BIOSTATISTICS

The Bare Essentials

Second Edition

A NOTE ON THE FRONT COVER

The cover depicts the famous "Study of Human Proportion in the Manner of Vitruvius" by Leonardo da Vinci, drawn about 1490, and done to death 500 years later in 2000. Those with a classical bent may wish to know the origin of the idea. According to Renaissance notions, the "Perfect Man" was based on geometric principles. The arms outstretched, the top of the head, and the tip of the feet defined a square, and the tips of the arms and legs outstretched in a fanlike position inscribed a circle centered on the navel.

What da Vinci failed *to notice is that the legs fit precisely on a normal curve, with the mean between the two heels and the apex at the crotch, one standard deviation falling exactly on the two kneecaps, and the asymptotes at the corners of the inscribed square. The centers of the two feet, at the point where they intersect the arc of the circle, then determine the conventional criterion for statistical significance at ± two standard deviations from the mean.*

Leonardo da Vinci can be forgiven, however. Statistics hadn't been invented yet in 1492.

BIOSTATISTICS
The Bare Essentials

Second Edition

Geoffrey R. Norman, PhD

Professor, Department of Clinical Epidemiology and Biostatistics
McMaster University
Hamilton, Ontario, Canada

David L. Streiner, PhD

Assistant Vice-President, Research
Director, Kunin-Lunenfeld Applied Research Unit
Baycrest Centre for Geriatric Care
Professor, Department of Psychiatry
University of Toronto
Toronto, Ontario, Canada

with 122 illustrations

2000
B.C. Decker Inc.
Hamilton • London

B.C. Decker Inc.
4 Hughson Street South
P.O. Box 620, L.C.D. 1
Hamilton, Ontario L8N 3K7
Tel: 905-522-7017
Fax: 905-522-7839
E-mail: info@bcdecker.com
Website: http//www.bcdecker.com

00 01 02 03 04/ PC / 9 8 7 6 5 4 3 2 1

ISBN 1-55009-123-9

Printed in Canada

Sales and Distribution

United States
B.C. Decker Inc.
P.O. Box 785
Lewiston, NY 14092-0785
Tel: 905-522-7017 / 1-800-568-7281
Fax: 905-522-7839
E-mail: info@bcdecker.com
Website: www.bcdecker.com

Canada
B.C. Decker Inc.
4 Hughson Street South
P.O. Box 620, L.C.D. 1
Hamilton, Ontario L8N 3K7
Tel: 905-522-7017 / 1-800-568-7281
Fax: 905-522-7839
E-mail: info@bcdecker.com
Website: www.bcdecker.com

UK, Europe, Scandinavia, Middle East, India, Asia, Africa
Harcourt Publishers Limited
Customer Service Department
Foots Cray High Street
Sidcup, Kent
DA14 5HP, UK
Tel: 44 (0) 208 308 5760
Fax: 44 (0) 181 308 5702
E-mail: cservice@harcourt_brace.com

Japan
Igaku-Shoin Ltd.
Foreign Publications Department
3-24-17 Hongo
Bunkyo-ku, Tokyo, Japan 113-8719
Tel: 3 3817 5680
Fax: 3 3815 6776
E-mail: fd@igaku.shoin.co.jp

Foreign Rights
John Scott & Company
International Publishers' Agency
P.O. Box 878
Kimberton, PA 19442
Tel: 610-827-1640
Fax: 610-827-1671

To two people whose hard work, patience, diligence,
and, most important, unflagging good humor,
have made it possible:

Geoff R. Norman
and
David L. Streiner

Too many people confuse being serious with being solemn.

John Cleese

One of the first symptoms of an approaching nervous breakdown is the belief that one's work is terribly important.

Bertrand Russell

Most researchers use statistics the way a drunkard uses a lamp-post—more for support than illumination.

Winifred Castle

PREFACE TO THE SECOND EDITION

We have been extremely pleased by the positive comments we have received about the first edition of *Biostatistics: The Bare Essentials.*[1] It is very gratifying to get e-mail messages out of the blue telling us that, for the first time, people really understand what statistics are all about and are having fun learning it—almost as gratifying as getting royalty checks in the snail-mail. We debated for a long time whether we should write a second edition. Our hesitation was due to two considerations. First, if the half-life of medical knowledge is about 5 years, then it must be longer than the life-time of an elephant for statistical knowledge. After all, we are still using the correlation coefficient that was proposed by Galton (and he died in 1911), and the work done by Ronald Fisher in the early 1900s continues to provide the basic core of statistics. Second, there were other things we wanted to do with our lives, such as eating, sleeping, and seeing our families.

[1] *And even more pleased by the reaction to the back cover. In response to many inquiries about it, yes, Geoff really does have four arms.*

So, what made us decide to do a second edition? For one thing, statistics *have* changed. Path analysis and structural equation modeling have been around for about a quarter of a century, but with the recent introduction of programs that do these easily on a desktop computer, their use has proliferated over the past few years. It is almost impossible to read any journal in fields such as psychology without seeing at least one example in each issue. The same is true for other computer-intensive techniques, such as logistic regression and multivariate analysis of variance; so, we have added chapters on all of these subjects. Now, after nearly a half century of debate, we are starting to reach some consensus about the best way to measure change, and this deserves its own chapter.

Writing a second edition has also allowed us to correct the mistakes that we and others have discovered over the years. However, writing new chapters has also offered us the opportunity to make new ones, so keep your eyes open and the e-mails coming. In closing, we would like to thank three people who have been especially diligent in pointing out our mistakes and in reading drafts of some of the new chapters: Bill Marks from Villanova University, Kathleen Wyrwich from St. Louis University, and Jose Luis Saiz from Universidad de la Frontera.

GRN
DLS

PREFACE TO THE FIRST EDITION

Are congratulations in order? Have you finally overcome those years of denial about your ignorance of statistics, those many embarrassing incidents at scientific meetings, those offhand comments at drug company receptions when someone dropped tidbits like "analysis of covariance" into the conversation and you had to admit your bewilderment? Are you prepared to recognize your condition and deal with your problem? Face it, you are a **photonumerophobic!**[1]

Now that you have come out of the closet (clinic), we are here to help. To begin, it would be useful for you to understand that all statisticians are not created equal, and as a result all statistics books are not equal.[2] An analogy with home renovation might help. Three basic types of folks are involved in home renovation. First there are architects, who design houses that no one except dermatologists can afford—they worry about concepts, esthetics, and design at the theory level. Next there are carpenters who *do* home renovations, are highly specialized and skilled,[3] and have a special language consisting of terms such as plates, sills, rafters, sheathing, R28, and the like that describe goings on at the practical level.[4]

Finally, there are the do-it-yourselfers (DIYers), who have the temerity to sally forth in blissful ignorance and make their own additions. Now, the fact of the matter is that it isn't all that difficult to put a nail into a 2 × 4, or to do anything else related to foundations, walls, ceilings, plumbing, and wiring. But a frustration for accomplished DIYers is that the books on do-it-yourselfing are written either by the architects, or by carpenters, but not by really good DIYers, and they all miss the mark. So, you either get pieces about the esthetic considerations involved in a $200,000 bathroom renovation, or a DIY book that starts and stops with "How to change a fuse."

Unfortunately, the same conventions hold in statistics. There are the architects of statistics—card-carrying PhDs who contribute to the theory of statistics and publish journal articles in *Biometrika* or little monographs to be read only by other members of this closed community. Then there are the carpenters—the most common species. They usually have a PhD in statistics, but they don't actually contribute to the discipline base of statistics—they just *do* statistics. They don't usually publish articles in statistics journals, beyond the cookbook recipes. Then there are the DIYers—folks like us who have arrived at statistics by the back door through disciplines such as psychology or education. With the advent of modern statistical packages and PCs, nearly anyone can be a do-it-yourself statistician—even you. Note that we are assuming in this book, unlike many other statistics books, that you will *not* actually *do* statistics. No one except students in statistics courses has done an analysis of variance for 20 years. If God had meant people to do statistics, He wouldn't have invented computers.

This description reveals two problems with the present state of affairs. First, doing statistics really is easier now than doing plumbing, but unfortunately errors are much better hidden—there is no statistical equivalent of a leaky pipe. Also, there is no building inspector or building code in statistics, although journal editors wish there were.

Secondly, most Do-It-Yourself stats books are written by tradesmen (oops, that should "tradespersons"). They are a possessive lot and likely feel a little guilty that they, too, don't publish in *Biometrica*.[5] So, they commit two fundamental errors. First, they cannot resist dazzling you with the mysteries of the game and subliminally impressing you with the incredible intelligence that they must have had to master the field. This is achieved by sprinkling technical lingo throughout the book, doing lots and lots of derivations and algebra to make it look like science, and, above all, writing in a stilted, formal, and ultimately unreadably boring prose, as if this is a prerequisite for credibility. That is one type of statistics book—until recently, in the majority.

There is a second strategy, however. Recognizing that no one in possession of his or her senses would actually lay out hard-earned cash to buy such a book,[6] a number of carpenters have begun to publish little thin books, with lively prose and with a sincere hope of demystifying the field and making good royalties. The only problem is that they usually presume that the really contemporary stuff of statistics is much too complicated for the average DIYer to comprehend. As a result, these books begin, and end, with statistical methods that were popular around the turn of the last century. An argument used to justify such books goes like, "We have carefully surveyed the biomedical literature, and contemporary and powerful methods like factor analysis are used only rarely, so we are just teaching

[1] *Photonumerophobia: fear that one's fear of numbers will come to light (thanks to Dave Sackett).*

[2] *Most statisticians who write statistics books don't understand this distinction, which is why most statistics books are so boring.*

[3] *Always the optimists, aren't we?*

[4] *Damn fools. If they had the good sense to put Graeco-Latin names on these things they could have tripled their salaries. Admit it, you can charge more for making a diagnosis of acute nasopharyngitis than for snotty nose.*

[5] *Norman can sympathize. He has a PhD in physics, which he never used. He was recently introduced at a meeting as a "fallen physicist," a term which Streiner calls a redundancy.*

[6] *Unless, of course, it was assigned reading in a course taught by another statistical carpenter.*

methods that appear commonly." The circular nature of this argument somehow escapes them.[7]

We have news for you. Contemporary statistics are not all that complicated; in fact, now that computers are around to do all the dirty work, it's much less painful than in yesteryear. Certainly compared to physiology or physics, it's pain free. But an author has to approach it with a genuine desire to try very hard to explain it. Let us just return to the DIY analogy one last time. There are really two types of activities that accomplished DIYers get involved in. For some chores on the house, they want to be sufficiently informed that they can hire a professional and feel confident that they will recognize when it is done well or poorly. That is, they know they can't do it all on their own, but they know enough to be able to tell shoddy workmanship when they see it. Other tasks they may decide to complete themselves. Again, for the biomedical researcher confronted with statistics, both avenues are open. On the one hand, it is a prerequisite, in examining the analyses conducted by others, to be able to understand when it was done well or poorly, even though one may choose to not do it oneself. On the other hand, with the flexibility and ease of many contemporary statistics packages, just about anyone can now get involved in the doing of statistics.

Our first book, *PDQ Statistics* (Norman and Streiner, 1986), was written to satisfy consumers of statistics. We found that it was possible to explain most of contemporary statistics at the conceptual level, with little recourse to algebra and proofs. However, it does take somewhat more knowledge and skill to do something—plumbing, wiring, or statistics—than it does to recognize when others are doing it well or poorly. That, then, is the intent of this book. If you never intend to do statistics, save a few bucks and buy *PDQ*. However, if you are actually involved in research, or if you have had your appetite whetted by *PDQ* or some other introductory book, pay the salesperson for this book and carry on.

Some comments about the format of the book. A perusal of the contents reveals that it is laid out much as any other traditional stats book. We contemplated doing it in problem-based fashion, both because we come from a problem-based medical school and also because it would sound contemporary and sell more books (we never said we were in it for altruism). But this would constitute, in our view, a debasement of the meaning of problem-based learning (PBL). This book is a resource, not a curriculum. By all means, we urge the reader to consult it when there is a statistical problem around, thereby doing PBL. But PBL does not dictate the format of the resources—all medical students, wherever they are, still engorge Harrison and the Merck Manual. We felt that we could better explain the conceptual underpinnings by following the traditional sequence.

Some differences go beyond style. Most chapters begin with an example to set the stage. Usually the examples were dreamt up in our fertile imaginations and are, we hope, entertaining. Occasionally we reverted to real-world data, simply because sometimes the real world is at least as bizarre as anything imagination could invent. Although many reviews of statistics books praise the users of real examples and castigate others, we are unapologetic in our decision for several reasons: (1) the book is aimed at all types of health professionals, and we didn't want to waste your time and ours explaining the intricacies of podiatry for others; (2) the real world is a messy place, and it is difficult, or well nigh impossible, to locate real examples that illustrate the pedagogic points simply;[8] and (3) we happen to believe, and can cite good psychological evidence to back it up, that memorable (read "bizarre") examples are a potent ally in learning and remembering concepts.

There are far more equations here than in *PDQ*, although we have still tried to keep these to a minimum. Our excuse is simply that this is the language of statistics; if we try to avoid it altogether, we end up with such convoluted prose that the message gets lost in the medium. But we continue to try very hard to explain the underlying concept, instead of simply dropping a formula in your lap.

There are a few other distinctive features. We have retained the idea of C.R.A.P. Detectors[9] from *PDQ* as a way to help you see the errors of other's (and your own) ways. We have included computer notes at the end of most chapters[10] to help you with one of the more common and powerful statistical programs—SPSS (Statistical Program for the Social Sciences). Finally, we acknowledge that many clinical investigators use most of their skills to get grants so that they can hire someone else to do statistics. Also, it is impossible to squeeze money out of most federal, state, or provincial agencies without an impressive sample size calculation.[11] That means, of course, that the only analysis many biomedical researchers do is the sample size calculations in their grant proposals. Recognizing this harsh reality, every chapter has a section devoted to sample size calculations (when these are available) so you will be as good as the next person at befuddling the grant reviewers.

On the issue of format, you will already have noticed that the book has an excessively wide outside margin. This is not a publisher's error or an attempt to salvage the pulp and paper industry. Instead, it accomplishes two things: (1) we can use the margin for rubrics,[12] expanding on things of slightly peripheral interest, or inflicting our base humor on the reader; and (2) you can use it to make your own notes if you don't like ours.

Finally, on the issue of style. You might have already noticed that we have cultivated a somewhat irreverent tone, which we will proceed to apply as we see fit to all folks who have the misfortune to appear in these pages—statisticians, physicians, administrators, nurses, physiotherapists, psycholo-

[7] *This is an argument for maintaining the status quo despite much discussion of the inadequacy of reporting statistics in the biomedical literature. It's analogous to saying that we have studied primary care clinics and we found that most visits (about 80%) are related to acute respiratory infections, hypertension, depression, and chronic pulmonary disease, so that is all we will teach our medical students.*

[8] *Every time we get on an airplane, we are grateful that the pilots practiced landing the 747 with both starboard engines blown on a simulator so (a) they would know what to do if it happened, and (b) they wouldn't have to practice on us.*

[9] *Lest we be accused of profane language, this stands for "Convoluted Reasoning and Anti-intellectual Pomposity Detectors." Ernest Hemingway likely thought so too—he coined the phrase.*

[10] *See the note at the end of this preface.*

[11] *Most sample size calculations are based on exact analysis of impossibly wild guesses, resulting in an illusion of precision. As Alfred North Whitehead said, "Some of the greatest disasters of mankind were inflicted by the narrowness of men with a sound methodology."*

gists, and social workers. We recognize that we run a certain risk of offending the "allied"[13] health professionals, who have historically felt somewhat downtrodden, with good reason, by folks with MD after their name. However, we felt the risk was greater if we omitted them altogether. Fear no evil, all ye downtrodden—our intent is not racist, sexist, or otherwise prejudiced. We will attempt, as much as possible, to insult all professions equally.[14]

Notes on the Computer Notes

We are of the firm belief that our mothers didn't raise us to waste our time doing calculations by hand; that's why we have computers and computerized statistical packages. However, learning the arcane code words demanded by many of these programs can be as intimidating as learning statistics itself. So, in our never-ending quest to be as helpful as possible, we've supplied the commands necessary to make one of these programs bow to your wishes.

A few years ago, it would have been a simple job to choose which programs to include; because there were only three or four that could be run on desktop computers, we could have included all of them and be seen as comprehensive and erudite. Now, though, it seems as if a new, "better," package is introduced every month, forcing us to make some choices.[15] When we wrote the first edition, there were a bunch of popular and powerful programs which stood out from the rest — SPSS, BMDP, SAS, and Minitab. So, we obligingly included some hints on how to run three of the four. Well, things have changed considerably in the past decade or so. SPSS (Statistical Package for the Social Sciences) has done to statistical software what Microsoft did to operating systems — it swallowed them whole for breakfast. While you can still buy SAS and Minitab, SPSS bought out BMDP had then let it wither on the vine. "Real" statisticians still use SAS, but you'll need a separate bookcase just to house all of its manuals. The reality is that wherever you look in the social and medical sciences, folks are running SPSS. It's never been the best at everything, but it's good at many things, and it's pretty well created a monopoly. Since no manual or Help directory can ever compete with a knowledgeable friend, and friends knowledgeable in SPSS are far more common than friends of the other ilks[16], it makes no sense for us to buck a trend. Accordingly, this time around we've only included instructions for the Windows version of SPSS (Version 9 point something or other).

Good luck (and don't call us if your machine blows up).

Acknowledgments

Many of our students have waded through early drafts of this book, giving us valuable advice about where we were going astray. Unfortunately, they are too numerous to mention (and we have forgotten most of their names). However, special thanks are due to Dr. Marilyn Craven, who patiently (and sometimes painfully) helped us with our logic and English. So, any mistakes you find should be blamed on them; we humbly accept any praise as due to our own efforts.

On a serious note (which we hope will be the last), we would like to express our thanks to Brian C. Decker, who dreamt up the idea of this book and who encouraged us from the beginning.

GRN
DLS

[12] *No doubt you wonder what a* **rubric** *is. Literally, it is the note written in red in the margin of the Book of Common Prayer telling the preacher what to do next. That's why these are red.*

[13] *We don't like the term either, but it's shorter than spelling out all the allies.*

[14] *We forget whether it was Lenny Bruce or Mort Sahl who ended every routine with the line, "Is there anyone in the audience whom I haven't insulted yet?" In either case, he was our inspiration.*

[15] *And thereby resulting in some people castigating us for not including the best statistical package (i.e., the one they have on their machine). Such are the perils of authorship.*

[16] *Not that we recommend, "Hi there. Do you know SPSS?" as an ice-breaker at a singles bar. Chapter 8 to the contrary notwithstanding, sex and stats make poor bedfellows.*

CONTENTS

SECTION THE FIRST **THE NATURE OF DATA AND STATISTICS**

1 The Basics 2

2 Looking at the Data 6
A First Look at Graphing Data

3 Describing the Data with Numbers 16
Measures of Central Tendency and Dispersion

4 The Normal Distribution 27

5 Probability 33

6 Elements of Statistical Inference 42

C.R.A.P. DETECTORS 57

SECTION THE SECOND **ANALYSIS OF VARIANCE**

7 Comparing Two Groups 62
The *t*-Test

8 More than Two Groups 68
One-Way ANOVA

9 Factorial ANOVA 79

10 Two Repeated Observations 89
The Paired *t*-Test and Alternatives

11 Repeated-Measures ANOVA 94

12 Multivariate ANOVA (MANOVA) 103

C.R.A.P. DETECTORS 114

SECTION THE THIRD **REGRESSION AND CORRELATION**

13 Simple Regression and Correlation 118

14 Multiple Regression 127

15 Logistic Regression 139

16 Advanced Topics in Regression and ANOVA 145

17 Measuring Change 155

18 Principal Components and Factor Analysis 163
Fooling Around with Factors

19 Path Analysis and Structural Equation Modeling 178

C.R.A.P. DETECTORS 197

SECTION THE FOURTH **NONPARAMETRIC STATISTICS**

20 Tests of Significance for Categorical Frequency Data 203

21 Measures of Association for Categorical Data 217

22 Tests of Significance for Ranked Data 224

23 Measures of Association for Ranked Data 230

24 Life-Table (Survival) Analysis 236

C.R.A.P. DETECTORS 250

SECTION THE FIFTH **REPRISE**

25 Screwups, Oddballs, and Other Vagaries of Science 256
Locating Outliers, Handling Missing Data, and Transformations

26 Putting It All Together 265

Test Yourself (Being a Compendium of Questions and Answers) 271

Answers to Chapter Exercises 276

References and Further Reading 287

Unabashed Glossary 293

Appendix 295

Index 317

SECTION

THE FIRST

THE NATURE OF DATA AND STATISTICS

In this chapter, we will introduce you to the concepts of *variables* and to the different types of data: *nominal, ordinal, interval,* and *ratio.*

CHAPTER THE FIRST

The Basics

STATISTICS: SO WHO NEEDS THEM?

The first question most beginning students of statistics ask is, "Why do we need it?" Leaving aside the unworthy answer that it is required for you to get your degree, we have to address the issue of how learning the arcane methods and jargon of this field will make you a better person and leave you feeling fulfilled in ways that were previously unimaginable. The reason is that the world is full of variation, and sometimes it's hard to tell real differences from natural variation. Statistics wouldn't be needed if everybody in the world were exactly like everyone else;[1] if you were male, 172 cm tall, had brown eyes and hair, and were incredibly good looking,[2] this description would fit every other person.[3] Similarly, if there were no differences and we knew your life expectancy, or whether or not a new drug was effective in eliminating your dandruff, or which political party you'd vote for in the next election (assuming that the parties finally gave you a meaningful choice, which is doubtful), then we would know this for all people.

Fortunately, this is not the case; people are different in all of these areas, as well as in thousands of other ways. The downside of all this variability is that it makes it more difficult to determine how a person will respond to some newfangled treatment regimen or react in some situation. We can't look in the mirror, ask ourselves, "Self, how do you feel about the newest brand of toothpaste?" and assume everyone will feel the same way.

DESCRIPTIVE AND INFERENTIAL STATISTICS

It is because of this variability among people, and even within any one person from one time to another, that statistics were born. As we hope to show as you wade through this tome, statistics allow us to describe the "average" person, to see how well that description fits or doesn't fit other people, and to see how much we can generalize our findings from studying a few people[4] to the population as a whole. So statistics can be used in two ways: to describe data, and to make inferences from them.

> *Descriptive statistics* are concerned with the presentation, organization, and summarization of data.

The realm of descriptive statistics, which we cover in this section, includes various methods of organizing and graphing the data to get an idea of what they show. Descriptive statistics also include various indices that summarize the data with just a few key numbers.

The bulk of the book is devoted to inferential stats.

> *Inferential statistics* allow us to generalize from our sample of data to a larger group of subjects.

For instance, when a dermatologist gives a new cream, attar of eggplant, to 20 adolescents whose chances for true love have been jeopardized by acne, and compares them with 20 adolescents who remain untreated (and presumably unloved), he is not interested in just those 40 kids. He wants to know whether all kids with acne will respond to this treatment. Thus he is trying to make an inference about a larger group of subjects from the small group he is studying. We'll get into the basics of inferential statistics in Chapter 6; for now, let's continue with some more definitions.

[1] *We also wouldn't need dating services because it would be futile to look for the perfect mate; he or she would be just like the person sitting next to you. By the same token, it would mean the end of extramarital affairs, because what's the use? But that's another story.*

[2] *Coincidently, this perfectly describes the person writing this section.*

[3] *Mind you, if everybody in the world were male (or female), we wouldn't need statistics (or anything else) in about 70 years.*

[4] *As we'll see later, "a few" to a statistician can mean over 400,000 people, as in the Salk polio vaccine trial. So much for the scientific use of language.*

VARIABLES

In the first few paragraphs, we mentioned a number of ways that people differ: gender,[5] age, height, hair and eye color, political preference, responsiveness to treatment, and life expectancy. In the statistical parlance you'll be learning, these factors are referred to as *variables.*

A *variable* is simply what is being observed or measured.

Variables come in two flavors: independent and dependent. The easiest way to start to think of them is in an experiment, so let's return to those acned adolescents. We want to see if the degree of acne depends on whether or not the kids got attar of eggplant. The outcome (acne) is the dependent variable, which we hope will change in response to treatment. What we've manipulated is the treatment (attar of eggplant), and this is our independent variable.

The *dependent variable* is the outcome of interest, which should change in response to some intervention.

The *independent variable* is the intervention, or what is being manipulated.[6]

Sounds straightforward, doesn't it? That's a dead giveaway that it's too simple. Once we get out of the realm of experiments, the distinction between dependent and independent variables gets a bit hairier. For instance, if we wanted to look at the growth of vocabulary as a kid grows up, the number of different words would be the dependent variable and age the independent one. That is, we're saying that vocabulary is dependent on age, even though it isn't an intervention and we're not manipulating it. So, more generally, if one variable changes in response to another, we say that the dependent variable is the one that changes in response to the independent variable.

Both dependent and independent variables can take one of a number of specific **values**: for gender, this is usually limited to either male or female; hair color can be brown, black, blonde, red, gray, artificial, or missing; and a variable such as height can range between about 50 cm for premature infants to about 200 cm for basketball players and coauthors of statistics books.

TYPES OF DATA

Discrete versus Continuous Data

Although we referred to both gender and height as variables, it's obvious that they are different from one another with respect to the type and number of values they can assume. One way to differentiate between types of variables is to decide whether the values are **discrete** or **continuous.**

Discrete variables can have only one of a limited set of values. Using our previous examples, this would include variables such as gender, hair and eye color, political preference, and which treatment a person received. Another example of a discrete variable is a number total, such as how many times a person has been admitted to hospital; the number of decayed, missing, or filled teeth; and the number of children. Despite what the demographers tell us, it's impossible to have 2.13 children—kids come in discrete quantities.

Discrete data have values that can assume only whole numbers.

The situation is different for continuous variables. It may seem at first that something such as height, for example, is measured in discrete units: someone is 172 cm tall; a person slightly taller would be 173 cm, and a somewhat shorter person would measure in at 171 cm. In fact, though, the limitation is imposed by our measuring stick. If we used one with finer gradations, we may be able to measure in ½-cm increments. Indeed, we could get really silly about the whole affair and use a laser to measure the person's height to the nearest thousandth of a millimeter. The point is that height, like weight, blood pressure, serum rhubarb, time, and many other variables, is really continuous, and the divisions we make are arbitrary to meet our measurement needs. The measurement, though, is artificial; if two people appear to have the same blood pressure when measured to the nearest millimeter of mercury, they will likely be different if we could measure to the nearest tenth of a millimeter. If they're still the same, we can measure with even finer gradations until a difference finally appears.

Continuous data may take any value, within a defined range.

We can illustrate this difference between discrete and continuous variables with two other examples. A piano is a "discrete" instrument. It has only 88 keys, and those of us who struggled long and hard to murder Paganini learnt that A-sharp was the same note as B-flat. Violinists (fiddlers to y'all south of the Mason-Dixon line), though, play a "continuous" instrument and are able to make a fine distinction between these two notes. Similarly, really cheap digital watches display only 4 digits and cut time into 1-minute chunks. Razzle-dazzle watches, in addition to storing telephone numbers and your bank balance, cut time into 1⁄100-second intervals. A physicist can do even better, dividing each second into 9,192,631,770 oscillations of a cesium atom. Even this, though, is only an arbitrary division. Only the hospital administrator, able to buy a Patek Phillipe analogue chronometer, sees time as it actually is: as a smooth, unbroken progression.[7]

Many of the statistical techniques you'll be learning about don't really care if the data are discrete or continuous; after all, a number to them is just a number. There are instances, though, when the distinction is important. Rest assured that we will point these out to you at the appropriate times.

[5] *Formerly referred to as "sex."*

[6] *These are different from the definitions offered by one of our students, who said that, "An undependable variables keeps changing its value, while a dependable variable is always the same."*

[7] *Actually, the escapement mechanism makes the second hand jump, but if you can afford a Patek, you'll ignore this.*

Nominal, Ordinal, Interval, and Ratio Data

We can think about different types of variables in another way. A variable such as gender can take only two values: male and female. One value isn't "higher" or "better" than the other;[8] we can list them by putting male first or female first without losing any information. This is called a **nominal** variable.

[8]*Although male chauvinist pigs and radical feminists would disagree, albeit for opposite reasons.*

> A *nominal* variable consists of named categories, with no implied order among the categories.

The simplest nominal categories are what Feinstein (1977) calls "existential" variables—a property either exists or it doesn't exist. A person has cancer of the liver or doesn't have it; someone has received the new treatment or didn't receive it; and, most existential of all, the subject is either alive or dead. Nominal variables don't have to be dichotomous; they can have any number of categories. We can classify a person's marital status as Single/Married/Separated/Widowed/Divorced/Common-Law (six categories); her eye color into Black/Brown/Blue/Green/Mixed (five categories[9]); and her medical problem into one of a few hundred diagnostic categories. The important point is that you can't say brown eyes are "better" or "worse" than blue. The ordering is arbitrary, and no information is gained or lost by changing the order.

[9]*"Bloodshot" is usually only a temporary condition and so is not coded.*

Because computers handle numbers far more easily than they do letters, researchers commonly code nominal data by assigning a number to each value: Female could be coded as 1 and Male as 2: or Single = 1, Married = 2, and so on. In these cases, the numerals are really no more than alternative names, and they should not be thought of as having any quantitative value. Again, we can change the coding by letting Male = 1 and Female = 2, and the conclusions we draw will be identical (assuming, of course, that we remember which way we coded the data).[10]

[10]*Other examples of numbers really being nominal variables and not reflecting measured quantities would be telephone numbers, social insurance or social security numbers, credit card numbers, and politicians' IQs.*

A student evaluation rating consisting of Excellent/Satisfactory/Unsatisfactory has three categories. It differs from a variable such as hair color in that there is an ordering of these values: "Excellent" is better than "Satisfactory," which in turn is better than "Unsatisfactory." However, the difference in performance between "Excellent" and "Satisfactory" cannot be assumed to be the same difference as exists between "Satisfactory" and "Unsatisfactory." This is seen more clearly with letter grades; there is only a small division between a B+ and a B, but a large one, amounting to a ruined summer, between a D– and an F+. This is like the results of a horse race; we know that the horse who won ran faster than the horse who placed, and the one who showed came in third. But there could have been only a 1-second difference between the first two horses, with the third trailing by 10 seconds. So letter grades and the order of finishing a race are called ordinal variables.

> An *ordinal variable* consists of ordered categories, where the differences between categories cannot be considered to be equal.

Many of the variables encountered in the health care field are ordinal in nature. Patients are often rated as Much improved/Somewhat improved/Same/Worse/Dead; or Emergent/Urgent/Elective.[11] Sometimes numbers are used, as in Stage I through Stage IV cancer. Don't be deceived by this use of numbers; it's still an ordinal scale, with the numbers (Roman, this time, to add a bit of class) really representing nothing more than ordered categories. Use the difference test: is the difference in disease severity between Stage I and Stage II cancer the same as exists between Stages II and III or between III and IV? If the answer is No, the scale is ordinal.

[11]*This is similar to the scheme used to evaluate employees: Walks on water/ Keeps head above water under stress/ Washes with water/ Drinks Water/ Passes water in emergencies.*

If the distance between values *is* constant, we've graduated to what is called an **interval** variable.

> An *interval variable* has equal distances between values, but the zero point is arbitrary.

Why did we add that tag on the end, "the zero point is arbitrary," and what does it mean? We added it because, as we'll see, it puts a limitation on the types of statements we can make about interval variables. What the phrase means is that the zero point isn't meaningful and therefore can be changed. To illustrate this, let's contrast intelligence, measured by some IQ test, with something such as weight, where the zero is meaningful. We all know what zero weight is.[12] We can't suddenly decide that from now on, we'll subtract 10 kilos from everything we weigh and say that something that previously weighed 11 kilos now weighs 1 kilo. It's more than a matter of semantics; if something weighed 5 kilos before, we would have to say it weighed –5 kilos after the conversion—an obvious impossibility.

[12]*It's a state aspired to by "high-fashion" models.*

An intelligence score is a different matter. We say that the average IQ is 100, but that's only by convention. The next world conference of IQ experts can just as arbitrarily decide that from now on, we'll make the average 500, simply by adding 400 to all scores. We haven't gained anything, but by the same token, we haven't lost anything; the only necessary change is that we now have to readjust our previously learned standards of what is average.

Now let's see what the implications of this are. Because the intervals are equal, the difference between an IQ of 70 and an IQ of 80 is the same as the difference between 120 and 130. However, an IQ of 100 is not twice as high as an IQ of 50. The point is that if the zero point is artificial and moveable, then the differences between numbers are meaningful, but the ratios between them are not.

If the zero point is meaningful, then the ratios between numbers are also meaningful, and we are dealing with (not surprisingly) a **ratio** variable.

A *ratio variable* has equal intervals between values and a meaningful zero point.

Most laboratory test values are ratio variables, as are physical characteristics such as height and weight. A person who weighs 100 kilos is twice as heavy as a person weighing 50 kilos; even when we convert kilos to pounds, the ratio stays the same: 220 pounds to 110 pounds.

That's about enough for the difference between interval and ratio data. The fact of the matter is that, from the viewpoint of a statistician, they can be treated and analyzed the same way.

Notice that each step up the hierarchy from ordinal data to ratio data takes the assumptions of the step below it and then adds another restriction:[13]

Variable type	Assumptions
Nominal	Named categories.
Ordinal	Same as nominal plus *ordered categories.*
Interval	Same as ordinal plus *equal intervals.*
Ratio	Same as interval plus *meaningful zero.*

Although the distinctions among nominal, ordinal, interval, and ratio data appear straightforward on paper, the lines between them occasionally get a bit fuzzy. For example, as we've said, intelligence is measured in IQ units, with the average person having an IQ of 100. Strictly speaking, we have no assurance that the difference between an IQ of 80 and one of 100 means the same as the difference between 120 and 140; that is, IQ most likely is an ordinal variable. In the real world outside of textbooks, though, most people treat IQ and many other such variables as if they were interval variables. As far as we know, they have not been arrested for doing so, nor has the sky fallen on their heads.

Despite this, the distinctions among nominal, ordinal, interval, and ratio are important to keep in mind because they dictate to some degree the types of statistical tests we can use with them. As we'll see in the later chapters, certain types of graphs and what are called "parametric tests" can be used with interval and ratio data but not with nominal or ordinal data. By contrast, if you have nominal or ordinal data, you are, strictly speaking, restricted to "nonparametric" statistics. We'll get into what these obscure terms mean later in the book. So, with that as background, on to statistics!

[13]*A good mnemonic for remembering the order of the categories is the French word NOIR. Of course, this assumes you know French. Anglophones will just have to memorize the order.*

EXERCISES

1. For the following studies, indicate which of the variables are *dependent* (DVs), *independent* (IVs), or neither.
   a. ASA is compared against placebo to see if it leads to a reduction in coronary events.
   The IV is______ The DV is______
   b. The relationship between hypocholesterolemia and cancer.
   The IV is______ The DV is______
   c. We know that members of religious groups that ban drugs, alcohol, smoking, meat, and sex (because it may lead to dancing) live longer than the rest of us poor mortals, but is it worth it? How do they compare with us on a test of quality of life?
   The IV is______ The DV is______
   d. One study (a real one, this time) found that bus drivers had higher morbidity rates of coronary heart disease than did conductors.
   The IV is______ The DV is______
2. State which of the following variables are *discrete* and which are *continuous.*
   a. The number of hair-transplant sessions undergone in the past year.
   b. The time since the last patient was grateful for what you did.
   c. Your anticipated before-taxes income the year after you graduate.
   d. Your anticipated after-taxes income in the same year.
   e. The amount of weight you've put on in the last year.
   f. The number of hairs you've lost in the same time.
3. Indicate whether the following variables are *nominal, ordinal, interval,* or *ratio.*
   a. Your income (assuming it's more than \$0).
   b. A list of the different specialties in your profession.
   c. The ranking of specialties with regard to income.
   d. Bo Derek was described as a "10." What type of variable was the scale?
   e. A range of motion in degrees.
   f. A score of 13 out of 17 on the Schmedlap Anxiety Scale.
   g. Staging of breast cancer as Type I, II, III, or IV.
   h. ST depression on the ECG, measured in millimeters.
   i. ST depression, measured as '1' ≤ 1 mm, '2' = 1 to 5 mm, and '3' > 5 mm.
   j. ICD-9 classifications: 0295 = Organic psychosis, 0296 = Depression, and so on.
   k. Diastolic blood pressure, in mm Hg.
   l. Pain measurement on a seven-point scale.

Here we look at different ways of graphing data, how to make the graphs look both accurate and esthetic, and how not to plot data.

CHAPTER THE SECOND

Looking at the Data

A First Look at Graphing Data

WHY BOTHER TO LOOK AT DATA?

Now that you've suffered through all these pages of jargon, let's actually do something useful: learn how to look at data. With the ready availability of computers on every desk, there is a great temptation to jump right in and start analyzing the bejezus out of any set of data we get. After all, we did the study in the first place to get some results that we could publish and prove to the Dean that we're doing something. However, as in most areas of our lives (especially those which are enjoyable), we must learn to control our temptations in order to become better people.

It is difficult to overemphasize the importance and usefulness of getting a "feel for the data" before starting to play with them. If there isn't a Murphy's Law to the effect that "There will be errors in your data," then there should be one. You do not look at the data just in case there are errors; they are there, and your job is to try to find as many as you can. Sometimes the problem isn't an error as such; very often, a researcher may use a code number such as 99 or 999 to indicate a missing value for some variable, then forget to tell you this little detail when he asks you to analyze his data. As a result, you may find that some people in his study are a few years older than Methuselah. Graphing the data beforehand may well save you from one of life's embarrassing little moments.

A second purpose for looking at the data is to see if they can be analyzed by the statistical tests you're planning to use. For example, some tests require the data to fit a given shape, or that a plot of two variables follow a straight line. Although there are specific tests of these assumptions, the power of the "calibrated eyeball test" should not be underestimated. A quick look often gives you a better sense of the data than does a bunch of numbers.

HISTOGRAMS, BAR CHARTS, AND VARIATIONS ON A THEME

The Basic Theme: The Bar Chart

Perhaps the most familiar types of graphs to most people are **bar charts** and **histograms** (we'll tell you what the difference is in a little bit). In essence, they consist of a bar whose length is proportional to the number of cases. To illustrate it, let's conduct a "gedanken experiment."[1] Imagine we do a study in which we survey 100 students and ask them what their most boring course was in college. We can then tabulate the data as is shown in Table 2–1.

[1]This is a German term, popularized by Albert Einstein, meaning "thought experiment." It is used here simply for purposes of pretentiousness.

The first step is to choose an appropriate length for the *Y*-axis, where we'll plot (at least for now) the number of people who chose each alternative. The largest number in the table is 42, so we will choose some number somewhat larger than this for the top

TABLE 2–1

Responses of 100 students to the question, "What was your most boring introductory course?"

Course	Number of students
Sociology	25
Economics	42
History	8
Psychology	13
Calculus	12

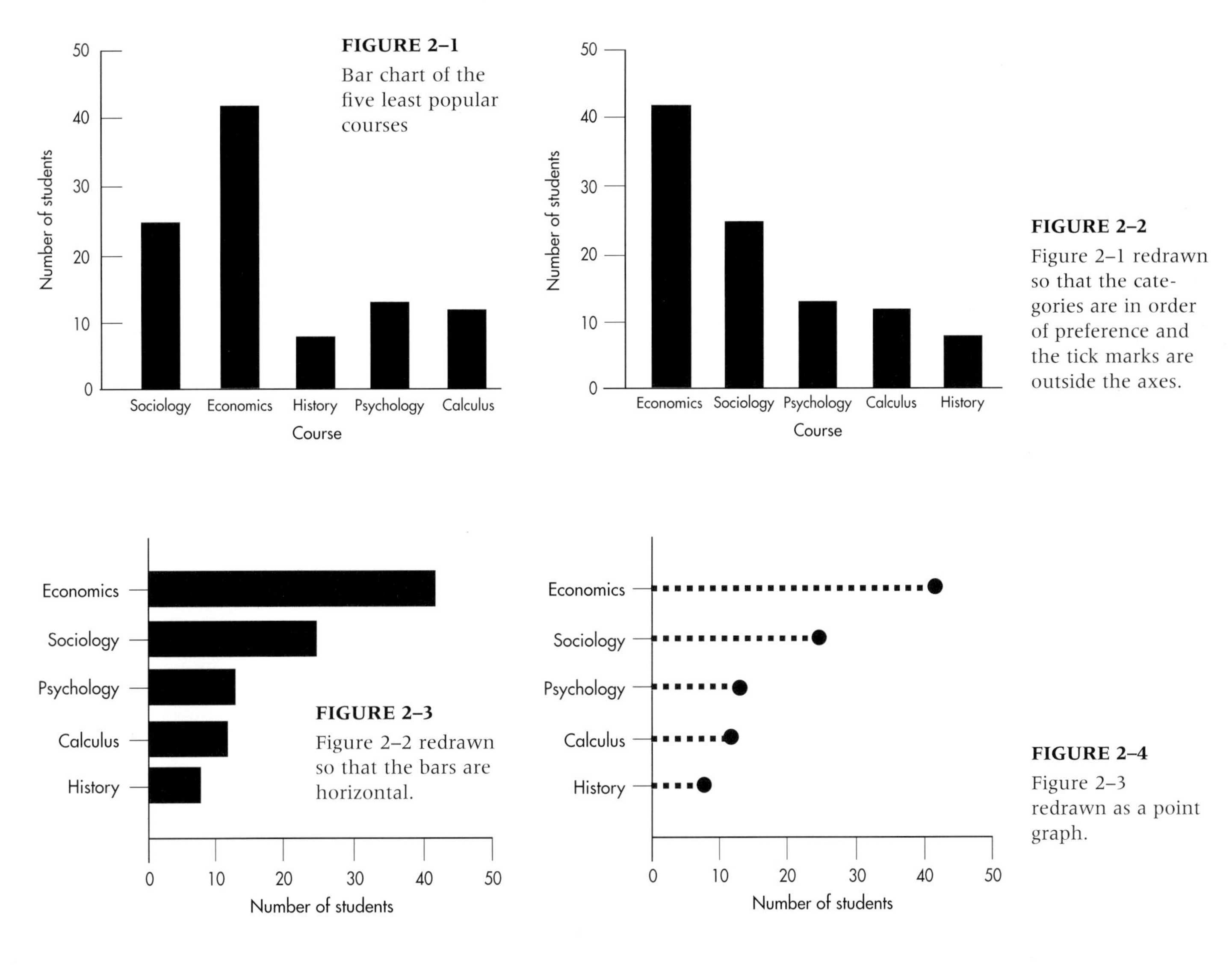

FIGURE 2–1
Bar chart of the five least popular courses

FIGURE 2–2
Figure 2–1 redrawn so that the categories are in order of preference and the tick marks are outside the axes.

FIGURE 2–3
Figure 2–2 redrawn so that the bars are horizontal.

FIGURE 2–4
Figure 2–3 redrawn as a point graph.

of the axis. Because we'll label the tick points every 10 units, 50 would be a good choice. If we had used the number 42, we would have had to label the axis either every 7 units (which are somewhat bizarre numbers[2]), or every even number, which would make the axis look too cluttered. So, our graph would look like Figure 2–1.

At first glance, this doesn't look too bad! However, we can make it look even better. It's obvious that the data are nominal; the order is arbitrary, so we can change the categories around without losing anything. In fact, we gain something if we rank the courses so that the highest count is first and the lowest one is last. Now the relative standing of the courses is more readily apparent. (As a minor point, it's often better to put the tick marks outside the axes rather than in. When the data fall near the *Y*-axis, a tick mark inside the axis may obscure the data point, or vice versa.) Making these two changes gives us Figure 2–2.

This is the way most bar charts of nominal data looked until recently. Within recent years, though, things have been turned on their ear—literally. If the names of the categories are long, things can look pretty cluttered down there on the bottom. Also, some research (Cleveland, 1984) has shown that people get a more accurate grasp of the relative sizes of the bars if they are placed horizontally. Adding this twist (pun intended), we'll end up with Figure 2–3.

[2]*Fast! Count by sevens, starting at 1 and ending at 64. See what we mean?*

Variation 1: Dot Plots

Another variant of the bar chart that is particularly useful when there are many categories is the dot plot, as shown in Figure 2–4. Instead of a bar, just a heavy dot is placed where the end of the bar would be. When there are many labels, smaller dots that extend back to the labeled axis are often used to make the chart easier to read.

Graphing Ordinal Data

The use of histograms isn't limited to nominal data; it can be used with all four types. However, a few other considerations should be kept in mind when using them with ordinal, interval, and ratio data. The first, which would seem obvious, is that because the values are ordered, you can't blithely move the categories around simply to make the graph look prettier. If you were graphing the number of students who received Excellent/Satisfactory/Unsatisfactory ratings, it would confuse more than help if you put them in the order: Satisfactory/Excellent/Unsatisfactory just because most students were in the first category.

Graphing Interval and Ratio Data

A few other factors have to be considered in graphing interval and ratio data. Let's say we had some data on the number of tissues dispensed each day by a group of 75 social workers. We look at our data, and we find that the lowest number is 10 and the highest is 117. The difference between the highest and lowest value is 107. (This difference is called the **range**. We'll define it a bit more formally later in the next chapter.) If we have one bar for each value, we'll run into a few problems. First, we have more possible values than data points, so some bars will have a "height" of zero units, and many others will be only one or two units high. This leads to the second problem, in that it will be hard to discern any pattern by eyeballing the data. Third, the *X*-axis is going to get awfully cluttered. For these reasons, we try to end up with between 10 and 20 bars on the axis.[3]

To do this, we make each bar represent a range of numbers; what we refer to as the **interval width**. If possible, use a width that most people are comfortable with: 2, 5, 10, or 20 points. Even though a width of 6 or 7 may give you an esthetically beautiful picture, these don't yield multiples that are easily comprehended. Let's use an example.

If we took 100 fourth-year nursing students and asked them how many bedpans they emptied in the last month, we'd get 100 answers, as in Table 2–2. The main thing a table like this tells us is that it's next to impossible to make sense of a table like this. We're overwhelmed by the sheer mass of numbers, and no pattern emerges. In fact, it's very hard even to figure out what the highest and lowest numbers are; who's been working like a Trojan and who's been goofing off. To make our lives (and all of the next steps) easier, the first thing we should do is to put the data in rank order,[4] starting with the smallest number and ending with the highest. Two notes are in order. First, you can go from highest to lowest if you wish, it makes no difference. Second, most computers have a simple routine, usually called SORT, to do the job for you. Once we do this, we'll end up with Table 2–3.

With this table we can immediately see the highest and lowest values and get at least a rough feel for how the numbers are distributed; not too many between 1 and 10 or between 60 and 70, and many in the 20s and 30s. We also see that the range is (66 – 1) = 65; far too many to graph when letting each bar stand for a unique number. An interval width of 10 would give us 7 boxes (not quite enough for our esthetic sense), whereas a width of 2 would result in 33 boxes (which is still too many). A width interval of 5 yields 14 boxes (which is just right). To help us in drawing the graph, we could make up a summary table, such as Table 2–4, which gives the interval and the number of subjects in that interval.

There are a few things to notice about this table. First, there are two extra columns, one labeled *Midpoint* and the other labeled *Cumulative Total*. The first is just what the name implies: it is the middle of the interval. Because the first interval consists of the numbers 0, 1, 2, 3, and 4, the midpoint is 2. If there were an even number of numbers, say 0, 1, 2, and 3, then the midpoint would again be in the middle. This time, though, it would fall half way between the 1 and 2, and we would label it 1.5. The other added column, the *Cumulative Total*, is simply a running sum of the number of cases; the first interval had 1 case, and the second 4, so the cumulative total at the second interval is (1 + 4) = 5. The 9 cases in the third interval then produce a cumulative total of (5 + 9) = 14. This is very handy because, if we didn't end up with 100 at the bottom, we would know that we messed up the addition somewhere along the line. The other point to notice is the interval. The first one goes from 0 to 4, the second from 5 to 9, and so on. Don't fall into the trap of saying an interval width of 5 covers the numbers 0 to 5; that's actually 6 digits.

Another point to notice is that we've paid a price for grouping the data to make it more readable, and

[3] *Note that this dictum is based on esthetics, not statistics.*

[4] *No pun is intended; it really is called "rank" order, even when the data aren't as smelly.*

TABLE 2–2

Number of bedpans emptied by 100 fourth-year nursing students in the past month

Student	Data				
1–5	43	45	16	37	33
6–10	41	24	11	34	51
11–15	14	29	55	9	25
16–20	31	24	24	28	16
21–25	35	36	14	15	18
26–30	36	27	42	7	43
31–35	45	32	42	46	14
36–40	57	16	27	34	12
41–45	32	30	26	25	1
46–50	58	26	19	17	35
51–55	32	17	31	22	16
56–60	42	28	26	36	51
61–65	52	28	15	38	49
66–70	21	11	66	37	35
71–75	37	41	11	27	43
76–80	7	20	56	13	61
81–85	38	38	39	54	31
86–90	26	24	43	38	33
91–95	34	12	52	47	25
96–100	26	17	7	22	54

that price is the loss of some information. We can tell from Table 2–4 that 1 person emptied between 0 and 4 bedpans, but we don't know exactly how many. In the next interval, we see that 4 people emptied between 5 and 9 pans, but again we're not sure precisely how many future nurses dumped what number of bedpans. *The wider the interval, the more information is lost.*

So, with these points in mind, we're almost ready to start drawing the graph. There's one last consideration, though: how to label the two axes. Looking at the count column in Table 2–4, we can see that the maximum number of cases in any 1 interval is 15. We would therefore want the *Y*-axis to extend from 0 to some number over 15. A good choice would be 20, because this would allow us to label every fifth tick mark. Notice that on the *X*-axis, we've labeled the middle of the interval. If we labeled every possible number, the axis would look too cluttered; the midpoint cuts down on the clutter, and (for reasons we'll explore further in the next chapter) is the best single summary of the interval. Our end product would look like Figure 2–5.

This figure differs from Figure 2–2 in a subtle way. In the earlier figure, because each category was different from every other one, we left a bit of a gap between bars. In Figure 2–5, the data are continuous, so it makes both statistical as well as esthetic sense[5] to have each bar abutting its neighbors. Now we can finally tell you the difference between bar charts and histograms:

Bar charts: *there are spaces between the bars.*
Histograms: *the bars touch each other.*

TABLE 2–3

Data from Table 2–2 put in rank order

1	17	26	35	43
7	17	27	35	43
7	17	27	35	43
7	18	27	36	45
9	19	28	36	46
11	20	28	36	46
11	21	28	37	47
11	22	29	37	49
12	22	30	37	51
12	24	31	38	51
13	24	31	38	51
14	24	31	38	52
14	24	32	38	52
14	25	32	39	54
15	25	33	41	56
15	25	33	41	56
16	26	33	42	57
16	26	34	42	58
16	26	34	42	61
16	26	34	43	66

TABLE 2–4

A summary of Table 2–3, showing the intervals, midpoints, counts, and cumulative total

Interval	Midpoint	Count	Cumulative total
0–4	2	1	1
5–9	7	4	5
10–14	12	9	14
15–19	17	11	25
20–24	22	8	33
25–29	27	15	48
30–34	32	12	60
35–39	37	14	74
40–44	42	9	83
45–49	47	5	88
50–54	52	6	94
55–59	57	4	98
60–64	62	1	99
65–69	67	1	100

[5] *Is "esthetic sense" an oxymoron?*

STEM-LEAF PLOTS AND RELATED FLORA

All these variants of histograms and bar charts are the traditional ways of taking a mess of data such as we found in Table 2–2 and transforming them into a graph such as Figure 2–5. The steps were:

1. Rank order the data.
2. Find the range (the highest value minus the lowest).
3. Choose an appropriate width to yield about 10 to 20 intervals.
4. Make a new table consisting of the intervals, their midpoints, the count, and a cumulative total.
5. Turn this into a histogram.
6. Lose some information along the way, consisting of the exact values.

Tukey (1977) devised a way to eliminate steps 1 and 6 and to combine 4 and 5 into one step. The resulting diagram, called a **Stem-and-Leaf Plot**, thus consists of only three steps:

1. Find the range.
2. Choose an appropriate width to yield about 10 to 20 intervals.
3. Make a new table that looks like a histogram and preserves the original data.

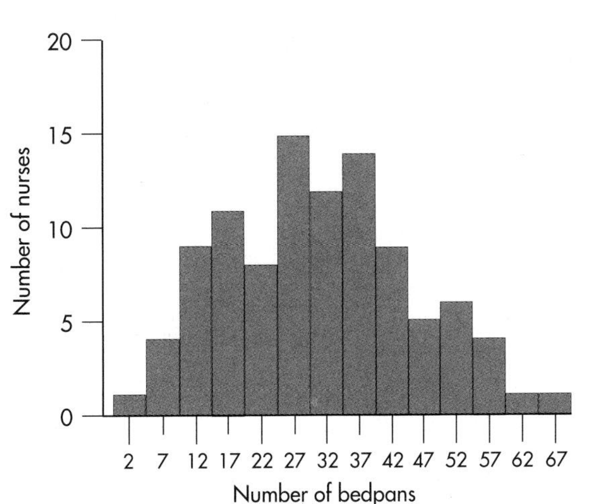

FIGURE 2–5

Histogram showing the number of bedpans emptied during the past month by each of 100 nursing students.

TABLE 2–5

First step in constructing a Stem-and-Leaf Plot: writing the stems

Stem	Leaf														
0															
0															
1															
1															
2															
2															
3															
3															
4															
4															
5															
5															
6															
6															

TABLE 2–6

Stem-and-Leaf Plot of the first 10 items of Table 2–2

Stem	Leaf														
0															
0															
1	1														
1	6														
2	4														
2															
3	3	4													
3	7														
4	3	1													
4	5														
5	1														
5															
6															
6															

TABLE 2–7

Stem-and-Leaf Plot of all the data in Table 2–2

Stem	Leaf														
0	1														
0	9	7	7	7											
1	1	4	4	4	2	1	1	3	2						
1	6	6	5	8	6	9	7	7	6	5	7				
2	4	4	4	2	1	0	4	2							
2	9	5	8	7	7	6	5	6	8	6	8	7	6	5	6
3	3	4	1	2	4	2	0	2	1	1	3	4			
3	7	5	6	6	5	6	8	7	5	7	8	8	9	8	
4	3	1	2	3	2	2	1	3	3						
4	5	5	6	9	7										
5	1	1	2	4	2	4									
5	5	7	8	6											
6	1														
6	6														

Let's take a look and see how this is done, at the same time explaining these somewhat odd-sounding terms. The "leaf" consists of the least significant digit of the number, and the "stem" is the most significant. So, for the number 94, the leaf is "4" and the stem is "9." If our data included numbers such as 167, we would make the "16" the stem. Using the data from Table 2–3 and the same reasoning we did for the histogram, we would again opt for an interval width of 5. We then write the stems we need, vertically, as in Table 2–5 (it's best to do this on graph paper, for reasons that will be readily apparent if you'll just be patient).

No, you are not seeing double. Table 2–5 really does have two 0s, two 1s, and so on. The reason is that, because we've chosen an interval width of 5, the first 0 will contain the numbers 0 to 4. Strictly speaking, the 0 is the stem of the numbers 00 (zero) to 04 (four). The second interval covers the numbers 5 (05) to 9 (09); the first 1 is the stem for the numbers 10 to 14, and the second for the numbers 15 to 19; and so on. Now, we go back to our original data and write the leaf of each number next to the appropriate stem. For example, the first number in Table 2–2 is 43, so we put a 3 (the leaf) next to the first 4. The second number, reading across, is 45 so we put a 5 next to the second 4, because this stem contains the intervals 45 to 49. If you did what we told you to earlier, and used graph paper, each leaf would be put in a separate and adjacent horizontal box. Table 2–6 shows a plot of the first 10 numbers, and Table 2–7 is the stem-and-leaf plot of all 100 numbers.

If you turn Table 2–7 sideways, you'll see it has exactly the same shape as does Figure 2–5. Moreover, the original data are preserved. Let's take the third line down, the first stem with a 1. Reading across, we can see that the actual numbers were 11, 14, 14, 14, 12, 11, 11, 13, and 12. If we want to be a bit fancier, we can actually rank order the numbers within each stem. Computer programs that product stem-leaf plots (see the end of this chapter) do this for you automatically. Most journals still prefer histograms or bar charts rather than stem-leaf plots, but this is slowly changing. In any case, it's simple to go from the plot to the more traditional forms.

FREQUENCY POLYGONS

Another way of representing interval or ratio types of data is called a **frequency polygon**. Let's start off by looking at one, and then we'll describe it. Now, look at Figure 2–6. This shows the same data as does Figure 2–5. However, instead of a bar that spans each interval, we've put a dot at the midpoint of the interval and then connected the dots with straight lines. There are a few other differences between histograms and frequency polygons.

First, as we've said, polygons should not be used with nominal or ordinal data because joining the dots makes the assumption that there is a smooth

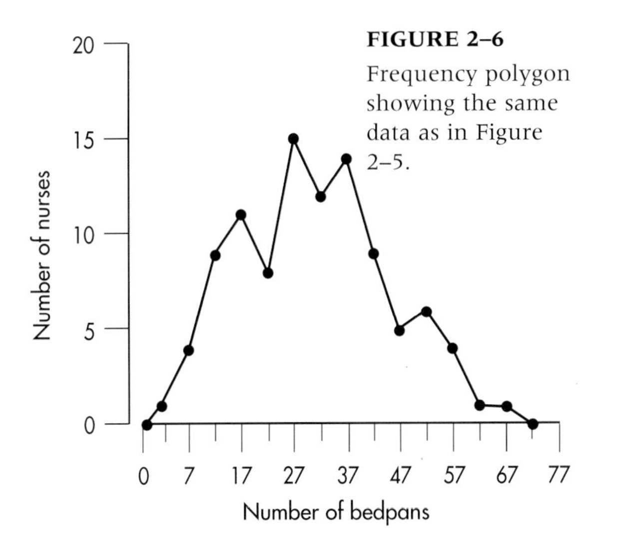

FIGURE 2–6 Frequency polygon showing the same data as in Figure 2–5.

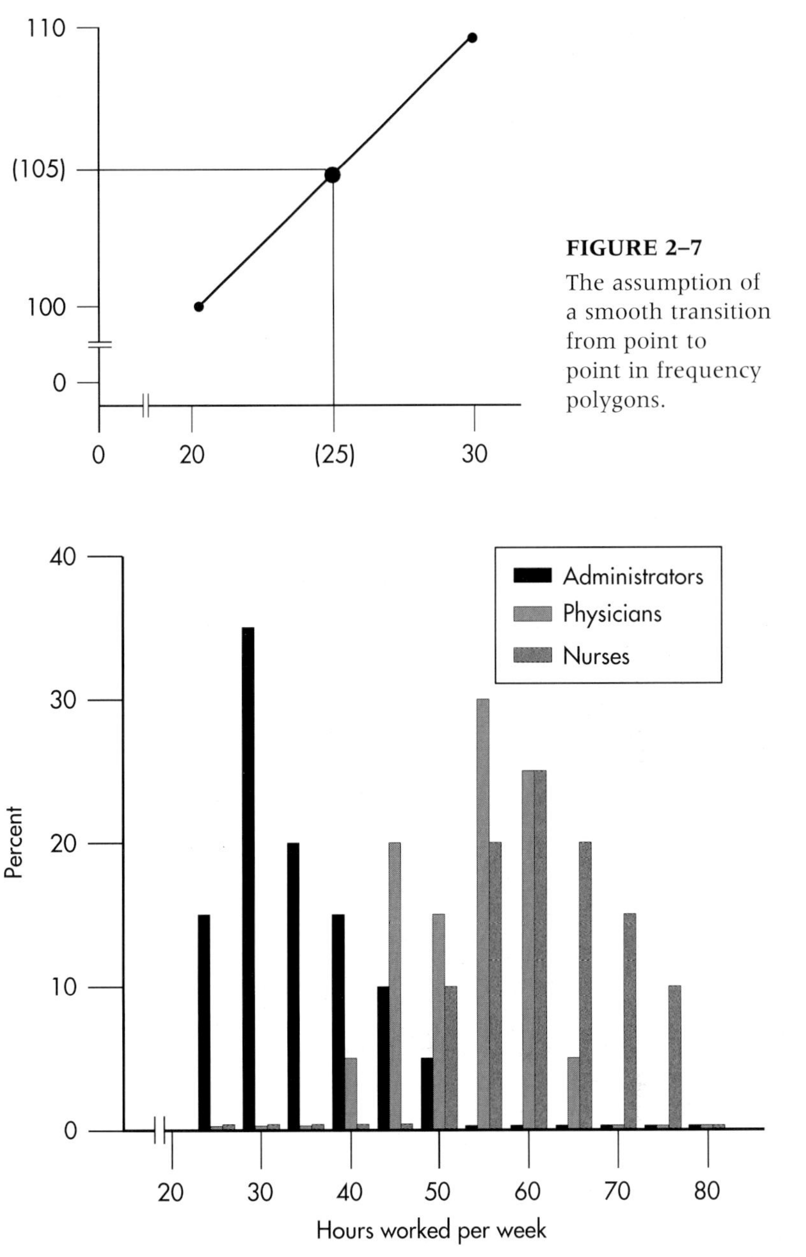

FIGURE 2–7 The assumption of a smooth transition from point to point in frequency polygons.

FIGURE 2–8 Data for three groups displayed as bar graphs.

transition from one datum point to another. For example, imagine that we have a polygon with just two points, as in Figure 2–7. The first point, at a midpoint of 20, shows 100 units on the *Y*-axis, and the second point, which falls at a midpoint of 30, shows 110 units. Even though we may not have gathered any data that correspond to an *X*-axis value of 25, we assume they fall on the line, half way between 20 and 30. In this case, they would correspond to 105 units (where the dot is). We can make this assumption only because we're using an interval or ratio level of data; if the distances between intervals are variable or unknown, as they are with ordinal data, we couldn't make this assumption.

A second difference is that bar charts seem to imply that the data are spread equally over the interval. For instance, if we had an interval width of 5 units spanning the numbers 20 through 24, and 10 cases were in that interval, it would appear (and we would assume) that 2 cases fell at 20, 2 at 21, 2 at 22, and so on. With a frequency polygon, we assume all the cases had the value of the midpoint. This is a closer representation of what we actually do in statistics; if we don't know the exact value of some variable, we usually use some midpoint as an approximation.

A third difference is that, by convention, frequency polygons begin and end with the line touching the *X*-axis. To accomplish this, we've added an extra interval at the upper end, which had a frequency count of zero. At the low end, it doesn't make sense in this case to add another interval because it would cover the numbers –1 to –5, so we just continue the line to the origin. If we were plotting data that did not include a value of zero, such as blood pressure, IQ, or height, we would have added an extra "empty" interval at the lower end.

So, when do we use a histogram and when a polygon? For nominal and ordinal data, you don't have a choice; you're limited to a histogram. If you're dealing with interval or ratio data and are showing the data for only one or two groups, it really doesn't matter; it's more a matter of personal preference, esthetics, and whatever your plotting package can manage. However, if you have more than two groups, then it's often better to use frequency polygons, with each group represented by a different line. The advantage is that all the data for any one group are joined; with a histogram, the values for one group are often broken up by the bars for the other groups. We've shown an example of this in Figure 2–8. Figure 2–9 then shows the same data with a polygon, which we feel is easier to follow.

When you're plotting two or more lines, they should be noticeably distinct from one another—different symbols representing the data points and different types of lines joining the points. If you're showing the graph at a meeting, you can also use

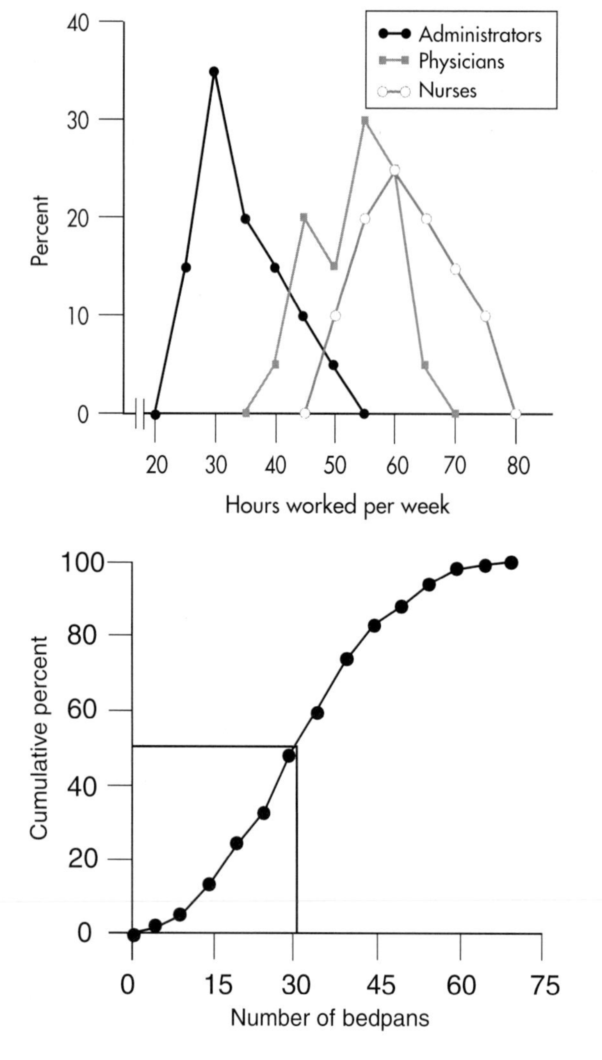

FIGURE 2–9

The same data as in Figure 2–8, but displayed as frequency polygons. The lines are differentiated by color, symbol type, and line type.

FIGURE 2–10

Cumulative frequency polygon of data in Figure 2–6.

different colors; however, most publications are in black and white, so this isn't an option.[6]

CUMULATIVE FREQUENCY POLYGONS

Before leaving the topic of graphing for a while, we'll mention one more variant, a **cumulative frequency polygon**. Cast your mind back, if you will, to our discussion of the emptying of bedpans. When we drew up Table 2–4, we added another column, labeled the Cumulative Total, and mentioned that one reason for using it was as a check on our addition. Now we'll mention another purpose; it helps us draw cumulative frequency polygons. With them, we plot not the raw count within each interval, but the cumulative count. You can also convert the cumulative total at each interval into a percentage of the total count and plot the cumulative *percents*, as we've done in Figure 2–10. In our example, because the total number of data points was 100, each cumulative total is also the percent, but you'll rarely be in the fortunate position of having exactly 100 subjects. Figure 2–10 again shows the data in Table 2–4, but this time as a cumulative polygon. The only difference in drawing a regular frequency polygon and a cumulative one is where we put the point: in the former case, it was at the midpoint; with cumulative polygons, we put the mark at the upper end of the interval, for reasons that will soon be apparent.

In Figure 2–10 we've drawn a horizontal line at 50%, starting at the *Y*-axis and extending to the curve, then dropped a vertical line to the *X*-axis. This shows us that 50% corresponds to 31 bedpans; that is, half of the people emptied fewer than 31 and half emptied more. We can also draw lines at other percentages, or even work backwards (i.e., draw a vertical line up from, say 40 bedpans, and see what percentage of people dumped more or fewer).

This is the reason the data are plotted at the end of the interval, rather than at the midpoint. As we've mentioned, we have lost some information by grouping the data, so we don't know exactly where within the interval the raw data actually occurred. We do know, though, how many cases there were, up to and including everyone within the interval. The difference may be small, but statisticians pride themselves on being accurate.[7]

Graphs of this sort are very common in plotting all sorts of anthropometric features, especially for kids—height, weight, head circumference, and other vital statistics. Then, after the doc takes the kid off the scale, she can look at a graph appropriate for age and sex and determine in what percentile this particular kid is.

MAKING BETTER TABLES

So far, we've been showing you different ways of presenting data in graphs, as if this were the only way that data can be portrayed. Indeed, graphs are excellent for displaying one or two variables at a time. There are times, though, when only a table of numbers will do—when we have many variables to show at the same time, or when we want the reader to see the actual numbers. It may seem at first glance as if tables were the simplest thing in the world to construct: just write the names of the variables as the headings of the columns, the subjects along the left to indicate the rows, and fill in the blanks. Table 2–8 is such a table, and it is typical of many you'll see. The countries are listed alphabetically, and the numbers are given with as much accuracy as possible. Now, quickly—which is the largest country? The smallest? The one with the highest GNP? The lowest infant mortality rate? If you think that was hard, imagine how hard it would be if we had listed all of the countries in Africa.[8]

Why was such a seemingly easy task so hard? The main reason is that there are too many numbers; not that there are too many columns but that we have "unnecessary" accuracy. Don't get us wrong; accuracy is good but, like a child, only in its place. If the exact numbers are important for archival purposes then, fine, maintain as many significant digits as you can come up with, but stick the table in an appendix. For most purposes, however, so many digits

[6] *Our publisher is a very generous guy and doesn't mind doing things in color.*

[7] *Even when working with inaccurate data.*

[8] *Don't bother to check; there are 55 of them.*

TABLE 2–8

Demographic characteristics of the countries in South America

Country	Area (km²)	Population	Per capita GNP (US$)	Births/ 1,000	Deaths/ 1,000	IMR*
Argentina	2,776,661	28,438,000	2,134	20	9	32
Bolivia	1,098,582	5,600,000	600	36	13	123
Brazil	8,506,663	110,098,992	2,434	27	7	67
Chile	756,946	11,275,400	1,979	21	6	18
Colombia	1,138,339	27,520,000	1,190	27	7	54
Ecuador	283,561	8,354,000	1,040	31	7	63
Guyana	214,970	820,000	324	31	12	49
Paraguay	406,752	2,973,000	1,180	36	6	49
Peru	1,285,215	17,031,221	1,850	29	8	69
Suriname	142,823	354,860	3,020	27	5	40
Uruguay	186,925	2,899,000	2,736	17	10	34
Venezuela	912,050	14,313,000	2,058	30	6	38

*IMR = infant mortality rate/1,000 live births.

gives an illusion of accuracy that is often misleading. For example, the population of Brazil is given as 110,098,992.[9] By the time you finish reading that number, it's already wrong. Even assuming that the census was correct when it was taken (a dubious assumption at best in developed countries, and most likely a myth in developing ones), it was out of date almost as soon as it was recorded. If the population increases by 3% a year, then there are nearly seven additional people every minute, or almost 10,000 a day. Between the time the census was taken (and don't forget it was probably taken over a period of weeks or months), recorded by the central government, reported in an official document, reproduced in the atlas, and read by you, years may have elapsed. That number is no longer correct—if it ever was to begin with—but the last three digits give the illusion of precision.

"Inaccurate precision" can be found all over the place. If we report that the average age of one group is 43.02 years, and for another group is 44.76 years, is that last decimal place really meaningful? Bear in mind that .01 years represents less than four days. Making the problem worse, we probably asked people their age to the nearest year, so they were introducing a loss of accuracy from the very outset (assuming that they didn't lie about their age[10]). Finally, without glancing up the page or at the table, do you remember the exact population of Brazil? Probably not; if you're like most people, you'll remember that it was somewhat over 110 million, but that final "90,992" has gone by the board. The moral of this story is to round, and then round again—keep enough digits to highlight important differences, and no more.

In Table 2–9, we've done some rounding. Now let's try the same exercise again: Which is the largest country? The smallest? The one with the highest GNP? The lowest infant mortality rate? That was much easier, wasn't it? Getting rid of unnecessary digits made the table much easier to comprehend. However, we can go even further. Keeping the countries in alphabetical order makes sense if this table is referred to often, or for a number of different purposes. But if there is one major point you want to make, such as focusing on the infant mortality rate (IMR), it would be even better to list the countries in order, ranging from the one with the highest IMR

[9] *All of whom squeeze into one stadium every Sunday to watch the soccer match.*

[10] *Isn't is somewhat passing strange that people will lie about their age, but not about their year of birth?*

TABLE 2–9

Data from Table 2–8, with rounding introduced

Country	Area (100,000 km²)	Population (1,000,000)	Per capita GNP (US$100)	Births/ 1,000	Deaths/ 1,000	IMR*
Argentina	28	28	21	20	9	32
Bolivia	11	6	6	36	13	123
Brazil	85	110	24	27	7	67
Chile	8	11	20	21	6	18
Colombia	11	28	12	27	7	54
Ecuador	3	8	10	31	7	63
Guyana	2	0.8	3	31	12	49
Paraguay	4	3	12	36	6	49
Peru	13	17	19	29	8	69
Suriname	1	0.3	30	27	5	40
Uruguay	2	3	27	17	10	34
Venezuela	9	14	21	30	6	38

*IMR = infant mortality rate/1,000 live births.

TABLE 2–10

Data from Table 2–9, reordered and with spaces

Country	IMR*	Births/ 1,000	Deaths/ 1,000	Per capita GNP (US$100)	Popula-tion (1,000,000)	Area (100,000 km²)
Bolivia	123	36	13	6	6	11
Peru	69	29	8	19	17	13
Brazil	67	27	7	24	110	85
Ecuador	63	31	7	10	8	3
Colombia	54	27	7	12	28	11
Guyana	49	31	12	3	0.8	2
Paraguay	49	36	6	12	3	4
Suriname	40	27	5	30	0.3	1
Venezuela	38	30	6	21	14	9
Uruguay	34	17	10	27	3	2
Argentina	32	20	9	21	28	28
Chile	18	21	6	20	11	8

*IMR = infant mortality rate/1,000 live births.

to the lowest (or vice versa). Then, ask yourself, Are the other columns really necessary? If you want to relate the IMR to the size of the country or to other indices of health such as the birth and death rates, then keep them; otherwise, out they go (or into an appendix). This also means that it may be worthwhile to reorder the columns; if IMR is the most important point of the table, list it first. If you want to relate it primarily to other health indices, then the birth and death rates are next, followed by the country's per capita income and size.

Finally, use spaces to highlight clusters, as in Table 2–10 in which information is arranged logically for you. For example, Bolivia seems to be in its own class, with an IMR that is quite a bit higher than that of the next country. Then, there appears to be a group of countries with a gradation of similar IMRs, and Chile is by itself, with the lowest IMR.[11] Those divisions are totally arbitrary; if you feel that there should be a break between Paraguay and Suriname, for instance, you can put one in; we're flexible.

If you *really* want to learn how to make good tables, look at the article and book by Wainer listed in the "To Read Further" section; that's where we got all these good ideas.

[11] It's probably no coincidence that they also produce the finest wine in South America. On the other hand, have you ever tasted Bolivian wine? We rest our case.

EXERCISES

Let's take another look at some of the variables we used in the exercises for Chapter 1, as well as a few others to minimize boredom. This time, though, indicate what type of graph you'd use to present the data (*bar chart, histogram, frequency polygon,* or something else). Just to keep you on your toes, there is sometimes more than one correct answer.

1. Number of hair transplant sessions per person.
2. Time since the last patient indicated his/her gratitude.
3. The number of patients with 0, 1, or 2+ vessels with >75% stenosis.
4. Before-taxes income.
5. Income for the different specialties in your profession.
6. Range of wrist motion for 100 patients.
7. Schmedlap Anxiety Inventory scores for 128 people.

How to Get the Computer to Do the Work for You

Histograms

- From **Graphs**, choose **Bar**
- The default is **Simple**; keep it and then click the **Define** button
- Click the name of the variable you want to graph from the list on the left, and click the arrow next to **Category Axis**
- You can name the axes by clicking the button marked **Titles**
- Click **OK**

Frequency Polygons

- From **Graphs**, choose **Line**
- The default is **Simple**; keep it and then click the **Define** button
- Click the name of the variable you want to graph from the list on the left, and click the arrow next to **Category Axis**
- You can name the axes by clicking the button marked **Titles**
- Click **OK**

Cumulative Frequency Polygons

- From **Graphs**, choose **Line**
- The default is **Simple**; keep it and then click the **Define** button
- In the **Line Represents** area, check **Cum. n of cases**
- Click the name of the variable you want to graph from the list on the left, and click the arrow next to **Category Axis**
- You can name the axes by clicking the button marked **Titles**
- Click **OK**

Stem-and-Leaf Plots

- From **Analyze**, choose **Summarize** → **Explore**
- Click the **Plots** button
- Choose **Stem-and-Leaf** and click the **Continue** button
- Click the name of the variable you want to graph from the list on the left, and click the arrow next to **Dependent List**
- Click **OK**

In this chapter, we discuss how to summarize the data with just a few numbers: measures of central tendency (such as the mode, median, and mean), and measures of dispersion (such as the range and standard deviation).

CHAPTER THE THIRD

Describing the Data with Numbers

Measures of Central Tendency and Dispersion

Graphing the data is a necessary first step in data analysis, but it has two limitations. First, if someone asks you to describe the essence of what you found, all you can do is find a spare napkin (preferably unused), and draw a graph. Second, there's not much we can do with the results, except show them; we can't easily compare the results of two or more different groups or see if they differ in important ways.[1] It would be helpful if we could summarize the results with just a few numbers. Not surprisingly, those numbers exist. The two most important are measures of *central tendency* and of *dispersion*. (We will later discuss two other indices, called skewness and kurtosis.)

[1] *Even more important, there wouldn't be any work for statisticians, and they'd have to find an honest profession.*

However, before we introduce these two terms, a brief diversion is in order to introduce some of the shorthand notation that is used in statistics.

A SLIGHT DIGRESSION INTO NOTATION

A specific data point—that is, the value of a variable for one subject—is represented by the capital letter X. The small letter x is used to denote something different, which we'll get to later in this chapter. In Table 2–2, for subject 1, $X = 43$. We denote the *mean* (see below for definition) of a variable by putting a bar over the capital letter X: $\bar{X}$. When speaking to another statistician, we can say either "the mean" or "X bar."[2]

[2] *"X bar" means "the arithmetic mean (AM)"; it is not the name of a drinking place for divorced statisticians (see the glossary at the end of the book).*

In recent years, we've seen a new abbreviation for the mean: M. This is because most word processing packages can't handle $\bar{X}$. In other words, we have to adapt our practices to accommodate the needs of the computer; does that strike you as bizarre as it does us?

The number of subjects in the sample is represented by N. There is no convention on whether to use uppercase or lowercase, but most books use a lowercase n to indicate the sample size for a group when there are two or more and use the upper case N to show the entire sample, summed over all groups. If there is only one group, take your pick and you'll find someone who'll support your choice. If there are two or more groups, how do we tell which one the n refers to? Whenever we want to differentiate between numbers, be they sample sizes, data points, or whatever, we use subscript notation. That is, we put a subscript after the letter to let us know what it refers to—n_1 would be the sample size for group 1, X_3 the value of X for subject 3, and so on.

To indicate adding up a series of numbers, we use the symbol Σ, which is the uppercase Greek letter sigma. (The lowercase sigma, σ, has a completely different meaning, which we'll discuss shortly.) If there is any possible ambiguity about the summation, we can show explicitly which numbers are being added, using the subscript notation:

$$\sum_{i=1}^{N} X_i \tag{3–1}$$

We read this as, "Sum over X-sub-i, as i goes from 1 to N." This is just a fancy way of saying "Add all the Xs, one for each of the N subjects."

X refers to a single data point. X_i is the value of X for subject i. n_j is the number of subjects (sample size) in group j. N is the total sample size. $\bar{X}$ is the arithmetic mean. Σ means to sum.

Later in this book, we'll get even fancier, and even show you some more Greek. But for now, that's enough background and we're ready to return to the main feature.

MEASURES OF CENTRAL TENDENCY

The Mean

Just to break the monotony, let's begin by discussing interval and ratio data and work our way down through ordinal to nominal. Take a look at Figure 3–1, where we've added a second group to the bedpan data from the previous chapter. As you can see, the shape of its distribution is the same as the first group's, but it's been shifted over by 15 units. Is there any way to capture this fact with a number?[3] One obvious way is to add up the total number of bedpans emptied by each group. For the first group, this comes to 3,083.[4] Although we haven't given you the data, the total for the second group is 4,583. This immediately tells us that the second group worked harder than the first (or had more patients who needed this necessary service).

However, we're not always in the position where both groups have exactly the same number of subjects. If the students in the second group worked just as hard, but they numbered only 50, their total would be only 2,291 or so. It's obvious that a better way would be to divide the total by the number of data points so that we can directly compare two or more groups, even when they comprise different numbers of subjects. So, dividing each total by 100, we get 30.83 for the first group and 45.83 for the second. What we've done is to calculate the *average* number of bedpans emptied by each person. In statistical parlance, this is called the **arithmetic mean (AM)**, or the **mean**, for short.

The reason we distinguish it by calling it the arithmetic mean is because there are other means, such as the harmonic mean and the geometric mean, both of which we'll touch on (very briefly) at the end of this chapter. However, when the term *mean* is used without an adjective, it refers to the AM. If there is any room for confusion (and there's *always* room for confusion in this field), we'll use the abbreviation. Using the notation we've just learned, the formula for the mean is:

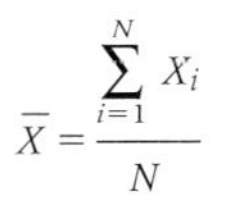

$$\bar{X} = \frac{\sum_{i=1}^{N} X_i}{N} \quad (3\text{–}2)$$

We spelled out the equation using this formidable notation for didactic purposes. From now on, we'll use conceptually more simple forms in the text unless there is any ambiguity. Because there is no ambiguity regarding what values of X we're summing over, we can simplify this to:

The Arithmetic Mean

$$\bar{X} = \frac{\Sigma X}{N} \quad (3\text{–}3)$$

The mean is the measure of *central tendency* for interval and ratio data.

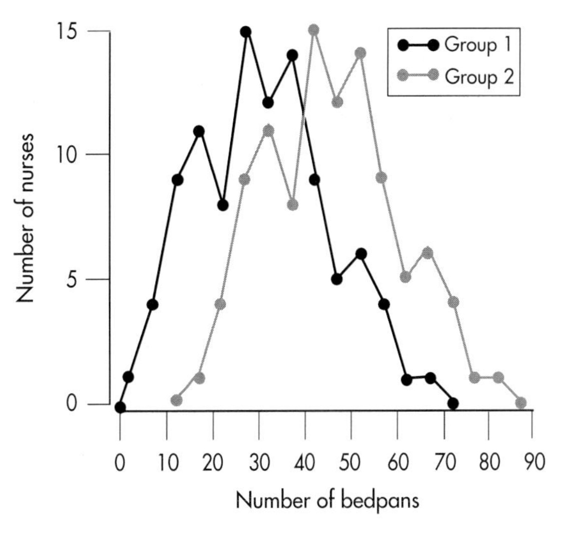

FIGURE 3–1
Graphs of two groups, with the second shifted to the right by 15 units.

A *measure of central tendency* is the "typical" value for the data.

One of the ironies of statistics is that the most "typical" value, 30.83 in the case of Group 1 and 45.83 for Group 2, never appears in the original data. That is, if you go back to Table 2–2, you won't find anybody who dumped 30.83 bedpans, yet this value is the most representative of the group as a whole.[5]

The Median

What can we do with ordinal data? It's obvious (at least to us) that, because they consist of ordered categories, you can't simply add them up and divide by the number of scores. Even if the categories are represented by numbers, such as Stage I through Stage IV of cancer, the "mean" is meaningless.[6] In this case, we use a measure of central tendency called the **median**.

The *median* is that value such that half of the data points fall above it and half below it.

Let's start off with a simple example; we have the following 9 numbers: 1, 3, 3, 4, 6, 13, 14, 14, and 18. Note that we have already done the first step, which is to put the values in rank order. It is immaterial whether they are in ascending or descending order. Because there is an odd number of values, the middle one, 6 in this case, is the median; four values are lower and four are higher.

If we added one more value, say 17, we'd have an even number of data points, and the median would be the AM of the 2 middle ones. Here, the middle values would be 6 and 13, whose mean is (6 + 13) ÷ 2 = 9.5; this would then be taken as the median. Again, half of the values are at or below 9.5 and half located at or above 9.5. (On a somewhat technical level, this approach is logically inconsistent. We're calculating the median because we're not supposed to use the mean with ordinal data. If that's the case, how can we then turn around and calculate this mean of the middle values? Strictly speaking, we can't, but yet we do.)

[3] *By now, you should have learned that we never ask a question unless we know beforehand what the answer will be.*

[4] *If you don't believe us, you can add up the numbers in Table 2–2!*

[5] *This is like the advice to a nonswimmer, to never a cross a stream just because its average depth is four feet.*

[6] *It also seems ridiculous to write that the mean stage is II.LXIV (that's 2.64, for those of you who don't calculate in Latin).*

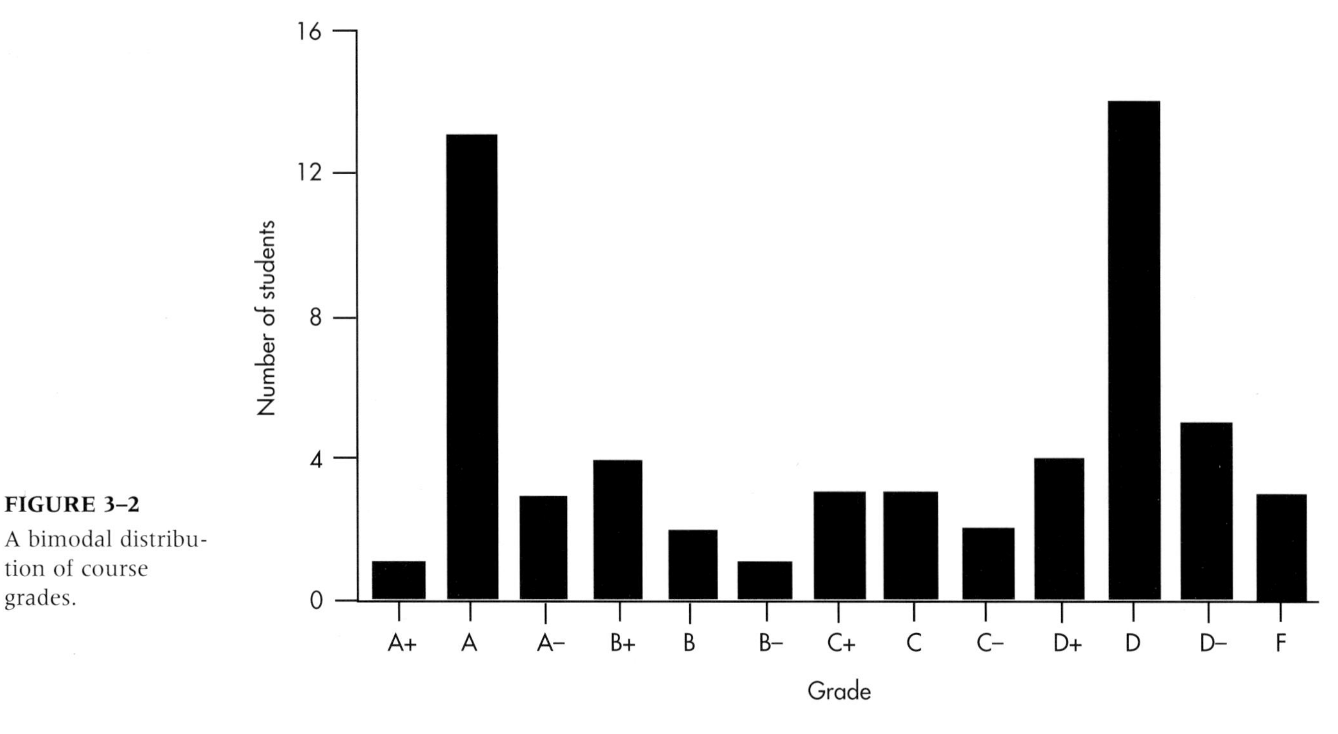

FIGURE 3–2
A bimodal distribution of course grades.

FIGURE 3–3
Two groups, differing in the degree of dispersion.

If the median number occurs more than once (as in the sequence: 5 6 7 7 7 10 10 11), some purists calculate a median that is dependent on the number of values above and below the dividing line (e.g., there are two 7s below and one above). Not only is this a pain to figure out, but the result rarely differs from our "impure" method by more than a few decimal places.

As we've said, the median is used primarily when we have ordinal data. But there are times when it's used with interval and ratio data, too, in preference to the mean. If the data aren't distributed symmetrically, then the median gives a more representative picture of what's going on than the mean. We'll discuss this in a bit more depth at the end of this chapter, after we introduce you to some more jargon; so be patient.

The Mode

Even the median can't be used with nominal data. The data are usually named categories and, as we said earlier, we can mix up the order of the categories and not lose anything. So the concept of a "middle" value just doesn't make sense. The measure of central tendency for nominal data is the **mode**.

> The *mode* is the most frequently occurring category.

If we go back to Table 2–1, the subject that was endorsed most often was Economics, so it would be the mode. If two categories were endorsed with the same, or almost the same[7] frequency, the data are called bimodal. This happened in one course I had in differential equations: if you understood what was being done, the course was a breeze; if you didn't, no amount of studying helped. So, the final marks looked like those in Figure 3–2—mainly As and Ds, with a sprinkling of Bs, Cs, and Fs. If there were three humps in the data, we could use the term *trimodal*, but it's unusual to see it in print because statisticians have trouble counting above two. However, you'll sometimes see the term *multimodal* to refer to data with a lot of humps[8] of almost equal height.

MEASURES OF DISPERSION

So far we've seen that distributions of data (i.e., their shape) can differ with regard to their central tendency, but there are other ways they can differ. For example, take a look at Figure 3–3. The two curves have the same means, marked $\bar{X}$, yet they obviously do not have identical shapes; the data points in Group 2 cluster closer to the mean than those in Group 1. In other words, there is less **dispersion** in the second group.

> A *measure of dispersion* refers to how closely the data cluster around the measure of central tendency.

[7] *The quantity "almost the same" is mathematically determined by turning to your neighbor and asking, "Does it look almost the same to you?"*

[8] *Another technical statistical term.*

This time, we'll begin with nominal data and work through to interval and ratio data. In fact, making our task even easier, we can dispense entirely with a measure of dispersion for nominal data; there isn't one. About all we can do is state how many categories were used. However, this is a fixed number in many situations; there are only two sexes, a few political parties,[9] and so on.

The Range

Having dispensed with nominal data, let's move on to ordinal data. When ordinal data comprise named, ordered categories, then they are treated like nominal data; you can say only how many categories were used. However, if the ordinal data are numeric, such as the rank order of students within a graduating class, we can use the **range** as a measure of dispersion.

> The *range* is the difference between the highest and lowest values.

If we had the numbers 102, 109, 110, 117, and 120, then the range would be (120 − 102) = 18. Do not show your ignorance by saying, "the range is 102 to 120," even though we're sure you've seen it in even the best journals. The range is always one number. The main advantage of this measure is that it's simple to calculate. Unfortunately, that's about the only advantage it has, and it's offset by several disadvantages. The first is that, especially with large sample sizes, the range is unstable, which means that its value can change drastically with more data or when a study is repeated. That means that if we add new subjects, the range will likely increase. The reason is that the range depends on those few poor souls who are out in the wings—the midgets and the basketball players. All it takes is one midget or one stilt in the sample, and the range can double. It follows that the more people there are in the sample, the better are the chances of finding one of these folks. So, the second problem is that the range is dependent on the sample size; the larger the number of observations, the larger the range. Last, once we've calculated the range, there's precious little we can do with it.

However, the range isn't a totally useless number. It comes in quite handy when we're describing some data, especially when we want to alert the reader that our data have (or perhaps don't have) some oddball values. For instance, if we say that the mean length of stay on a particular unit is 32 days, it makes a difference if the range is 10 as opposed to 100. In the latter case, we'd immediately know that there were some people with very long stays, and the mean may not be an appropriate measure of central tendency, for reasons we'll go into shortly.

The Interquartile Range

Because of these problems with the range, especially its instability from one sample to another or when new subjects are added, another index of dispersion is sometimes used with ordinal data, the **interquartile range** (sometimes referred to as the **midspread**). To illustrate how it's calculated, we'll use some real data for a change. Table 3–1 shows the length, width, breadth, and gonad grade for 35 littleneck clams, *Protothaca staminea*, harvested in Garrison Bay. These data were taken from a book by Andrews and Herzberg (1985), called, simply, *Data*. Although our book is intended as family reading, we had to include the data on the gonad grade of these clams because we will be using them later on in this section.[10] If any reader is under 16 years of age, please read the remainder of this section with your eyes closed. And yes, we know the data are ratio, but you can use this technique with ordinal, interval, and ratio data.

For this part, we'll focus on the data for the width; to save you the trouble, we've rank ordered the data on this variable and indicated the median and the upper and lower quartiles.[11] Remember that the median divides the scores into two equal sections, an upper half and a lower half. There are 35 numbers in Table 3–1, so the median will be the

TABLE 3–1 Vital statistics on 35 littleneck clams

Clam number	Length (mm)	Width (mm)	Breadth (mm)	Gonad grade	
31	91	77	42		
30	169	141	81	0	
29	305	264	172		
28	330	268	188	3	
23	420	282	265	3	
18	335	288	193	3	
26	393	333	209	2	
16	394	338	253	3	
24	402	340	216		Q_L = 340
25	410	349	253	3	
27	389	356	249	3	
15	455	385	269	3	
11	459	394	282	2	
9	452	395	282	3	
12	449	397	278	3	
14	471	401	271	2	
22	465	402	299	2	
5	487	407	286	2	Median = 407
7	485	408	298	3	
13	472	408	281	3	
4	512	413	302	2	
21	474	414	317	3	
10	468	417	272	3	
17	475	422	287	3	
6	481	427	315	3	
8	479	430	314	2	
34	509	433	284	3	Q_U = 433
20	486	436	275	3	
35	511	447	285	3	
33	519	456	312	3	
19	508	464	298	3	
3	505	471	338	3	
2	517	477	334	3	
1	530	494	337	3	
32	537	498	345	3	

[9] *Except in Italy and Israel, where the number of parties is variable and equal to one more than the sum of the total population.*

[10] *The data, not the gonads.*

[11] *These data are kosher, although the subject matter isn't. However, we couldn't find any data on hole sizes in bagels or the degree of heartburn following Mother's Friday night meal.*

TABLE 3–2 Calculation of the mean deviation

Column 1 number of coffee breaks X	Column 2 raw deviation $X - \overline{X}$	Column 3 absolute deviation $\lvert X - \overline{X} \rvert$	Column 4 squared deviation $(X - \overline{X})^2$
1	−8	8	64
3	−6	6	36
4	−5	5	25
7	−2	2	4
9	0	0	0
9	0	0	0
11	2	2	4
12	3	3	9
16	7	7	49
18	9	9	81
$\Sigma X = 90$	$\Sigma(X - \overline{X}) = 0$	$\Sigma \lvert X - \overline{X} \rvert = 42$	$\Sigma(X - \overline{X})^2 = 272$

eighteenth number, which is 407. Now let's find the median of the lower half, using the same method. It's the ninth number, 340, and this is the lower quartile, symbolized as Q_L. In the same way, the upper quartile is the median of the upper half of the data; in Table 3–1, Q_U is 433. So, what we've done is divide the data into four equal parts (hence the name **quartile**).

> The *interquartile range* is the difference between Q_L and Q_U and comprises the middle 50% of the data.

Because the interquartile range deals with only the middle 50% of the data, it is much less affected by a few extreme scores than is the range, making it a more useful measure. We'll meet up with this statistic later in this chapter, when we deal with another way of presenting data, called **box plots**.

Variations on a Range

The interquartile range, which divides the numbers into quarters, is perhaps the best known of the ranges, but there's no law that states that we have to split the numbers into four parts. For example, we can use *quintiles* that divide the numbers into five equally sized groups, or *deciles* that break the data into 10 groups. Having done that, we can specify a range that includes, for example, the middle 30% ($D_6 - D_4$), 50% ($D_7 - D_3$), or 70% ($D_8 - D_2$) of the data. The choice depends on what information we want to have: narrower intervals (e.g., $D_6 - D_4$) contain less of the data but fall closer to the median; wider ranges (e.g., $D_7 - D_3$) encompass more of the data but are less accurate estimates of the median.

The Index of Dispersion

The final index of dispersion for nominal and ordinal data is called, with an amazing degree of originality, the *index of dispersion*. It is defined as:

$$D = \frac{k\,(N^2 - \Sigma f_i^2)}{N^2\,(k - 1)} \qquad \textbf{(3–4)}$$

where k is the number of categories, f_i the number of ratings in each category, and N the total number of ratings. If all of the ratings fall into one category, then D is zero; whereas, if the ratings were equally divided among the k categories, D would be equal to 1. If we go back to the data in Table 2–1, $k = 5$ (Sociology, Economics, History, Psychology, and Calculus); $f_1 = 25$, $f_2 = 42$, and so on; $\Sigma f_i^2 = 2{,}766$; and $N = 100$. Therefore,

$$D = \frac{5\,(100^2 - 2766)}{100^2\,(4)} = \frac{36{,}170}{40{,}000} = 0.904 \qquad \textbf{(3–5)}$$

showing a nice spread of scores across the courses. However, if the eight people who rated History as their most boring course changed their responses to Calculus, meaning that only four of the five categories were used, then D would drop to 0.370.

The Mean Deviation

An approach that at first seems intuitively satisfying with interval and ratio data would be to calculate the mean value and then see how much each individual value varies from it. We can denote the difference between an individual value and the mean either by $(X - \overline{X})$ or by the lowercase letter, x. Column 1 of Table 3–2 shows the number of coffee breaks taken during 1 day by 10 people;[12] their sum, symbolized by ΣX, is 90. Dividing this by N, which is 10, yields a mean of 9. Column 2 shows the results of taking the difference between each individual value and 9. The symbols at the bottom of Column 2, $\Sigma(X - \overline{X})$, signify the sum of the differences between each value and the mean. We could also have written this as Σx. Adding up these 10 deviations results in—a big zero. This isn't just a fluke of these particular numbers; by definition, *the sum of the deviations of* any *set of numbers around its mean is zero.* So clearly, this approach isn't going to tell us much. We can get around this problem by taking the absolute value of the deviation; that is, by ignoring

[12] *Judging from the numbers, obviously civil servants.*

the sign. This is done in Column 3, where taking the absolute value of a number is indicated by putting the number between the vertical bars: $|+3| = 3$, and $|-3| = 3$. The sum of the absolute deviations is 42. Dividing this by the sample size, 10, we get a mean deviation of 4.2; that is, the average of the absolute deviations. To summarize the calculation:

$$\text{Mean deviation (MD)} = \frac{\Sigma|X - \overline{X}|}{N} = \frac{\Sigma|x|}{N} \tag{3-6}$$

This looks so good, there must be something wrong, and in fact there is. Mathematicians view the use of absolute values with the same sense of horror and scorn with which politicians view making an unretractable statement. The problem is the same as with the mode, the median, and the range; absolute values, and therefore the mean deviation (MD), can't be manipulated algebraically, for various arcane reasons that aren't worth getting into here.

The Variance and Standard Deviation

But all is not lost. There is another way to get rid of negative values: by squaring each value.[13] As you remember from high school, two negative numbers multiplied by each other yield a positive number: $-4 \times -3 = +12$. Therefore any number times itself must result in a positive value. So, rather than taking the absolute value, we take the square of the deviation and add these up, as in Column 4. If we left it at this, then the result would be larger as our sample size grows. What we want, then, is some measure of the average deviation of the individual values, so we divide by the number of differences, which is the sample size, N. This yields a number called the **variance**, which is denoted by the symbol s^2.

$$s^2 = \frac{\Sigma(X - \overline{X})^2}{N} = \frac{\Sigma x^2}{N} \tag{3-7}$$

This is more like what we want, but there's still one remaining difficulty. The mean of the 10 numbers in Column 1 is 9.0 coffee breaks per day, and the variance is 27.2 squared coffee breaks. But what the #&$! is a squared coffee break? The problem is that we squared each number to eliminate the negative signs. So, to get back to the original units, we simply take the square root of the whole thing and call it the **standard deviation**, abbreviated as either SD or s:

The Standard Deviation

$$s = \sqrt{\frac{\Sigma(X - \overline{X})^2}{N}} = \sqrt{\frac{\Sigma x^2}{N}} \tag{3-8}$$

The result, 5.22 (the square root of 27.2), looks more like the right answer. So, in summary, the SD is the square root of the average of the squared deviations of each number from the mean of all the numbers, and it is expressed in the same units as is the original measurement. The closer the numbers cluster around the mean, the smaller s will be. Going back to Figure 3–3, Group 1 would have a larger SD than would Group 2.

Do NOT use the above equation to actually calculate the SD. To begin with, you have to go through the data three times: once to calculate the mean, a second time to subtract the mean from each value, and a third time to square and add the numbers. Moreover, because the mean is often a decimal that has to be rounded, each subtraction leads to some rounding error, which is then magnified when the difference is squared. Computers use a different equation which minimizes these errors. Finally, this equation is appropriate only in the highly unlikely event that we have data from every possible person in the group in which we're interested (e.g., all males in the world with hypertension). After we distinguish between this situation and the far more common one in which we have only a sample of people (see Chapter 6), we'll show you the equation that's actually used.

Let's look for a moment at some of the properties of the variance and SD. Say we took a string of numbers, such as the ones in Table 3–2, and added 10 to each one. It's obvious that the mean will similarly increase by 10, but what will happen to s and s^2? The answer is, absolutely nothing. *If we add a constant to every number, the variance (and hence the SD) does not change.*

[13] *Erasing the minus sign is not considered to be good mathematical technique.*

SKEWNESS AND KURTOSIS

We've seen that distributions can differ from each other in two ways: in terms of their "typical" value (the measure of central tendency), and in how closely the individual values cluster around this typical value (dispersion). With interval and ratio data, we can use two other measures to describe the distribution: **skewness** and **kurtosis**. As usual, it's probably easier to see what these terms mean first, so take a look at the graphs in Figure 3–4. They differ from those in Figure 3–3 in one important respect. The curves in Figure 3–3 were symmetric, whereas the ones in Figure 3–4 are not; one end (or tail, in statistical parlance) is longer than the other. The distributions in this figure are said to be **skewed**.

Skew refers to the symmetry of the curve.

The terminology of skewness can be a bit confusing. Curve A is said to be **skewed right**, or to have a **positive skew**; Curve B is **skewed left**, or has a **negative skew**. So, the "direction" of the skew refers to the direction of the longer tail, not to where the bulk of the data are located. We're not going to give you the formula for computing skew

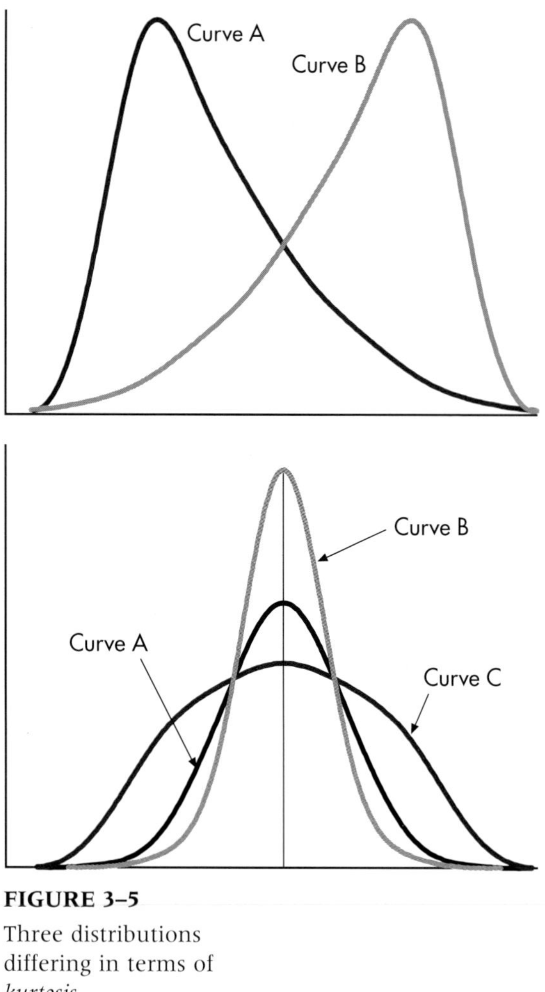

FIGURE 3–4
Two curves, one with positive and one with negative *skew.*

FIGURE 3–5
Three distributions differing in terms of *kurtosis.*

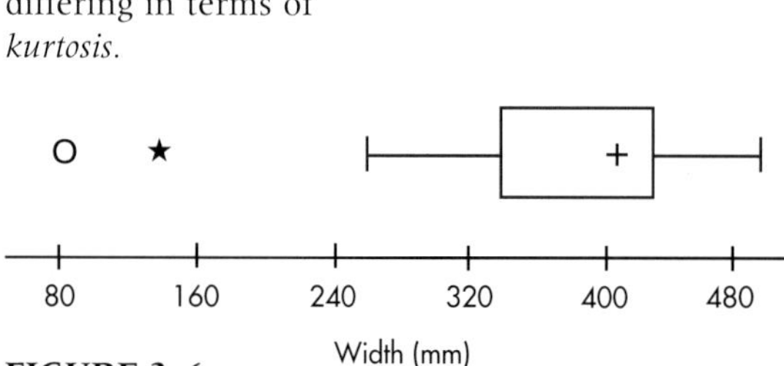

FIGURE 3–6
Box plot of widths of littleneck clams.

because we are unaware of any rational human being[14] who has ever calculated it by hand in the last 25 years. Most statistical computer packages do it for you, and we've listed the necessary commands for a few of them at the end of this chapter. A value of 0 indicates no skew, a positive number shows positive skew (to the right), and a negative number reflects negative, or left, skew.[15]

The three curves in Figure 3–5 are symmetric (i.e., their skew is 0), but they differ with respect to how flat or peaked they are, a property known as **kurtosis**. The middle line, Curve A, shows the classical "bell curve," or "normal distribution," a term we'll define in a short while. The statistical term for this is **mesokurtic**. Curve B is more peaked; we refer to this distribution as **leptokurtic**. By contrast, Curve C is flatter than the normal one; it's called **platykurtic**. The formula for calculating kurtosis, as for skew, would be of interest only to those who believe that wading through statistical text books makes them better people; such people are probably related to those who buy *Playboy* just for the articles. Again, most statistical computer packages figure out kurtosis for you. The normal distribution (which is mesokurtic) has a kurtosis of 3. However, many computer programs subtract 3 from this, so that it ends up with a value of 0, with positive numbers reflecting leptokurtosis, and negative numbers, platykurtosis. You'll have to check in the program manual[16] to find out what yours does.

> *Kurtosis* refers to how flat or peaked the curve is.

Although kurtosis is usually defined simply in terms of the flatness or peakedness of the distribution, it also affects other parts of the curve. Distributions that are leptokurtic also have heavier tails; whereas platykurtic curves tend to have lighter tails. However, kurtosis doesn't affect the variance of the distribution.

BOX PLOTS

Now that we've introduced some of these numeric ways of summarizing data, we will return to the realm of descriptive statistics and talk about one more type of graph. One of the most powerful graphing techniques, called the **box plot**, comes from the fertile brain of John Tukey (1977), who has done as much for exploring the beauty of data as Marilyn Monroe has done for the calendar.[17] Again, the best way to begin is to look at one (a box plot, not a calendar), and then describe what we see. Figure 3–6 shows the data for the width of those delightful littleneck clams we first encountered in Table 3–1.

Let's start off with the easy parts. The "+" in the middle represents the **median** of the distribution.[18] The ends of the box fall at the upper and lower **quartiles**, Q_U and Q_L , so the middle 50% of the cases fall within the range of scores defined by the box. Just this central part of the box plot yields a lot of information. We can see the variability of the data from the length of the box; the median gives us an estimate of central tendency; and the placement of the median tells us whether or not the data are skewed. If the median is closer to the upper quartile, as is the case with these numbers, the data are negatively skewed; if it is closer to the lower quartile, they are positively skewed.

The long lines coming out the sides are called **whiskers**. To fully understand them and their usefulness, we're going to have to introduce a bit more of Tukey's jargon. Remember that the interquartile range (IQR) was defined as $Q_U - Q_L$. A **step** is 1.5 times this value; that is, 1.5 box lengths. The end of the whisker (which may or may not have that small perpendicular line at the end of it) corresponds to

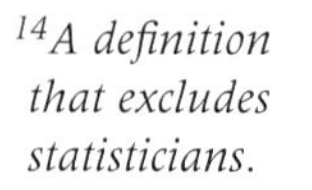

[14] *A definition that excludes statisticians.*

[15] *At least some things in statistics make sense.*

[16] *Usually something we do only as a last resort, when everything else has failed.*

[17] *Unfortunately, he has also done more to confuse people than did Abbott and Costello doing "Who's on First," by making up new terms for old concepts. For example, Tukey refers to something almost like the upper and lower quartiles as "hinges." As much as possible, we'll try to use the more familiar terms.*

[18] *Actually, there's no fixed convention for this. Some computer programs use a plus sign, others an asterisk. Tukey himself drew a solid line across the width of the box. But, because there's little ambiguity, this really doesn't matter too much.*

the **inner fence**. For simplicity's sake, let's talk about the upper whisker first. If an actual datum point falls exactly at one step, then the inner fence is drawn one step above the upper quartile. However, if a datum point doesn't happen to be there, then the fence is drawn to the largest observed value that is still less than one step away from Q_U. The same thing is done for the lower whisker. If a lot of data are about and the distribution is roughly symmetrical, then both whiskers will be about the same length. However, if the data points are relatively sparse on one side, it's possible that one whisker may be considerably shorter than the other, simply because no datum point is near the step. The **outer fence**, which is not usually drawn, is two steps beyond the quartile; that is, it's 3.0 times the interquartile range.

A logical question that arises (or should arise, if you're paying attention) is why the fences are chosen to be 1.5 and 3 times the IQR. These values actually make a lot of sense. If the data are normally distributed, then 95% of the data points would fall within the range defined by the inner fences, and 99% are encompassed by the outer fences.

Any data points that fall between the fences are called **outliers**, and any beyond the outer fence are called **far outliers**. Most computer packages that produce box plots differentiate between them, using different symbols for near and far outliers.[19] In Figure 3–6, there is one outlier and one far, or extreme, outlier, both falling at the lower end of the distribution.

Just to pull things together, Figure 3–7 labels the various parts of a box plot. Notice that we've drawn it vertically rather than horizontally. It can be drawn either way, but when we use box plots to compare two or more groups, they're probably easier to read in the vertical orientation.

FIGURE 3–7
Anatomy of a box plot.

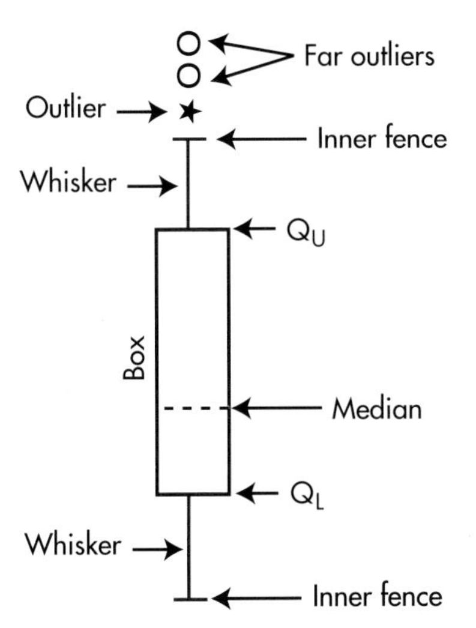

WHEN DO WE USE WHAT (AND WHY)?

Now that we have three measures of central tendency (the mode, the median, and the mean), and three measures of dispersion (the range, the interquartile range, and the SD), when do we use what? Under ideal circumstances, we can use the guidelines shown in Table 3–3.

For each listing, the most appropriate measures are listed first. If we have interval data, then our choice would be the mean and SD. Whenever possible, we try to use the statistics that are most appropriate for that level of measurement; we can do more statistically with the mean (and its SD) than with the median or mode, and we can do more with the median (and the range) than with the mode.

Having stated this rule, let's promptly break it. The mean is the measure of central tendency of choice for interval and ratio data when the data are symmetrically distributed around the mean, but not when things are wildly asymmetric; a synonym is "if the data are highly skewed." Let's see why. If the data are symmetrically distributed around the mean, then the mean, median, and mode all have the same value, as in Figure 3–8.

This isn't true for skewed distributions, though. Figure 3–9 shows some data with a positive skew, like physicians' incomes. As you can see, the median

[19] *For example, SPSS/PC uses an O for outliers and an E for extreme (i.e., far) outliers; Minitab uses an asterisk (*) for near outliers and an O for far outliers. So much for computers simplifying our lives.*

TABLE 3–3
Guidelines for use of central tendency and measure of dispersion

Type of data	Measure of central tendency	Measure of dispersion
Nominal	Mode	—
Ordinal	Median	Range
	Mode	Interquartile range
Interval	Mean	SD*
	Median	Range
	Mode	Interquartile range
Ratio	Mean	SD
	Median	Range
	Mode	Interquartile range

*SD—standard deviation.

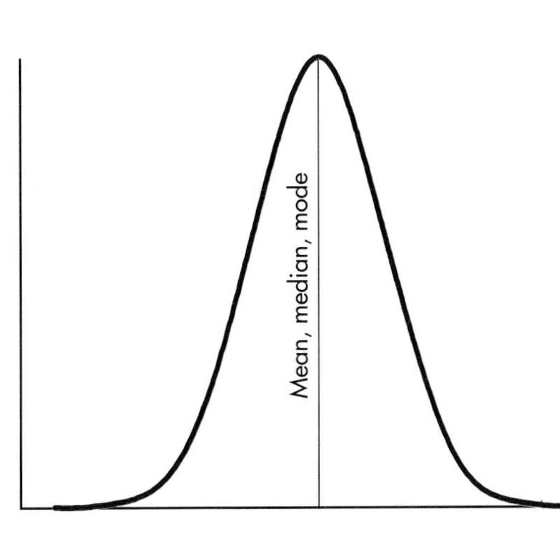

FIGURE 3–8
The mean, median, and mode in a symmetric distribution.

is offset to the right of the mode, and the mean is even further to the right than the median. If the data were skewed left, the picture would be reversed: the mode (by definition) would fall at the highest point on the curve, the median would be to the left of it, and the mean would be even further out on the tail. The more skewed the data, the further apart these three measures of central tendency will be from one another.

Another way data can become skewed is shown in Figure 3–10. If we ignore the oddball off to the right,[20] both the mode and the median of the 17 data points are 4, and the mean is 3.88. All these estimates of central tendency are fairly consistent with one another and intuitively seem to describe the data fairly well. If we now add that eighteenth fellow, the mode and median both stay at 4, but the mean increases to 6.06. So the median and the mode are untouched, but the mean value is now higher than 17 of the 18 values.

Similarly, the range of the 17 data points on the left is 5, and their SD is 1.41. After adding that one discrepant value, the range shoots up to 42 and the SD up to 9.32.

The moral of the story is that the median is much less sensitive to extreme values than is the mean. If the data are relatively well behaved (i.e., without too much skew), then this lack of sensitivity is a disadvantage. However, when the data are highly skewed, it becomes an advantage; for skewed-up data, the median more accurately reflects where the bulk of the numbers lie than does the mean.

[20]From our political perspective, most people off on the right are a bit odd.

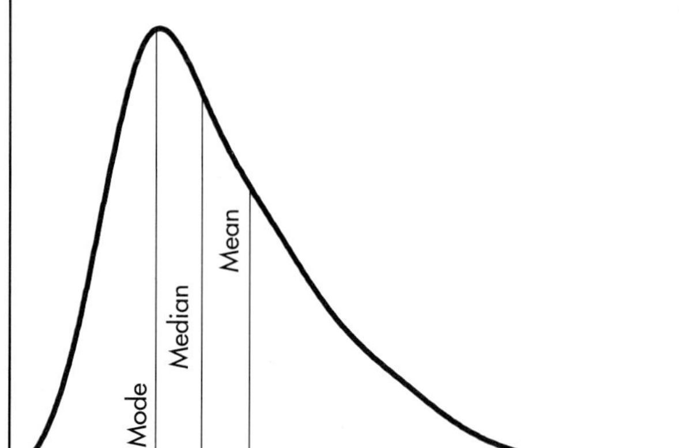

FIGURE 3–9
The mean, median, and mode in a skewed distribution.

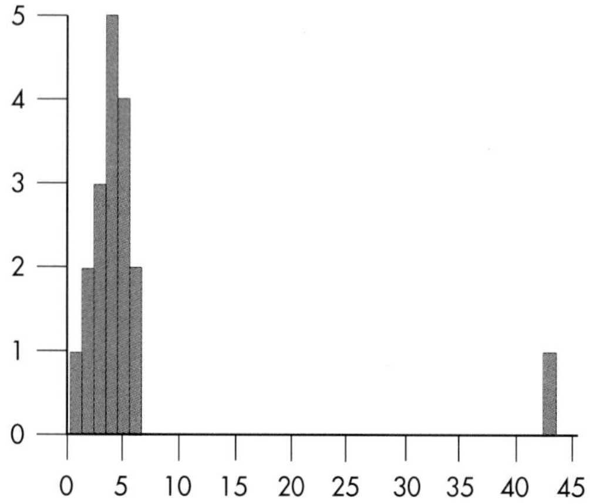

FIGURE 3–10
Histogram of highly skewed data.

OTHER MEASURES OF THE MEAN

Although the arithmetic mean is the most useful measure of central tendency, we saw that it's less than ideal when the data aren't normally distributed. In this section, we'll touch on some variants of the mean and see how they get around the problem.

The Geometric Mean

Some data, such as population growth, show what is called exponential growth; that is, if we were to plot them, the curve would rise more steeply as we move out to the right, as in Figure 3–11. Let's assume we know the value for X_8 and X_{10} and want to estimate what it is at X_9. If the value of X_8 is 138, and it is 522 for X_{10}, then the AM is (138 + 522) ÷ 2 = 330. As you can see in the graph, this overestimates the real value. On the other hand, the dot labeled *Geometric mean* seems almost dead on. The conclusion is that when you've got exponential or growth-type data, the geometric mean is a better estimator than is the AM.

The formula for the geometric mean is:

The Geometric Mean

$$GM = \sqrt[n]{\prod_{i=1}^{n} X_i} \qquad \textbf{(3–9)}$$

FIGURE 3–11
The difference between the arithmetic and geometric means.

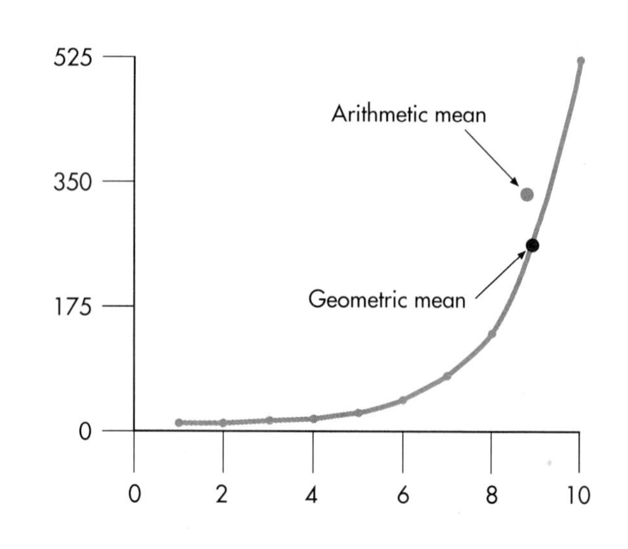

This looks pretty formidable, but it's not really that bad. The Greek letter π (pi) doesn't mean 3.14159; in this context, it means *the product* of all those *Xs*, So:

$$\sum_{i=1}^{3} X_i = X_1 + X_2 + X_3 \quad \text{(3-10)}$$

$$\prod_{i=1}^{3} X_i = X_1 \times X_2 \times X_3 \quad \text{(3-11)}$$

Then, the n to the left of the root sign ($\sqrt{\ }$) means that if we're dealing with two numbers, we take the square root; if there are three numbers, the cube root; and so on. In the example we used, there were only two numbers, so the geometric mean is:

$$GM = \sqrt[2]{138 \times 522} = \sqrt{72036} = 268.4 \quad \text{(3-12)}$$

Most calculators have trouble with anything other than square roots. So you can use either a computer or, if you're really good at this sort of stuff, logarithms. If you are so inclined, the formula using logs is:

$$GM = \text{antilog} \frac{1}{n} \sum_{i=1}^{n} \log X_i \quad \text{(3-13)}$$

Be aware of two possible pitfalls when using the GM, due to the fact that all of the numbers are multiplied together and then the root is extracted: (1) if any number is zero, then the product is zero, and hence the GM will be zero, irrespective of the magnitude of the other numbers; and (2) if an odd number of values are negative, then the product will be negative and the computer will have an infarct when it tries to take the root of a negative number.

The Harmonic Mean

Another mean that we sometimes run across is the *harmonic mean*; its formula is:

$$HM = \frac{n}{\sum_{i=1}^{n} \frac{1}{X_i}} \quad \text{(3-14)}$$

So, the harmonic mean of 138 and 522 is:

$$HM = \frac{2}{\frac{1}{138} + \frac{1}{522}} = 218.29 \quad \text{(3-15)}$$

Despite its name, it is rarely used by musicians (and only occasionally by statisticians). Usually, the only time it is used is when we want to figure out the average sample size of a number of groups, each of which has a different number of subjects. The reason for this is that, as we can see in Table 3–4, it gives the smallest number of the three means, and the statistical tests are a bit more conservative.[21] When all of the numbers are the same, the three means are all the same. As the variability of the numbers increases, the differences among the three means also increases, and the AM is always larger than the GM, which, in turn, is always larger than the HM.

[21] *Although why you'd want to be more conservative in this (or any) regard escapes us.*

TABLE 3–4 Different results for arithmetic, geometric, and harmonic means

Data	Arithmetic mean	Geometric mean	Harmonic mean
10, 10, 10, 10	10	10	10
9, 10, 10, 11	10	9.97	9.95
5, 10, 10, 15	10	9.31	8.57
1, 2, 18, 19	10	5.11	2.49

EXERCISES

1. Coming from a school advocating the superiority (moral and otherwise) of the SG-PBL approach (that stands for Small Group—Problem-Based Learning and is pronounced "skg-pble"), we do a study, randomizing half of the stats students into SG-PBL classes and half into the traditional lecture approach. At the end, we measure the following variables. For each, give the best *measure of central tendency* and *measure of dispersion*.
   a. Scores on a final stats exam.
   b. Time to complete the final exam (there was no time limit).
   c. Based on a 5-year follow-up, the number of articles each person had rejected by journals for inappropriate data analysis.
   d. The type of headache (migraine, cluster, or tension) developed by all of the students during class (i.e., in both sections combined).
2. Just to give yourself some practice, figure out the following statistics for this data set (we deliberately made the numbers easy, so you don't need a calculator): 4 8 6 3 4
   a. The *mean* is _______.
   b. The *median* is _______.
   c. The *mode* is _______.
   d. The *range* is _______.
   e. The *SD* is _______.
3. A study of 100 subjects unfortunately contains 5 people with missing data. This was coded as "99" in the computer. Assume that the true values for the variables are:

   $\bar{X} = 45.0$ SD = 5.6
   Minimum = 16   Maximum = 65

   If the statistician went ahead and analyzed the data as if the 99s were real data, would it make the following parameter estimates *larger, smaller,* or *stay the same?*
   a. The mode
   b. The median
   c. The mean
   d. The standard deviation
   e. The range.

How to Get the Computer to Do the Work for You

- From **Analyze**, choose **Summarize** → **Descriptives**
- Click on the variables you want to graph, and click the arrow next to the **Variable(s)** box
- Click the **Options** button
- Click the boxes you want (will likely include **Mean, Standard Deviation, Skewness,** and **Kurtosis**)
- Click **OK**

How to Get the Computer to Do the Work for You

- From **Graphs**, choose **Boxplot**
- Click the box marked **Simple** (unless you changed it before; it's the default option)
- In the box labeled **Data in Chart Are**, click **Summaries of separate variables**
- Click the **Define** button
- Click on the variable you want from the list on the left, and click the arrow to move it into the box labeled **Boxes Represent**
- Click **OK**

CHAPTER THE FOURTH

The Normal Distribution

The normal distribution is ubiquitous in statistics. Here, we discuss what it is, why it's useful, and how to use it.

SETTING THE SCENE

A survey of contraceptive practices found that the most widely used method is the phrase, "Not tonight, dear, I've got a headache," uttered by one or the other partner. Based on a survey of 2,000 people, it was found to be used an average of 100 times a year, with an SD of 15. Can we determine what proportion of the public uses this reason at least 115 times a year, fewer than 70 times a year, or between 106 and 112 times annually?

Before you can answer these important questions, you'll need to have some more information, starting with what we mean by a "normal distribution." We've made passing mention to it in the earlier chapters without really defining what it is. Now the moment of truth has come, and we'll tell you what is meant by a normal distribution and why you really want to know about it.

The normal curve has appeared in several previous figures, such as Figure 3–8, although it wasn't explicitly labeled as such. It's often referred to by a couple of other names, such as a **bell curve** or a **Gaussian distribution**. The term "bell curve" comes from its shape;[1] "Gaussian" from its discoverer.[2,3] So the alternative terms make sense and reflect attributes of the curve—its shape and history. Unfortunately, the standard term doesn't make sense; there's nothing inherently "normal" about this distribution, nor "abnormal" about other types.

WHY WE CARE ABOUT THE NORMAL DISTRIBUTION

There are several reasons why the normal curve is important. First, many of the statistical tests we'll be discussing in this book assume that the data come from a normal distribution. Second, with normally distributed data, the mean and variance aren't dependent on each other; if we increase the mean of a normal distribution, its variance should remain the same. This isn't true for many other types of distributions. Third, it's held that many natural phenomena are in fact approximately normally distributed. That is, if we were to measure the height, weight, blood pressure, or urine dehydroepiandrosterone level in a large number of people ("large" meaning at least 1,000) and make frequency polygons of our findings, they would each approximate the normal curve. Each measure, naturally, would have a different mean, but all of the curves would be roughly symmetric around their means and resemble that general shape. The only fly in the ointment is that the resemblance may be more illusory than real. Lippman (in Wainer and Thissen, 1976) put it well; he said, "Everybody believes in the theory of errors (the normal distribution). The experimenters because they think it is a mathematical theorem. The mathematicians because they think it is an experimental fact." On an empirical level, Micceri (1989) looked at the distributions of scores from well over 400 widely used psychological measures, such as achievement and aptitude tests, and found that distributions that were strictly normal were as rare as hen's teeth.[4]

[1] *And has led to the "gong phenomenon" —ask a statistician any question, and the first thing he or she will do is draw bell curve.*

[2] *Although rumor has it that, when lying on his back, Karl Friedrich Gauss himself resembled a Gaussian curve.*

[3] *A pity Alexander Graham Bell spent all his time on the phone. If he had discovered this curve, we would have only one name to remember.*

[4] *Thus you can say that, in some sense, normal curves are abnormal.*

The fourth reason that the normal distribution is important is that, whatever the distribution of the data, if we drew a large number of samples of reasonable size (we'll define "reasonable" shortly), then the distribution of the means of those samples will always be normally distributed. Now for the real heart of the matter—the data don't have to be normally distributed for this to be true because of what's called the **Central Limit Theorem**.

The *Central Limit Theorem* states if we draw equally sized samples from a **non-normal** distribution, the distribution of the means of these samples will *still* be normal, as long as the samples are large enough.

How large is "large?" Again, it all depends. If the shape of the population is pretty close to normal, then "large" can be as small as 2. If the population is markedly different from normal, then 10 to 20 may be large enough. To play it safe, though, we usually say that anything over 30 is enough under almost all circumstances.

We can illustrate this with another *gedanken* experiment. Imagine that we had a die that we rolled 600 times, and we recorded the number of times each face appeared. If the die wasn't loaded (and neither were we), no face would be expected to appear more often than any other. Consequently, we would expect that each number would appear one-sixth of the time, and we would get a graph that looks like Figure 4–1. This obviously is not a normal distribution; because of its shape, it's referred to as a **rectangular** distribution.

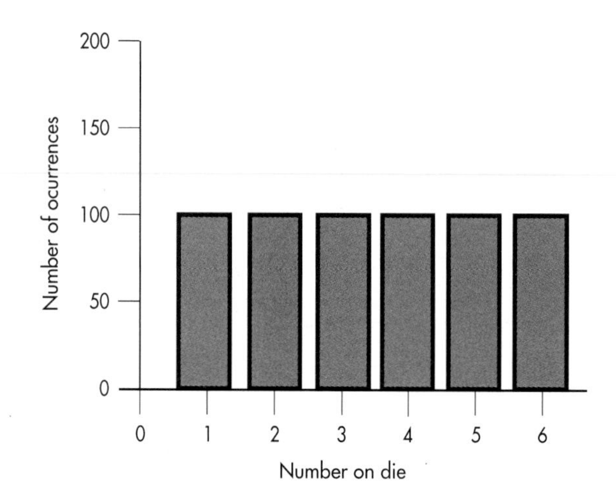

FIGURE 4–1 Theoretic distribution from rolling a die 600 times.

Now, let's roll the die twice and add up the two numbers. The sums could range from a minimum of 2 to a maximum of 12. But this time, we wouldn't expect each number to show up with the same frequency. There's only one way to get a 2 (roll at 1 on each throw) or a 12 (roll a 6 each time), but two ways to roll a 3 (roll a 1 followed by a 2, or a 2 followed by a 1), and five ways to roll a 6. So, because there are more ways to get the numbers in the middle of the range, we expect that they will show up more often than do those at the extremes. This tendency becomes more and more pronounced as we roll the die more and more times.

We did a computer simulation of this; the results are shown in Figure 4–2. The computer "rolled" the die twice, added the numbers, and divided by 2 (i.e., took the mean for a sample size of 2) 600 times; then it "rolled" the die four times, added the numbers, and divided by four (the mean for a sample size of 4) for 600 trials; and again it rolled the die eight times, added the numbers, and divided by eight. Notice that rolling the die even twice, the distribution of means has lost its rectangular shape and has begun to look more normal. By the time we've rolled it eight times, the resemblance is quite marked. This works with any underlying distribution, no matter how much it deviates from normal. So, the Central Limit Theorem guarantees that, if we take enough even moderately sized samples ("enough" is usually over 30), the means will approximate a normal distribution.

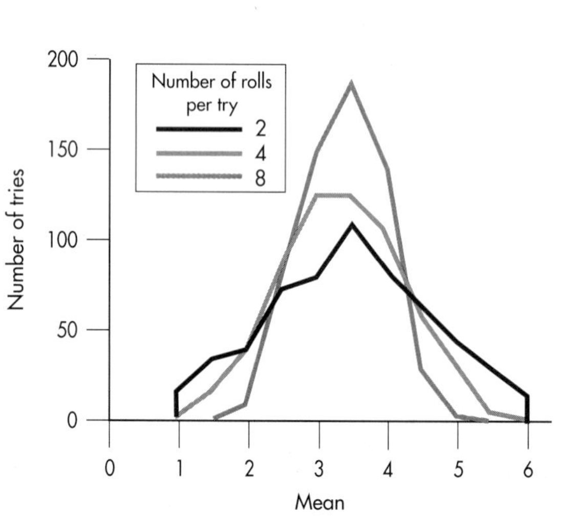

FIGURE 4–2 Computer simulation of averaging the sum of rolling the die 2, 4, and 8 times, each done 600 times.

STANDARD SCORES

Before we get into the intricacies of the normal distribution, we have to make a minor detour. If hundreds of variables were normally distributed, each with its own mean and SD, we'd need hundreds of tables to give us the necessary specifications of the distributions. This would make publishers of these tables ecstatic but everyone else mildly perturbed. So statisticians have found a way to transform all normal distributions so that they (the distributions, not the statisticians) use the same scale. The idea is to specify how far away an individual value is from the mean by describing its location in **standard deviation (SD) units**. When we transform a raw score in this manner, we call the results a **standard score**.

A *standard score*, abbreviated as z or Z, is a way of expressing any raw score in terms of SD units.

TABLE 4–1

Data in Table 3–2 transformed into standard scores

X	z
1	−1.53
3	−1.15
4	−0.96
7	−0.38
9	0
9	0
11	0.38
12	0.57
16	1.34
18	1.72

The standard score

$$z = \frac{(X - \bar{X})}{s} = \frac{x}{s} \tag{4–1}$$

Adding a bit to the confusion, Americans pronounce this as "zee score," whereas Brits and Canadians say "zed score."[5] A standard score is calculated by subtracting the mean of the distribution from the raw score and dividing by the SD. Just to try this out, let's go back to the data in Table 3–2; we found that civil servants took an average of 9.0 coffee breaks per day, with an SD of 5.22. A raw score of 1 coffee break a day corresponds to:

$$z = \frac{(1 - 9)}{5.22} = -1.53 \tag{4–2}$$

that is, −1.53 SD units, or 1.53 SD units below the mean. We can do the same thing with all of the other numbers, and these are presented in Table 4–1.

In addition to allowing us to compare against just one table of the normal distribution instead of having to cope with a few hundred tables, *z*-scores also have other uses. They allow us to compare scores derived from various tests or measures. For example, several different scales measure the degree of depression, such as the *Beck Depression Inventory* (BDI; Beck et al, 1961) and the *Self-Rating Depression Scale* (SDS; Zung, 1965). The only problem is that the BDI is a 21-item scale, with scores varying from a minimum of 0 to a maximum of 63; whereas the SDS is a 20-item scale with a possible total score between 25 and 100. How can we compare a score of, say, 23 on the BDI with a score of 68 on the SDS? It's a piece of cake, if we know the mean and SD of both scales. To save you the trouble of looking these up, we've graciously provided you with the information in Table 4–2.

TABLE 4–2

Means and standard deviations of two depression scales

	Beck Depression Inventory	Self-Rating Depression Scale
Mean	11.3	52.1
SD	7.7	10.5

What we can now do is to transform each of these raw scores into a *z*-score. For the BDI score of 23:

$$z = \frac{23 - 11.3}{7.7} = 1.52 \tag{4–3}$$

Similarly, for the SDS score of 68:

$$z = \frac{68 - 52.1}{10.5} = 1.51 \tag{4–4}$$

So, these transformations tell us that the scores are equivalent. They each correspond to *z*-scores of about 1.5; that is, 1½ SD units above the mean. Let's just check these calculations. In the case of the BDI, the SD is 7.7, so 1½ SD units is (1.5 × 7.7) = 11.6. When we add this to the mean of 11.3, we get 22.9, which is (within rounding error) what we started off with, a raw score of 23. This also shows that if we know the mean and SD, we can go from raw scores to *z*-scores, and from *z*-scores back to raw scores. Isn't science wonderful?

There are a few points to note about standard scores that we can illustrate using the data in Table 4–1. First, the raw score of 9, which corresponds to the mean, has a *z*-score of 0.0; this is reassuring, because it indicates that it doesn't deviate from the mean. Of course, not every set of data contains a score exactly equal to the mean; however, to check your calculations, any score that is close to the mean should have a *z*-score close to 0.0. Second, if we add up the *z*-scores, their sum is 0 (plus or minus a bit of rounding error). This will always be the case if we use the mean and SD from the sample to transform the raw scores into *z*-scores. It is the same reason that the mean deviation is always 0; the average deviation of scores about their mean is 0, even if we transform the raw scores into SD units (or any other units). A third point about standard scores is that if you take all the numbers in the column marked *z* in Table 4–1 and figure out their SDs, the result will be 1.0. By definition, when you convert any group of numbers into *z*-scores, they will always have a mean of 0.0 and a standard deviation of 1.0 (plus or minus a fudge factor, for rounding error).

However, we don't have to use the mean and SD of the sample from which we got the data; we can take them from another sample, or from the population. We do this when we compare laboratory test results of patients against the general (presumably healthy) population. For instance, if we took serum rhubarb levels from 100 patients suffering from hyperrhubarbemia[6] and transformed their raw scores into *z*-scores using the mean and SD of those 100 scores, we would expect the sum of the *z*-scores to be 0. But if we used the mean and SD derived from a group of normal[7] subjects, then it's possible that all of the patients' *z*-scores would be positive.

[5] *This further confirms Churchill's statement that the United States and Britain are two countries separated by a common language. Canada is one country divided by two languages.*

[6] *A nonfatal disorder that makes people long and green and turns their hair red.*

[7] *Here, "normal" means healthy, not bell-shaped.*

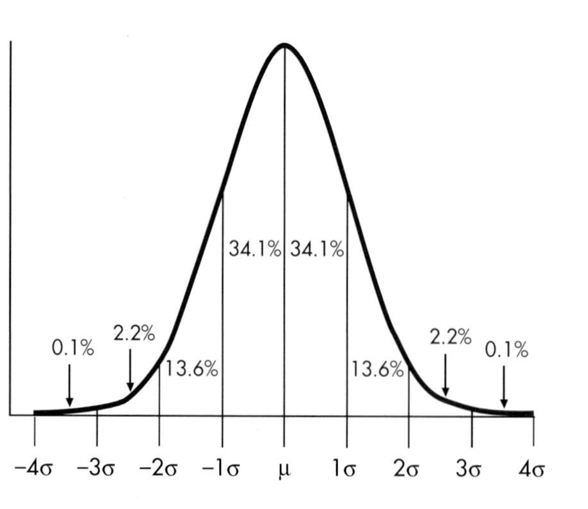

FIGURE 4–3
The normal curve.

TABLE 4–3
A portion of the table of the normal curve

z	Area below
0.00	.0000
0.10	.0398
0.20	.0793
0.30	.1179
0.40	.1554
0.50	.1915
0.60	.2257
0.70	.2580
0.80	.2881
0.90	.3159
1.00	.3413
1.00	.3413
1.10	.3643
1.20	.3849
1.30	.4032
1.40	.4192
1.50	.4332
1.60	.4452
1.70	.4554
1.80	.4641
1.90	.4713
2.00	.4772

THE NORMAL CURVE

Now armed with all this knowledge, we're ready to look at the normal curve itself, which is shown in Figure 4–3. Notice a few properties:

1. The mean, median, and mode all have the same value.
2. The curve is symmetric around the mean; the skew is 0.
3. The kurtosis is also 0, although you'll have to take our word for this.
4. The tails of the curve get closer and closer to the *X*-axis as you move away from the mean, but they never quite reach it, no matter how far out you go. In mathematical jargon, the curve approaches the *X*-axis **asymptotically.**
5. For reasons we'll discuss in Chapter 6, we've used μ (the Greek *mu*) for the mean and σ (lower-case *sigma*) for the SD.

These properties are true for *theoretical* normal curves; that is, those which exist only in the imaginations and dreams of statisticians. Reality deviates from this to some degree; any set of real numbers will show a slight degree of skew and kurtosis, and the mean, median, and mode will not be exactly the same. Most importantly, the curve will eventually touch the *X*-axis, unless we have an infinite set of data points.[8] For all intents and purposes, though, most of the action takes place between the lines labeled −3σ and +3σ, so the discrepancy between theoretical and real normal curves bothers only the purists.[9]

Let's now take a look at the numbers inside the curve. What they tell us is that 34.1% of the area under the normal curve falls between the mean (μ) and one SD above the mean (+1σ); because the curve is symmetric, it follows that another 34.1% falls between μ and −1σ. So, roughly two-thirds of the area (actually 68.2%) is between +1σ and −1σ. Going a bit further, 13.6% of the area is between +1σ and +2σ (and between −1σ and −2σ); therefore, 47.7% of the area is between the μ and +2σ,[10] and just slightly over 95% of the curve falls between +2σ and −2σ.

All this raises two questions: first, who really cares about the area under this odd-looking curve; and second, how did we get these numbers? To answer the first question, we'll return to those intrepid nurses and their never-empty bed pans. If you remember Figure 2–5, the distribution wasn't quite normal, but it's close enough.[11] We calculated the mean to be 30.83, and if you go through the calculations, you'll find that the SD = 14.08. So, putting this together with the numbers in Figure 4–3, we know that 68% of the nurses emptied between (30.83 − 14.08) and (30.83 + 14.08), or between 16.75 and 44.91 (let's say 17 and 45) bedpans. The vast majority—95% of them—emptied between (30.83 − [2 × 14.08]) and (30.83 + [2 ×14.08]), or between about 3 and 59 pans. Anyone who dumped fewer was really slacking off, and those who cleaned 60 or more were working harder than about 97% of their mates.

The important point is that, if our data are relatively close to being normally distributed, the properties of the normal curve apply to our data. So the normal curve can give us information about the data we've collected, not just about some theoretical line on a piece of paper.

[8]*However, to misquote Albert Einstein, "There are only two things that are infinite—the universe and human stupidity—and I'm not sure about the universe."*

[9]*By now, you should know that "purist" is one term that will never be assigned to us.*

[10]*That's 34.1 + 13.6 for those of you whose calculator batteries died.*

[11]*Another one of those precise statistical terms.*

The second question is about where those numbers came from. That's easy; look at Table A in the back of the book, titled **Area of the Normal Curve**. Where *those* numbers came from is a bit more difficult. There is an equation, which we won't bother you with, that can be solved to give the area between the mean and any value of σ. We "simply" solved this a few hundred times and put the results in the table. To simplify your life yet again, we've reproduced a part of it in Table 4–3. Now, how to read it.

Table 4–3 has two columns, one labeled "z" and one labeled "Area Below." There are a few things to notice about the table: first, the z is in SD units. Tables in other books may refer to it as x/σ or as σ. They all mean the same thing; 0.1 is one-tenth of an SD. Second, Table A starts at 0.00 and goes up to 4.00 (we've given only a few values up to 2.00 in Table 4–3); because the curve is symmetric, it does not make sense to waste ink and paper going from 0.00 to −4.00. We'll show you how to deal with negative z values in a minute. Last, be careful reading tables of the normal curve in other books. Many show the curve the same way it is here, giving the area between mean ($\mu = 0.00$) and the value of z (or σ, or x/σ, or however it's labeled). Other books give the area to the *left* of z; these are easy to spot because the area equivalent to $z = 0.00$ is 0.5000 rather than 0.0000, as it is here. Finally, a few tables give the area to the *right* of z. So be sure to check which type of table you are using.[12]

Now, let's start using it. Notice that the number next to a value of z of 1.00 is .3413; not coincidentally, it's the same number as in Figure 4–3, showing the percent of the area between μ and $+1\sigma$. This shows first, how we got the number, and second, that the total area under the curve is 1.0000 units, so that an area of .341 is 34.1% of the total area.

To really given the normal curve a good workout, let's return to the problem posed in *Setting the Scene*, and try to determine how many times the phrase, "Not tonight, dear, I've got a headache," has been used.

1. **How many people used this excuse up to 115 times?** First, we have to transform 115 to a z-score, using the format of Equation 4–1:

$$z = \frac{115 - 100}{15} = 1.00 \qquad \textbf{(4–5)}$$

Table 4–3 tells us that the area of the curve between the mean and +1.00 SD is .3413. This means that 34.13% of the people use this delightful phrase between 100 and 115 times. But we're interested in all of the people who said it 115 times and less, so we'll have to add the 50% of the area that falls below the mean, as in Figure 4–4. So the answer is 84.13% of 2,000, or 1,683 people.[13]

2. **How many people said this fewer than 70 times in 1 year?** Again, we start off by converting this to a z-score:

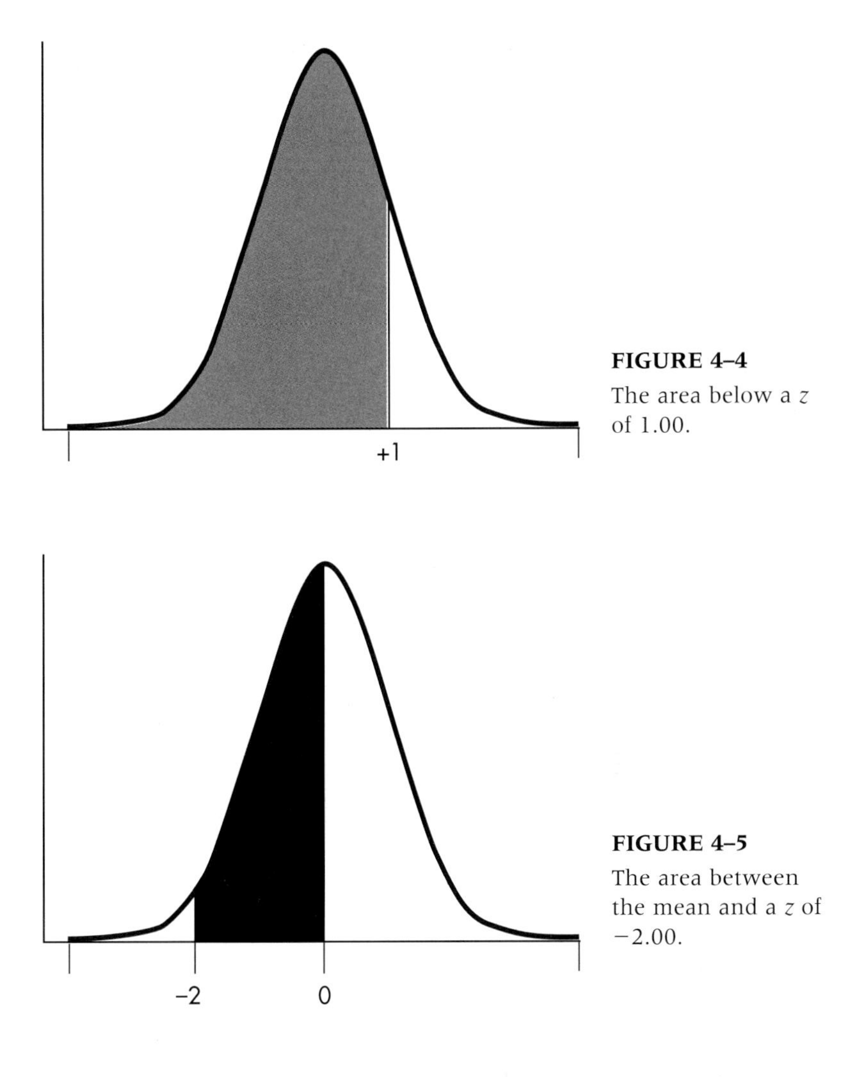

FIGURE 4–4 The area below a z of 1.00.

FIGURE 4–5 The area between the mean and a z of −2.00.

$$z = \frac{70 - 100}{15} = -2.00 \qquad \textbf{(4–6)}$$

As we mentioned, the table does not include negative z-scores. What we do is ignore the sign, but keep it in our minds. Looking up 2.00 in the table, we find .4772. This is the area between the mean and +2.00, and also between the mean and −2.00; because the sign was negative, we use this latter figure. It corresponds to the shaded area in Figure 4–5. But this isn't the area we're interested in; we want to know the area below 70. Because the total area between the extreme left and the mean is half the area of the curve, or 0.5000, the area to the left of the shaded portion is (0.5000 − .4772), or .0228; that is, 2.28% of the people.

What this also shows is that it is very helpful to draw a rough sketch of the normal curve and the area that the table shows; it helps clarify in our mind the portion that we're interested in. This isn't just for neophytes; us oldtimers do it all the time.[14] Just one more for practice.

[12] *Although we couldn't begin to imagine why you would want to look at, much less own, any other statistics book.*

[13] *So that's why the U.S. birthrate is falling!*

[14] *Perhaps a reflection of our increasing decreptide. We have also been told that the correct phrase is "we oldtimers." But, us oldtimers prefer "us oldtimers."*

3. **How many people use this phrase between 106 and 112 times a year?** As usual, we begin by changing the raw numbers into *z*-scores, which in this case are +0.40 and +0.80, and making a rough sketch (Figure 4–6). Table 4–3 tells us that the area between the mean and +.80 is .2881, and the mean and +.40 is .1554. We're interested in the area between these; the difference is .1327, or 13.3% of the 2,000 people.

This finishes our discussion of the normal distribution. It is not the only one used in statistics; there are many others with names such as *Poisson, exponential, Gompertz,* and the like. However, we're not going to discuss them for two reasons. First, unless you plan on doing some very fancy stuff with statistics, the normal curve will get you through almost everything. Second, we don't know how to use them, so why should you?

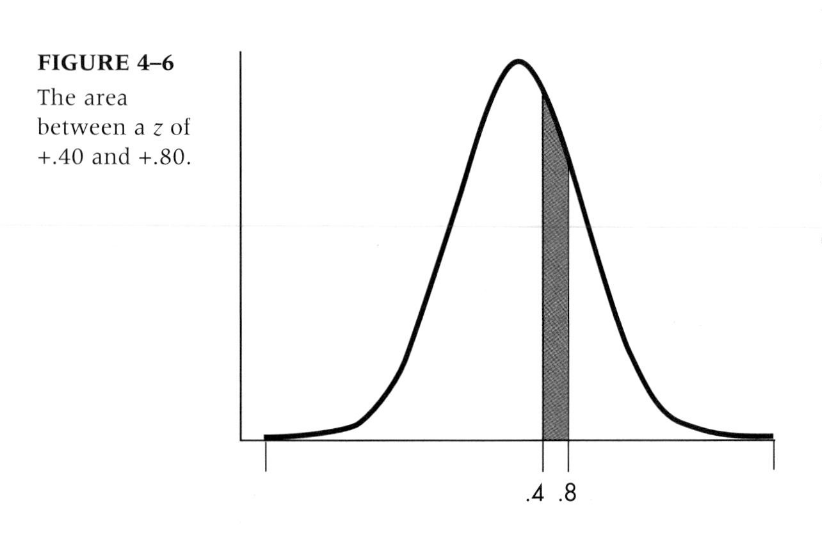

FIGURE 4–6
The area between a *z* of +.40 and +.80.

EXERCISES

The entire first-year class in Billing Practices 101 takes the Norman-Streiner Test of Real or Imagined Licentiousness (the NoSTRIL; often referred to as the NoSE). The results were:

	Males	Females
Mean	60	40
SD	12	10
N	138	97

Unlike the students, the scores were fairly normal for both men and women. Based on these data, figure out the following:

1. If a male gets a score of 70, what's his *z*-score?
2. What's the *z*-score for a female with a score of 35?
3. What score for females is equivalent to a male's score of 78?
4. What proportion of women get scores between 30 and 45?
5. What proportion of men get scores over 68?
6. What score demarcates the upper 10% of women?

CHAPTER THE FIFTH

Probability

This chapter introduces the basics of probability theory, the binomial theorem, and the relationship between the binomial and the normal distributions.

SETTING THE SCENE

Imagine you have an urn with 73 white marbles and 136 black marbles. What is the probability that, if you took out 12 marbles (replacing each one after you took it out), you'd have seen exactly 5 white ones and 7 black ones?

A DISCLAIMER

Open up just about any other book on statistics, and you'll find a long section on probability theory. It usually consists of examples such as the one above in *Setting the Scene*. The answer given by most students to problems such as this, "Who cares?" To us, they got the right answer. The two of us have been messing around with statistics for a total of about 50 years now,[1] and we can't remember when we've ever had to figure out a problem like this—except when we were wading through statistics texts, trying to solve problems in probability theory. Much of what's covered in such chapters is undoubtedly of great interest to those who are so inclined, but are of little direct value to the clinician. Instead, this chapter gives you what we think are the necessary survival skills to understand and deal with probabilities in situations you're likely to encounter; anybody who wants to figure out the correct answer to this and other such problems should be reading another book (we would recommend anything by the Count Sacher von Masoch[2]).

What Do We Mean by "Probability?"

This is not as easy to answer as it sounds. But, rather than getting bogged down in philosophical discussions, we'll rely for now on your intuitive understanding.

> Probability deals with the relative *likelihood* that a certain event will or will not occur, relative to some other events.

We can derive probabilities in one of two ways: empirically or theoretically.

The Empirical Way

Each of us, in our youth (or second childhood), has probably asked out for a date a number of people of the opposite (or the same) sex. We've been accepted by some and rejected by others; now we want to look back and see how we've done, possibly as a guide to trying out our old skills. To keep things nonsexist and simple, let's say we can categorize our askees into four mutually exclusive types based on what it was that first attracted us to them: their *Body*, their *Mind*, their *Wallet*, and our *Desperation*. In Table 5–1, we put down how we did (allowing for some degree of poetic license). What the *Percent Success* column tells us is how well we did in the past with each of these four types and gives us the probability of success in the future, *assuming nothing has changed.* The key point is that the probability, based on past performance, holds true now and in the future only under similar circumstances. If the circumstances *have* changed (e.g., we haven't been able to see our toes for the past decade, or those who had been the *Body* class now have moved to the *Desperate*), then the probabilities no longer apply.

The classic example of empirically derived probabilities is the tout sheet sold for horse races. The odds they give (another way of expressing probabilities) are based on how well the horse did in races of the same length, under the same track conditions, ridden by the same jockey (and these days, under the influence of the same drugs). Almost all of the probabilities we encounter in the health field are derived empirically. For instance, the probability of survival for a cancer patient is based on the known survival rates of similar patients who have the same

[1] *That's combined, not* each.

[2] *From whom we get the term masochism.*

TABLE 5-1

Our batting average

Attraction	Number asked	Number accepted	Percent success
Body	10	3	30.00
Mind	12	5	41.67
Wallet	5	1	20.00
Desperation	23	21	91.30
TOTALS	50	30	60.00

stage of disease and have undergone the same treatment regimen. Our definition also assumes that if any of these factors change, such as admitting patients at an earlier stage or changing what we do to them, these empirical probabilities go out the window.

The empirical way is also the basis for the old diagnostic dictum that if you hear hoof beats, it's more likely to be coming from a horse than from a zebra; horses are more common here than zebras.[3] Analogously, it's more likely (or probable) that the patient has a more common disease than a rare one.

The Theoretical Way

When we've gotten tired of losing our money on the nags and want to lose it some other way, we can always shoot craps at Las Vegas or Atlantic City. If you've ever been there, you'll have noticed that the craps tables are covered with a green baize cloth, stating the pay-offs for various throws.[4] The odds given for rolling a 7 or 11 on the first throw, or hitting a certain number, are not based on the experience of the croupier; they're figured out based on the theory of probability. To take a simple example, let's roll a die. Each of the six sides has an equal likelihood of ending up on top, and which one actually appears is a purely random event. Consequently, the probability of rolling a 3 on this one toss of the die is one in six, or .1667; we don't have to do this experiment 1,000 times to find this out. We can get even fancier and calculate the probability of getting a 10 with one roll of two dice, of drawing an inside straight, or of rolling a 3 at roulette.[5,6]

All of these calculations are based on our knowledge of the likelihood of occurrence of various chance events, which is the essence of probability theory, and that's what we'll be concerning ourselves with in this chapter. We want to emphasize that all of this is to make you better clinical researchers, not to lead you down the road of corruption by making you better gamblers.

MUTUALLY EXCLUSIVE AND CONDITIONALLY DEPENDENT EVENTS

To understand probability theory, it is necessary to differentiate between two types of events: those which are **mutually exclusive** and those which are **conditionally dependent**.

> Two events, *X* and *Y*, are *mutually exclusive* if the occurrence of one precludes the occurrence of the other.

The simplest example of this is flipping a coin; heads and tails are mutually exclusive in that, if the head side appears, the tail side won't, and vice versa. Which party a person votes for in an election for a specific office is a mutually exclusive event.[7] However, how a person voted in the last election for *all* candidates may not be mutually exclusive; the voter may have voted for one party for some offices and for another party for other offices.

Closer to home, respiratory acidosis and alkalosis are mutually exclusive events; if you have one, you can't simultaneously have the other. On the other hand, cardiac disease and esophageal reflux are *not* mutually exclusive. If a person has some chest pain, and the ECG confirms the presence of an infarct, it doesn't necessarily mean that the person can't have reflux at the same time.

> Two events, *X* and *Y*, are *conditionally dependent* if the outcome of *Y* depends on *X*, or *X* depends on *Y*.

Returning to the gaming tables, the probability of throwing a 5 with a single toss of two dice is 11.11%—there are 36 possible combinations, four of which yield a 5 (1 and 4, 2 and 3, 3 and 2, and 4 and 1). However, if we throw the dice one after the other, and the first die comes up a 1, then the probability that the sum will be 5 is one in six, or 16.67%. That is, the probability of a 5, **conditional** on the first die being a 1, is .1667. Using the example in Table 5-1, our overall, or **unconditional**, success rate was 30 out of 50, or 60%. However, our hit rate with *Bodies* was 30%; that is, our success (*Y*), *conditional on the person having being chosen for his or her body (X)*, was .30, or 30%.

Turning to more mundane examples, we've all heard that the life expectancy of a person is somewhere around 74 years. But this doesn't tell the whole story. Women live longer than men; 78.1 years for white females born in 1980 as opposed to 70.7 years for white males born in the same year. Black people's life expectancies are about 7 years less than this, and all the figures are about 3 years more than for people born in 1970. So the probability of a person living to be 80 is conditional on several factors, such as gender, race, and year of birth.

The difference between mutually exclusive and conditionally dependent events is important because we have to figure out probabilities differently for each of them.

Mutually Exclusive Events and the Additive Rule

To illustrate the difference between mutually exclusive and conditionally probable events, let's assume

[3] *What do they tell medical students in Kenya, "It's more likely coming from a zebra than a horse?"*

[4] *Our knowledge of such matters is derived solely from movies and the reports of others, not from personal experience.*

[5] *As a free bit of information, the payoff is always less than the calculated odds. With the exception of Donald Trump, has any casino owner ever gone bankrupt?*

[6] *And as a second bit of free information, the payoff at the tables is light-years better than in state or provincial lotteries. Casinos pay an average of 50 to 80 cents on the dollar; lotteries pay, at most, 10 cents on the dollar.*

[7] *Except in Chicago, where they adhere to the motto, "Vote early, and vote often."*

that the unit we work on admits only those patients with one of three mutually exclusive disorders: cryptogenic tinea pedis (CTP),[8] idiopathic hangnail syndrome (IHS), and iatrogenic systemic degeneration (ISD). However, these conditions don't occur with equal frequency; CTP is relatively rare, and only 10% of our patients suffer from this, as opposed to 30% from IHS and 60% from ISD. Moreover, the proportion of males and females is different for each disorder; these are given in the third column of Table 5–2.

TABLE 5–2 Relative frequencies and gender differences for three disorders

Disorder	Relative frequency	Male/female ratio
Cryptogenic tinea pedis	.10	50/50
Idiopathic hangnail syndrome	.30	30/70
Iatrogenic systemic degeneration	.60	80/20

Now, what is the probability that the next person through the door has *either* CTP or IHS? These are mutually exclusive events, so if the patient has one, he or she can't have the other. Thus the probability is .10 *plus* .30; that is, there is a 40% probability that the person has either CTP or IHS, and, by extension, a 60% probability that he or she has ISD.

Why do we add the probabilities (rather than, say, multiplying them or taking their square roots)? It may help if we think for a moment in terms of bodies instead of proportions. If there were a total of 100 patients on our ward, 10 would have CTP, 30 would suffer from IHS, and 60 from ISD. So the condition would be satisfied (i.e., the next person through the door has either CTP or IHS) if he or she were 1 of the 10 from the first group or 1 of the 30 from the second; in other words, 40 of the 100 patients would satisfy the condition, and 60 would not. We can summarize what we've said by the additive rule:

> If X and Y are *mutually exclusive* events, then the probability of X or Y is the probability of X *plus* the probability of Y.

For obvious reasons, this is called the **additive law**. Put into formal jargonese:

$$\Pr(X \text{ or } Y) = \Pr(X) + \Pr(Y) \tag{5-1}$$

where "Pr" is statistical shorthand for *probability*. Needless to say, we're not limited to just two events; the same law holds with as many mutually exclusive events as we want.

Conditionally Probable Events and the Multiplicative Law

Now, let's change the question a bit. What's the probability that the next patient will be a male and have ISD? These are not mutually exclusive events; a person with ISD can be either male or female. However, we know from experience that ISD is more common in males. This is a case of **conditional probabilities** because the probability that the patient has the diagnosis is conditional on the probability that the patient is a male (and vice versa). We know from Table 5–2 that 80% of patients with ISD are male, 50% of patients with CTP are male, and 30% of IHS patients are male. One way to answer this question is to redraw the table, giving the number of males and females with each of the diagnoses, as we've done in Table 5–3. We've based this on having 100 patients so that we're working with whole numbers, but this will work with any number. We see that 48 of the 100 patients on our ward are males with ISD, so the answer is that there is a 48% chance that the next patient admitted to our ward will be a male and have ISD.

TABLE 5–3 Actual number of patients with the three disorders

Disorder	Males	Females	Total
Cryptogenic tinea pedis	5	5	10
Idiopathic hangnail syndrome	9	21	30
Iatrogenic systemic degeneration	48	12	60
TOTALS	62	38	100

[8]*For those of you who are not fluent in Latin and Greek, "cryptogenic tinea pedis" means "athlete's foot of unknown origin."*

We can get at the answer another way, by looking only at the row and column labeled *Total*. In statistical parlance, we say that we're looking at the **marginals**. The probability of having ISD is 60/100, or .60, whereas the probability of being a male, if the diagnosis ISD, is 48/60, or .80. So, the probability that both events occur together (i.e., a male with ISD) is .60 × .80, or 48%, which is exactly what we got before. We multiplied in this case because we're looking at a part of a part. That is, some of the people are male (the others are female); and, looking at the patients from the other perspective, some of the total have ISD (the remainder have the other disorders).

Using this technique of multiplying probabilities means that we can figure out the conditional probabilities by simply knowing the individual probabilities that certain events will happen, and we don't have to make up a table such as Table 5–3. So this rule reads:

> If X and Y are *conditionally probable*, then the probability that both will occur is the probability of X *times* the probability of Y, given X has occurred.

TABLE 5–4 The eight possible outcomes, and their probabilities, of three tests with false-positive rates of 5%

Test 1	Test 2	Test 3	Probability
P	P	P	$(.05) \times (.05) \times (.05) = .000125$
P	P	N	$(.05) \times (.05) \times (.95) = .002375$
P	N	P	$(.05) \times (.95) \times (.05) = .002375$
P	N	N	$(.05) \times (.95) \times (.95) = .045125$
N	P	P	$(.95) \times (.05) \times (.05) = .002375$
N	P	N	$(.95) \times (.95) \times (.05) = .045125$
N	N	P	$(.95) \times (.95) \times (.05) = .045125$
N	N	N	$(.95) \times (.95) \times (.95) = .857375$

It goes without saying that this is referred to as the **multiplicative law**, which is written in statisticalese as:

$$\Pr(X \text{ and } Y) = \Pr(X) \times \Pr(Y \mid X) \tag{5–2}$$

where the symbol $\Pr(Y|X)$ means the probability of Y given X, in our example, the probability of being a male, given that the patient has ISD. So, translating this equation from statistics into English, it reads, "The probability that the patient has ISD [X] and is a male [Y] is the probability of having ISD [$\Pr(X)$] times the probability of being male, given that ISD is present [$\Pr(Y|X)$]."

Just for practice, let's run through a few other examples. The probability that the patient is a female with CTP is the probability of CTP (.10) times the probability of being female, given a diagnosis of CTP (.50), or 5%. The probability of a female with IHS is 21%; you figure it out for yourself.

Independent Events

Many events are neither mutually exclusive nor conditionally probable; they are *independent* of one another.[9] A problem arises when events that are independent of one another are mistakenly assumed by some people to be conditional. Let's say you're back in the casino, standing over the roulette wheel. You've seen that the last five numbers have all been red. Now, you know that, assuming the wheel is honest, red and black have the same probability of appearing, so half the numbers should turn up red and half black. What's the probability that the next number will be black?

The "gambler's fallacy" is thinking that the sixth roll is conditional on the previous five, that after a long run of red the probability of a black is higher, so as to make the overall proportion of reds and blacks closer to 50%. However, the ball does not have a memory and has never studied probability theory; it doesn't "know" what the previous results were, and the probability of black is 50% (ignoring the 0 and 00 slots for the moment), exactly what it would be if the previous five rolls had also been black. That is, the outcome is not conditional on the previous run; they are independent events. However, it's been rumored that casino owners' dreams are filled with fantasies of having a room full of people who believe in the gambler's fallacy, rather than with images of girls from the chorus line.

[9] *We were going to use the example of fitness for office and actually being elected to office. However, we quickly realized that this is more likely an example of mutually exclusive events.*

The Law of "At Least One"

Let's assume that 5% of the time, a lab test from a perfectly normal person comes back labeled "abnormal"; that is, if we tested the serum rhubarb level of 100 eurhubarbic individuals, five people would have results that indicate either hyper- or hyporhubarbemia.[10] What is the probability that, if you order an SMA 12 on a completely healthy person, one of these 12 test results will be a false-positive?

[10] *We'll see later why this is a fairly safe assumption to make.*

To make things simpler, let's consider the case of a healthy person who has been given three different lab tests, each of which has a 5% chance of yielding a false-positive result. Eight combinations of positive and negative results are possible; these various alternatives, with the probability that each will occur, are given in Table 5–4. Now, the probability of *any* test being positive includes all but the last line (N-N-N). We can add up all of the lines up to this point, but the sum of all the outcomes has to be 1.0; there has to be a 100% probability that *one* of these eight alternatives will occur, so it's easier to take $(1.0 - .95^3) = .1426$.

What we have done is turn things around. We are saying that the probability of "at least one event" is the complement of the probability of "no events"; that is:

$$\Pr(\text{At Least One}) = [1 - \Pr(\text{None})] \tag{5–3}$$

So, returning to our SMA 12 test examples, if each of the 12 component tests has a false-positive rate of 5%, the probability of at least one false-positive out of 12 is:

$$1 - .95^{12} = 45.96\% \tag{5–4}$$

For your edification and amazement, Figure 5–1 shows the probability that at least one test will be abnormal in a perfectly healthy individual. As you can see, it increases with the number of tests done. We've shown this for three false-positive rates: 1%, 5%, and 10%. You can see that changing the false-positive rate moves the curve up or down, but the basic relationship between the number of tests and the probability of at least one being abnormal stays the same.

Just to recapitulate: to figure out the probability of *at least one* event occurring, we first determine the probability of no events occurring, and then subtract this number from 1. So, in addition to learning some stats, you've also learned a lesson in clinical care; don't order more tests than you really need!

THE BINOMIAL DISTRIBUTION

Question: What do these statements (taken from Bloch, 1979) have in common? Circle the correct answer:

"Any wire cut to length will be too short."

"Any error in any calculation will be in the direction of most harm."

"If you miss one issue of any magazine, it will be the issue that contained the article, story, or installment you were most anxious to read."

"For a bike rider, it's always uphill and against the wind."

Answers:

a. They're all cynical.
b. They're all correct.
c. They all express the probabilities of dichotomous events.
d. All of the above.

In case you didn't know, the correct answer is d, "All of the above." I was first introduced to this apparent breakdown of the laws of probability when my kids were small and learning to put on their shoes. You would expect that if they didn't know right from left, and put their shoes on at random, they'd get it wrong only half the time. This is not what happened; it seemed that they put their left shoes on their right feet at least 89% of the time. Now, is there some way to tell how often this deviation from chance would be expected to occur?

Again, a give-away question; of course there is. What we're dealing with here is called the **binomial distribution.**[11]

What is the Binomial Distribution?

As you no doubt recall, the normal curve describes how a continuous variable (such as blood pressure or IQ) would be distributed if we measured it in a large number of people. The curve can also be used to give us the probability of a given event, such as a diastolic blood pressure of 95 mm or greater. However, the examples we just gave are not continuous, but have only two possible outcomes: the wire either will be too short, or it won't be too short; the missing issue either will be the one containing the last installment of the mystery story, or it won't be; and so on. What we would like to have is something equivalent to the normal distribution, but that can be used to both describe and give us the probabilities for dichotomous events. Not surprisingly, we have such an animal; it's called the **binomial distribution**.

The *binomial distribution* shows the probabilities of different outcomes for a series of random events, each of which can have only one of two values.

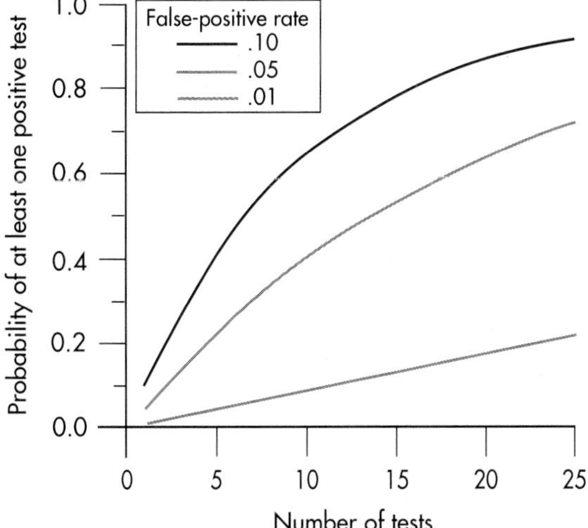

FIGURE 5–1
Probability that at least one test will be positive in a healthy individual, given false-positive rates of 1%, 5%, and 10%.

Let's start off with the easiest case, where each of the two values is equally likely. The usual example, used in every other textbook, is flipping a coin and seeing how many times it comes up heads in 10 flips. For that reason, we'll avoid that example assiduously and stick with a kid putting on his shoes.

If we let the kid try to put his shoes on once, there are two possible outcomes: right (R) or wrong (W), each of which should occur 50% of the time.[12] If there are two attempts at getting shod, then the possible outcomes are: (1) R on both tries; (2) W on both tries; (3) R on the first and W on the second; and (4) W on the first and R on the second. It's easy enough in this instance to figure out the probability of getting it wrong both times: there are four equally possible outcomes, one of which is the combination W-W, so the chances are 1 in 4. The other way to figure it out is to use the multiplicative law: the probability of W on the first try is .50, as it is on the second (i.e., the probability of getting it wrong on the second try, conditional that the first try was wrong). Consequently, the probability of W on both trials is $.50 \times .50 = .25$, which is what we got before.

We could do the same thing for 3, 10, or 100 tries, but these methods are laborious. For example, we could ask the question, If a kid puts his shoes on 10 times, what's the probability that he will get it wrong on exactly 7 of those tries? If we tried to solve this by making a table of the possible outcomes, we'd quickly get bogged down. On the first try, there are two possible results—right or wrong. For each of these outcomes, there are two possible results for the second try—again, right or wrong, yielding the four different patterns we just discussed. On each trial, the number of possibilities doubles, so that by the time we reach 10 trials, there are 2^{10}, or 1,024 possibilities.

However, there's an easier way to figure things out, which is called the **binomial expansion**. Although we're trying to avoid equations as much as possible, this one comes in quite handy, so bear

[11] *Of, if you prefer, contrary children—your choice.*

[12] *This assumes the kid really doesn't know right from left, and the attempts are truly random. It doesn't apply if the kid does know, but does it wrong to get you annoyed; that is, it doesn't apply about 97% of the time.*

with us. Let's define a few terms and symbols first, and then get into answering the question of putting on shoes.

n is the number of tries (10 in our example);
r is the number of favorable outcomes (7 in this case);[13]
p is the probability on each try of the outcome of interest (0.5 in this example) occurring; and
q is $(1 - p)$.

[13] *As you can see, the term "favorable" is just an expression meaning "the outcome of interest"; from a parent's point of view, the outcome is anything but that. This terminology becomes particularly disconcerting when the outcome of interest is death.*

Now, the formula for the binomial expansion is:

$$\frac{n!}{r!\,(n-r)\,!}\,p^r\,q^{n-r} \tag{5-5}$$

The symbol $n!$ does not mean "emphatically n"; it means "***n* factorial**," which is defined as:

$$n! = n \times (n-1) \times (n-2) \times \cdots \times 1 \tag{5-6}$$

For instance, $5! = 5 \times 4 \times 3 \times 2 \times 1 = 120$. (By definition, $0! = 1$). Equation 5–5 can also be written as:

$$\left(\frac{n}{r}\right) p^r q^{(n-r)} \tag{5-7}$$

because the term $(\frac{n}{r})$ is simply a shorthand way of writing:

$$\frac{n!}{r!(n-r)!} \tag{5-8}$$

These equations may look fairly scary, but actually they're not hard to handle. The only difficult part is calculating the factorials, but nowadays, many pocket calculators can do it for you. Putting the numbers from our example into Equation 5–5 gives us:

$$\frac{10!}{7!\,(10-7)!} \times .5^7 \times .5^{(10-7)}$$

$$= \frac{10!}{7!\,3!} \times .5^7 \times .5^3 = .1172 \tag{5-9}$$

So, the probability is just under 12% that the kid would get it wrong 7 times out of 10, if he were really putting the shoes on at random.

Now, let's get a bit fancier. What's the probability that he does it wrong at *least* 7 times out of 10 (instead of *exactly* 7 out of 10)? This means getting it wrong 7, 8, 9, or 10 times out of 10 trials. To calculate the cumulative probability of any of these outcomes, we figure out the individual probabilities and then add them up. We already figured out the probability of 7 out of 10. Next, 8 out of 10 looks like:

$$\frac{10!}{8!\,(10-8)!} \times .5^8 \times .5^{(10-8)} = .0439 \tag{5-10}$$

9 out of 10 is:

$$\frac{10!}{9!\,(10-9)!} \times .5^9 \times .5^{(10-9)} = .0098 \tag{5-11}$$

and for 10 out of 10

$$\frac{10!}{10!\,0!} \times .5^{10} \times .5^0 = .0010 \tag{5-12}$$

Adding these up gives us .1719, or just over 17%.

So, the binomial expansion has allowed us to figure out that the kid has a 12% chance of putting his shoes on wrong in 7 out of 10 tries and a 17% chance that he'll get it wrong 7 or more times out of 10.

So far, we've dealt with situations that have a 50:50 chance of happening, but we're not limited to this. For example, let's say that the bug committee at the hospital has really been effective and has knocked the incidence of nosocomial infections down to 20% following abdominal surgery. If we have 15 of these hapless abdominal surgery patients on our wards, what's the probability that 5 of them will develop an infection from the hospital? In this case, $n = 15$, $r = 5$, $p = .20$, and $q = .80$. Putting these into the equation gives us:

$$\frac{15!}{5!\,(15-5)!} \times .2^5 \times .8^{(15-5)} = .1032 \tag{5-13}$$

So the probability that 5 of the 15 patients will develop a hospital-acquired infection is 10.32%.

What we've learned in this section is how to extend the binomial expansion beyond the case where each alternative has a 50% chance of occurring to the more general situation where the two outcomes have different probabilities.

Learning a Bit More about the Binomial Distribution

Staying with this example for a minute, how many people with nosocomial infections would we expect to see on our 15-bed unit? It is almost intuitive that, given 15 patients and an incidence of .20, we would expect that, most of the time, 3 infected patients would be on the unit simultaneously (i.e., 20% of 15). In Figure 5–2, we've plotted the probabilities of having anywhere between 0 and 15 nosocomial patients on the ward at the same time. This was done using Equation 5–5 by setting $r = 0$, then $r = 1$, up through $r = 15$. This figure, then, shows the binomial distribution for $p = .2$ and $n = 15$.

What happens when we change the probability and the number of trials (in this case, each patient can be thought of as one trial)? In Figure 5–3, we've kept n at 15, but we changed p from 0.2 to 0.3. You would expect that the average number on the ward at any one time would increase (30% of 15 = 4.5), and sure enough the graph has shifted to the right a bit. It also looks as if the data are spread out some more.

If we keep p at .2 but increase n from 15 to 30, we would again expect a shift to the right, with an expected average of 6 (Figure 5–4). Mirabile dictu,[14] the data behave just as we predicted, and again, there seems to be a greater spread in the scores.

So let's summarize what we've seen so far. First, as p gets closer to .5, the graph becomes more symmetric. When it is exactly equal to .5, the graph is perfectly symmetric. When p is less than .5, the distribution is skewed to the right; it's skewed left when p is greater than .5 (we haven't shown that, but trust us). Second, the closer p is to .5, the greater the variability in the scores. Third, there isn't just one "binomial distribution"; there's a different one for every combination of n and p.

We learned in the previous chapters how to figure out the mean, SD, and variance of continuous data. We can do the same for binomial data, and thus numerically describe the properties of the binomial distribution that we just saw graphically. As we would expect from the graphs, these properties depend on n and p (and therefore also on q, which you remember is $1 - p$). What we have, then, is:

Properties of the Binomial Distribution

$$\text{Mean} = np$$

$$\text{Variance} = npq$$

$$\text{SD} = \sqrt{npq}$$

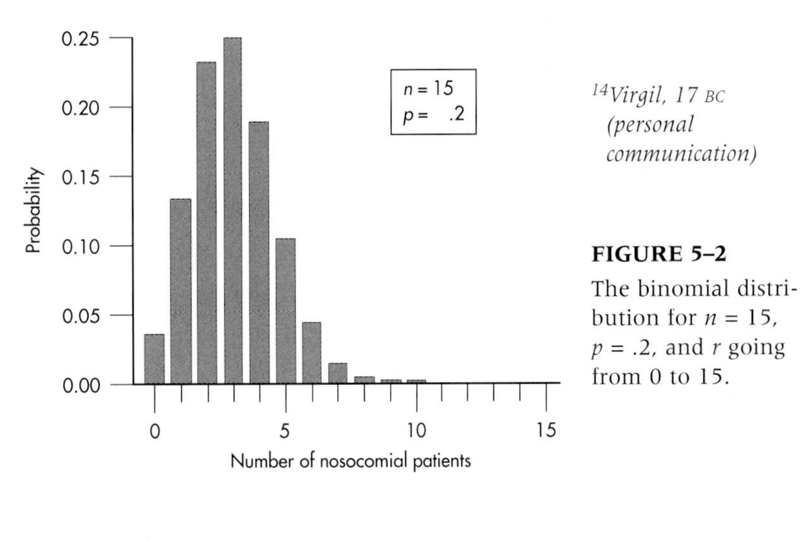

[14] *Virgil, 17 BC (personal communication)*

FIGURE 5–2 The binomial distribution for $n = 15$, $p = .2$, and r going from 0 to 15.

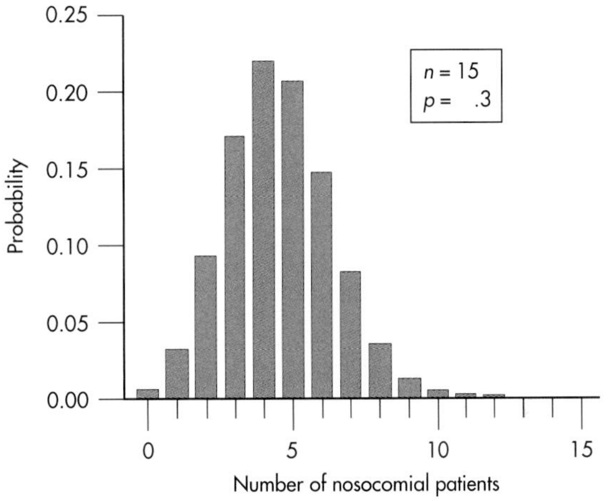

FIGURE 5–3 Changing p from .2 to .3.

The Binomial and Normal Distributions

If we go back and compare Figure 5–2 with Figure 5–4, it looks as though increasing the sample size with the same value of p makes the graph seem more normally distributed. Yet again, your eyes don't deceive you; as n increases, the binomial distribution looks more and more like the normal distribution. Let's pursue this a bit further. In Figure 5–5, we show a binomial distribution with $p = q = .5$. The left graph is for $n = 5$, the middle shows $n = 10$, and the right part shows $n = 20$. As you can see, the graph looks more and more normal as n increases; by the time $n = 30$, the figure is virtually indistinguishable from the normal distribution. What this means is that, if we're dealing with binomial distributions where n is 30 or more, we don't have to worry about using Equation 5–3 to figure out probabilities; we can approximate the binomial distribution by using the normal curve. In fact, when $p = .5$, we can use the normal curve when n is as low as 10; however, the more p deviates from .5, the worse the approximation to the normal distribution, so using the normal curve only when n is at least 30 is fairly safe.

To illustrate how we can use the normal distribution to approximate the binomial one, let's stick with the example of patients who leave the OR minus an appendix but with an infection, and we'll figure out how likely it will be that we'd have five such people on our unit at one time. Now, one difference between the normal and binomial distribu-

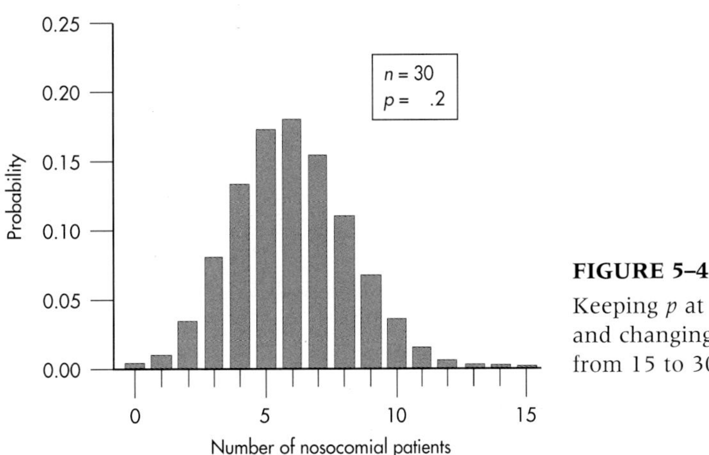

FIGURE 5–4 Keeping p at .2, and changing n from 15 to 30.

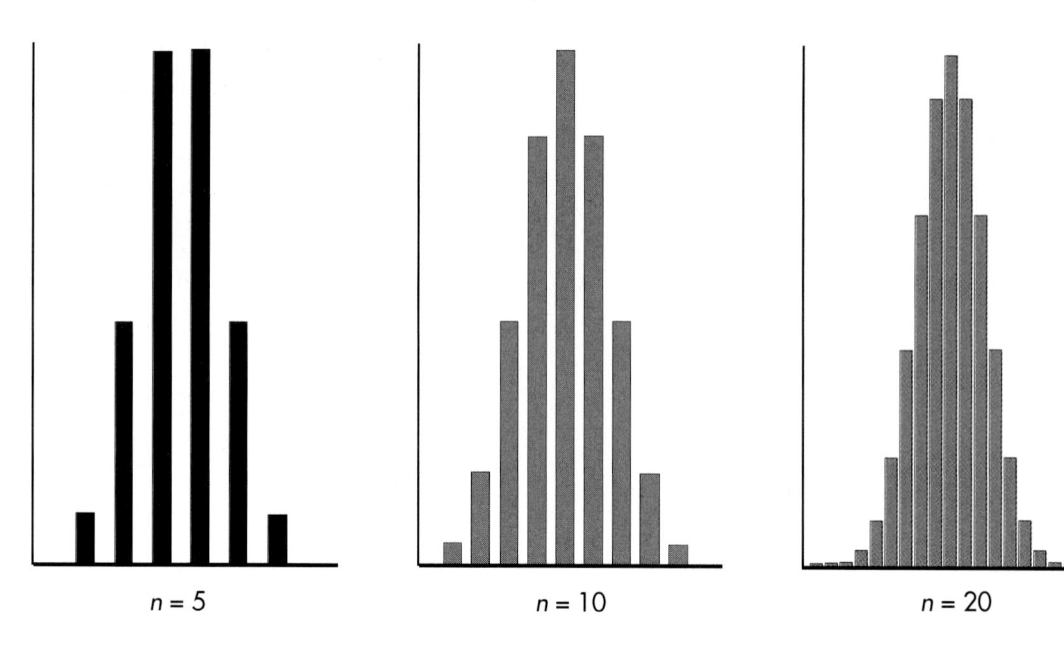

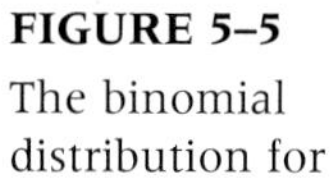

FIGURE 5–5 The binomial distribution for $n = 5$, $n = 10$, and $n = 20$, with $p = q = .5$.

tions is that the former is intended to be used with continuous variables (those which can assume any value between the highest and lowest ones), and the latter with discrete variables. Consequently, we have to consider the discrete value of 5 people as actually covering the exact limits of 4.5 to 5.5.[15] The next step is to convert these two numbers (4.5 and 5.5) to standard scores, using the formula we encountered in Chapter 4. Remember that the mean for a binomial distribution is np, and its standard deviation is $\sqrt{npq}$. This means that in our case, the mean is $15 \times .2 = 3.0$, and the SD is $\sqrt{15 \times .2 \times .8} = 1.55$. Plugging these values into the equation, we get:

$$z_1 = \frac{5.5 - 3.0}{1.55} = 1.61 \quad z_2 = \frac{4.5 - 3.0}{1.55} = 0.97$$

(5–14)

We look these two numbers up in a table of the normal distribution and find that $z_{1.61} = .4463$, and $z_{0.97} = .3340$. The difference between them is .1123, meaning that the probability of finding five nosocomial patients on the ward at the same time is about 11%. This approximation isn't bad, especially considering that in this case, p deviates from .5 quite a bit and n is less than 30; it's fairly close to what we found before, .1032.

RECAP

In this chapter, we've looked at the nature of probability, and explored[16] figuring out probabilities of events with two outcomes. We also saw that when n is over 30, the binomial distribution shades into the normal one, which is easier to use.

(By the way, the answer to the problem of 7 black and 5 white balls drawn from the urn is 20.32%. Because there are 73 white balls and 136 black ones, the probability of drawing a white one is 73/209 = 34%, and is 65.1% for pulling a black one. So,

$$p = \frac{12!}{7!\,5!} \times .349^5 \times .651^7$$

(5–15)

We just thought you'd like to know.)

[15] *We'll ignore the fact that no one but a gross anatomist has ever seen 4.5 or 5.5 people and simply remind you that we did the same thing in Chapter 3 when we were discussing the median.*

[16] *Possibly in more depth than you cared to go.*

EXERCISES

1. According to the Office of Technology Assessment, it will require about 30 space shuttle flights to build a proposed space station. They state that, even if the reliability of the shuttle could be increased to 98%, there is an 8-in-9 chance that a shuttle will fail while building the station (Friedman, 1990). How did they get this figure?
2. There's a pot on the table of $750, and you're holding three aces. If you discard your other two cards, what's the probability of drawing that fourth ace?
3. Two health trends have swept the country over the past few years: one concerned with diet, and one with exercise. Assume that these two fads (oops, that should read "beliefs") are independent, in that people who keep a healthy diet are no more or less likely to exercise than those who don't eat raw fish and Granola bars. If 40% of people exercise, and 10% are wheat germ addicts, then:
   a. What proportion both jog and eat health food (call them Type 1)? What proportion jog but don't eat health food (Type 2)? Eat health food but don't jog (Type 3)? Neither jog nor eat health food (Type 4)?
   b. If we choose three people at random, what is the probability that all will be health food addicts?
   c. What is the probability that none of the three people will be addicted to these behaviors?
   d. What is the probability of finding only Type 2s in a sample of 1 person; 2 people; 3 people?
   e. What is the probability of finding either Type 1s or Type 3s in a sample of 1 person; 2 people; 3 people?
4. According to the weather report, the probability of rain is 10% each day for the next 7 days. If you go camping for 3 days, what is the probability that it will rain every day?

In this chapter, we discuss the problem of comparing a sample to a population of values with a known mean and standard deviation.

CHAPTER THE SIXTH

Elements of Statistical Inference

SETTING THE SCENE

For some time, you have noticed that a sample *of hospital administrators just doesn't seem like other folks. You decide to put it to a test, and you begin with the stuff you know best—lab data. You run an electrolyte screen on a bunch of them and find that their mean serum sodium is 138. Published values for serum Na in the* population *have a mean of 140 and a standard deviation of 2.5. Is this difference statistically significant?*

BASIC CONCEPTS

When you approach the average man on the street and ask what statistics means to him, the answer is simple. If he is less than 30, the only statistic of interest is 900 − 600 − 900 (36 − 24 − 36 before metric); between ages 30 and 60, statistics are the inflation rate and the Dow Jones averages; and over 60, it's vital statistics and mortality rates that count.

However, in research, these descriptive statistics, the type we discussed in Chapter 3, count for little. What we spend the most time on is the stuff of **inferential statistics:** *t*-tests, chi-squares, ANOVAs, life-tables, and their ilk. The basic goal of these statistics is not to describe the data—that's what the previous statistics do—but to determine the likelihood that any conclusion drawn from the data is correct.

Inferential statistics are used to determine the probability (or likelihood) that a conclusion based on analysis of data from a sample is true.

The fly in the ointment that leads to all sorts of false conclusions and keeps all us statisticians employed is **random error**. Any measurement based on a sample of people, even though they are drawn at random from the **population** (more on this later) of all individuals of interest, will differ from the true value by some amount as a result of random processes. So, whenever you compare two treatments, or look for an association between two variables, some differences or association will be present purely by chance. As a result, unless you take the role of chance into account, every experiment will conclude that one treatment was better or worse than another.

To explore how chance wrecks things, imagine trying to determine the average height of all statisticians. It would be difficult and unfundable to try to measure all of us, so you would likely sample us somehow—perhaps by sending a letter to the department heads at some northwestern colleges or steering delegates at the annual statisticians' conference into your booth with offers of beer and pizza. If you were unlucky enough to get one of your esteemed authors in your sample (a good possibility with a six-pack for bait), we guarantee that your estimate will be in trouble. You see, Streiner is about 5′8″, a bit on the short side, whereas Norman is 6′5″, a basketball reject. That doesn't matter too much unless you want to make an **inference** that the height you measured is an accurate reflection of all statisticians. If you got Streiner, your estimate would be too low; if you got Norman, you'd be too high. If you pick us both, you'll likely be about right. If you wanted to generalize from the sample to the population of statisticians, there is a good chance that your estimate may be too high or too low just as a result of the operation of chance in determining who walks through the door of the hospitality suite.

The goal of inferential statistics is to be highly specific about these chances. Instead of saying, as we just did, that there is some chance that the estimate will be a bit off, we want to do just like Gallup and state that "the true height will lie within plus or minus 2 inches of what we measured 95% of the time."

SAMPLES AND POPULATIONS

The Difference between Them

In part, this generalization is strengthened by the methods of sampling. It is clear that if we confine our interest to only those patients who are in the hospital at the time of the study, we will miss all those who (1) have less severe illness and were not referred to the hospital, and (2) have different manifestations of illness and thus were not referred to the particular clinicians at our hospital. But, if we make an honest attempt to reach all individuals fitting our criteria, by a process of **random sampling**, the chances that the generalization will be successful are enhanced.

Of course, from the previous paragraph, it is obvious that *no one* has ever made a truly random sample from a list of everyone of interest, if for no other reason than because some of those to whom the results will hopefully apply have not actually been born yet. Also, many more of them are too long a plane trip away. Nevertheless, the notion of defining a *population* consisting of all folks of interest to you in the particular experiment, and then drawing a *sample*, hopefully at random, from the population is at the root of most experimentation and all inferential statistics. Note that some of the methods of sociology, particularly ethnography, are deliberately not intended to generalize beyond the situation under study. For more details about this idea, try *PDQ Epidemiology* (Streiner and Norman, 1996).

The *sample* describes those individuals who are in the study; the *population* describes the hypothetical (and usually) infinite number of people to whom you wish to generalize.

The Implications for Statistics

Inferential statistics emerge at the point when the data from the sample are then analyzed and you wish to draw some conclusions, proceeding with some degree of confidence that they will apply to the hypothetical population from which you began. The dilemma is that the sample data and their means and SDs will always differ from the true value obtained by analyzing all the individuals in the population, simply because of the role of chance. If we are looking at height or IQ, we may have, just by chance, picked someone in the sample who was particularly tall or short, or smart or dumb, and this will throw our estimate off by a little. Even if no unusual character was in the sample, there is still reason to suspect that the estimate would be a little different from the true value. The point of inferential statistics is to *quantify* the degree of imprecision in the estimate. Thus, at a philosophical level, we are able to determine the confidence we can have in our generalizations, just like Mr. Gallup.

That seems like a truly magical feat. How can you, without knowledge of the true value, estimate how far you might be away from it? But it really isn't all that mysterious. It depends on only two variables—the extent to which individual values differ from the average, often expressed as a *standard deviation* (SD); and the sample size. If relatively little variation is found about the mean of the sample, it is likely that the sample mean will lie fairly close to the true value. Also, if we have a large sample size, regardless of the variation, all the differences in individual values will tend to cancel themselves out, and our estimate will be close to truth.

A Bit More on Nomenclature

Of course, as we start inferring, we have suddenly doubled the number of variables we have on hand. We now have *sample* means and *population* means, *sample* SDs and *population* SDs, *sample* variances and *population* variances, and so on. As one strategy to keep things straight, statisticians, a long time ago, created two sets of labels. Sample values are labeled with the usual Roman letters, as we have been doing all along, and population values are labeled with Greek letters. Undoubtedly this was a good idea back in those wondrous days of yore when every school person had to survive courses in Latin and Greek. Nowadays, the only people who know Greek are Greek scholars, Greek fraternity members, and Greeks, so the convention confuses. However, one of us had the benefit of a Greek fraternity[1] (but thankfully no Greek course), so all will now be enlightened. Below is a small sprinkling of Greek and Roman letters, and their names:

Greek letter	Name	Roman letter	Statistical term
α	alpha	a	Type I error (see below)
β	beta	b	Type II error (see below)
δ	delta	D	difference
π	pi	p	proportion
μ	mu	M	mean
σ	sigma	s	standard deviation

So, the little squiggles aren't all that mysterious; most stand for the same quantity in the sample and the population. Sample means begin with *M*, population means begin with Greek *m*, or *mu* (μ) . . . and so on.

As yet, we haven't said anything about how one goes about calculating these mystical quantities. In fact, you don't, because only God has access to the entire population.[2] What one does is use the calculated sample *statistic*—the mean or standard deviation calculated from the sample to **estimate** the population *parameter*.

[1] *The other author will be happy to furnish Hebrew equivalents on request.*

[2] *This isn't entirely correct. You may actually have access to the population. For example, Yugo Motors has access to the entire population of 1993 Yugos, at least until they are both sold. So, when they say that the average gas mileage for 1990 Yugos is 23.4 mpg, they may well mean just that. No estimation of error exists, and inferential statistics are not required.*

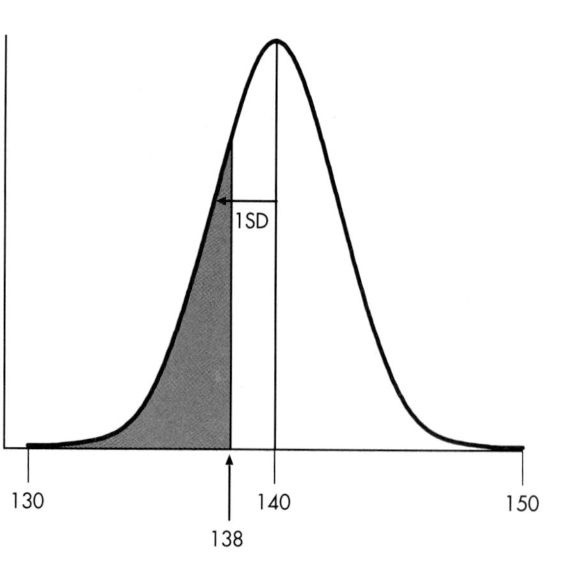

FIGURE 6–1
Range of normal for serum sodium.

ELEMENTS OF STATISTICAL INFERENCE

Enough of philosophy. Now, let's return to the relatively real world and examine a slightly atypical (in its simplicity) problem in statistical inference.

As a consultant in psychiatric biochemistry, you have become suspicious that individuals who are inclined to an obsessive-repulsive personality disorder are hyponatremic (low sodium), causing them to want to compulsively rub salt into everyone else's wounds. The Clinical Chemistry lab in our local hospital states the range of normal values of serum sodium are from 135 to 145 mmol/L. By convention, the normal range is ±2 SDs, so about 95% of all people fall within the normal range. This implies that the mean value is 140 and that the SD is (140 − 135) ÷ 2 = 2.5. You sample a total of 25 people from the hospital administration area (reasoning that they would have a particularly high incidence of obsessive-repulsive disorder) and discover that their mean serum sodium is 138.0. Is this evidence that they are hyponatremic, supporting your hypothesis?

A first approach to understanding the problem conceptually is to graph the distribution of normal values and indicate the sample mean, as shown in Figure 6–1. This would appear to show that the sample mean of 138.0, although somewhat out on the wing of the distribution, is not all that unusual. By inspection, it would seem that about 21% of people have values more extreme (in this case, lower) than 138.

THE CONVENTION OF HYPOTHESIS TESTING

Statisticians are a cynical lot. Although their bread and butter is proving that effects, however small, are statistically significant and therefore worthy of attention, they always start out the other way, by assuming that there is no effect. They frame a **null hypothesis** (abbreviated as H_0) that looks like:

> H_0: There is *no* difference between the serum sodium of hospital administrators and normal people.[3]

Of course, if this is true, then the administrators are like everyone else (fat chance!), the syndrome is unsupported, and the paper gets rejected.[4] So, we want to beat up on (i.e., reject) the null hypothesis to make our reputation. The alternative is called, to no one's surprise, the **alternative hypothesis**, and it is labeled H_1. This hypothesis states that the sample and population *are* different.

[3] *Note that we are not claiming that hospital administrators are a random sample of the general population in all their characteristics. Even statisticians are not that thickheaded!*

[4] *Lest you think these are the ravings of a mad author, there is good evidence, from a variety of fields that it is easier to publish results that show a difference than results that don't.*

THE STANDARD DEVIATION AND THE STANDARD ERROR

The distribution in Figure 6–1 displays how *individual values* fall about the mean. But this is not really what interests us. What we really care about is how the *mean* value of the sample compares with the population mean. We are not dealing with individual values any more; we are dealing with the mean from a sample of 25 people. Instead of dealing with the original distribution of values, we must consider what would happen if we repeatedly sampled 25 people and measured their serum sodium, that is to say, suppose we did the study a zillion times, using 25 subjects each time, calculated the mean, and then displayed all these means.

It should seem evident that these mean values for a sample size 25 would be more tightly distributed about the true mean than would the original individual values. If this is not evident to you, imagine what happens if you vary the sample size. If we use a sample size of one (i.e., we simply sample individuals and plot their values), we will, of course, reproduce the original distribution. If we use a sample size of two, taking two people and averaging their sodium levels, we would expect that the means would fall a little closer to the true mean than would either considered alone because the chance deviation of one person from the population mean may cancel out that of the other. If we go to 10 values, it would seem plausible that the mean values would be quite a bit closer to the true mean, so the distribution for a sample size of 10 would be quite a bit narrower than would the original distribution. As we go up the sample size ladder, things get closer to the truth, so that a sample size of 100 should yield a mean value very close indeed to the true (i.e., population) value.

Recognizing that things get tighter to the mean as the sample gets larger, the issue is now, "How much tighter?" It would seem that the SD of these means would be directly related to the original SD and somehow inversely related to the sample size. As it turns out, there is a simple relationship between the sample size and the SD of the sample means (now called the **Standard Error of the Mean**, or SE_M), as shown:

Standard Error of the Mean

$$SE_M = \frac{SD}{\sqrt{\text{Sample Size}}} = \frac{s}{\sqrt{N}} \quad (6\text{–}1)$$

So, the *SD reflects how close individual scores cluster around their mean, whereas the SE shows how close mean scores from repeated samples will be to the true (population) mean.* All this discussion is predicated on the notion that the sample we have chosen is a random sample of the population of all sodiums; that is, the hospital administrators are simply a random sample of the general population, at least insofar as their serum sodiums go. This is the **null hypothesis** we mentioned before.

CALCULATING THE z-TEST

In the present sample, the normal range, which went from –2 SD to +2 SD, was equal to 10 mmol/L. So, 1 SD is 10 ÷ 4 = 2.5 mmol/L. Thus the SE of the mean for a sample size of 25 is equal to 2.5 ÷ $\sqrt{25}$ = 0.5 mmol/L. This then signifies that samples of size 25 repeatedly drawn from the "normal" population would have a mean of 140 mmol/L and an SE (i.e., an SD of the means) of 0.5 mmol/L. We can now have a second look at what our sample mean of administrators looks like, in Figure 6–2. Now we have a different picture. The sample mean is well out on the curve; in fact, it is

$$z = \frac{(\bar{X} - \mu)}{\sigma/\sqrt{n}} = \frac{(138 - 140)}{0.5} = -4.0 \quad (6\text{–}2)$$

SDs below the mean. If we now look this up in Table A in the book appendix, which displays the area corresponding to different places on the normal curve starting from the mean, we see that the area corresponding to a z of +4.0 is .4999. This means that about 1/1000 of the area under the curve is to the right of 4.0. Similarly, less than 1/1000 of the area of the curve falls to the left of −4.0. The probability of observing a difference between the sample and population means this large or larger, **under the null hypothesis**, is vanishingly small. As a result, the null hypothesis (that there is no difference between administrators and normal people) seems rather unlikely, and we **reject the null hypothesis** in favor of the alternative hypothesis that we really wanted all along, that administrators have lower sodiums than you or me. That is, we have determined that the probability of arriving at a sample mean of 138 or less, from a sample size of 25 drawn at random from the population with a mean of 140, was sufficiently small (namely .0001, or 1 chance in 10,000) that we reject the hypothesis that this was where the sample originated. We have achieved our first *statistically significant* result.[5]

The z-Test and the z-Score

As you'll find when you get further into this book, statisticians have an annoying habit of using the same letter to indicate different things (e.g., Σ can mean either to sum everything up, or the variance-covariance matrix; π can be either 3.14159 or the product of some numbers). Now we've used the letter z in Chapter 4 to refer to the standard score, and in this chapter to indicate a statistical test between a sample and a population mean. Just to confuse you a bit more, there actually *is* a connection between the two. In the equation for the z-test (6–2), the denominator is $\sigma/\sqrt{n}$. The formula for the z-score is exactly the same, except that n = 1, so we don't bother to write the square root term. In other words, the z-score is a z-test for a single number. See, we're not always irrational.

THE RATIONALE BEHIND "SIGNIFICANCE TESTING"

So, what we've found is that the population mean is 140, and the mean for 25 hospital administrators is 138. Why can't we stop right there and conclude that people who don't work have a lower mean than those who toil for a living? This goes to the heart of hypothesis testing. In Figure 6–1, we drew the distribution of serum Na scores in the population. It has a population mean (μ) of 140, with a population SD (σ) of 2.5. If we were to draw a large sample of people at random from this population and draw a graph of their scores, what would we find? Another normal distribution with a sample mean ($\bar{X}$) of about 140 and a sample SD (s) of 2.5.

But we're interested in the mean of a sample, so we'll draw a sample of 25 and figure out their mean, then repeat this for a few hundred random samples of N = 25 each. If we now draw a graph of these few hundred *mean* values, what will it look like? Based on what we just went over, we should again get a nor-

[5] *If you want to become a real statistician when you grow up or grow old, this is the point where you throw your pencil in the air (some of us high-tech types throw our programmable calculators in the air, but it's a bit hard on them), bounce up and down in your chair, emit squeals of joy, and rush out and embrace the first young member of the opposite sex you see. So, to help you in learning the rituals of the culture, we strongly suggest that you take a moment before reading further to throw something in the air, squeal or chirp a bit, and embrace your dog or budgie. They won't mind the eccentricity— they're probably used to it.*

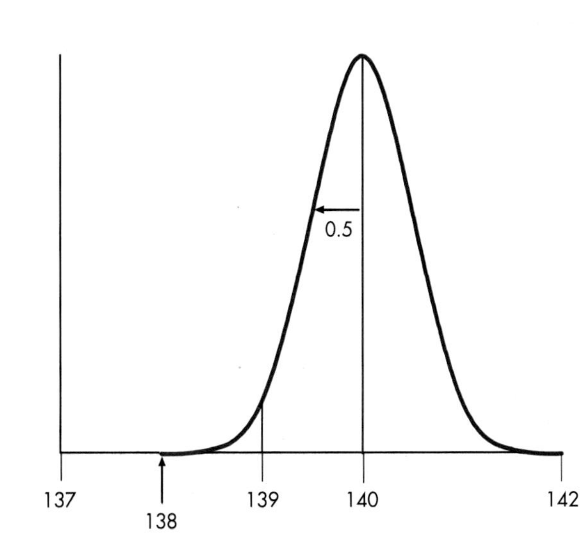

FIGURE 6–2
Standard error of serum sodium.

mal distribution with a mean of 140, but its SD would be equal to the SE based on 25 subjects, or 0.5.

Now let's go one step further. We'll go back to our random samples, and subtract 140 from all the computed sample means. How will the *differences* between the sample means and the population mean be distributed? Effectively, this amounts to simply moving the whole curve 140 units to the left, so it's centered on 0. Once again, this results in a normal distribution, but this time the mean is 0 because, on the average, there is no difference between the sample mean and the population mean. Also, the differences will have a standard error of 0.5, as we computed above, which means that if we do a study with 25 folks randomly sampled from the population, 2.5% of the time their mean will be more than 2 standard errors to the left of 140, or less than 139; 2.5% of the time it will fall more than one unit to the right of 140, or 141 or more, purely by chance.

The problem is when we do a study, such as the one with serum Na, and find a difference between the means, what can we conclude? It *may* be caused by the fact that the two groups are different, or it simply may result from sampling (like ending up with either Norman or Streiner in your group of statisticians). What we do now is play a game; we say that if a difference as large as the one we found, given our values of SD and *N*, can occur by chance more than 5 times in 100, there's too great a likelihood that it was caused by chance only. But if the probability was less than 5%, we say that the difference was caused by the fact that the two samples actually are different.

Where Did that 5% Come from?

Changing the subject a moment to statistical sociology, we might as well explore the mysteries of statistical significance a bit further. Long before you laid down your hard-earned cash for this gem, you knew that statistical significance meant $p < .05$; you just didn't know what $p < .05$ meant. Now you do—but why, says you, .05?

It turns out that this is really a historical issue. One day, Sir Ronnie Fisher (the granddaddy of statistics, and not to be confused with Ronnie Corbett, the little British comedian) was having tea with his cronies, and mused that, "If the probability of such an event were sufficiently small—say, 1 chance in 20—then one might regard the result as significant." And the emperor spake, and that was that.

Lest this seem somewhat arbitrary, try this out on your friends. Imagine you're betting, by throwing a coin in the air. If it comes up heads, they'll pay you $1.00; if tails, you'll pay them. You keep tossing it, and it keeps coming up heads. How many tosses before your friends will think it's rigged?

If we were doing it, our friends would say "1 or fewer." For you, though, it will probably be about 4 or 5. Now, if we assume that chance is operative, then the probability on the first coin is 50%. Three more tosses corresponds to 1 chance in 2^4, or 1 in 16. Four tosses is 0.5^5, or 1 in 32. One in 20 falls nicely in between. Maybe Sir Ronnie wasn't that far off after all! Bear in mind, though, that 5% is only an agreed-upon convention, and not some absolute criterion of truth. An effect that exists at $p = .05$ doesn't suddenly disappear at $p = .051$. This was expressed best by Rosnow and Rosenthal (1991), who said, "Surely, God loves the .06 nearly as much as the .05."

STATISTICAL INFERENCE AND THE SIGNAL-TO-NOISE RATIO

The essence of the *z*-test (and as we will eventually see, the essence of all statistical tests) is the notion of a **signal**, based on some observed difference between groups, and a **noise**, which is the variability in the measure between individuals within the group. If the signal—the difference—is large enough as compared to the noise within the group, then it is reasonable to conclude that the signal has some effect. If the signal does not rise above the noise level, then it is reasonable to conclude that no association exists. The basis of all inferential statistics is to attach a probability to this ratio.

> Nearly all statistical tests are based on a signal-to-noise ratio, where the *signal* is the important relationship and the *noise* is a measure of individual variation.

To bring home the concept of signal-to-noise ratio, we'll make a brief diversion into home audio. As the local electronics shops and our resident adolescents continue to remind us, the stereo world has undergone yet another revolution. The last one in recent memory was the audio cassette, which had the advantage of portability so it would fit into the Walkmen (Walkpersons?) of us on-the-move yuppies, and also would continue to blare music out of our BMWs without skipping a beat as we rounded corners at excessive speed. The cost of all this miniaturization was lots and lots of hiss that no amount of Dolbyizing would resolve. But now we have CDs—compact discs—which deliver all that rap noise at a zillion decibels, completely distortion-free.

All that hissing and wowing was **noise**, brought about by scratches and dents on the album or random magnetization on the little tape. This was magically removed by digitizing the signal and implanting it as a bit string on the CD, letting the **signal**—the original music (or rap noise or heavy metal noise)—come booming on through. In short, 2 decades of sound technology can be boiled down to a quest for higher and higher signal-to-noise ratios so worse and worse music can be played louder and louder without distortion.

Although we are referring to music, we are simply using this as one example of a small signal detected above a sea of noise. When it comes to receiving the radio signal from Voyager 2 as it rounds the bend at Uranus, signal-to-noise ratio of the radio receiver is not just an issue of entertainment value;

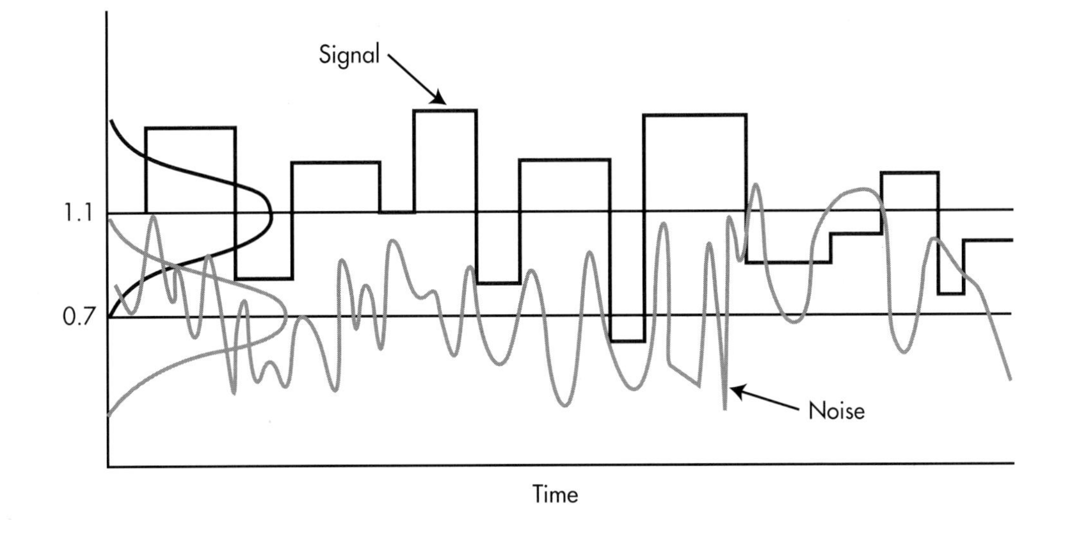

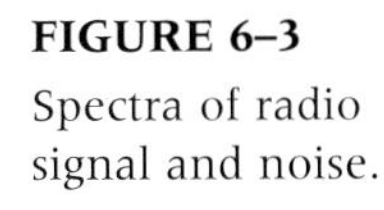

FIGURE 6–3
Spectra of radio signal and noise.

it's a measure of whether any information will be detected and whether all those NASA bucks are being well spent.

You might imagine the signals from Voyager 2 whistling through the ether as a "blip" from space. This is superimposed on the random noise of cosmic rays, magnetic fields, sunspots, or whatever. The end result looks like Figure 6–3. Now, if we project these waves onto the *Y*-axis, we get a distribution of signals and noises remarkably like what we have already been seeing. The signals come from a distribution with an average height about +1.1 microvolts (μV), and the noises around another distribution at +0.7 μV. If we now imagine detecting a blip in our receiver and trying to decide if it is a signal or just a random squeak, we can see that it may come from either distribution. Of course, if it is sufficiently high, we then conclude that it is definitely unlikely to have occurred by chance. Conversely, if it is very low, we do not hear it at all above the noise, and we falsely conclude that no signal was present. That is, there are always four possibilities: (1) concluding we heard a signal when there was none, (2) concluding there was a signal when there was, (3) concluding no signal when there was one, and (4) concluding there was no signal when there was none. Two of these are correct decisions (2 and 4), and two are wrong ones (1 and 3). Our problem is to determine which our decision is.

TYPE I, TYPE II, ALPHA, BETA, AND CONCLUSION ERRORS

Let's return to the serum Na example and complete the analogy. When we left off, we had determined that our sample size of 25, with a mean of 138 mmol/L, was sufficiently far away from the population mean of 140 that the difference was statistically significant.

For the moment, we must recognize that we have gone only partway in the logic of the inference. We have concluded that it was unlikely that our sample came from the population of *normal* people; that is, *we rejected the null hypothesis* of no difference between our sample and the reference population. But we have not, as yet, made any claims about the alternative population that they might have come from. It is clear that if they didn't come from the population we started with, they must have come from somewhere else.[6] In other words, we have rejected the null hypothesis, H_0, in favor of the alternative hypothesis, H_1, that the sample was drawn from a different population with a different mean, μ_1. Most of the time we don't worry too much about this alternative because achieving statistical significance is equivalent to stating that the experiment worked. Who cares how much it worked?

Paradoxically, the alternative hypothesis does matter a lot when you don't achieve significance. If you don't reject the null hypothesis, then you are in Never-Never Land, where it is unclear whether there really was no difference or whether there was a difference but your sample was too small to detect it. The philosophical dilemma is that you can never prove the **non**existence of something.

Suppose for the moment that the administrators actually did come from a different population, with a mean of 137.5 mmol/L and the same SD. (Of course, we have no way of actually knowing what this value is.) Then the two distributions would look like Figure 6–4. Now we have two overlapping distributions. The bell curve on the right was the one we started with, based on the null hypothesis that no difference existed between administrators and everyone else. This is H_0, or the Null Hypothesis. The bell curve on the left is the one based on the hypothesis that a difference does exist between the population of administrators and the normal population.[7] As we said before, this is the Alternative Hypothesis, or H_1.

[6] *"Howdy stranger, y'all ain't frum these hyar parts." "Nope, ah drifted down frum Somewhere Else."*

[7] *For no apparent reason, every other statistics book in the world always makes the difference positive, putting the H_1 curve to the right of the H_0 curve. If it really bothers you, use a wall mirror and read this over your shoulder.*

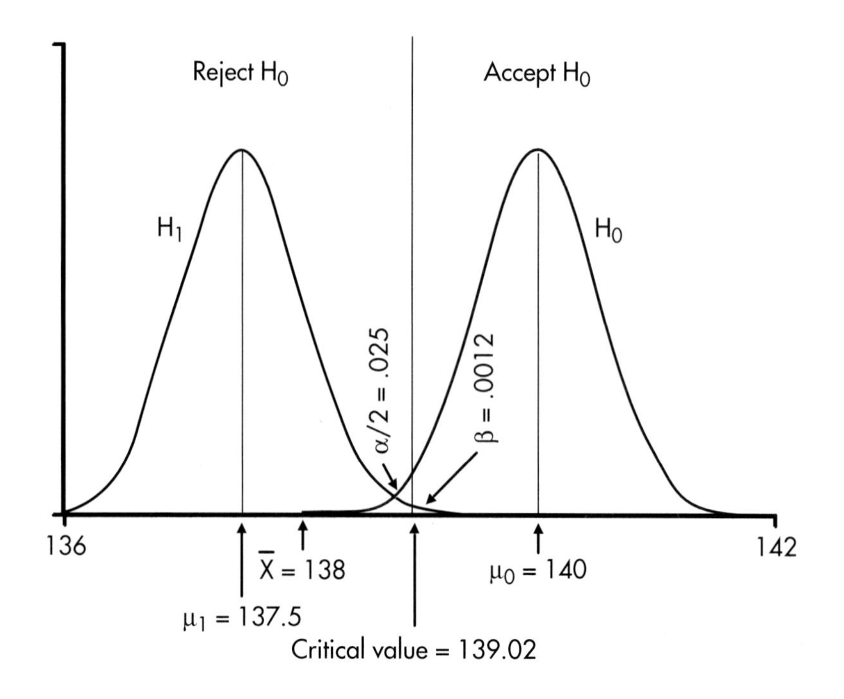

FIGURE 6–4
Null and alternate hypotheses for serum sodium.

The Type I Error

If the difference between the sample mean and the reference population were big enough to yield a small probability to the left on the null hypothesis distribution, then we were prepared to say that the difference was unlikely to have arisen by chance. Specifically, using the conventional level of significance of .05, we know from the normal distribution that 5% of the values fall outside ±1.96 standard deviations. So, if the sample mean fell any more than 1.96 standard errors to the left of 140, or anything less than $(140 - (1.96 \times 0.5) = 139.02)$, we would "reject the null hypothesis" and declare that the difference is unlikely to have arisen by chance; hence, it is statistically significant. This part is old hat. But what we are implying is that we are ready to conclude that the sample actually comes from the alternative distribution on the left (it has to be from one distribution or the other). There is a danger that we are wrong in this decision, captured in the tail from H_0 that we have been talking about. This error is called a **Type I** error, the error of concluding that there was a difference when, in fact, there was none; the associated probability in the tail is called, for no particular reason, **alpha** or α. When we choose to use a *p* level of .05 for statistical significance, we accept the fact that this error will occur 5% of the time.

> **Alpha** (α) is the probability of concluding that the sample came from the H_1 population (i.e., there is a significant difference) when in fact it didn't (making a **Type I** error).

The Type II Error

Things are symmetrical, and there is an opposite danger lurking in the wings. The distribution of H_1 also stretches out to the right, into the H_0 distribution. As a result, there is a small but finite probability that for any value of the difference which arose from the experiment and was too close to the normal mean of 140.0, we may well wrongly conclude that there was no difference when, in fact, there was one (i.e., that the sample came from the H_0 distribution), with a probability corresponding to the tail of the H_1 distribution. It is called, we suppose for the sake of uniformity, a **Type II error**,[8] and the associated probability is called **beta** or ***β***.

> **Beta (*β*)** is the probability of concluding there was **no** difference when, in fact, there **was** one (making a **Type II** error).

Since we have already decided that any value to the left of 139.02 will be declared significant, then it follows that any value to the *right* of 139.02 will be declared *nonsignificant*. What's the chance we could make a Type II error?

To get at this, we compute the other *z*-test, corresponding to values on the right hand tail of the H_1 distribution, which equals $(139.02 - 137.5)/0.5 = 3.04$. (Remember that we have assumed that the population mean of administrators was 137.5.) Looking this up in Table A, we find that this corresponds to a value of .4988, so the probability to the right of the critical value is .0012. Our chances of committing a Type II error is only 12 in 10,000.

Reviewing the underlying logic, then, we begin with the null hypothesis of no difference. We declare our alpha level (most of the time it's .05) and then compute the standard error. The combination of the alpha level and the standard error determines a **critical value** (in this situation, 1.96 standard errors to the left of the population mean). Any sample value which falls to the left of the critical value will be

[8] A Type III error is getting the right answer to a question nobody asked.

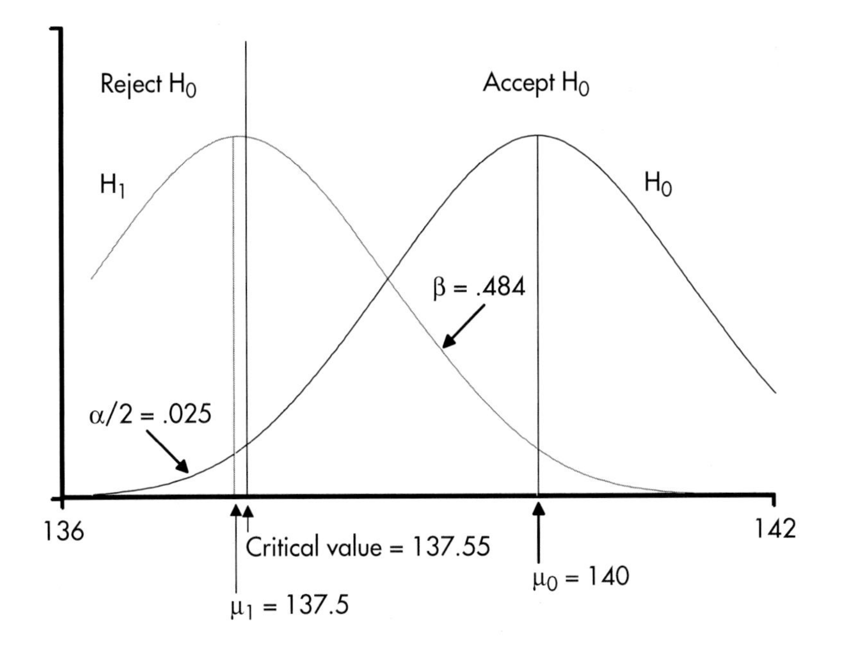

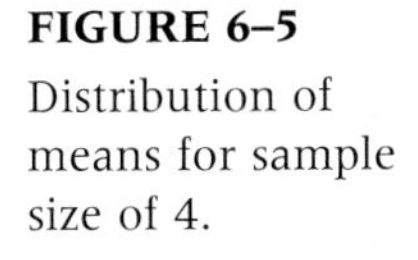
FIGURE 6–5
Distribution of means for sample size of 4.

declared significant and we will reject the null hypothesis; any sample value which falls to the right will be declared nonsignificant, and we will fail to reject the null hypothesis. In doing so, we run the risk of both a Type I error, corresponding to the values of H_0 to the left of the critical value, and a Type II error, corresponding to the values of H_1 to the right of the critical value.

At least, that's what we're supposed to do. Of course, we're also supposed never to exceed the speed limit, to look both ways before crossing the street, and to brush and floss after every meal. What actually happens in practice is that you do the experiment, you get some difference between one group and the other (in this case, between the sample and the population, which equals 2.0), you compute the appropriate test (in this case, a z-test, which equals $[138 - 140]/0.5 = 4.0$), you look up the p value in the back of the book (or more commonly, the computer doing the calculation for you also computes the p value; in this case, $p = .00001$) and, if it's less than .05, you close down the lab and celebrate.

Although the logic is perverted, the end result is the same, and the notion of a critical value with Accept and Reject regions is critical to interpreting things like power and Type II errors.

The Relationship between ß and *N*

To review the situation, let's have another run at the data, only using a smaller sample size. You have probably been admonished by researchers and statisticians on one occasion or another that using too small a sample means less chance of showing a statistically significant difference. Let's see why.

If we used a small sample, then standard error of the mean will be larger, so the two distributions may overlap a lot. In the present case, for the same value of the sample mean, and a sample size of, say, 4, the standard error would be $2.5/\sqrt{4} = 1.25$, and the two distributions would look like Figure 6–5.

In this case, since there is considerably more overlap in the two distributions, it is less likely that we will reject the null hypothesis. The critical value to reject H_0 is now $(140.0 - 1.96 \times 1.25) = 137.55$. Since the sample mean is 138.0, this falls to the right of the critical value, and we would conclude this time that there was no significant difference.[9] Now, with the benefit of hindsight derived from the previous calculation based on the first, bigger experiment, it is a pretty safe bet that this really was the wrong conclusion. But how safe a bet? After all, it seems that statistical inference is a game of putting probabilities on such things. To see, have a closer look at Figure 6–5. The critical value determined from this "study" is indicated, and the probability of making a mistake, as we just did, is the area of the left hand (or H_1) curve to the *right* of the critical value, 137.55. This is the probability of concluding that there was no difference, when there was, in fact, a difference.

[9] Note that all that has changed is the sample size. There is a message in this that we will return to later.

This time, the z is equal to $(137.55 - 137.50)/1.25 = 0.04$, and the associated probability is .4841; see if you can figure it out from Table A.

Power

There is one final quantity to be extracted from this pretty picture. Since experiments are usually done to demonstrate differences, often in the face of some risk that this won't happen, statisticians are often interested in the probability of detecting a true difference. Mindful of the personal consequences of continued success, this probability is called the **power** of the test.[10]

[10] Since this is directly related to publication, power, and prestige, it can be referred to by the symbol $, or "money."

TABLE 6–1

The relationship between α and β

Called	Truth	
	No difference	Difference
Accept H_0	$(1 - \alpha)$	β
Reject H_0	α	$(1 - \beta)$

Power is the probability of concluding that there was a difference when, in fact, there was one (power = $1 - \beta$).

As we can see from the diagram, it is the area to the *left* of the critical value on the left curve, and is equal simply to $(1 - \beta)$, or .5159. In particular, in the circumstances of this last example in which we were "unable to reject the null hypothesis," a natural questions is, "Was there really no difference, or were we just unable to detect it with the sample size we had?" By tradition, we like the power of a statistical test to be at least 80%.[11] In this case, since the power was only 51.6%, we're left in the uncomfortable position of having to say that the null hypothesis wasn't rejected, but by the same token, we didn't have enough power in any case, given our sample size.[12] So the experiment is over, our dreams have been shattered, the Nobel prize eludes us once again, and we sift through the ashes to see what went wrong.

[11] *Why 80%? Because Cohen surveyed the literature and found that the average power was barely 50%. His hope was that, eventually, both α and β would be .05 for all studies, so he took β = .20 as a compromise, and thought that over the years, people would adopt more stringent levels. It never happened.*

[12] *It doesn't make any sense to calculate the power you had to detect a difference if you have already detected a difference; you obviously had enough power!*

Putting It All Together

One other way to look at the four types of conclusions we can draw is to cast an analogy with diagnostic tests. Epidemiologists, and for that matter, many clinicians, are always concerned with false-positive and false-negative results. If we go along with this, we might say that a false-positive result comes from calling a conclusion significant when it isn't, and a false-negative result comes from calling our answer nonsignificant when it is. Thus a correspondence is seen between (1) our call, (2) the truth, and (3) the probabilities we have been messing with, as shown in Table 6–1. So, α is the probability of saying there is a difference when there isn't, β is the probability of not saying there is a difference when there is, and power $(1 - \beta)$ is the chance of detecting a difference when there is one.

Another way to get this Greek tragedy figured out is to review the logic of an experiment. After the analysis is completed, there are only two possibilities; either you conclude there **is** a difference, or you conclude there **isn't** one.

If you conclude there is a difference, then a natural concern is the likelihood that you have made an error; that is, the probability that there was actually no difference, and the sample you observed came from the null hypothesis distribution. This is captured in the α error. By and large, only studies that show differences get published anyway, explaining why the α probability is quoted all the time.

Conversely, if you conclude that no difference exists, then the opposite error arises; namely, the likelihood that a difference really did exist and the sample you studied came from the alternative hypothesis distribution. This is expressed in the β error, but to achieve this, you have to make a guess at how big a difference there might have been because the probability of missing small differences is higher than the probability of not detecting large ones. So you hazard a guess at a "clinically important difference" (10%, 25%, or whatever) and then calculate the β error. This can also be reported as $(1 - \beta)$, the **power** to detect a difference of such and such.

There is one design implication. Sometimes the situation arises where you really want to show that no difference exists; for example, comparing generic with brandname drugs. In this case, you **really** don't want to conclude there is a difference if there isn't. The strategy is to reduce the α probability, say to .01 or .001. Looking at Figure 6–5, we see that this amounts to moving the critical value further out, thereby increasing the β error and reducing power. To avoid an α error while keeping the power to detect a difference, if there is one, the only solution is to increase the sample size.

TWO TAILS VERSUS ONE TAIL

You will have noticed that we have been preoccupied with the left side of the pictures up to now. We had set out to show that administrators had a sodium deficiency, leading to a predilection for rubbing salt in. For obvious reasons, this is therefore called a **one-tailed** test.

A *one-tailed test* specifies the direction of the difference in advance.

Given that this particular hypothesis is a bit far-fetched anyway, it might have been equally interesting to simply ask whether administrators' serum sodium levels are *different from*, not higher or lower, than that of normal folks. Now, the *different from* hypothesis implies that we would be equally pleased if the administrators' level were either higher than or lower than those of the normals. If this were so, then we would have to consider the tails of the distribution on both sides, therefore conducting a **two-tailed** test.

A *two-tailed test* is a test of any difference between groups, regardless of the direction of the difference.

That is, for a one-tailed test:

$$H_0: \mu_A \geq \mu_N; \qquad H_1: \mu_A < \mu_N; \qquad \textbf{(6–3)}$$

or

H_0: $\mu_A \leq \mu_N$; H_1: $\mu_A > \mu_N$;

(6–4)

and for a two-tailed test:

H_0: $\mu_A = \mu_N$; H_1: $\mu_A \neq \mu_N$

(6–5)

where μ_A is the population mean of the administrators and μ_N is the population means of normal people.

Aside from the philosophy, it is not immediately evident what difference all this makes. But remember that the significance or nonsignificance of the test is predicated on the probability of reaching some conventionally small criterion (usually 0.05). If this occurs only on one side of the distribution, then from Table A in the book appendix, we see that this probability occurs at a z value of 1.645 (i.e., 1.645 SDs from the mean). By contrast, if we want the *total* probability on both sides to equal 0.05, then the probability on one side is 0.025, which corresponds to a z value of 1.96. So, to achieve significance with a one-tailed test, we need only achieve a z of 1.645; if it is a two-tailed test, we must make it to 1.96. Clearly the two-tailed test is a bit more stringent. (If you look back at page 47, you'll see we actually **did** use a two-tailed test, looking for a difference of ±1.96 standard errors.)

You would think that one-tailed tests would be the order of the day. When we test a drug against a placebo, we don't usually care to prove that the drug is worse than the placebo.[13] If we want to investigate the effects of high versus low social support, we wouldn't be thrilled to find that folks with high support are more depressed. In fact, except for the circumstance where you are testing two equivalent treatments against each other, it is difficult to find circumstances where a researcher isn't cheering for one side over the other.

However, there is a strong argument against the use of one-tailed tests. We may well begin a study hoping to show that our drug is better than a placebo, and we expect, for the sake of argument, a 10% improvement. Taking the one-tail philosophy to heart, imagine our embarrassment when the drug turns out to have lethal, but unanticipated, side effects, so that it is 80% worse. Now we are in the awkward situation of concluding that an 80% difference in this direction is *not significant*, where a 10% difference in the other direction was. Strictly speaking, in fact, we don't even have the right to analyze whether this difference was statistically significant; we would have to say it resulted from chance. Oops![14]

So that is the basic idea. One-tailed tests are used to test a *directional* hypothesis, and two-tailed tests are used when you are *indifferent* as to the direction of the difference. Except that everybody uses two-tailed tests all the time.

CONFIDENCE INTERVALS

There is an alternative, but related, approach to the yin-yang strategy of hypothesis testing. We could say, "Okay, we did the experiment, and this is what we found. There is some error inherent in our estimate, but we are pretty confident that the true value falls between *X* and *Y*." Mind you, by now you will have realized that words such as "pretty confident" send shivers down statisticians' spines. How confident is "pretty confident?" Are you 95% certain that the truth is somewhere in the interval? In other words, what is the **95% confidence interval** (CI)?

Over the past few years, George Gallup's successors have adopted this strategy as a matter of routine. Every poll proclamation is now issued with the disclaimer that ". . . this poll is estimated to be accurate within 2.4 percentage points, 95 times out of 100."[15]

Now, if we return to the administrator example and attempt to follow through the logic, it would go something like this. Remember we found they had a serum sodium with a sample mean of 138 mmol/L and an SD of 2.5 mmol/L based on a sample size of 25. What we are attempting to do is establish an upper and lower bound in such a way that there is a 95% probability that the true population mean falls within it.

Let's look at the lower bound first. We want to find out where the population mean would have to be so that the distribution of sample means for a sample size of 25 would end up with 2.5% above 138. The SE of the mean, as we calculated before, is $s \div \sqrt{n}$, or $2.5 \div \sqrt{25} = .5$. Two SDs is $1.96 \times .5 = .98$. So if the true mean was $(138 - .98) = 137.02$, there is a 2.5% probability of observing a sample mean of 138 or greater. Similarly, looking at the upper bound, if the true mean was $(138 + 0.98) = 138.98$, there is a 2.5% chance of observing a sample mean of 138 or less. So, putting it all together, there is a 5% chance that the truth is outside the range, or a 95% chance that the true population mean falls within the range.

Another way to see this is to look at Figure 6–6.

The 95% CI is such that there is a 2.5% chance that the population mean falls below the interval, shown as the shaded area of the lefthand curve to the right of 138, and a 2.5% chance that the population mean is above the interval, shown as the shaded area of the righthand curve to the left of 138.

To formalize all this into an equation, the $(1 - \alpha)$ CI, where α is, as before, the level of statistical significance, is:

Confidence Interval Around a Mean

$$CI = \bar{X} \pm z_{\alpha/2} \frac{s}{\sqrt{N}}$$

(6–6)

From the equation, it is evident that a relationship exists between the CI and the sample size and

[13]*And trying too hard to prove this is a surefire way to cut onself off from the filthy lucre of the drug companies.*

[14]*This is not as far-fetched as it may sound. Nobody expected pure oxygen to produce blindness in neonates, or that clofibrate would kill more people with high cholesterol than it saved, but that's what happened.*

[15]*We have often wondered what the average reader of the Des Moines New Dealer does with such information. Perhaps, before you read on, you could send us a postcard and let us know.*

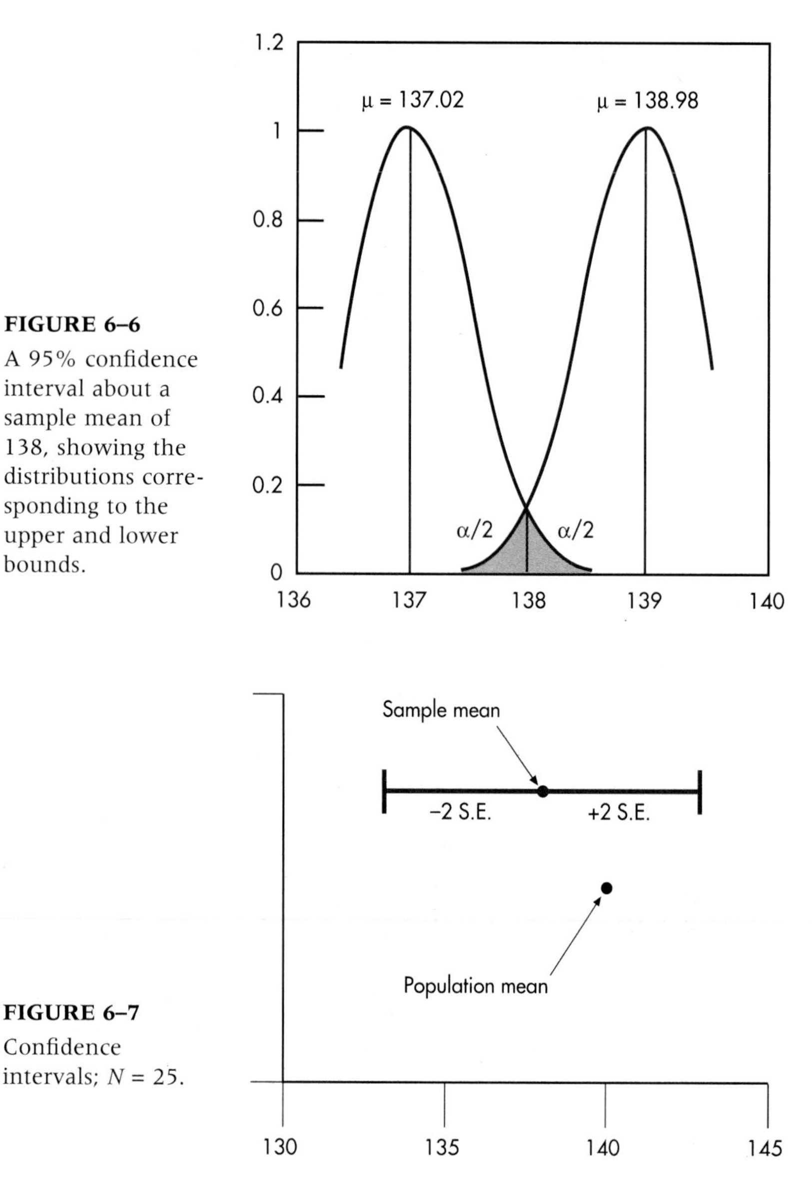

FIGURE 6–6
A 95% confidence interval about a sample mean of 138, showing the distributions corresponding to the upper and lower bounds.

FIGURE 6–7
Confidence intervals; $N = 25$.

SD. The smaller the sample size, the larger the CI. If the original SD is large, the CI will be as well.

It is not quite so obvious, but a relationship also exists between the CI and statistical significance. To explore this, let's return to the second experiment on the administrators, done with a sample size of 4. Here the CI would be:

$$\text{CI} = 138.0 \pm 1.96 \times 5.0 \div 2 = 138 \pm 4.90 = 133.1 \textit{ to } 142.9 \quad \text{(6–7)}$$

In particular, the 95% CI includes the original population mean of 140. So, clearly, the likelihood that the difference between the two means is 0 is something greater than .05. This can be seen in Figure 6–7. Putting it another way, if the 95% confidence intervals of two means overlap, then the difference is not statistically significant; if they do not overlap, the difference is significant. This can be awfully useful if a graph of means contains SEs. All you do is visually double the SEs on the graphs, then announce to your friends which differences are and are not significant. They rush to their computers, crank out the data, and return full of admiration for your amazing magical powers.[16]

STATISTICAL SIGNIFICANCE VERSUS CLINICAL IMPORTANCE

It may have dawned on you by now that statistical significance is all wrapped up in issues of probability and in tables at the ends of books. Whatever actual differences were observed were left far behind. Indeed, this is a very profound observation. Statistical significance, if you read the fine print once again, is simply an issue of the probability or likelihood that there *was* a difference—*any* difference of *any* size. If the sample size is small, even huge differences may remain non- (not in-) significant. By the same token, with a large sample size, even tiddly little differences may be statistically significant. As our wise old prof once said, "Too large a difference and you are *doomed* to statistical significance."

As one example, imagine a mail-order brochure offering to make your rotten little offsprings smarter so they can go to Ivy League colleges, become stockbrokers or surgeons, and support you in a manner to which you would desperately like to become accustomed. This is what the insurance companies call "Future Planning." Suppose the brochure even contains relatively legitimate research data to support its claims that the product was demonstrated to raise IQs by an amount significant "at the .05 level."

How big a difference is this? We begin by noting that IQ tests are designed to have a mean of 100 and an SD of 15. Suppose we did a study with 100 RLKs (rotten little kids) who took the test. Just like the earlier example, we know the distribution of scores in the population if there is no effect. Under the null hypothesis, our sample of RLKs would be expected to have a mean of 100 and an SD of 15. How would the means of a sample size of 100 be distributed?

The SE is equal to:

$$SE_M = \frac{SD}{\sqrt{N}} = \frac{15}{\sqrt{100}} = 1.5 \quad \text{(6–8)}$$

Now, the z value corresponding to a probability of .05 (two-tailed, of course) is 1.96. So, if the difference between the RLK mean and 100 is δ, then:

$$\frac{\delta}{1.5} = 1.96 \quad \text{(6–9)}$$

so $\delta = (1.96 \times 1.5) = 2.94$ IQ points. That is, for $N = 100$, a difference of only 3 points would produce a statistically significant difference.

[16] *Actually, this works only if the data are well behaved and normally distributed; otherwise, it's a good "eyeball" test, but it's not exact. If the distance between the lines is small, you may amaze your friends with your wrong conclusion.*

This is not the thing of which carefree retirement, supported by rich and adoring offspring, is made! Working the formula out for a few more sample sizes, it looks like Table 6–2. It would seem important, before finding that little cottage in the Florida swampland, to investigate how large the sample was on which the study was performed.

Of course, like everything else, "large" and "small" in terms of sample size are relative terms. By and large (and small), if the study deals with measured quantities such as blood sugar, clinical ratings, aptitude tests, or depression scores, any difference worth worrying about can be attained with about 30 to 50 subjects in each group. By contrast, with relatively rare events such as death,[17] it may take depressingly large samples.[18] For example, the first large-scale sample of cholesterol-lowering drugs screened 300,000 men to get 4,000 who fit the inclusion criteria. They were followed for 7 to 10 years, then analyzed. There were 38 heart-related deaths in the control group and 30 in the treatment group—just significant at the .05 level.

It would seem important to clearly outline the difference between statistical significance and clinical importance. As we have shown (we hope), statistical significance simply addresses the likelihood that the observed difference is, in truth, not actually zero. Statistical significance *says nothing about the actual magnitude or the importance of the difference*. The importance of the difference, often called **clinical significance** or **clinical importance**,[19] is a separate issue, and it can be decided only by judgment, not by any whiz-bang mathematics. It's a pity that statistical significance has assumed such magical properties, because it really is addressing a pretty mundane idea.

Note, however, that the two concepts are not unrelated. Although statistical significance makes no claims to the importance of a difference, it is a necessary precondition for clinical significance. If a difference is not statistically significant, it might as well be zero, or, for that matter, it might as well be in the opposite direction. Trying to argue that a difference that is not statistically significant (i.e., may be equal to zero) is still clinically important is illogical and, frankly, dumb.

Statistical significance is a necessary precondition for a consideration of *clinical importance* but says nothing about the actual magnitude of the effect.

SAMPLE SIZE ESTIMATION

As we already indicated, a lot of clinical research is horrendously expensive. To keep the cost of doing the study down, it has become *de rigueur* to include a sample size calculation in the grant proposal. Essentially, this begins with the clinicians guessing the amount of the minimum *clinically* significant difference worth detecting. Then the statistics are messed around so that this minimum clinical difference corresponds to the statistical difference at $p = .05$.

TABLE 6–2 Relation between sample size and the size of a difference needed to reach statistical significance when SD = 15

Sample size	Difference
4	14.70
9	9.80
25	5.88
64	3.68
100	2.94
400	1.47
900	0.98

Returning to the example of the RLKs, suppose we decide, about the time the encyclopedia salesman is shoving his foot further into the door, that the minimum difference in IQ we would shell out for is 5 points. How big a sample would *Encyclopedia Newfoundlandia (E.N.)* need to prove that its books will raise IQ levels by 5 points?

Now the picture is like Figure 6–8. We know where the mean of the null distribution is, at 100 points. We know where the mean of the population of RLKs who had the dubious benefit of *E.N.* is—a 5-point gain, at IQ 105. Finally, we must keep in mind that the normal curves we have drawn in the figure correspond to the distribution of *means* for repeated experiments, where the values are distributed about either 100 (if *E.N.* had no effect) or 105 (if *E.N.* had an effect). Of course, we don't know what distribution our *E.N.*-exposed RLKs come from; that's the point of the experiment. Either way, we know how wide the normal curves are—they correspond to an SD of $15 \div \sqrt{n}$. The challenge is to pull it all together and solve for n.

Imagine that the experiment was completed in such a way that it just achieved statistical significance at the .05 level, by the skin on its chin. Then the **critical value** (CV) corresponding to this state is 1.96 SEs to the right of the null mean.[20] We will call this distance z_α, the z value corresponding to the alpha error. Now we have to decide how much we want to risk a Type II error, the area of the H_1 curve to the left of this point. Suppose we decide that we will risk a beta error rate of .10; this, then, puts the critical value at 1.28 SEs to the left of the alternative hypothesis mean. By analogy, this will be called z_β, the z value on the alternative curve corresponding to the beta error. IMPORTANT NOTE: The z-value for β is always based on a one-tailed test. This doesn't contradict what we said about two-tailed tests because that applies only to the α level. The reason can be seen in Figure 6–8, where the tail of the H_1 distribution overlaps that of H_0 on only one side.

We can formalize this with a couple of equations:

[17] *Yes, we know, death has a 100% prevalence. But in a follow-up period sufficiently short that the investigators themselves have some certainty of survival, death can be relatively rare.*

[18] *There is an up side. With large samples, there is a need for multicenter trials, resulting in a need for international collaborative meetings in exotic locales.*

[19] *Presumably to make clinicians feel that there is a role for them just about the time that they are totally intimidated by the whole thing.*

[20] *Note: This time, we are putting the alternative hypothesis where everyone else has it. If you are still looking over your shoulder at the wall mirror, you can sit down now.*

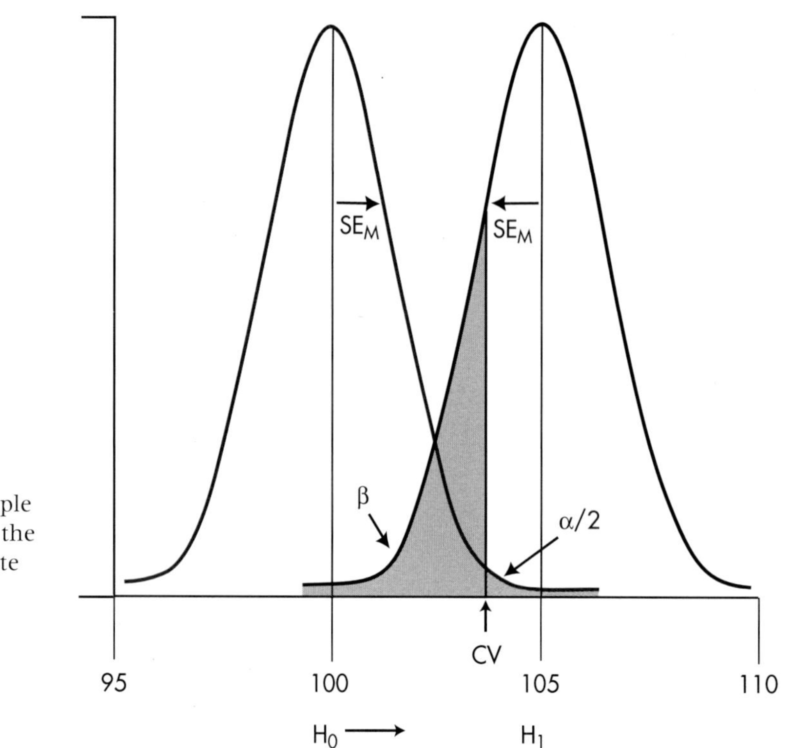

FIGURE 6–8
Mean IQ of sample of RLKs against the null and alternate hypotheses.

$$\frac{(CV - 100)}{s/\sqrt{n}} = z_\alpha = 1.96 \qquad \textbf{(6–10)}$$

Similarly:

$$\frac{(105 - CV)}{s/\sqrt{15}} = z_\beta = 1.28 \qquad \textbf{(6–11)}$$

where CV is the critical value between the H_0 and H_1 curves. Adding the two equations together, we get rid of CV.

$$\frac{(105 - 100)}{s/\sqrt{15}} = z_\alpha + z_\beta = 3.24 \qquad \textbf{(6–12)}$$

If, for the sake of generality, we call (105 – 100) the difference Δ, the algebra becomes:

$$\frac{\Delta}{s/\sqrt{15}} = (z_\alpha + z_\beta) \qquad \textbf{(6–13)}$$

so that

$$\sqrt{n} = \frac{(z_\alpha + z_\beta)\, s}{\Delta} \qquad \textbf{(6–14)}$$

and squaring everything up:

$$n = \left[\frac{(z_\alpha + z_\beta)\, s}{\Delta}\right]^2 \qquad \textbf{(6–15)}$$

We should put this equation in big, bold type because it, and variations on it, are the things of which successful grant proposals are made. The same strategy will be used in subsequent chapters to derive sample size estimates for a variety of statistical tests. To save you the agony of having to work out this formula every time you want to see how many subjects you need to compare two means, we've given you these in Table B in the appendix at the end of the book. Obviously, we couldn't do this for every possible value of σ and Δ. What we've done is to present *N* for different *ratios* of σ ÷ Δ.

Note that the ratio of the difference between groups to the SD is called the **effect size** (ES). The effect size is like a *z*-score, and it tells you how big the difference is in SD units. If the difference you're looking for is 5 points and the SD is 15 points, then the ES is 5 ÷ 15 = .33. So the ratio in the sample size equation, σ/Δ, is the inverse of the effect size.

For completeness, we'll put the numbers of Figure 6–8 back in:

$$n = [(3.24 \times 15) \div 5]^2 = 95 \text{ subjects} \qquad \textbf{(6–16)}$$

What is the distinction between the αs and βs in this calculation and the one before? Really only one of timing. In the previous example, the experiment was finished and did not show a difference. In this case, we are in the position of designing a trial, and so we based our calculations on a critical value for the sample mean that corresponded to the difference required to just reject the null hypothesis. If the experiment had turned out at that critical value, we then would have been able to determine exactly the probability of rejecting the null hypothesis when it was true (α, the Type I error) and the probability of rejecting the alternative hypothesis (accepting the null hypothesis) when it was true (β, the Type II error). It was these values that were used in the sample size calculation.

SUMMARY

You can use a *z*-test to determine the statistical significance of the difference between a sample and a population with known mean and SD. The *z*-test, like all statistical tests, relates the magnitude of an observed difference to the probability that such a difference might occur by chance alone. The notion of statistical significance is embodied in this probability. But statistical significance does not, of itself, reveal anything about the importance of the observed difference.

EXERCISES

1. A report of a clinical trial of a new anticocaine drug, Snortstop, versus a placebo, noted that the new drug gave a higher proportion of successes than did the placebo. The report ended with the statement that the statistical test was significant ($p < .05$). In light of this information we may conclude:
  a. Fewer than 1 patient in 20 will fail to benefit from the drug.
  b. The chance that an individual patient will fail to benefit is less than .05.
  c. If the drug were effective, the probability of the reported finding or one more extreme is less than 1 in 20.
  d. If the drug were ineffective, the probability of the reported finding or one more extreme is less than .05.
  e. The power of the test exceeds 0.95.

2. In a small, randomized, double-blind trial of a new treatment in patients with acute myocardial infarction, the mortality in the treated group was half that in the control group, but the difference was not significant. We can conclude that:
  a. The treatment is useless.
  b. There is no point in continuing to develop the treatment.
  c. The reduction in mortality is so great that we should introduce the treatment immediately.
  d. We should keep adding cases to the trial until the Normal test for comparison of two proportions is significant.
  e. We should carry out a new trial of much greater size.

3. Consider two randomized trials of the effect of anabolic steroids on commuters' times in the "100 meter train dash." Both studies used the same populations and experimental design. The only difference is that the first study used a total of 10 office workers per group, whereas the second used 100 per group. For the first study, the means (SDs) of the two groups were 12.0 (3.0) seconds for the placebo group and 16.0 (3.0) seconds for the group that received anabolic steroids. Answer the following questions regarding the expected results of the second study:

	Larger	Smaller	Stay the same	Can't tell from the data
SD	_____	_____	_____	_____
SE of mean	_____	_____	_____	_____
Statistical test	_____	_____	_____	_____
p-value	_____	_____	_____	_____

4. In a two-group design comparing the effects of diet restriction and exercise on quality of life of obese patients, researchers used a quality-of-life instrument, the CPQ (Couch Potato Questionnaire), with 5 subscales (Emotional Function, Social Function, Physical Function, Self-Esteem, Eating Attitudes). Because of concern about the use of multiple tests, the alpha level (probability of declaring a difference under the null hypothesis) was set at .01 instead of the usual .05. What effect will this have on the power to detect a true difference between the two groups on the Eating Attitudes subscale?
  a. Increase power.
  b. Decrease power.
  c. Stay the same
  d. Insufficient data to tell.

5. Second only to terminal zits, the biggest concern of every nubile adolescent in the 2000s is "Quality of Life." So the local teener's health office developed a questionnaire to assess satisfaction with social interactions, depression, self-esteem, mirror avoidance, and time spent in closets. Because of concerns about using multiple *t*-tests, the investigators used a Bonferroni correction; α was divided by 5, so only *p* levels less than .01 were considered significant. What effect will this have on:
  a. The Type I error rate?
  b. The Type II error rate?
  c. Power?
  d. Degrees of freedom?

6. You have just completed a study of a patent medicine for basketball players, designed to make them jump higher, spin around faster, and fool the opposition by looking like they're going backwards and forwards at the same time. It's called MJ^3 Elixir and is endorsed by Magic Johnson, Michael Jackson, and Michael Jordan. Testing the first part only, you find that a sample of 16 collegiate players fed the elixir for 2 weeks can jump an average height of 56 cm. Population data gathered by university phys-ed coaches across the country show a normal jump height of 50 cm, SD 15 cm.

a. What is the probability that this difference could have occurred by chance?
b. Suppose the true benefit was 10 cm. What is the power of the study to detect this difference?
c. How large a sample would you need to have a 90% power of detecting this difference (using alpha = .05 as a critical value)?

I–1. Hospital administrators used a graph like the one shown in Figure I–1, which shows the number of hours worked each week between 1970 and 1985, to justify their request for a large pay increase. They argued that this graph showed their workload jumped about 500% between 1980 and 1985.[1] Can they use this to justify a 500% increase in their salary?

No, for three reasons. First, they already get paid too much. Second, they never needed any justification in the past to award themselves increases, so why start now? The third reason, though, is that from our perspective as unbiased, disinterested scientists, this graph distorts the data. The problem is the **missing zero**. The *Y*-axis does not start at zero, but at some arbitrary point (in this case, 30 hours per week), so that increases look magnified. Also, this is equivalent to taking ratio data and making it into interval data; this means that we can't calculate ratios, even mentally, from the graph.

> C.R.A.P. DETECTOR I–1
>
> **The *Y*-axis should start at 0, unless there are compelling reasons why it should not (see C.R.A.P. detector I–3).**

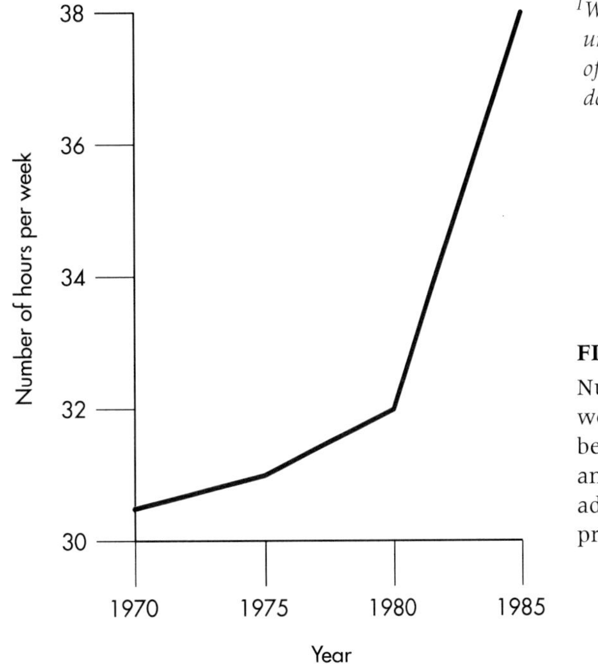

FIGURE I–1
Number of hours worked per week between 1970 and 1985 by administrators, as presented by them.

[1] *We won't ask the unworthy question of what they were doing prior to 1980.*

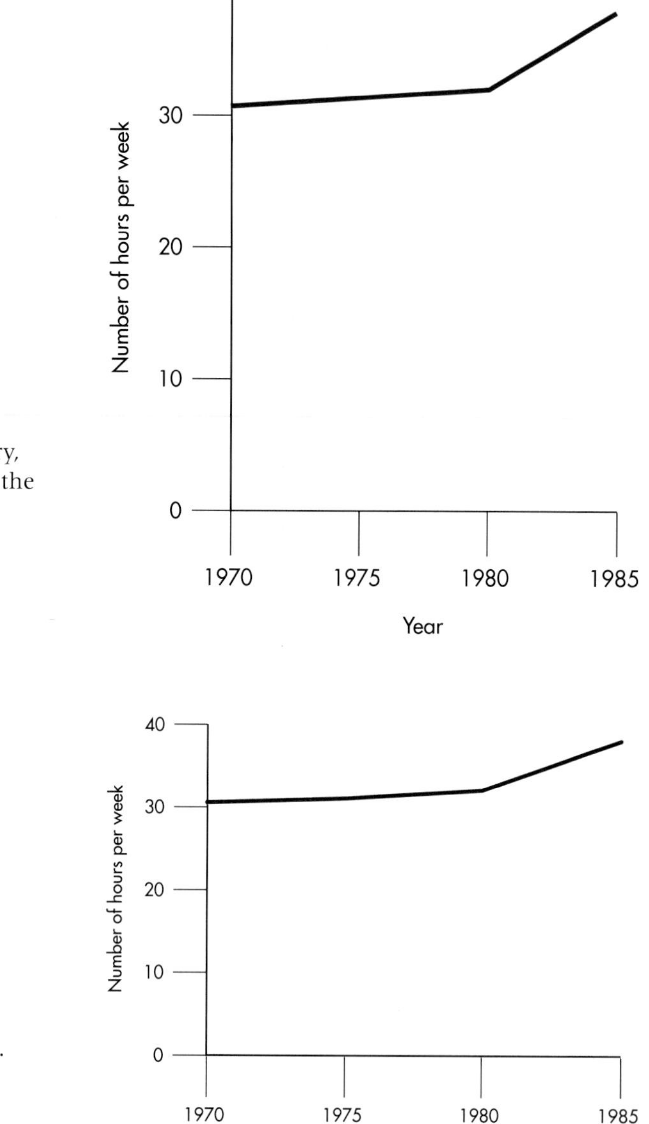

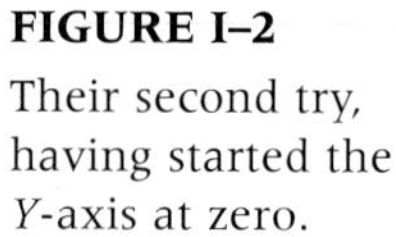

FIGURE I–2
Their second try, having started the *Y*-axis at zero.

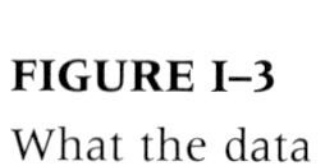

FIGURE I–3
What the data really look like.

I–2. Foiled in their dastardly attempt to flummox the Board of Directors of the hospital, the administrators brought in a second graph, Figure I–2, which they said corrected the problem of the missing zero, and which still showed a marked increase in hours worked per week. Have they learned the error of their ways and turned to the path of righteousness?

Are you kidding? If you look over the graphs we've presented so far, you'll notice that the vast majority of them are oriented horizontally. Figure I-2 is turned so that the *Y*-axis is parallel to the long side of the paper. Although the numbers displayed in the graph are correct, squeezing the data displayed along the *X*-axis tends to magnify vertical differences in our mind. The data should really have been displayed as it is in Figure I-3.

C.R.A.P. DETECTOR I–2

The graph should not magnify the visual effect of relatively small changes.

I–3. Claiming that they have repented, the contrite administrators[2] show up at the next board meeting with a graph showing the hours of work per week for the epidemiologists and statisticians. They maintain that Figure I–4, which starts at zero and is arranged horizontally, shows that these people have barely increased what they do since 1970 and so should not get any increase at all. Have we been falsely maligned?

Of course! The problem here is the converse of the missing zero; the zero *should* be missing because the bottom 80% of the graph is blank, thus all the action takes place in the upper 20%. The effect of this is to squeeze any changes into a very small range, making it look as if nothing is happening. A better way of presenting these data would be as in Figure I–5. Note that we statisticians, pure of heart, showed that the *Y*-axis did not start at zero by breaking the axis and putting in those two short lines. This lets the reader know that the graph shouldn't be read as reflecting ratio data.

So, when is not starting at zero a cardinal sin, and when is including it an offense? The clear, unambiguous answer is, "It all depends." When starting at zero would result in the bottom 75% or so of the graph being blank, as in Figure I–4, it's best to start somewhere else; otherwise, begin at zero.

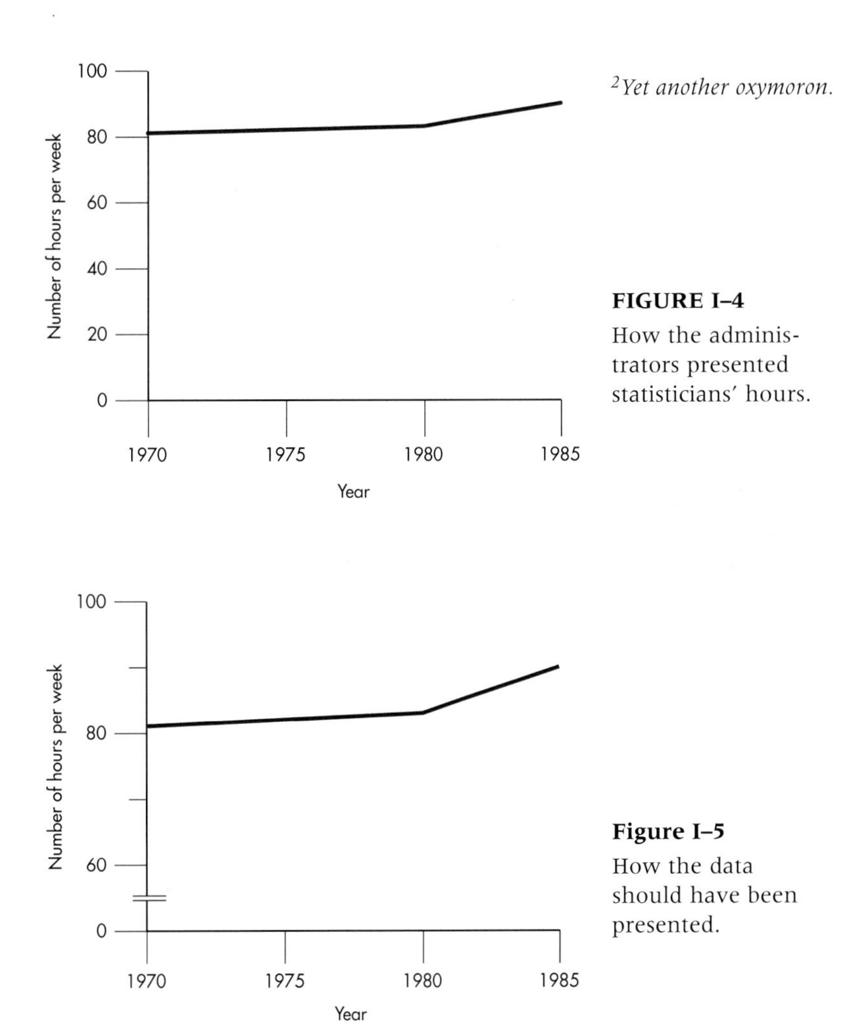

FIGURE I–4
How the administrators presented statisticians' hours.

Figure I–5
How the data should have been presented.

> **C.R.A.P. DETECTOR I–3**
> **The *Y*-axis should *not* start at 0 if it means that most of the graph is blank or if it visually distorts the data.**

[2] *Yet another oxymoron.*

TABLE I–1

Salaries (in thousands) of nine male and nine female administrators

Males	Females
55	47
50	43
70	43
60	45
75	175
40	35
58	48
50	43
65	42

I–4. Now the administrators were faced with a problem from within their own ranks. The women said that they were being paid less than were the men, and presented Table I–1. The CEO said that she (the one earning $175,000) did some calculations and found that the mean for both groups was exactly the same, $58,111. Do the women have a case?

Yes they do. Notice that the data for females are highly skewed by one very high number. Under these conditions, it would be better to use the median (refer to Chapter 3). This would show that the median salary for men is $58,000 and for women is $45,000, probably a more accurate representation of the bulk of the data.

C.R.A.P. DETECTOR I–4

If the data have a few outliers, or are seriously skewed, the median should be used as the estimate of central tendency rather than the mean.

SECTION

THE SECOND

ANALYSIS OF VARIANCE

The *t*-test is used for comparing the means of two groups and is based on the ratio of the difference between groups to the standard error of the difference.

CHAPTER THE SEVENTH

Comparing Two Groups

The *t*-Test

SETTING THE SCENE

To help young profs succeed in academia, you have devised an orientation course where they learn how to use big words when little ones would do. And, to help yourself survive in academia, you decide to do some research on it. So, you randomize half your willing profs to take the course and half to do without, then measure all the obscure words they mutter. How can you use these data to tell if the course worked? In short, how can you determine how much of the variation in the scores arose from differences ***between*** *groups and how much came from variation* ***within*** *groups?*

Perhaps the most common comparison in all of statistics is between two groups—cases vs. controls, drugs vs. placebos, boys vs. girls. Reasons why this comparison is ubiquitous are numerous. First, when you run an experiment in biomedicine, in contrast to doing an experiment in Grade 7 biology, you usually do something to some poor souls and leave some others alone so that you can figure out what effect your ministrations may have had. As a result, you end up looking at some variable that was measured in those lucky folks who benefited from your treatment and also in those who missed out.[1]

Note that we have implied that we measure something about each hapless subject. Perhaps the most common form of measurement is the FBI criterion—dead or alive. There are many variations on this theme: diseased or healthy; better, same, or worse; normal or abnormal x-ray; and so on. We do *not* consider this categorical type of measurement in this section. Instead, we demand that you measure something more precise, be it a lab test, a blood pressure, or a quality-of-life index, so that we can consider means, SDs, and the like. In the discussion below, we examine **Interval** or **Ratio** variables.

AN OVERVIEW

As we indicated in Chapter 6, all of statistics comes down to a signal-to-noise ratio. To show how this applies to the types of analyses discussed in this section, consider the following example.

A moment's reflection on the academic game reveals certain distinct features of universities that set them apart from the rest of the world. First, there is the matter of the dress code. Profs pride themselves on their shabbiness. Old tweed jackets that the rest of the world gardens in are paraded regularly in front of lecture theatres. The more informal among us, usually draft dodgers with a remnant of the flowerchild ethos, tramp around in old denim stretched taut over ever-expanding derrières.

But even without the dress code, you can tell a prof in a dark room just by the sound of his voice. We tend, as a group, to try to impress with obscure words in long, meandering sentences.[2] It's such a common affliction that one might be led to believe that we take a course in the subject, and foreigners on the campus might do well to acquire a Berlitz English-Academish dictionary.

Imagine if you will a course in Academish 1A7 for young, contractually limited, tenureless, assistant profs. As one exercise, they are required to open a dictionary to a random page, pick the three longest words, and practice and rehearse them until they roll off their lips as if Mummy had put them there.

Of course, not wanting to pass up on a potential publication, the course planners design a random-

[1] *Or maybe the lucky folks who missed out, and the poor souls who "benefited" from your treatment.*

[2] *The long, obscure word for that is sesquipedalianism, which literally means a foot and a half.*

ized trial; graduate students are required to attend a lecture from one of the graduands and some other prof from the control group and count all the words that could not be understood. After the data are analyzed, the graduands ($n_1 = 10$) used a mean of 35 obscure words. A comparable group ($n_2 = 10$) who didn't take the course used a mean 27 such words in their lectures. Did the course succeed? The data are tabulated in Table 7–1.

TABLE 7–1 Number of incomprehensible words in treatment and control groups

	Participants	Controls
	35	22
	31	25
	29	23
	28	29
	39	30
	41	28
	37	30
	39	33
	38	21
	33	29
Sum	350	270
Mean	35.0	27.0
Grand mean	31.0	

It is apparent that some overlap occurs between the two distributions, although a sizeable difference also exists between them. Now the challenge is to create some method to calculate a number corresponding to the **signal**—the difference between those who did and did not have the course, and to the **noise**—the variability in scores among individuals within each group.

The simplest method to make this comparison is called **Student's *t*-test**. Why it is called Student's is actually well known. It was invented by a statistician named William Gossett, who worked at the Guinness brewery in Dublin around the turn of the century. Perhaps because he recognized that no Irishman, let alone one who worked in a brewery, would be taken seriously by British academics, he wrote under the pseudonym "Student." It is less clear why it is called the "*t*"-test. There is some speculation that he did most of his work during the afternoon breaks at the brewery. Student's Stout test probably didn't have the same ring about it, so "tea" or "t" it became.[3]

[3] *Actually, all Guinness employees were forbidden to publish. Too bad Guinness doesn't run universities.*

EQUAL SAMPLE SIZES

To illustrate the *t*-test, let's continue to work through the example. From the table, the profs who made it through Academish 1A7 had a mean of 35 incomprehensible words per lecture; the control group only 27. One obvious measure of the signal is simply the difference between the groups or (35 − 27) = 8.0. More formally:

$$\text{Numerator} = \bar{X}_1 - \bar{X}_2 \tag{7–1}$$

Under the null hypothesis, we are presuming that this difference arises from a distribution of differences with a mean of zero and a standard deviation that is, in some way, related to the original distributions.

There are two differences between the *t*-test and the *z*-test. The first is that, with the former test, we focus on the distribution of **differences** between the two groups, so that we are testing a null hypothesis:

$$H_0\text{: } \mu_1 - \mu_2 = 0; \qquad H_1\text{: } \mu_1 - \mu_2 \neq 0 \tag{7–2}$$

rather than:

$$H_0\text{: } \mu_1 = \mu_2; \qquad H_1\text{: } \mu_1 \neq \mu_2 \tag{7–3}$$

We therefore calculate the mean and SD of the *differences.*

The second difference is that the SD is not provided. In the case of the *z*-test, discussed in Chapter 6, the SD of the population, σ, was furnished to us (remember we were dealing with serum sodium levels, where we were given the mean and SD of the population).

This is not the case here, so the next challenge is to determine the SD of this distribution of differences between the means: the amount of variability in this estimate that we would expect by chance alone. Because we are looking at a difference between two means, one strategy would be to simply assume that the error of the difference is the sum of the error of the two estimated means. The error in each mean is the *standard error (SE)*, $s \div \sqrt{n}$, as we demonstrated in Chapter 6. So, a first guess at the error of the difference would be:

$$\text{Standard error}_{\text{difference}} = SE_d = \frac{s_1}{\sqrt{n_1}} + \frac{s_2}{\sqrt{n_2}} \tag{7–4}$$

This is almost right, but as we mentioned many times, statisticians like to square and add things. So, the SE (squared) of the difference between the two means is the sum of the two squared SEs, and the SE is the square root of the whole thing:

$$SE_d = \sqrt{\frac{s_1^2}{n_1} + \frac{s_2^2}{n_2}} \tag{7–5}$$

Because the sample sizes are equal (i.e., $n_1 = n_2$), this equation simplifies a bit further to:

$$SE_d = \sqrt{\frac{s_1^2 + s_2^2}{n}} \tag{7–6}$$

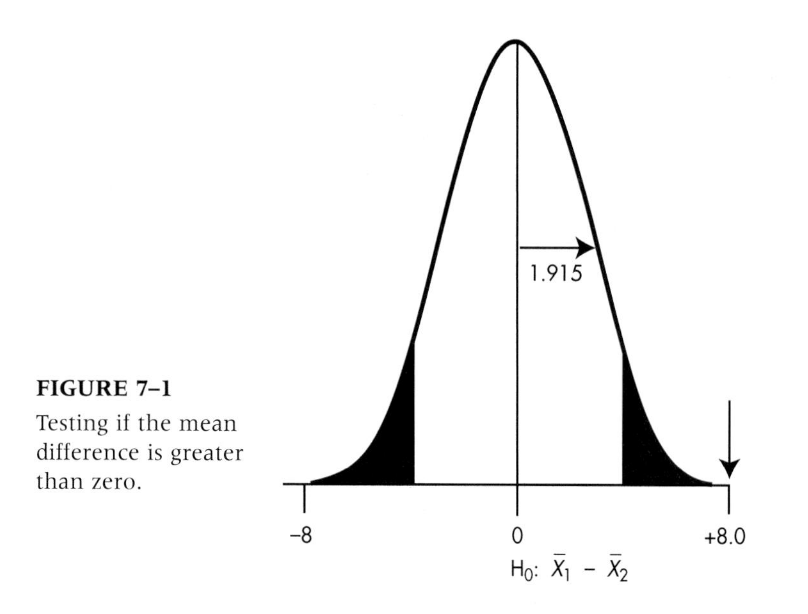

FIGURE 7–1

Testing if the mean difference is greater than zero.

In the present example, then, we can calculate the variances of the two groups separately, and these are equal to:

$$s_1^2 = \frac{(35 - 35)^2 + (31 - 35)^2 + \ldots + (33 - 35)^2}{10 - 1}$$

$$= \frac{186}{9} = 20.67$$

$$s_2^2 = \frac{(22 - 27)^2 + (25 - 27)^2 + \ldots + (29 - 27)^2}{10 - 1}$$

$$= \frac{144}{9} = 16.0$$

(7–7)

Then the denominator of the test is equal to $\sqrt{[(20.67 + 16.0)] \div 10} = 1.915$.

We can see what is happening by putting the whole thing on a graph, as shown in Figure 7–1. The distribution of differences is centered on zero, with an SE of 1.915. The probability of observing a sample difference large enough is the area in the right and left tails. If the difference is big enough (i.e., sufficiently different from zero), then we can see that it will achieve significance. The *t*-test is then obtained by simply taking the signal-to-noise ratio:

$$t = \frac{(\bar{X}_1 - \bar{X}_2)}{\sqrt{\frac{s_1^2 + s_2^2}{n}}} = \frac{8}{1.915} = 4.178$$

(7–8)

We can then look this up in Table C in the appendix and find a whole slew of numbers we don't know how to handle. The principal problem is that, unlike the situation with the *z*-test, there is a different *t* value for every degree of freedom, as well as for every α level. Instead of finding that, if $\alpha = .05$, then $t = 1.96$, as we could expect if it behaved like a *z*-test, we find that now, *t* can range anywhere from 1.96 to 12.70. The problem is that, because we have estimated both the means and the SDs, we have introduced a dependency on the degrees of freedom. As it turns out, for large samples, *t* converges on *z*—they are both equal to 1.96 when $\alpha = 0.05$. However, *t* is larger for small samples, so we require a larger relative difference to achieve significance.

Of course, we don't as yet know how to identify this magical quantity. We began with 20 data points, so we had 20 *df*. But we lost one when we calculated $\bar{X}_1$ and another when we calculated $\bar{X}_2$, leaving us with 18. (In general, $df = n_1 + n_2 - 2$). We can now look up the critical *t* for our situation (18 df) at the 0.05 level, which is 2.10. So our calculated *t*, which is equal to 4.178, is wildly significant. If we were presenting the results in a paper, we'd write $t(18) = 4.178$, $p < .05$.

The Formulae for the SD

If you were paying attention, you would have noticed that the equation we just used to calculate the variance (Equation 7–7) differs slightly from the equation we used in Chapter 3, when we introduced the concept of variance and the SD; in particular, the denominator is $N - 1$, rather than just *N*. Equations 3–7 (for the variance) and 3–8 (for the SD) are used to calculate these values for the **population**; whereas the equivalent formulae with $N - 1$ in the denominator are used for **samples** (the usual situation). Why the difference? The answer is based on three factors, all of which we've already discussed. First, the purpose of inferential statistics is to estimate the value of the parameter in the population, based on the data we have from a sample. Second, we expect that the estimated sample statistic (the mean, in this case) will differ from the population value to some degree, small as it may be. Finally, any set of numbers will deviate less from their own mean than from any other number. Putting this all together, the sample data will deviate less from the sample mean than they will from the population mean, so that the estimate of the variance and SD will be biased downward. Dividing the squared deviations by $N - 1$ rather than by *N* compensates for this, and leads to an **unbiased** estimate of the population parameter.

TWO GROUPS AND UNEQUAL SAMPLE SIZES: EXTENDED *t*-TEST

If there are unequal sample sizes in the two groups, the formula becomes a little more complex. To understand why, we must again delve into the philosophy of statistics. In particular, when we used the two sample SDs to calculate the SE of the difference, we were actually implying that each was an equally good estimate of the population SD, σ.

Now, if the two samples are different sizes, we might reasonably presume that the SD from the larger group is a better estimate of the population value. Thus it would be appropriate, in combining the two values, to **weight** the sum by the sample sizes, like this:

$$\sigma^2(est.) = \frac{n_1 s_1^2 + n_2 s_2^2}{n_1 + n_2}$$

(7–9)

This is close, but by now you have probably gotten into the habit of subtracting 1 every time you see an n. This is not the place to stop, so:

$$\sigma^2(est.) = \frac{(n_1 - 1)\, s_1^2 + (n_2 - 1)\, s_2^2}{n_1 + n_2 - 2}$$

(7–10)

This is the best guess at the SD of the difference. But we actually want the SE, which introduces yet another $1 \div n$ term. In this case, there is no single n; there are two n terms. Instead of forcing a choice, we take them both and create a $(1/n_1 + 1/n_2)$ term. So, the final denominator looks like:

$$Denominator = \sqrt{\frac{(n_1 - 1)\, s_1^2 + (n_2 - 1)\, s_2^2}{n_1 + n_2 - 2} \times \left(\frac{1}{n_1} + \frac{1}{n_2}\right)}$$

(7–11)

and the more general form of the t-test is:

$$t = \frac{(\overline{X}_1 - \overline{X}_2)}{\sqrt{\frac{(n_1 - 1)\, s_1^2 + (n_2 - 1)\, s_2^2}{n_1 + n_2 - 2} \times \left(\frac{1}{n_1} + \frac{1}{n_2}\right)}}$$

(7–12)

Although this looks formidable, the only conceptual change involves weighting each SD by the relevant sample size. And of course, the redeeming feature is that computer programs are around to deal with all these pesky specifics, leaving you free. From here we proceed as before by looking up a table in the appendix, and the relevant df is now $(n_1 + n_2 - 2)$.

Pooled versus Separate Variance Estimates

The whole idea of the t-test, as we have talked about it so far, is that the two samples are drawn from the same population and hence have the same mean *and* SD. If this is so, then it makes good sense to pool everything together to get the best estimate of the SD. That's why we did it; this approach is called a **pooled estimate**.

However, it might not work out this way. It could be that the two SDs are wildly different. At this point, one might rightly pause to question the whole basis of the analysis. If you are desperate and decide to plow ahead, some computer packages proceed to calculate a new t-test that doesn't weight the two estimates together. The denominator now looks like:

$$SE = \sqrt{\frac{s_1^2}{n_1} + \frac{s_2^2}{n_2}}$$

(7–13)

This looks very much like our original form and has the advantage of simplicity. The trade-off is that the df are calculated differently and turn out to be much closer to the smaller sample of the two. The reason is not all that obscure. Because the samples are now receiving equal weight in terms of contributing to the overall SE, it makes sense that the df should reflect the relatively excess contribution of the smaller sample. This strategy, as we mentioned in Chapter 3, is called the **harmonic mean** (abbreviated as $\overline{n}_h$), and comes about as:

$$\overline{n}_h = \frac{2}{\frac{1}{n_1} + \frac{1}{n_2}}$$

(7–14)

In short, if n_1 was 4 and n_2 was 20, the arithmetic mean would be 12; the harmonic mean would be $2 \div (1/4 + 1/20) = 6.67$, which is closer to 4 than to 20. So the cost of the separate variance test is that the df are much lower, and it is appropriately a little harder to get statistical significance.

SAMPLE SIZE AND POWER

Sample size estimates for the t-test closely follow the formalism developed in Chapter 6. However, note one small wrinkle. Because there are two groups, a factor of 2 sneaks into the equation. So the new formula for the sample size requirements for a two-group comparison looks like:

$$n = 2\left[\frac{(z_\alpha + z_\beta)\, \sigma}{\Delta}\right]^2$$

(7–15)

For example, if we wanted to compare a clam juice group and a placebo group, and our dependent variable was the misery of psoriasis, measured as percent of body area, we would proceed as follows:

1. *What is known about the extent of psoriasis in my patient population?*
   For the sake of argument, let's assume that the mean extent is 42% and the SD is 15%.[4]
2. *How big a treatment effect do I think I will get?*
   This is never known. If it were, you wouldn't need to do the study. So, make it up. If the sample size is more than you can manage in a year, double the treatment effect. If it's too small, and you can't justify enough funding,

[4] *If these data are not available, make them up. For the sake of the granting agency, though, try to back it up with some data from the literature.*

halve the treatment effect. Usually, though, it's the smallest difference that you would say is *clinically important*. Even if a smaller difference were statistically significant, you wouldn't change your practice because of it. So, for the sake of argument, let's say 20% in relative terms, so .20 × 42 = 8.4% in absolute terms.

3. *How big a Type I and Type II error do you want?* Unfortunately, you can never diddle with the α level (unless you try one-tailed tests, but this should be used only as a last resort when all else fails). However, you can pick out β levels of .05, .10, or .20, or even .50 if you are really desperate. So, for the sake of argument, let's say α = .05, so z_α = 1.96; and β = .10, so z_β = 1.28.

Now we put it all in the old sausage machine, and crank:

$$n = 2\left[\frac{(1.96 + 1.28)\ 15}{8.4}\right]^2$$

$$= 66.94 \text{ (say 67) per group} \tag{7–16}$$

If 67 per group is too large or too small, diddle away.

To save you the agony of having to buy batteries for your calculator, Table D in the appendix gives you the sample sizes you need. The first column is labeled *d*, which is the ratio of Δ ÷ σ. That's upside down from the way it appears in the formula, but it's the standard way of expressing differences in SD units; the formal term is the **effect size.** In this case, 8.4 ÷ 15 is about .5. So, looking up a two-sided α of .05 and β of .20, you'll find 63 subjects per group, which is pretty close.

Table E goes the other way. If you've stumbled across a study that reports a nonsignificant *t*-test, you can check if the groups really were equivalent or if a high probability of a Type II error existed. Use the article to find out the sample size (if the two groups are different, use the harmonic mean), the difference between the means that they actually found (Δ), and the SD (σ). Then, with an α of your choosing, you can look up the power of the test. For example, if the previous study was done with only 30 subjects per group, look across the row with 30 in column 1 until you get to the two-tailed α = .05. There's one column with *d* = .4, and one for *d* = .6, so we'll use a number half way between. For *d* = .4, the power is .346; for *d* = .6, power is .648. The mean of the two is .497, so for an effect size of 0.5, there was only a 50% probability that the study would have found a difference if it were actually there. This is too low for our blood (we usually want power to be at least .80), so we'd conclude that this study was too small and that the negative results were probably a Type II error.

The moral of the story is that a sample size calculation informs you about whether you need 20 or 200 people to do the study. Anyone who takes it more literally than that, unless the data on which it is based are very good indeed, is suffering from delusions.

One last word about power. The *t*-test and all related statistics such as the *z*-test and the analysis of variance (which will be discussed in later chapters), are based on the assumption that the data are normally distributed. The concern is that if the data aren't normally distributed, the value of the test will still be correct, but the significance level wouldn't be accurate. Well, put your mind at ease. Sawilowsky and Hillman (1992) found that even with a "radically nonnormal distribution,"[5] the tabled significance levels are accurate, except when the *n*s are small, and the groups differ in sample size.

[5] *That is, one skewed to the left; if it's skewed to the right, we presume it's referred to as a "conservatively nonnormal distribution."*

SUMMARY

The *t*-test is the easiest approach to the comparison of two means. The distinction between the *t*-test and the *z*-test, discussed in the previous chapter, is that the *t*-test estimates both the means and the SD, which introduces a dependency on sample size. Despite its computational ease, the *t*-test is *not* appropriate when there are more than two groups or when individuals in one group are matched to individuals in another.

EXERCISES

1. Answer True or False: When comparing the means of two samples using the *t*-test:
   a. the null hypothesis is that the means are equal
   b. the null hypothesis is that the means are not significantly different
   c. the sample sizes must be equal
   d. the SEs of the means must be equal
   e. the data must be normally distributed
2. Let's look at hair loss, the last bastion of male vanity (and a personal issue with your intrepid authors). Till recently, most patent hair restorers contained ethyl alcohol as the main active ingredient, presumably to ease the anguish. Now, a legitimate drug has changed all that. But does it really work? We take 10 chrome-domes, randomize them to two groups, and have them rub the active drug or a placebo into the affected part for 6 weeks. A blind (technically, not literally) observer counts hairs per cm^2 on the dome, and we calculate the means and SDs. The data look like this:

Drug		Placebo	
Subject	Hairs	Subject	Hairs
1	12	5	5
2	14	7	10
3	28	8	20
4	3	9	2
5	22	10	12
Mean	15.8		9.8
SD	8.59		6.21

   Calculate the following quantities:
   a. Difference between the means
   b. SE of the difference
   c. *t*-test
   d. Is this result significant?
3. Okay, so you tried and failed to grow hair. Maybe the sample wasn't big enough (and you can get even more money to do a bigger and better study).
   a. How much power did you have to detect a difference of 100% (i.e., the treatment mean is 19.6, the control mean is 9.8)?
   b. How big a sample size would you need to detect a true difference of 50% with α of .05 and β of .10?

How to Get the Computer to Do the Work for You

- From **Analyze**, choose **Compare Means → Independent-Samples *T* Test**
- Click on the variable(s) to be analyzed from the list on the left, and click the arrow to move it (them) into the box marked **Test Variable(s)** [Note: One *t*-test will be performed for each variable listed.]
- Click on the grouping variable from the list on the left, and click the arrow to move it into the box marked **Grouping Variable**
- Click the **Define Groups** button
- Enter the value which defines the first group into **Group 1** and press the <**Tab**> key
- Do the same for **Group 2** and press <**Tab**>
- Click **Continue**
- Click **OK**

CHAPTER THE EIGHTH

More than Two Groups

One-Way ANOVA

One-way ANOVA deals with statistical tests on more than two groups. We create a sum of squares representing the differences between individual group means and a second sum of squares representing variation within groups. There are also methods (called pairwise, planned, or post-hoc comparisons) to examine specific comparisons among individual means.

SETTING THE SCENE

To further the goal of "Safe Sex for Sinners," you decide to investigate which is the most cost-effective condom. You are rapidly discouraged by the challenge, as a visit to the local pharmacy reveals an overwhelming array of choices. What you really want to do is select a few brands and determine if any difference overall exists among the group means, then try to find out what affects these differences.

In the last chapter, we discovered a neat way to compare two means, the *t*-test. Why go further? Well, ponder if you will what happens when you have more than two groups. How, as a conscientious researcher, do you deal with the problem that assaults consumers daily when they must choose among dozens of apparently identical products to deal with every aspect of life from brushing their teeth in the morning to knocking them out at night?

As an example, consider condoms.[1] Leaving aside the exotica, which come in all the colors, shapes, and sizes under the sun and are apparently only dispensed in men's rooms of sleazy bars, there are dozens of brands, all promising to lift you to new heights of erotic pleasure, dispensed by every drugstore in the land. Interestingly, almost all are made by Julius Schmid,[2] who probably took a cue from the beer companies in finding the advantages of producing multiple brands from the same vat.

Those of us with an empirical bent might wish to put the promise to the test and determine if there really was any difference in pleasure derived from different brands. We certainly wouldn't do it two brands at a time, one study for Brand A versus Brand B, a second study for A versus C, another for A versus D, etc.—think of all the extra effort our subjects would have to put in and all the extra pleasure they would have to put up with. It would be far easier to do the study with a number of different brands all at once; get a bunch of willing volunteers (which shouldn't be too difficult), randomize them to various brands (all delivered for experimental purposes in plain brown wrappers), do IT, then provide a rating on a 10-point scale.[3]

Suppose we test four brands, Ramses (R), Sheiks (S), Trojan (T), and unknown house brand (U), with 10 subjects each.[4] Now, what of the hypothesis? Going in armed with the knowledge that the condoms all likely came off the same production line, we might really be interested in whether any difference is discernible among the brands. If there isn't, we would stop right there. If there is, then we might like to find out which is best. Formalizing it a bit, the null hypothesis is:

$$H_1: \mu_R = \mu_S = \mu_T = \mu_U$$

and our alternative hypothesis is simply:

H_1: *Not* all the μs are equal.

Now, if we were to set about comparing the means[5] with a *t*-test, a problem would arise. We can do only two at a time, so we end up comparing *R* with *S*, *R* with *T*, *R* with *U*, *S* with *T*, *S* with *U*, and *T* with *U*. There are 6 possible comparisons, each of which has a .05 chance of being significant by chance, so the overall chance of a significant result, even when no difference exists, approaches .30.[6] In any case, we really don't care about the specific differences in the first round.

[1] *When* PDQ Statistics *was written, we couldn't consider them. However, now every Grade 5 student knows all the arcane details, so we view this as an opportunity to bring the adults up to speed.*

[2] *Actually his kids and grandkids. So much for practising what you preach. The remainder are made by Ortho, which clearly likes to cover both bases, as it were.*

[3] *What we (Streiner and Norman, 1995) have previously called a "Bo Derek scale."*

TABLE 8–1

Satisfaction ratings with 4 brands of condoms, 10 subjects each

Subject	Ramses	Sheiks	Trojans	Unnamed
1	4	5	7	2
2	4	5	8	1
3	5	6	7	2
4	5	6	9	3
5	6	7	6	3
6	3	6	3	4
7	4	4	2	5
8	4	5	2	4
9	3	6	2	4
10	4	3	3	3
Mean	4.2	5.3	4.9	3.1
Grand mean =	4.375			

Based on a 10-point scale where 0 is the pits and 10 is ecstasy.

This is where the complicated formula comes in. Thinking in terms of signals and noises, what we need is a measure of the overall difference among the means of the groups and a second measure of the overall variability within the groups. We approach this by first determining the sum of all the squared differences between group means and the overall mean. Then we determine a second sum of all the squared differences between the individual data and their group mean. These are then massaged into a statistical test.

THE PARTS OF THE ANALYSIS

Sums of Squares

Let's just fake up some sex satisfaction data[7] to prove the point. They might look like Table 8–1.

Now the Sum of Squares (Between) is the sum of all the squared differences **between** the individual means and the grand mean. It looks like:

Sum of Squares (Between) = $10[(4.2 - 4.375)^2 + (5.3 - 4.375)^2 + (4.9 - 4.375)^2 + (3.1 - 4.375)^2] = 27.875$[8]

Algebraically, if that's your fancy:

$$SS(between) = n \sum (\bar{X}_{.k} - \bar{X}_{..})^2 \quad (8\text{–}1)$$

where n is the number of subjects in each group, $\bar{X}_{.k}$ is the group mean, and $\bar{X}_{..}$ is the overall (Grand) mean. If the groups have different sample sizes, you can use the harmonic mean of the sample sizes, as we discussed in Chapter 3.

Similarly, the Sum of Squares (Within) is the sum of all the squared differences between individual data and the group mean **within** each group. It looks like:

Sum of Squares (Within) = $(4 - 4.2)^2 + (4 - 4.2)^2 + \ldots + (4 - 4.2)^2 + (5 - 5.3)^2 + (5 - 5.3)^2 + \ldots + (3 - 5.3)^2 + (7 - 4.9)^2 + \ldots + (3 - 4.9)^2 + (2 - 3.1)^2 + \ldots + (3 - 3.1)^2$ [40 terms]

After much anguish, this turns out to equal 101.50. Again, the algebraic formula, for the masochists in the crowd, is:

$$SS(within) = \sum_i \sum_k (X_{ik} - \bar{X}_{.k})^2 \quad (8\text{–}2)$$

where the first summation sign, the one with the i under it, means to add over all of the subjects in the group, and the second summation sign means to add across all the groups.

Finally, the Sum of Squares (Total) is the difference between all the individual data and the grand mean. It is the sum of SS (Between) and SS (Within). But in longhand:

Sum of Squares (Total) = $(4 - 4.375)^2 + (4 - 4.375)^2 + \ldots + (5 - 4.375)^2 + (5 - 4.375)^2 + \ldots + (7 - 4.375)^2 + (8 - 4.375)^2 + \ldots + (2 - 4.375)^2 + (1 - 4.375)^2 + \ldots + (3 - 4.375)^2$ [40 terms] = 129.375

To check the result, this should be equal to the sum of the Between and Within Sums of Squares, 27.875 + 101.50 = 129.375; and the algebraic formula is:

$$SS(total) = \sum_i \sum_k (X_{ik} - \bar{X}_{..})^2 \quad (8\text{–}3)$$

There are two things to note. First, because summation is usually obvious from the equation, from now on, we'll just use a single Σ without a subscript, even when two or more are called for, and let you work things out. Second, the equations look a little bit different if there aren't the same number of subjects in each group, but we'll let the computer worry about that wrinkle.

Degrees of Freedom

The next step is to figure out the degrees of freedom (*df* or d.f.) for each term, preparatory to calculating the Mean Squares. There are four groups for the

[4] *We recognize that good sex, like good tangos, usually takes two (or more). For the moment, we will assume that the ratings were made by the male partners, not because of any sexist leanings, but simply because they are the ones who are always whining about the intrusion.*

[5] *The astute reader may well point out that we have no business comparing means of numbers from a rating scale. Indeed, there is no assurance that the distance between 9 and 10 on the scale is the same as the distance between 3 and 4, so it is not apparently interval level measurement. Debates have raged about this one for literally 50 years and we won't resolve it here (although some of the key references are at the end of the book).*

[6] *Actually it's not quite that. The correct formula to calculate the overall probability of a significant result by chance alone when there are* n *comparisons, as we outlined in Chapter 5, is* $1.0 - .95^k$, *in this case* $(1.0 - .95^6) = .26$.

[7] *A trick known to all consenting adults.*

[8] *We multiply by 10 because this comparison is actually based on 10 values.*

TABLE 8–2 The ANOVA summary table

Source	Sum of squares	*df*	Mean square	*F*
Between	27.875	3	9.292	3.296
Within	101.500	36	2.819	
Total	129.375	39		

Between-Groups Sums of Squares, but one *df* was lost in calculating the grand mean. So, the *df* (between) is equal to 4 – 1 = 3. More generally, for *k* groups,

$$df(between) = k - 1 \quad \textbf{(8–4)}$$

For the Within-Groups Sums of Squares, there are 40 terms (data points); 4 groups and 10 subjects per group. But we lose one *df* for each group mean, so we lose 4 overall. Thus the *df* (within) is 40 – 4 = 36. More generally, when you have *n* observations in each of the *k* groups, then:

$$df(within) = k(n - 1) \quad \textbf{(8–5)}$$

Finally, the total *df* is based on 40 terms and 1 lost *df* (the grand mean), for 39 *df*. Again, generally, this is equal to:

$$df(total) = nk - 1 \quad \textbf{(8–6)}$$

It's no coincidence that the *df* for the individual variance components (between and within) add up to the total *df*. This is always the case, and it provides an easy check in complex designs.

Mean Squares

Now we can go the next, and last, steps. First we calculate the Mean Square by dividing each Sum of Squares by its df. This is then a measure of the average deviation of individual values from their respective mean (which is why it's called a *Mean Square*), since the df is about the same as the number of terms in the sum. Finally, we form the ratio of the two Mean Squares, the *F*-ratio, which is a signal-to-noise ratio of the differences between groups to the variation within groups. This is summarized in an ANOVA table such as Table 8–2. We can then look up the calculated *F*-ratio to see if it is significant.

The critical values of the *F*-test at the back of the book are listed under the *df* for both the numerator and the denominator. When you publish this piece (good luck!!), the *F*-ratio would be written as $F_{3,36}$, or, if you can't afford the word processor, $F(3,36)$ or $F(3/36)$. Either way, the calculated ratio turns out to be significant because 3.296 is just a bit greater than the published *F*-value for 3 and 36 *df*, 2.86. So Julius may have taken them all out of the same latex vat, but all condoms are not created equal. At this point, we know that at least two are different from each other, but we can't say yet which they are; we'll get to that when we discuss *post-hoc comparisons*.

EXPECTED MEAN SQUARES AND THE DISTRIBUTION OF *F*

If you peruse the table of *F*-ratios in the back, one fact becomes clear—you don't see *F*-ratios anywhere near zero. Perhaps that's not a surprise; after all, we didn't find that any *t* values worth talking about were near zero either. But it actually should be a bit more surprising, if you consider where the *F*-ratio comes from. After all, the numerator is the signal—the difference between the groups—and the denominator is the differences within the groups. If no difference truly exists between the groups, shouldn't the numerator go to zero?

Surprisingly, no. Imagine[9] that there really was no difference among the condoms. All the μs are therefore equal. Would we expect the Sum of Squares (Between) to be zero? As you might have guessed, the answer is "No." The reason is because whatever variation occurred within the groups as a result of error variance would eventually find its way into the group means, and then in turn into the Sum of Squares (Between) and the Mean Square. As it turns out, in the absence of any difference in population means, the expected Mean Square (Between) [usually abbreviated as $E(MS_{bet})$] is exactly equal to the variance (within), σ^2_{err}.

Conversely, if absolutely no variance exists within groups, then the difference between sample means is equal to the difference between population means, and the expected Mean Square (Between) $= n \times \sigma^2_{bet}$.

Putting it together, then, the expected value of the Mean Square (Within) is just the error variance, σ^2; and the expected value of the Mean Square (Between) is equal to the sum of the two variances:

$$E(MS_{bet}) = \sigma^2_{err} + n\sigma^2_{bet} \quad \textbf{(8–7)}$$

Then, when there is no true variance between groups, the σ^2_{bet} drops out and the ratio (the *F*-ratio) equals 1.

As we go to hairier and hairer designs, the formulae for the expected mean squares will also become hairier (to the extent that this is the last time you will ever see the beast derived exactly), but one thing will always remain true: in the absence of an effect, we expect the relevant *F*-ratio to equal 1. Conversely, if we go to a really simple design and do a one-way ANOVA on just two groups, the calculated *F*-ratio is precisely the square of the *t*-test.

Does this mean that you'll never see an *F*-ratio less than 1? Again, the same answer, "No." Because of sampling error, it sometimes happens that when nothing is going on—there's no effect of group membership—you'll end up with an *F* that's just below 1. Usually it's in the high .90s.

[9] *If you would rather gedanken, feel free.*

MULTIPLE COMPARISONS

One could assume, in the above experiment, that finding the *F*-ratio concluded the analysis. The alternative hypothesis was supported, the null hypothesis was rejected, and so on. You don't really care which of the condoms resulted in the most satisfaction—or do you? There are certainly many occasions where one might, out of genuine rather than prurient interest, wish to go further after having rejected the null hypothesis to determine exactly which specific levels of the factor are leading to significant differences.

In fact, situations also occur when, although there may be more than two levels in the analysis, the previous hypothesis can be framed much more precisely than simply, "Not everything is equal." In the present example, if we were going up against Julius Schmid, our real interest is a comparison of Brand *U* (unnamed) against the average of brands *R*, *S*, and *T*. More commonly, a comparison of three or four drugs, such as a group of aspirin-based analgesics that includes a placebo, almost automatically implies two levels of interest—all analgesics against the placebo, and, if this works, comparisons among analgesics.

These two situations are described as **post-hoc comparisons**, occurring out of interest after the primary analysis has rejected the null hypothesis; and **planned comparisons**, which are deliberately engineered into the study before the conduct of the analysis.

> *Planned comparisons* are hypotheses specified before the analysis commences; *post-hoc comparisons* are for further exploration of the data after a significant effect has been found.

As you might have guessed, post-hoc comparisons are considered to be more like data-dredging, and thus inferior to the elegance of planned comparisons. However, they are much more common and also easier to understand, so we will start at the end and work forward.

POST-HOC COMPARISONS

All the post-hoc procedures we discuss—**Tukey's LSD (Least Significant Difference), HSD (Honestly Significant Difference),** the **Scheffé Method**, the **Newman-Keuls Test**, and **Dunnett's** ***t***—involve two comparisons at a time. Because we have only a limited number of ways to look at the difference between two means (subtract one from the other and divide by a noise term), they all end up looking a lot like a *t*-test.

Bonferroni Correction

Why not just do a bunch of *t*-tests? Two reasons: (1) it puts us back into the swamp we began in, of losing control of the α level; and (2) we can use the Mean Square (Error) term as a better estimate of the within-group variance. This does point to one of the simplest strategies devised to deal with multiple comparisons (of any type). Recognizing that the probability of making a Type I error on any one comparison is .05, one easy way to keep things in line is to set an alpha level that is more stringent. This is called a **Bonferroni correction**. All you do is count up the total number of comparisons you are going to make (say *k* comparisons), then divide .05 by *k*. If you have four comparisons, then the alpha level becomes $.05 \div 4 = .0125$.

It should more appropriately be called the Bonferroni over-correction because it does overcompensate. To see why, refer back to marginal note 6. So let's proceed to the more sophisticated (it's a relative term) methods—variations of the Bonferroni correction, the Scheffé method, Tukey's LSD and HSD, the Newman-Keuls, and Dunnett's *t*.

Modifications of the Bonferroni Correction

In an attempt to overcome the extremely conservative nature of the Bonferroni correction, a number of alternatives have been proposed. Some are only moderately successful,[10] so we won't waste your time (or ours) going over them. The most liberal ones, proposed by Holm (1979) and Hochberg (1988), use critical values that change with each test, rather than the fixed value of α/T (where *T* is the number of tests), which Bonferroni uses. Both of the tests start off by arranging the *p* levels in order, from the smallest (p_1) to the largest (p_T). In the Holm procedure, we start off with the smallest (most significant) *p* level, and compare it to α/T. If our *p* level is smaller than this, then it is significant, and we move on to the next *p*-value in our list, which is compared to $\alpha/(T - 1)$. We continue doing this until we find a *p*-value larger than the critical number; it and all larger *p*s are nonsignificant.

[10] *More like a middle-of-the-roader rather than an arch conservative, but still not in the liberal camp.*

The Hochberg variation[11] starts with the largest value of *p* and compares it to α. If it's significant, then so are all smaller values of *p*. If it's not significant, we move on to the next value, which is compared with its critical value, $\alpha/(T - 1)$. When we find a *p* level that is larger than the critical value, we stop; it and all subsequent *p*s are not significant.

[11] *Not to be confused with the Goldberg Variations.*

Let's run through an example to see how these procedures work. Assume we have the results of five tests, and their *p* values, in order, are:

$p_1 = .008$ $p_2 = .011$ $p_3 = .030$ $p_4 = .040$ $p_5 = .045$

In the Holm method, we would compare p_1 against $\alpha/5 = .010$. Because it is smaller, p_1 is significant, and we move on to p_2, which is compared against $\alpha/4 = .0125$. It too is significant, so we test p_3 against $\alpha/3 = .0167$. It's larger, meaning that p_3 through p_5 are not significant. With the Hochberg version, we start with p_5 and compare it to α. Because it's significant, then all of the other *p*-values are, too. Had we used the Bonferroni correction, and tested all of the *p*s against $\alpha/5 = .010$, then only p_1 would have reached significance. So, the Bonferroni correction is the most conservative, followed by the Holm method, and then the Hochberg variation.

Although we have been discussing these three tests in relation to ANOVA, they can be used in a variety of circumstances. For example, if we ran three separate *t*-tests or did 10 correlations within a study, we should use one of these corrections (Holm or Hochberg) to keep the experiment-wise α level at .05.

The Studentized Range

Common to all the remaining methods is the use of the overall Mean Square (Within) as an estimate of the error term in the statistical test, so we'll elaborate a bit on this idea. You remember in the previous chapter that we spent quite a bit of time devising ways to use the estimate of σ derived from each of the two groups to give us a best guess of the overall SE of the difference. In ANOVA, most of this work is already done for us, in that the Mean Square (Within) is calculated from the differences between individual values and the group mean across all the groups. Furthermore, as we showed already, the Mean Square (Within) is the best estimate of σ^2. So the calculation of the denominator starts with Mean Square (Within). We first take the square root to give an estimate of the SD. Finally, we must then divide by some *n*s to get to the SE of the difference. In the end, the denominator of the test looks like:

$$Denominator = \sqrt{MS_{within} \times \left(\frac{1}{n_1} + \frac{1}{n_2}\right)} \quad \textbf{(8–8)}$$

If the sample sizes are the same in both groups, this reduces further to:

$$Denominator = \sqrt{\frac{2 \times MS_{within}}{n}} \quad \textbf{(8–9)}$$

For reasons which may, with luck, become evident, we're going to call this quantity $\mathbf{q'}$. It really represents a critical range of differences between means resulting from the error in the observations. We could, for example, create a *t*-test using this new denominator:

$$t = \frac{\bar{X}_1 - \bar{X}_2}{q'} \quad \textbf{(8–10)}$$

So, in the present example, if we want to compare *U* to *S* (the reason for this choice will be obvious in a minute), we would first compute the denominator:

$$q' = \sqrt{\frac{2 \times 2.82}{10}} = 0.751 \quad \textbf{(8–11)}$$

Then, we could proceed with a *t*-test:

$$t = \frac{|3.1 - 5.3|}{0.751} = \frac{2.2}{0.751} = 2.93 \quad \textbf{(8–12)}$$

This is just an ordinary *t*-test, except that we computed the denominator differently, using the Mean Square (Within) instead of a sum of variances, which we called $\mathbf{q'}$. This does introduce one small wrinkle. Since we are using all the data (in this case, 40 observations) to compute the error, we modify the degrees of freedom to take this into account; so, we compare this with a critical value for the *t*-test on 36 df, which turns out to be 2.03.

In fact, almost all the post-hoc procedures we will discuss are, at their core, variants on a *t*-test. However, since by definition they involve multiple comparisons (between Group A and Group B, Group A and Group C, etc.), they go about it another way altogether, turning the whole equation on its head, and computing the difference required to obtain a significant result. That is, they begin with a range computed from the Mean Square (Within), multiply it by the appropriate critical value of the particular test, and then conclude that any difference between group means which is larger than this range is statistically significant. Sticking with our example of a *t*-test for the moment, we would compute the range by multiplying 0.751 by the critical value of the *t*-test (in this case, 2.03). Then, any difference we encounter which is larger than this, we'll call statistically significant by the *t*-test.

Regrettably, before we launch into the litany of post-hoc tests, one more historical diversion is necessary. It seems that, when folks were devising these post-hoc tests, someone decided early on that the factor of 2 in the square root was just unnecessary baggage, so they created a new range, called the **Studentized range.** This has the formula:

$$q = \sqrt{\frac{MS_{within}}{n}} = \sqrt{\frac{2.82}{10}} = 0.531 \quad \textbf{(8–13)}$$

And that's where $\mathbf{q}$ and $\mathbf{q'}$ come about. Why they bothered to create a new range, one will never know. After all, it doesn't take a rocket scientist to determine that q' is just $\sqrt{2} = 1.414$ times q. Since most of the range tests use q, we have included critical values of q in Table M.

Tukey's Least Significant Difference

Tukey's LSD is on the left wing in terms of conservatism, and it is actually nothing more than a computational device to save work; goodness knows how this got into the history books. You begin with the critical value of *t*, given the *df*. In this case, we have 36 *df*, so a significant *t* (at .05) is 2.03. We worked out before that the denominator of the calculated *t*-test is .751, so any difference between means greater than .751 × 2.03 = 1.52 would be significant. Sooo—1.52 becomes the LSD, and it is not necessary to calculate a new t for every comparison. Just compare the difference to 1.52; if it's bigger, it's significant. The formula for the LSD is therefore:

$$LSD = t_{n-2} \sqrt{\frac{2 \times MS_{within}}{n}} = t_{n-2} \times q' \quad (8\text{–}14)$$

And this time the $T=U$ comparison, at 1.80, is statistically significant. One would be forgiven if there was some inner doubt surfacing about the wisdom of such strategies. Tukey's LSD does nothing to deal with the problem of multiple comparisons because the critical value is set at .05 for each comparison; all it does is save a little calculation.

Tukey's Honestly Significant Difference

Perhaps because Tukey saw that the LSD would not win him fame and glory,[12] he reappraised the situation and came up with the **HSD** (**Honestly Significant Difference**).[13] This time the test statistic is changed to something closer to the square root of an F statistic. It has its own table at the back of some stats books (including this one). In the present example, with 4 and 36 df, the statistic q equals 3.79. Tukey then creates another critical difference, called the **Honestly Significant Difference**, or **HSD**:

$$HSD = q_{(k,M)} \sqrt{\frac{MS_{(within)}}{n}} \quad (8\text{–}15)$$

when n is, as before, the sample size in each group, k is the number of groups (4 in this case), and M is the df for the within term, equal to $k(n-1)$. This time around, then, the HSD equals:

$$HSD = 3.79 \sqrt{\frac{2.82}{10}} = 2.01 \quad (8\text{–}16)$$

and now the $T=U$ comparison is not significant.

Note that this test, like the one that follows, uses the q statistic. Second, these tests (including the LSD), don't start with a difference between means and then compute a test statistic, but, rather, start with a test statistic, and compute from it how big the difference between means has to be in order for it to be significant. This involves looking up the critical value of the q statistic in Table M, which turns out to involve not just the mean, SD, and sample size, but also the number of means, k.

This isn't at all unreasonable. After all, the more groups there are, the more possible differences between group means there are, and the greater the chance that some of them will be extreme enough to ring the .05 bell. So, the test statistic for the HSD and the next test, which we've called q up until now, takes all this into account.

TABLE 8–3

Differences between means for different steps, with critical values of q

	Mean	Differences		
		1 step	2 steps	3 steps
Sheiks	5.3			
		0.4		
Trojans	4.9		1.1	
		0.7		2.2
Ramses	4.2		1.8	
		0.9		
Unnamed	3.1			
Critical q		1.53	1.84	2.023

The Newman-Keuls Test

For reasons known only to real statisticians,[14] the **Newman-Keuls** test appears to have won the post-hoc popularity contest lately. It, too, is a minor variation on what is becoming a familiar theme. We begin by ordering all the group means from highest to lowest.

Now in order to apply the N-K test, we have to introduce a new concept, called the **step**. Basically, we need to know how many steps (means) we have to traverse to get from one mean of interest to the other mean of interest, and what the difference is between the means; we've worked this out in Table 8–3. So, to get from S to U, we have four steps; to go from R to U, only one. For $\alpha = .05$, four means, and 36 degrees of freedom, the tabled value equals 3.81, so any difference between means greater than a critical value of:

$$Critical\ (\bar{X}_i - \bar{X}_j) = 3.81 \times q = 3.81 \times \sqrt{\frac{2.82}{10}} = 2.023 \quad (8\text{–}17)$$

Since the S - U comparison is bigger than 2.023, it is declared significant. Now, to discuss two three-step comparisons. From the table, the critical q is 3.464; so any difference larger than 3.464 × .531 = 1.84 is significant. So, R - U is almost, but not quite, significant, and S - T is not significant. Since neither comparison is significant, there is no point in continuing, as none of the two-step comparisons would be significant. Just to prove the point, for the two-step comparisons, the critical q is 2.875; so, any comparison greater than 2.875 × .531 = 1.53 is significant. And none are.

Dunnett's t

Yet another t-test. We include it for two reasons only. The big one is that it is the right test for a particular circumstance that is common to many clinical studies, in which you wish to compare several treatments to a control group; and second, because Dunnett is an emeritus professor in our department, and the only person named in this book whom we've actually met.[15]

[12] *Unlike Timothy Leary, who found that LSD did win him fame and glory. Rest in Peace, Tim.*

[13] *Makes you wonder, doesn't it? If this one is "honestly" significant, what does that say about his other post-hoc tests?*

[14] *Thereby counting us out.*

[15] *Aside from Streiner and Norman, but they don't count since they're not real statisticians.*

To illustrate the point, returning yet again to the lurid example that got this chapter rolling, one clear difference emerges between *R*, *S*, *T*, and the last one. *U* you can get for free almost anywhere—student health services, counsellors, etc.; whereas, to get *R*, *S*, or *T*, you might have to shell out some hard-earned scholarship dough. Surely, the primary question must be whether the ones you pay for are really any better than the freebies.[16] That is, the study is now designed to compare multiple treatments with a control group.[17]

[16] *Not to be confused with frisbees, which is the other favourite undergraduate sport.*

[17] *We were tempted to call it a placebo group after the Latin meaning "I will please," but, on reflection, all the treatments are quite pleasing.*

Dunnett's test is the essence of simplicity, doing essentially what all the others did. We compute a critical value for the difference using a test statistic called, as you might expect, Dunnett's *t*, and another standard error derived from Mean Square (Within), only this time with the $\sqrt{2}$ back in—that is, a q' range. So the critical value looks like:

$$\textit{Critical}\ (\bar{X}_i - \bar{X}_c) = t_d \sqrt{\frac{2\ MS_{within}}{n}}$$

(8–18)

Dunnett's main contribution was that he worked out what the distribution of the *t* statistic would be (for reasons of space, you won't find it in the back of this book). Not surprisingly, like the Studentized range, it is dependent on the number of groups and the sample size. For the present situation, with four groups and a sample size of 10, the t_d is 2.31. So, with a Mean Square (Within) of 2.82 the critical value is:

$$\textit{Critical}\ (\bar{X}_j - \bar{X}_c) = 2.31 \times \sqrt{\frac{2(2.82)}{10}} = 2.31 \times .751 = 1.73$$

(8–19)

On this basis, brands *S* and *T*, with differences of 2.2 and 1.8, exceed this critical value and so, are declared significantly different from brand *U*; brand *R* doesn't make the grade, regrettably.

Scheffé's Method

The last procedure we'll discuss is also the oldest. Scheffé's method is intended to be very versatile, allowing any type of comparison you want (for example, A versus B, A versus C, A + B versus C, and so on). To conduct the procedure, you calculate a range, using the overall *F* test, the Mean Square (Within), and a complicated little bit of coefficients. It looks like:

$$\sqrt{S = (k-1)F_{\alpha,df}} \sqrt{MS_{within} \times \sum \frac{C_i^2}{n_j}}$$

(8–20)

For a simple contrast, it's not quite as bad as it looks. For example, in our present data set, the critical value of *F* at .05 for 3 and 36 degrees of freedom is 2.88. The number of groups, *k*, is 4. Mean Square (Within) is, as usual, 2.82. Now taking a simple contrast, if we just wanted to look at the difference between *R* and *U*, then C_1 is +1, C_2 is 0, C_3 is 0, and C_4 is −1. Putting it all together, the equation is:

$$S = \sqrt{(3)(2.88)}\ \sqrt{2.82 \times (1 + 0 + 0 + 1)} = 6.972$$

(8–21)

So, any comparison of means greater than 6.972 would be significant at the .05 level. This value is quite a bit larger than any of the other methods, which is the price you pay for versatility. In fact, Scheffé recommended using the tabled value for $\alpha = .10$, rather than .05, just to compensate for this conservativeness.

On balance, comparing these methods, it is evident that the LSD method is liberal—it is too likely to find a difference. The Scheffé method is too conservative. One reason for its conservativeness is that it was meant to test *all* possible comparisons, including highly weird ones. Even if we don't do all of these comparisons, Scheffé "protects" α from them. In general, the HSD and the Newman-Keuls methods are somewhere in the middle, although the Newman-Keuls test is a little less conservative (so we are told). It would seem that most folks these days are opting for the Newman-Keuls test when they are doing pairwise comparisons. Dunnett's test is in a class by itself and should be the method of choice when comparing multiple means to a control mean.

This pretty well sums up the most popular post-hoc procedures. We have not, by any means, exhausted the space. One general statistics book we examined listed 10 methods. If you have the fortitude to pursue this further, we have listed a couple of readable references at the back of the book.

PLANNED ORTHOGONAL COMPARISONS

In contrast to these bootstrap methods, planned contrasts are done with a certain *élan*. The basic strategy is to divide up the signal, the Sum of Squares (Between), among the various hypotheses or **contrasts.** The sum of squares associated with each is used as a numerator, and the Mean Square (Within) as a denominator, to calculate *F*-ratios for each test.

To accomplish this sleight of stat, it is necessary to devise the comparisons in a very particular way. If we just went ahead, as we do with post-hoc comparisons, taking differences among means as our whims dictate, then the Sum of Squares associated with all the contrasts would likely add up to greater than the Sum of Squares (Between). The reason is that the comparisons overlap—($\text{Mean}_1 - \text{Mean}_2$), ($\text{Mean}_2 - \text{Mean}_3$), and ($\text{Mean}_3 - \text{Mean}_1$) are, to some degree, capitalizing on the same sources of variance.

To avoid this state of affairs, the comparisons of interest must be constructed in a specific way so that they are nonoverlapping, or **orthogonal**.

Two things (contrasts, factors, or whatever) are said to be *orthogonal* if they do not share any common variance.

We ensure that this condition is met by first standardizing the way in which the comparison is written. We do this by introducing weights on each mean. So, each contrast among means is written like:

$$C = w_1\overline{X}_1 + w_2\overline{X}_2 + w_3\overline{X}_3 + w_4\overline{X}_4 \tag{8-22}$$

For example, those condom connoisseurs among the readers probably know that certain expensive classes can be found among condoms; spermicide present or absent, lubricated or not, and other more architectural differences, the details of which will be spared the reader. Suppose Brands *R* and *S* have one such a characteristic, and *T* and *U* do not. To see if it matters, we would make a comparison as shown:

$$C = \tfrac{1}{2}\overline{X}_R + \tfrac{1}{2}\overline{X}_S - \tfrac{1}{2}\overline{X}_T - \tfrac{1}{2}\overline{X}_U \tag{8-23}$$

In a similar manner we might like to compare Brand *R* with Brand *S*, ignoring *T* and *U*. This looks like:

$$C = 1\overline{X}_R - 1\overline{X}_S + 0\overline{X}_T + 0\overline{X}_U \tag{8-24}$$

And finally, making the same comparison within the other category ends up looking like:

$$C = 0\overline{X}_R + 0\overline{X}_S + 1\overline{X}_T - 1\overline{X}_U \tag{8-25}$$

Now comes the magic. How do we know that these are orthogonal? By multiplying the coefficients together, according to the equation:

$$\sum(w_i \times w_j) = 0 \tag{8-26}$$

where *i* refers to one contrast and *j* to the other.

How, you might ask, does this guarantee things are orthogonal? We asked the same question and decided that it was anything but self-evident. Try this, however. Suppose there are two dimensions, *X* and *Y*. If we imagine two lines, $(a_1X + b_1Y)$ and $(a_2X + b_2Y)$, they are at right angles (orthogonal) if the sum of the product of the weights is equal to zero.

In this case, the product of the first two sets equals $(\tfrac{1}{2})(1) + (\tfrac{1}{2})(-1) + (-\tfrac{1}{2})(0) + (-\tfrac{1}{2})(0) = 0$. So far, so good.

Similarly, we have to prove that contrasts 2 and 3 are orthogonal. The sum of weights is $(1)(0) + (-1)(0) + (0)(1) + (0)(-1) = 0$. We're getting tired of all this, so you can check the 1 and 3 contrast.

Now that we have established a set of contrasts equal to the number of *df*, it's almost easy.[18] We calculate the sum of squares for each contrast as follows:

1. First, calculate the actual contrast. From our data set, they look like:
   $C1 = \tfrac{1}{2} \times 4.2 + \tfrac{1}{2} \times 5.3 - \tfrac{1}{2} \times 3.1 - \tfrac{1}{2} \times 4.9 = 0.75$
   $C2 = 1 \times 4.2 - 1 \times 5.3 + 0 \times 3.1 + 0 \times 4.9 = 1.10$
   $C3 = 0 \times 4.2 + 0 \times 5.3 + 1 \times 3.1 - 1 \times 4.9 = -1.80$
2. Next, calculate the sum of $w_i^2 \div n$, and call it *W*:
   $W1 = (\tfrac{1}{2}^2 + \tfrac{1}{2}^2 + \tfrac{1}{2}^2 + \tfrac{1}{2}^2) \div 10 = 1 \div 10 = .10$
   $W2 = (1^2 + 1^2) \div 10 = 2 \div 10 = .20$
   $W3 = (1^2 + 1^2) \div 10 = 2 \div 10 = .20$
3. The sum of squares for each contrast is then, by some further chicanery, equal to $C^2 \div W$:
   $SS(C1) = .75^2 \div .10 = 5.625$
   $SS(C2) = 1.1^2 \div .20 = 6.05$
   $SS(C3) = 1.8^2 \div .20 = 16.2$

And these are all supposed to sum up to the Sum of Squares (Between), 5.625 + 6.05 + 16.2 = 27.875. So, the net effect of the creation of these planned comparisons is to parcel out the total Sum of Squares (Between) into three linear contrasts. In a similar manner, we noted above that we could have only as many contrasts as there were df between groups, so the df were divided among the contrasts. This is illustrated in Figure 8–1.

Finally, we can do a test of significance on each contrast. This is done by taking the ratio to the Mean Square (Within), which leads to an elaborate ANOVA table (Table 8–4).

Now the critical *F*-value for 1 and 36 df is between 4.08 and 4.17, so only the last of these individual comparisons is significant. In general, if the overall *F* test is significant, then at least one of the comparisons will be as well. Conversely, if the overall test is not significant, then none of the individual comparisons will be either.

The advantage of the method is twofold; first, concern about the individual comparisons being liberal or conservative are unnecessary—they are all exactly right. Second, the comparisons provide direct tests of the hypotheses of interest. Planned comparisons should probably be used more, but because they require a bit of creativity and some manual calculations (instead of simply pressing a button), they remain a quaint curiosity to most investigators.

THE STRENGTH OF RELATIONSHIP

The logic behind ANOVA is that we want to see if one variable (in this case, type of condom) is related to another one (here, satisfaction). The *F*-ratio tells us if the association is statistically significant, but it doesn't give us any information about the *strength*

[18] *If you really aren't interested in all those contrasts, what do you do? Make some up to fit the sum = 0 rule, calculate the sum of squares as below, then ignore the result.*

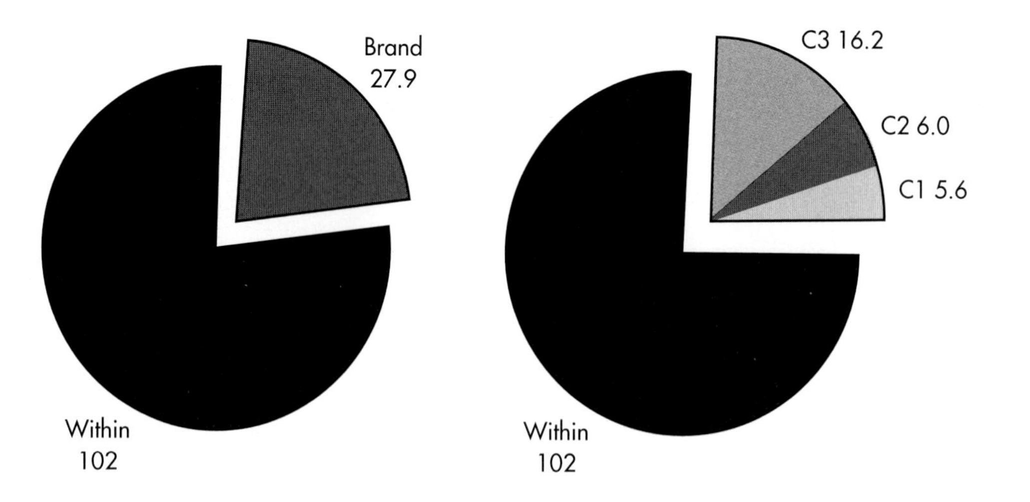

FIGURE 8–1
Parceling out the Total Sum of Squares into three linear contrasts.

TABLE 8–4

The ANOVA summary table for the orthogonal comparisons

Source	Sum of squares	df	Mean square	F
Between	27.875	3	9.29	
C1	5.625	1	5.625	1.995
C2	6.05	1	6.05	2.146
C3	16.20	1	16.20	5.746
Within	101.50	36	2.84	
Total	129.375	39		

of the relationship. As it happens, we can pull this information out from the ANOVA summary table. We can express the strength of the relationship in terms of a variable called **eta-squared** and written η^2.

$$\eta^2 = \frac{SS\ between}{SS\ total} = 1 - \frac{SS\ within}{SS\ total} \tag{8–27}$$

This will always yield a number between 0 and 1 and is interpreted as the *proportion* of the variance in the dependent variable that can be attributed to the independent variable. We discuss this concept in greater detail when we discuss correlation.

In our example,

$$\eta^2 = \frac{27.875}{129.375} = 0.2155 \tag{8–28}$$

so that almost 22% of the variance in satisfaction scores can be explained by condom brand; 78% of the variance results from other factors.[19]

[19] *As a homework assignment, make a list of what you think those other factors may be.*

SAMPLE SIZE AND POWER

The basic idea for sample size estimation developed in the preceding two chapters is made a little more complicated when we get to one-way ANOVA. Just to remind you, the formula for the sample size for a t-test was:

$$n = 2\left[\frac{(z_\alpha + z_\beta)\sigma}{\delta}\right]^2 \tag{8–29}$$

where δ is the difference between the two groups. If you reflect on the way this formula works, all the action is contained in the ratio of the difference between means to the SD. The rest of the stuff, the zs and such like, are just niceties related to the arbitrary choice of α and β levels. Putting it more directly, the **effect** of the group differences is contained in this ratio. For this reason, and none other, Cohen (1977), the granddaddy of sample size calculations, called this an **effect size**, symbolized by the letter d, which expresses the effect of the treatment in SD units.

But things get a bit hairier in the case of ANOVA, for two reasons. First, we have to worry about several means, not just two; and second, the means can be distributed in various ways, as we'll explain in a bit. This means that we have to make a couple of guesses; one about the average difference between means, and another about their probable distribution.

As before, let's call the distance between the highest and the lowest mean δ, and the effect size $(\delta \div s) = d$. Then, let's think about how the means can be spread out over this interval. One possibility arises when we have three groups; two fairly similar drugs and a placebo. Plausibly, the two drugs might be clustered together at one end of the distribution of means and the placebo at the other. However, if we had a whole bunch of treatments, a first guess is that they would be equally scattered along the line. A third variation may be that one treatment is a clear winner; another obviously does nothing; and the remaining ones are all bunched up in the middle.

Cohen (1977) then took the value of d and transformed it into the effect size for the ANOVA, which he called f. In essence, d is multiplied by some fancy formula, which varies depending on the distribution of means—minimum dispersion (Figure 8–2, A); maximum dispersion (Figure 8–2, C); or intermediate dispersion (Figure 8–2, B). The formulae that accompany these three patterns are:

Minimum dispersion: $d \times \sqrt{\frac{1}{2k}}$ **(8–30)**

Intermediate dispersion: $d \times \frac{1}{2} \sqrt{\frac{(k+1)}{3\,(k-1)}}$ **(8–31)**

Maximum dispersion: (k = odd): $d \times \sqrt{\frac{k^2 - 1}{2k}}$ **(8–32)**

Maximum dispersion (k = even): $d \times 1$ **(8–33)**

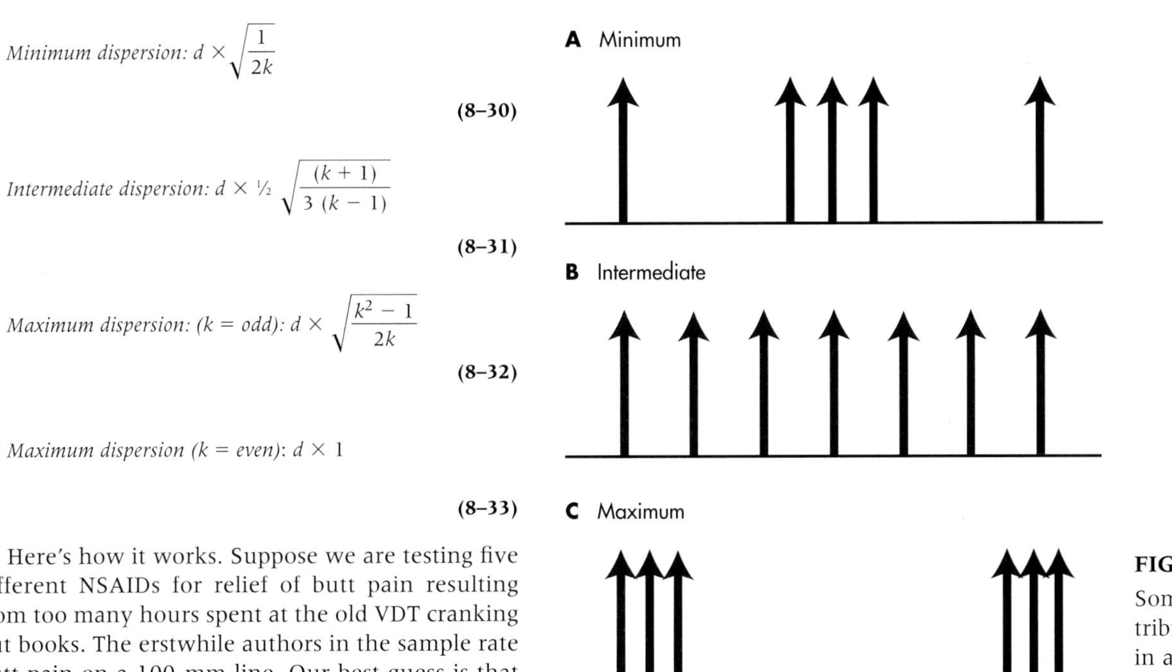

FIGURE 8–2 Some possible distribution of means in a one-way ANOVA.

Here's how it works. Suppose we are testing five different NSAIDs for relief of butt pain resulting from too many hours spent at the old VDT cranking out books. The erstwhile authors in the sample rate butt pain on a 100-mm line. Our best guess is that all the drugs are all the same, of course, but this is not the way to get drug company money. So, based on previous research or intuition or just plain imagination, we presume (1) a difference of 1 cm (10 mm) between the best and the worst, (2) that the individual means are distributed evenly along the 10 mm difference, and (3) the SD is 8 mm. How big a sample size do we need to detect this distribution of differences?

First, *d* is $10 \div 8 = 1.25$. The *f*, the effect size, for this intermediate distribution of means, is:

$$f = 1.25 \times \tfrac{1}{2} \sqrt{\frac{(5+1)}{3\,(5-1)}} = 1.25 \times \tfrac{1}{2} \sqrt{.5} = .442 \qquad \textbf{(8–34)}$$

Now, what do we do with this? We look it up in a table—more specifically, Table I in the book's appendix, which shows the sample size per group, having chosen the appropriate values of α and β. As usual, we've also made up a table that goes the other way; Table J gives you the power of the study for various values of *f*, α, *k*, and *N*.

SUMMARY

We have already indicated pretty strongly the reasons for using one-way ANOVA: it provides an exact test of the hypothesis for multiple groups and, in combination with planned comparisons, is an exact (and elegant) alternative to multiple *t*-tests. Actually, it is not an alternative—it is the *only* way to proceed when there are more than two groups.

But as we shall see in the next few chapters, one-way ANOVA is only one way (ho ho ho) to divide up the world, and the more complex ANOVA methods that build on this formalism are a powerful and elegant way to view the world of numbers. So, turn the page.

EXERCISES

1. Select the answer to each of the following statements from the list below. Note that each statement may have more than one answer.
   a. Sum of squares (between)
   b. Sum of squares (within)
   c. Mean square (between)
   d. Mean square (within)
   e. Degrees of freedom (between)
   f. Degrees of freedom (within)
   g. *F* ratio
   h. Probability of *F*

   A. Related to the size of the effect _____
   B. Related to the random variation within each group _____
   C. Increases with the number of groups _____
   D. Increases with the number of subjects in each group _____
   E. Decreases with the number of subjects per group _____
   F. Decreases as the signal-to-noise ratio gets bigger _____
2. One dilemma facing all lovers of fiery food is that different culinary establishments have different standards. "Suicide" wings in one joint don't rate more than a "Medium" in another—or so it seems.

   It's a slow day in the lab, so let's put this one to the test. We locate 3 different roadhouses and 12 fearless undergraduates. We randomize diners to diners (so to speak) and they sally forth, late at night, armed to the teeth with clipboards, Tums, and Pepto-

Bismol. They screw up their collective courages, order the platter of "Suicide," and then, if they remain conscious, rate fire on the ubiquitous 10-point scale.
The data look like this:

	Roadhouse		
Rater	A	B	C
1	4	7	7
2	4	8	9
3	7	6	10
4	5	7	10
Mean	5.0	7.0	9.0
SD	1.41	0.82	1.41

Now is your chance to flex your computational muscles.

a. Construct an ANOVA table and see if there is really a difference in suicide ratings among roadhouses.
b. Where does the difference lie? Do post-hoc comparisons using Scheffé and Tukey LSD methods.

How to Get the Computer to Do the Work for You

- From **Analyze**, choose **Compare Means** → **One-Way ANOVA**
- Click on the variable(s) to be analyzed from the list on the left, and click the arrow to move it (them) into the box marked **Dependent List** [Note: One ANOVA will be performed for each variable listed.]
- Click on the grouping variable from the list on the left, and click the arrow to move it into the box marked **Factor**
- Click the **Post Hoc** button and select those tests you want [we recommend **LSD, Tukey**, and **Tukey's-b**], then click **Continue**
- Click the **Options** button and click on **Descriptives** and **Homogeneity-of-Variance**
- Click **Continue**
- Click **OK**

CHAPTER THE NINTH

Factorial ANOVA

This chapter explores more complex forms of Analysis of Variance, involving multiple independent factors. The principle is the same: dividing the total Sum of Squares into components because of each factor. Additional information is derived from the interaction between factors.

SETTING THE SCENE

*The results of the condom experiment are in question. No account was taken of a second **factor**—circumcision status. Also, when the data are examined, it seems that uncircumcised males rate Brand T higher, whereas circumcised males rate Brand U higher, indicating a possible **interaction** between the two **factors**.*

We have now discovered one of the joys of ANOVA—we can compare multiple groups in a single test without losing track of the actual probability. But this doesn't seem such a big deal, let alone cause for joyous celebration. Surely there must be more than this?

Indeed there is. In Chapter 6, we introduced the notion of splitting up the total variance into components due to signal and noise. But nothing compels us to limit ourselves to only a single factor and a single noise term. We can easily introduce additional factors in the design, then examine the effects of each singly (**main effects**) and in combination (**interactions**).

Going back to our previous example, one other age-old question, which has been the subject of endless bits of folklore, is whether circumcised males have more—or less—fun than do uncircumcised males. It's difficult for any of us individually to provide evidence on the matter because few among us have had the opportunity to experience sex under both conditions. But an experiment such as the one we just did would provide an opportunity to put matters to the test.

We could let nature take its course and examine the ratings provided by males of both types after the fact, using a *t*-test. But the vast majority of men are circumcised, so there may well be a large imbalance in the two groups. Although this does not invalidate the test, it is less than optimal. A better approach would be to deliberately recruit equal numbers of males of both types so that we could eventually compare 20 circumcised to 20 uncircumcised men. It would be just another *t*-test. But as we shall see, there is a still better way.

Let's think about it a minute. When we compared the four brands, we contrasted the variance resulting from different brands against the variance within groups. This latter is called random error, but that is just a glib phrase to cover our ignorance of its cause. A better term would be "unexplained variance." Well, what we have been talking about is one possible cause of within-group variation. If circumcision does make a difference, then the presence of both types of men in the groups has led to some of the within-group variation. By explicitly dealing with this factor, we are accounting for some of this variance, and less is left over to go into the "error" term. So, as well as permitting an independent test of a second hypothesis, introducing a second factor (to the extent that it does contribute to the variance in the dependent variable) reduces the magnitude of the remaining error variance and thereby results in a more sensitive test of the first hypothesis.

There is one other boon to introducing additional factors—the possibility of uncovering interaction effects, such as, "Circumcised males prefer Brand *R*, uncircumcised males prefer Brand *S*," but we will leave this until later. The data would now look like Table 9–1. Now we proceed just about as we did before. In fact, the Sum of Squares (Brands) is exactly the same:

$$\text{Sum of Squares (Brands)} = 10[(4.2 - 4.375)^2 + (5.3 - 4.375)^2 + (4.9 - 4.375)^2 + (3.1 - 4.375)^2] = 27.875$$

TABLE 9-1

Ratings of satisfaction for different condom brands by circumcised and uncircumcised males

	Ramses	Sheik	Trojan	Unnamed	Mean
Uncircumcised	4	5	7	2	
	4	5	8	1	
	5	6	7	2	5.05
	5	6	9	3	
	6	7	6	3	
Group mean	4.8	5.8	7.4	2.2	
Circumcised	3	6	3	4	
	4	4	2	5	
	4	5	2	4	3.70
	3	6	2	4	
	4	3	3	3	
Group mean	3.6	4.8	2.4	4.0	
Brand mean	4.2	5.3	4.9	3.1	4.375

Algebraically:

$$SS_j = nI\sum(\overline{X}_{.j} - \overline{X}_{..})^2 \quad \textbf{(9-1)}$$

where j is the subscript for the columns (brand), I is the number of rows (in this case, 2), and n is the sample size in each cell (in this case, 5). It's just the squared differences between the column means and the grand mean (with a sample size diddle factor).

The Sum of Squares (Circumcised/Uncircumcised) is exactly analogous, involving a difference between the two group means and the grand mean, this time multiplying by the number of data in each circumcision group, 20:

$$\text{Sum of Squares } (C/UC) = 20[(5.05 - 4.375)^2 + (3.70 - 4.375)^2 = 18.225$$

And again the algebra looks like:

$$SS_i = nJ\sum(\overline{X}_{i.} - \overline{X}_{..})^2 \quad \textbf{(9-2)}$$

where now i is the subscript for the rows (circumcision status), and J is the number of columns (4). This is simply the squared difference between the two means and the grand mean (again, with a sample size diddle factor).

The Sum of Squares (Error) is conceptually the same as before, consisting of the difference between individual values and their group mean. This time, though, there are more group means to consider, and so it consists of terms such as:

$$\text{Sum of Squares (Error)} = (4 - 4.8)^2 + (4 - 4.8)^2 + (5 - 4.8)^2 + (5 - 4.8)^2 + (6 - 4.8)^2 + (5 - 5.8)^2 + \ldots + (7 - 5.8)^2 + \ldots + (3 - 2.2)^2$$
$$\text{[over all the top groups]} + (3 - 3.6)^2 + (4 - 3.6)^2 + \ldots + (4 - 4.0)^2 + (3 - 4.0)^2 \text{ [40 terms]}$$

This turns out to be equal to 24.80.

Once more, with feeling, the equation is:

$$SS_{within} = \sum(X_{ijk} - \overline{X}_{ij})^2 \quad \textbf{(9-3)}$$

This is the sum of the squared differences between all the individual data and their respective cell mean (with no diddle factor needed).

However, we have one more term in our bag of tricks—it's called an **interaction**.[1] As we indicated, it explores the idea that the value of the dependent variable (satisfaction) may relate in some nonadditive way to the value of both factors. Putting it more simply, circumcised males may express a strong preference for some brands and uncircumcised males for other brands.

It is almost easier to see what an interaction is by first considering the appearance of **non**interaction. But to illustrate the point, perhaps we can begin with some simpler data. Imagine an experiment similar in design to the present one. A sample of 30 boys and 30 girls is assigned to three different educational programs that teach algebra—lectures, small groups, and computers. There are 10 boys and 10 girls in each group. The goal is to determine what the expected average score in each cell would be if there were no interaction.

Now, if we knew only that the average score of all subjects was 50%, then our best guess at the expected mean score in each cell is just that, 50%, as we show in Table 9-2 under the first category. But suppose we have a bit more information, namely that girls score, on average, 10% above the mean, and boys, 10% below. We can now add this effect to the information and determine that the best estimate for the cell means in the top row is now 40% and in the bottom row is 60%, as shown in Table 9-2 under the second category.

[1] *Given the topic under discussion, interaction seems particularly apropos. However, this time there is only a dry, technical intent to the terminology.*

Now let's add some more information. Computers beat lectures by 10%, and lectures beat small groups by 10%.[2] If we add in these effects, we would guess that the expected values in the cells are as shown in Table 9–2 under the third category.

But so far there is still no interaction among factors. The extent to which the *actual* cell means depart from this picture of *expected* means is a measure of the *interaction* between teaching method and gender. So, for example, if boys did much better on computers and worse in groups, whereas girls did better in groups and worse on computers, the Boy-Computer mean would be higher than 50, the Boy-Group mean would be lower than 30, the Girl-Computer mean would be lower than 70, and the Girl-Group mean would be higher than 50. The data might look like that in Table 9–2 under the fourth category. This would constitute an *interaction* between gender and teaching method. Note that the marginal differences remain the same as in the third category.

The extent to which the *actual* cell means depart from this picture of *expected* means is a measure of the **interaction** between teaching method and gender. The calculation of expected means is also called an **additive model.**

TABLE 9–2

An example of predicting cell means from overall differences

Gender	Computer	Lecture	Group	
Knowing only overall mean				
Boys	50	50	50	
Girls	50	50	50	
				50
Knowing overall mean and row effects				
Boys	40	40	40	−10
Girls	60	60	60	+10
				50
Knowing overall mean, row effect, and column effect				
Boys	50	40	30	−10
Girls	70	60	50	+10
	+10	0	−10	50
Including interaction terms				
Boys	65	40	15	−10
Girls	55	60	65	+10
	+10	0	−10	50

The *interaction* between two variables is the extent to which the cell means depart from an expected value based on addition of the marginals.

Applying this logic to our present data, on the average, Brand *R* is a bit below par—4.2 versus 4.375, or 0.175 points. And on the average, uncircumcised men really do have more fun—5.05 versus 4.375, or 0.675 points better. So we would predict (if the effects simply added together) that uncircumcised males using Brand *R* would be up 0.675 from the mean, and down from it 0.175 points; so they would be (0.675 − 0.175), or 0.500 points above the overall average, which is (4.375 + 0.500) = 4.875. As we see, they actually average 4.8, which is pretty close to expectation. But if, for example, uncircumcised males scored Brand *R* at 5.5 when we expected 4.925, and circumcised males averaged 2.9 when we expected (4.375 + [−0.175 − 0.675]) = 3.525, we can suspect some suggestion of a relationship (or an **interaction**) between circumcision status and condom brand. Of course, taking the usual nonpartisan, noncommercial view favored by academics who haven't a ghost of a chance at making any entrepreneurial money, we are not specifically interested in the interaction only with Brand *R*; we want to show an overall interaction across all brands. So we create an interaction term, which is based on the difference between the observed cell means and that which we would expect based on the marginal means. The first and second terms are based on the expected values we have already calculated, and look like:

$$(4.8 - 4.875)^2 + (3.6 - 3.525)^2$$

As usual, though, these must be multiplied by the cell sample size, in this case, 5. In the end, the sum consists of 8 terms, the last of which is the squared difference between the observed value in the bottom right cell, 4.0, and its expected value [4.375 + (3.1 − 4.375) + (3.7 − 4.375)] = 2.425, so it all looks like:

$$\text{Sum of Squares (Interaction)} = 5[(4.8 - 4.875)^2 + (3.6 - 3.525)^2 + \ldots + (4.0 - 2.425)^2] = 58.475 \qquad (9\text{–}4)$$

And of course we feel duty-bound by now to furnish the masochists with yet another algebraic equation:

$$SS_{interaction} = n\sum(\overline{X}_{ij} - \overline{X}_{.j} - \overline{X}_{i.} + \overline{X}_{..})^2 \qquad (9\text{–}5)$$

This is, then, the sum of the differences between the individual cell means and what we would have expected if there were no interaction with one final diddle factor *n* for good measure.

The next step, as before, is to determine the *degrees of freedom*. This must be done for each Sum of Squares, and it is a bit more complicated than before. For brand, it's the same as before—four groups and one grand mean, so (4 − 1) = 3 *df*. For circumcision status, it's 2 groups and 1 mean, so we have 1 *df*. For the Error Sum of Squares, there are 8 groups and 5 data in each group, for 8 × 5 = 40; but we lose one *df* for each of the means in each group, so the actual degrees of freedom are (8 × 4) = 32.

[2] *This must be a hypothetical example. There has never been a convincing demonstration that any curriculum approach is any better than any other.*

TABLE 9–3

ANOVA table for two factors (brand and circumcision)

Source	Sum of squares	*df*	Mean square	*F*	*p*
Brands	27.88	3	9.29	11.99	<.0001
Circumcision status	18.23	1	18.23	23.52	<.0001
Brand × Circumcision	58.48	3	19.49	25.15	<.0001
Error	24.80	32	.78		
Total	129.38	39			

Once again, conceptual mind-bending surrounds the interaction term. The tortuous logic goes like this: we have four column means and two row means that are the data for the sum, but the overall row mean and column mean had to be estimated, so the *df* are $(4 - 1) \times (2 - 1) = 3$. We remain unconvinced by the logic too, but there is one way to check. The total *df* must equal the total number of data minus 1 (because the overall mean had to be estimated), or 39. From our above discussion we have

$$df\text{ (total)} = 3 + 1 + 3 + 32 = 39 \quad \textbf{(9–6)}$$

so the arcane logic above must be right.

Finally, after all the fooling around, we are ready to put it together into an ANOVA table. Obviously, the table (Table 9–3) has a few more lines in it than did the one-way table.

It is now evident that all the factors are significant. Uncircumcised males do have more fun. There is a difference in brands. Finally, the interaction between the two factors is significant (whatever that means; see below). Note that, although the Sum of Squares and Mean Square for brand is exactly the same as before, the *F* test has gone up to 11.99 and the probability has gone down correspondingly. Why? Because we have managed to move some of the variance that was previously contained in the error term into variance attributable to circumcision status and to the interaction between brand and circumcision. As a result, the error term has shrunk. The idea is illustrated in Figure 9–1. Because the Sums of Squares are additive, the sections in the figure have an area proportional to the relevant sum of squares.

Underlying the idea is a fundamental notion, which we mentioned in the beginning of this chapter. *Error variance* is not really error at all; it is simply variation for which we have no ready explanation. And the more explanatory variables that are introduced—to the extent that they really do explain variance—the smaller will be the unexplained, or error, variance.

It is subject to the law of diminishing returns, however. Because each variable costs at least one *df*, and usually more, if a variable is *not* accounting for a significant proportion of the variance, it can result in a less powerful test of the remaining factors. For this reason, some authors state that the term "error" is misleading and replace the term with "within" or "residual." However, in repeated-measures designs we describe in Chapter 11, we distinguish between "within-subject" and "between-subject" sources of variance. In deference to terminology, we call the variance term expressing variance not resulting from any of the identified factors in the design, "error."

SUMS OF SQUARES AND MEAN SQUARES FOR FACTORIAL DESIGNS

In the last chapter, we introduced you to the notion of an *Expected* Mean Square, a sum of variances that together represent the expected value of the calculated mean square. Last time around, it was almost straightforward: the expected mean square between groups was the sum of the variance between groups and the variance within groups, weighted by an *n* or two here and there; and the expected mean square within groups was the within-group variance.

In the present situation, we have many more possible variances that could enter the sum. As it turns out, the conceptual rule is as follows.

> The *Expected Mean Square* for a main effect or interaction of a variable contains other terms from interactions as well as the error term.

What that bit means is this: the expected mean square for the interaction between Circumcised/Uncircumcised and Brand contains σ^2 (Brand × Circumcised/Uncircumcised) and σ^2 (Error). The expected mean square for the main effect of Brands contains σ^2 (Brands), σ^2 (Brands × Circumcised/Uncircumcised) and σ^2 (Error). All are multiplied by *n*s here and there, using obscure rules that we will avoid.[3]

The effect of all this is that different effects require different error terms. The error term, MS (Error), contains only σ^2 (error). The interaction (Brand × Circumcised/Uncircumcised) contains only σ^2 (Brand × Circumcised/Uncircumcised) and the σ^2 (error), so that, if there is no interaction in the population, it contains only σ^2 (error). So MS (Error), which is equal to σ^2 (error), is the appropriate denominator for the *F* test of significance. By contrast, the main effect of brand is estimated to contain variance from the error term, the interaction, and the main effect. Then the appropriate denominator for the test of significance is the Mean Square (Brand × Circumcised/Uncircumcised).[4]

GRAPHING THE DATA (TAKE 1)

In our excitement to explore the delights of factorial ANOVA, we violated one cardinal rule of data analysis—*first, graph the data*. If we had done so, some of the mysteries of the analysis might have become clear. Look at Figure 9–2.

If we just squinted at Brands *R* and *S*, all is as expected. Everybody likes *S* a bit better, and uncircumcised males enjoy sex more. But the mean val-

[3] *If you can't resist exploring the rules more (masochist!), see Glass and Stanley (1970).*

[4] *Usually, except when the Circumcised/Uncircumcised effect is a fixed effect—see below.*

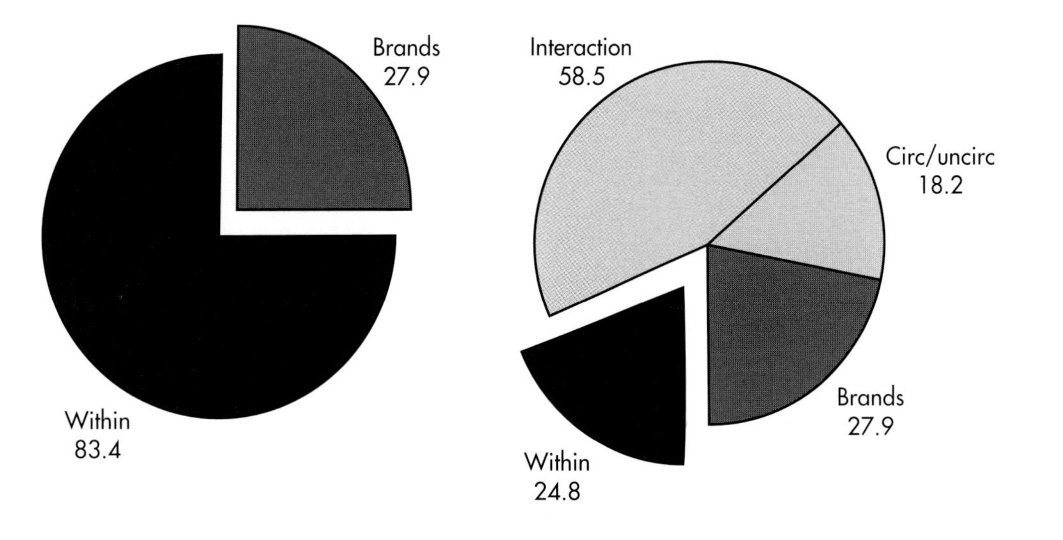

FIGURE 9–1
Sums of squares and interactions caused by factors and interactions.

ues of *T* and *U* present a very different picture. For some unexplained reason, uncircumcised males express a strong preference for the *U* brand and circumcised males for the *T* brand. Therein lies the explanation for the strong interaction term uncovered in the ANOVA. This is magnified in Figure 9–3.

This is only one of several possible types of interactions, some of which are shown in Figure 9–4. In the top left graph, the lines are parallel, but displaced, so the effect of circumcision is the same for both *T* and *U*. There are main effects of brand and circumcision status, but no interaction.

In the top right, if we take the average of the two points, one on top of the other, for *T* and then for *U*, they are the same, so there is no effect of brand. Similarly, the mean scores for circumcised and uncircumcised are the same, so there is no main effect of circumcision status. But a strong interaction is in evidence because the uncircumcised strongly prefer *T* and the circumcised prefer *U*.

Using the same kind of analysis on the lower left, the average for *T* and *U* is the same, so there is no effect of brand; but the uncircumcised are always above the circumcised, giving a main effect of circumcised status. Moreover, the lines are not parallel, so there is an interaction. Finally, the bottom right has everything going on—none of the means are the same as any other and the lines are not parallel—so there are both main effects and an interaction.

> The extent to which the lines are not parallel is an indication of the presence of an *interaction*.

If you are still having trouble conceptualizing the idea of interaction, it is synonymous with *synergy*; the whole is greater than (or less than) the sum of the parts. A match alone has little free energy; a gallon of gasoline alone has little free energy. Put them together, and suddenly you have a lot of energy (and synergy, too).

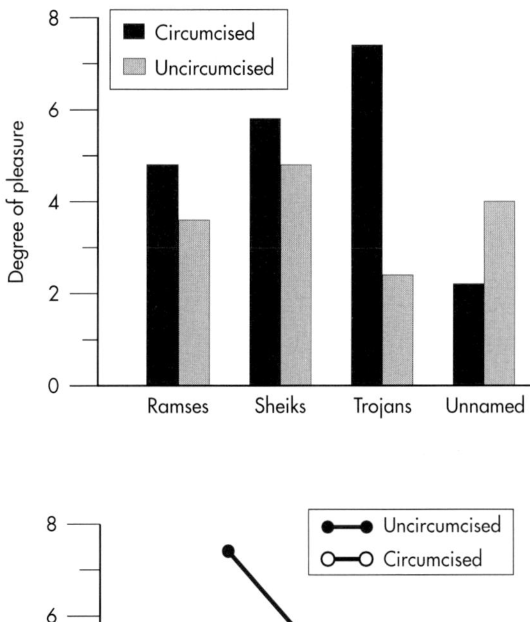

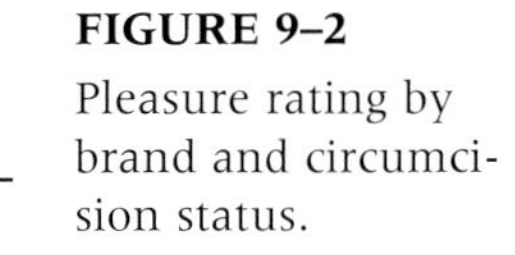

FIGURE 9–2
Pleasure rating by brand and circumcision status.

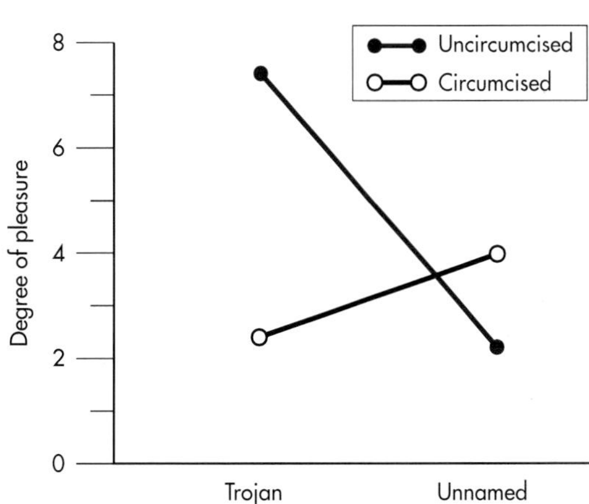

FIGURE 9–3
Interaction between brand and circumcision.

There is a divergence of opinion about interactions. Some folks hate 'em because if an interaction exists, then they cannot say that the effect of treatment is equal to such-and-such. One version, particularly prevalent in epidemiology, is that one should test only one hypothesis, such as "The drug works"—preferably with only two groups. Obvi-

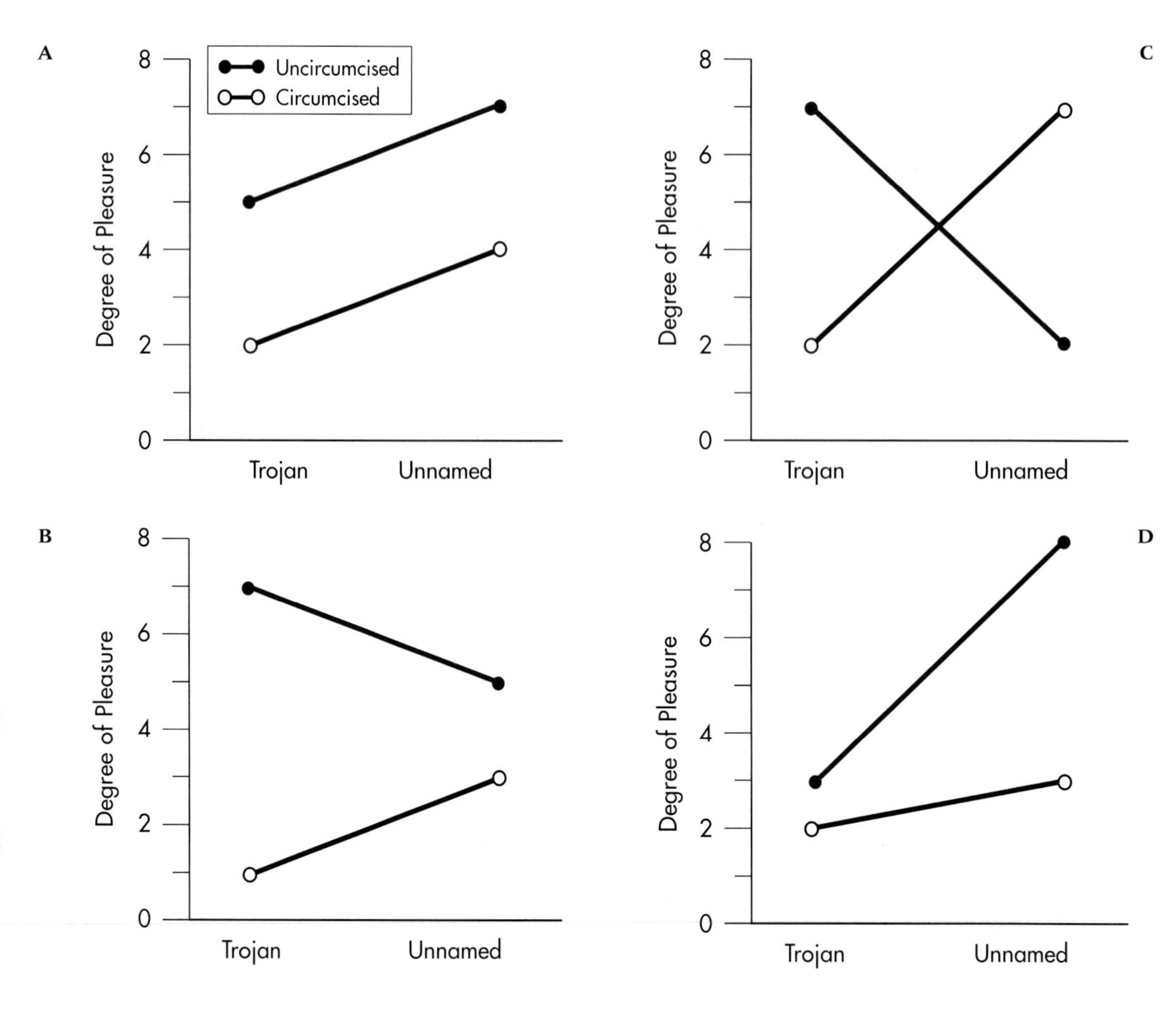

FIGURE 9–4
A, Main effect of brand and circumcision; no interaction. **B**, Main effect of circumcision; no effect of brand; significant interaction. **C**, No main effect of brand or circumcision; significant interaction. **D**, Main effect of brand and circumcision; significant interaction.

ously the drug companies like this approach because if you are testing their drug against a placebo, there is no chance that some other company's drug may come out better. This approach has one and only one virtue—simplicity.[5] But there are several reasons to contemplate including more than one variable. First, as we showed above, if you can account for some of the variance with another variable—in this case, circumcision status—then you can increase the power of the statistical test of the primary hypothesis.

[5]*It was Albert Einstein who said that "Everything should be made as simple as possible—and no simpler."*

Secondly, there is the glory of interactions. In designing our experiments, we actually often go looking for interactions. We believe that it provides much stronger information than a main effect. As an example, one study showed that if you take a group of patients with transient ischemic attacks, aspirin reduces the likelihood of a subsequent stroke by about 20%—but only for men. If the researchers had analyzed the data without including male/female as a factor in the design, they would have concluded that the effect was only about 10%, which in this study would have no longer been statistically significant. In addition, if the effect had been shown to be significant without the analysis by gender, the recommendation would have been to treat everyone with aspirin. The predictable result would have been a few more stomach ulcers and no benefit for the women.

Methodologic benefit is also gained from designing interactions into the study. Suppose we had reason to suspect a bias in the study. For example, perhaps physicians were unblinded and, being skeptical that aspirin could possibly work, put only the patients with a milder stroke episode on aspirin. Now, if all we had was an overall risk reduction of 10%, this bias might indeed explain the results. But if the conclusion is based on an interaction, we must now explain why unblinding of the physicians would result in a bias in assignment to treatment for only the males, which is much less plausible.

As a second example of a deliberately manipulated interaction, it is fair to say that all psychological studies of expertise date back to the studies of a single investigator. Adrian de Groot (1965) studied a group of chess masters, himself included, on a long voyage to America in the 1940s. As it turned out, the single best predictor of expertise in chess was the ability to recall a typical mid-game position. After a few seconds, experts could recall about 90% of the pieces; novices about 20%. Now if he had left it at that and done a *t*-test on two group means, post-hoc hypotheses would be hanging off every tree. After all, experts are not randomized, so maybe they are self-selected with better memories. Maybe chess playing results in biochemical changes that increase memory. Maybe experts are older, and age, up to a point, results in increased memory performance (this was the 1940s, and most psychologists were studying rat and pigeon memory, not human).

But de Groot didn't stop there. He also placed the pieces at random on the chess board, and did the same thing. This time there was *no* effect of expertise—everybody recalled about 20%. So he ended up with an interaction between expertise and real/random position, and the alternate hypotheses came tumbling down. Clearly, expertise in chess resulted in better memory performance *in chess*. He then went on to theorize that experts are able to "chunk" the data, using memory for previous positions, so as to reduce memory load. The result is that this one paper has directed the last 30 years of research in expertise.

GRAPHING THE DATA (TAKE 2)

We have a confession to make: when we described graphing interactions, we over-simplified things a bit. The problem is that plotting the observed means may give a misleading picture when one or more of the main effects is significant. What we should really do (and we'll show you how in this section) is to *adjust* the observed cell means to take these main effects into account. In essence, what we're doing is going back to the definition of an interaction, that it's "the extent to which the cell means depart from an expected value based on addition of the marginals,"[6] and subtracting the main effects from the cell means. Let's go back to the example in Table 9–2 and see what difference this makes.

In Part A of Figure 9–5, we've graphed the data from the last part of the table, showing the effects of different types of education in algebra for boys and girls. The figure shows us that there is an interaction (as we've said), with boys doing best when taught by computer, poorest when taught in a group, and middling when lectured to; for girls, the relationship goes the other way. The graph also tells us that girls do poorer than boys with computers, but better than boys with the other two teaching formats. The problem with trying to interpret the picture in this way is that there are main effects of both gender and teaching method. We remove these effects by subtracting the row and column means from each cell, and then adding in the grand mean. So, for boys taught by computer, the adjusted cell mean is $65 - 40 - 60 + 50 = 15$; for boys taught by lecture, it is $40 - 40 - 50 + 50 = 0$; and so on. The results are the deviations from the expected value for each cell. If we want to show the results so that the adjusted and unadjusted grand means are the same, we would then add the grand mean (which is 50) to each cell. We've done this, and graphed it in Part B of Figure 9–5.

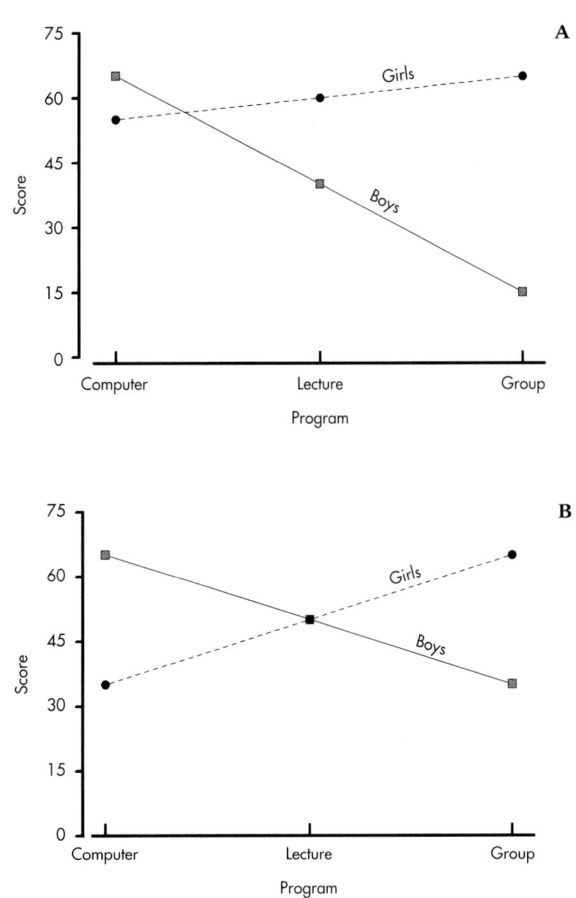

FIGURE 9–5
A, Data from Table 9–2, without adjusting for main effects. **B**, The same data after adjusting for main effects.

Once we've accounted for the fact that girls are better than boys in algebra, and that computers are better than lectures, which in turn are better than group instruction, a different (and more accurate!) interpretation of the interaction emerges: gender doesn't make a difference with lecture-based material; boys are better than girls with computerized teaching; and girls are better than boys with group instruction. The moral of the story is clear—don't try to interpret graphs of interactions or the cell means without first removing the lower-order effects (main effects for two-way interactions; main effects and two-way interactions for three-way interactions; and so on).

If none of the lower-order effects is significant, it's not necessary to do any adjustment. If only one main effect is significant in a factorial design, you need adjust only for that one, but you don't lose anything by adjusting for nonsignificant effects.

RANDOM AND FIXED FACTORS

Although it might not have been obvious when we began, there is a subtle difference between our two independent factors. The brand factor contained only a few of the possible "levels" of the factor. If you were to browse the shelves of the local drugstore or other sex shops, you would find dozens or hundreds of other brands. It is almost as if we randomly sampled the brands in the study from a population of possible brands. Nevertheless, our hope is that the results can be applied to other brands. Not

[6] *We just love it when we can quote ourselves.*

exactly of course; if we didn't study Rainbow Delights, we won't be able to make a statement about them. But if we don't find a difference across the four we chose, we presume that we wouldn't find a significant difference among any four brands. For this reason, brand is considered a **random** factor.

> A *random factor* contains only a sample of the possible levels of the factor, and the intent is to generalize to all other levels.

The same cannot be said for the circumcised/uncircumcised factor. Either a male is circumcised or he's not. We need not generalize beyond the two levels of the factor included in the study. For this reason, we call this a **fixed factor**. A factor can also be fixed if we have other levels of the factor but we do not wish to generalize to them. For example, a study done in the United States might include blacks, whites, and Hispanics. These are only a sample of all possible races, but if the results of the study are applied only to these three, then race remains a fixed factor. It comes down to the statistical notion of population, instead of the street definition.

> A *fixed factor* contains all levels of the factor of interest in the design.

Who cares about the distinction? Unfortunately, you have to when you move to more complex ANOVA designs. As we pointed out earlier, in complex designs the choice of error term becomes a bit complicated, and the choice is further complicated by the fixed versus random issue. In the present example, if brand is a fixed factor, then the denominator for brand is the within error term; for circumcised/uncircumcised it is the interaction term.

Having said all that, without bothering to tell you why it is so, the fact is that most of the time most computer programs never ask.

CROSSED AND NESTED FACTORS

We are not quite through with the generation of jargon yet, all to a worthwhile end, we hope. The design we used to address this question was only one of a number of possibilities. In particular, we ensured that both circumcised and uncircumcised subjects tried out every brand. This was not absolutely necessary because we could have had circumcised men use *R* and *S* and uncircumcised men use *T* and *U*. If we did, as long as there were equal numbers, we could still have made a perfectly legitimate statement about the differences among brands overall (the main effect of brand) and the effect of circumcision (the main effect of C/UC). However, it would not have been possible to state whether circumcised males preferred some brands and uncircumcised males preferred other brands. In the present design, both circumcised and uncircumcised men sampled all the levels of the brand factor. Thus the two factors are said to be **crossed.**

> Two factors are *crossed* if each level of one factor occurs at all levels of the other factor.

If we had used the other approach instead, we would have said that C/UC was partially "nested" in brand. A complete nesting would require that we test only two brands, with circumcised males using one, uncircumcised males the other.

> Two variables are *nested* if each variable occurs at only one level of the other variable.

One other variable in the present design is subject, which we chose to make nested; that is, we assigned individual subjects to only one cell or level of both factors. We could have crossed subject with brand (i.e., have each subject try out all brands) but chose not to so we wouldn't tire out the poor dears.[7]

[7] *Attempting to cross subjects with circumcision status would have led to severe problems in recruitment, especially among those who were already circumcised.*

Crossing and *nesting* are just technical terms, a shorthand way of communicating about experimental designs. But they describe differences that have profound implications for analysis. In general, crossed designs are more powerful because they create the possibility of examining interactions as well as main effects. Conversely, it is impossible or infeasible to cross some factors, and so, of necessity, we end up with nested factors. For example, we cannot have patients both have their appendices out and keep them; similarly, it would be hard to have hospitals doing cost containment 1 month and not the next.

It's easy, straightforward, and often very powerful to have crossover drug trials in which a subject gets one drug for a certain period and a second drug for an alternate period. This works nicely because there is a "washout" effect: after some period, the effect of the drug is gone and the subject is okay. Unfortunately, this doesn't generalize well. Curative drugs such as antibiotics are a one-shot affair. Education interventions hopefully have some lasting effect. And most surgery is a one-way street.

The factor most commonly "crossed" with other factors is the subject or patient. Chapter 11 is devoted to analysis of such designs, which involve using the same patient at various levels of the other factors. Here are some other examples of crossed and nested designs.

1. An intervention to convince obstetricians to reduce their rate of cesarean sections was conducted at a random sample of hospitals in the state. Rates of C-sections were determined for each physician in the treatment and control hospital:
   Hospital is nested in treatment
   Physician is nested in hospital
2. Patients with lupus, 50 males and 50 females, are treated in a randomized trial with cyclosporine. Each patient is randomized to receive either cyclosporine or steroids

for 6 weeks. At this point, there is a 2-week washout followed by 6 weeks on the alternative therapy:
Treatment is crossed with gender
Patient is nested in gender, crossed with treatment
3. An educational intervention involves completing several computer and paper problems on two organ systems. Three problems are given on each of cardiopulmonary and respiratory systems, represented as both computer and paper questions:
Format (computer/paper) is crossed with system
Problem (e.g., chest pain) is nested within system
Format is crossed with problem

As these examples illustrate, it is easy to go from one to two or more factors. Our primary example in the chapter involved only two factors, so it is called a **two-way factorial ANOVA**. All the other, more complex designs are simply called **factorial ANOVAs**, just because they involve many factors. Note that factorial ANOVA bears no relationship to factor analysis, except the similarity in names. Factor analysis is covered in Chapter 18. We won't attempt to do the analysis for these designs because it gets very hairy very fast. Winer (1971) covers many complex designs, and computer packages handle such complicated designs with ease.

SAMPLE SIZE CALCULATIONS FOR FACTORIAL ANOVA DESIGNS

You won't be blamed for rereading the following. The one-way ANOVA case covered in the last chapter led to all sorts of conditions and ramifications. Surely, now that we have really hairy designs, the sample size issue will be horrendous! Amazingly, no.

As it turns out, we use exactly the same strategies for sample size calculations related to *main effects* as we did in the one-way ANOVA case. Regardless of the design, pick the effect (or effects) you *really* care about, treat it as a difference among means, and bash off the sample size.

Interactions are more complicated, naturally. The concept is straightforward enough. You create an effect size, this time based on the difference between the cell means and an expected cell mean based on the main effects, divided by an estimate of the within-cell SD. Then you go to the table and look it up. However, calculating the numerator of the effect size means guessing a minimum of four cell means and four row and column means (for a simple 2×2 case), and the denominator requires even more guesswork. However plausible the exercise may be in theory, in practice, the situations where there is enough information available *a priori* are so limited that the exercise is one of futility. When we do it, we again reduce the comparison to a contrast between two means and use the basic formula.

ASSUMPTIONS AND LIMITATIONS

Factorial ANOVA seems to be the answer to all our dreams (or nightmares). One may rightfully ask why it isn't used all the time for all things. We have already described some of the limitations and assumptions of ANOVA in Chapter 7. Factorial ANOVA also rests on these assumptions, and then some. In particular, the issue of equal sample sizes or *balanced designs*, which was alluded to in Chapter 7, must now be dealt with. One form of balanced design is simply one in which there are equal numbers in each cell. But more generally, a design is balanced if proportionally the same number of individuals appears at each level of a factor. So in the present example, if we were having some difficulty recruiting uncircumcised males, we may decide to sample in a ratio of 1 (uncircumcised) to 2 (circumcised). As long as this ratio was maintained over all levels of the brand factor, the design is still balanced.

The reason balance is important is that, without it, it is possible to get biased estimates of means and variances when there are interactions about. What can you do about it if some souls depart the scene and your data are unbalanced? If the discrepancies are small, do nothing—it won't matter. If the discrepancies are large, say 15% or more difference, then either (1) throw out cases in the larger cells (but nobody wants to do this) (2) scrap ANOVA and do a complicated regression analysis, which is a bit beyond the scope of this book, or (3) threaten them with death beforehand if they choose to die.

EXERCISES

1. For the following studies, identify the independent and dependent variables, figure out the design, and decide which factors are crossed and which are nested. If you are up to it, draw the experimental design.
   a. Groups of laboratory mice from a particular ulcer-prone strain are assigned to different mazes; one with no barriers, and the other with many unsynchronized stop lights and slow-moving rats ahead of them. One third of the mice in each group get beta-blockers, one third get antacids, and one third get milk and digestive biscuits. After 2 weeks, they are all sacrificed and the size of the stomach lesions calculated.
   b. As above, only an additional factor is added. The mice are further subdivided, and two different brands of beta-blockers, antacid, and biscuit are tested.
   c. Five beer brands and five ales are each rated for quality by four engineering undergraduates on a scale from 1 equals slop to 9 equals super.
   d. A predictive validity study examined whether undergraduate grades predicted success or failure in podiatry school. Success/failure was classified as Honors = 3, Pass = 2, Fail = 1.
   e. At the beginning of a course of manipulation, patients with acute gluteitis maximus (pain in the butt) are rated by their chiropractor as to the likelihood of a successful outcome on a scale of 1 equals never to 10 equals a complete cure. Patients are further subdivided into lateral (one cheek) and bilateral (both cheeks).
2. Now go back over the list of factors and decide which are random and which are fixed effects.
3. Let's return to the roadhouse example of Chapter 8. In addition to different heat of "Suicide" wings, there may be systematic differences at other levels of heat. Suppose we extend the study to include two levels of heat—"Mild" and "Suicide." Same three roadhouses. We get a total of 24 undergraduates and send them into the assorted roadhouses. In the kitchen a sealed envelope tells the chef to dish out a Mild or Suicide. Each student rates the platter of wings for heat on a 10-point scale.
   Before going further, see if you can work out the design.
   Now the data look like this:

Heat	Roadhouse	Rater 1	2	3	4	Mean	
	A	4	4	7	5	5.0	
Suicide	B	7	8	6	7	7.0	7.0
	C	7	9	10	10	9.0	
	A	4	2	6	4	4.0	
Mild	B	3	3	4	2	3.0	3.0
	C	1	3	2	2	2.0	

   a. What are the factors in the design? Are they crossed or nested?
   b. Plot the data. By inspection, what do you think are the significant effects?
   c. Work out the ANOVA table. To ease the pain, we'll tell you in advance that the error term, SS (Within), equals 26.0.

How to Get the Computer to Do the Work for You

- From **Analyze**, choose **General Linear Model → GLM—General Factorial**
- Click on the variable to be analyzed from the list on the left, and click the arrow to move it into the box marked **Dependent Variable**
- If you have any fixed effect(s), click on it (them) from the list on the left, and click the arrow to move it (them) into the box marked **Fixed Factor(s)**
- Do the same for any random effects
- Click the **Post Hoc** button and select those you want [we recommend **LSD, Tukey,** and **Tukey's-b**], then click **Continue** [Note: You can run these only if there are more than two groups.]
- Click the **Options** button and click on **Descriptive Statistics**
- Click **Continue**
- Click **OK**

CHAPTER THE TENTH

Two Repeated Observations

The Paired t-Test and Alternatives

One common analysis problem results from situations where individuals are measured at the beginning of a period of time (e.g., at the start of treatment) and again later (at the end of treatment). This design requires a new test, the paired t-test, which explicitly accounts for systematic variance between subjects.

SETTING THE SCENE

In a blatant attempt to cash in on the North American preoccupation with girth, Dr. Casimir from Chittigong designs yet another diet plan. To add a dash of science to the whole affair, he does a study where he weighs a bunch of chubbies before and after they indulge in the plan. He dumps the data on your desk, promising endless riches if you analyze it right. Somehow it seems that you must pair up the beginning and ending observations on each patient. How do you proceed?

All this stuff about randomizing folks to groups, although now *de rigeuer* for medical research, goes against a lot of intuition. A much more natural experiment is to measure something, do something to make it better, and then measure it again. It seems nonsensical to do it to some folks and not to others, and then measure everybody only after it is all over.

For example, when we reach middle age, we tend to get most of our exercise stepping up and down on the bathroom scales each morning.[1] The point of the exercise is to compare today's weight with yesterday's. Our hope is that by resisting the third donut at coffee or walking to the mailroom, some magical transformation will take place so that the belt will move in a notch or two.

If we were serious about combating this growing girth, we might even consider enlisting in an experiment. One possibility is Marine Basic Training at Parris Island, but they wouldn't want middle-aged academics for all sorts of reasons, of which big bellies are the least. A more likely option is some local group, such as Stomach Starers or Girth Gazers. And there we would go once a week, to pay for our pounds of flesh with our pounds of cash, to suffer public humiliation inflicted by the sadistic scales.

The measure of the success or failure of this treatment is based entirely on the comparison of this week's measurement with last week's. Although we may derive some perverse pleasure out of comparing ourselves with other pathetic creatures in the group, the comparison is based on weight loss (or more likely, not lost), not absolute weight. It is small consolation to the formerly petite housewife of 70 kg (154 lb) that the football alumnus and current used car salesman beside her tops in at 140 kg (308 lb). Even if we were to enroll a bunch of these folks in an experiment where they were randomly assigned to a treatment and control group, no scientist (or for that matter, no 6-year old) in his or her right mind would simply weigh them all after the course of treatment.

Forgive us for being so pedantic, but *why* exactly is it so evidently right to measure change in weight within an individual instead of final weight between groups of individuals? In particular, in view of the inevitable statistical sleight of hand to be inflicted on the unsuspecting data in the search for the magical p, why is change better than terminal measure?

The reason is that, when it comes to weight, stable differences *between* individuals are far greater than any likely difference resulting from treatment *within* individuals. This is not simply a reflection that some of us are gaunt and some gross. Recall again that your esteemed authors differ somewhat in height. Stretch is 6′5″; Shrimp is 5′8″. Both have approximately the same size of self-induced life pre-

[1] It's not really that simple. You know you have become obsessed with the problem when you spend a few minutes each day exploring different positions of feet, arms, and so on to see what results in minimum weight. Actually, we have found that leaving one foot off the scale works better; leaving both feet off works best of all.

server about the midriff. But the big guy weighs about 200 lb and the little fella about 160 lb. Arguably, both could afford to lose about 15 lb.

Suppose, by some miracle, they achieved this lofty goal, whereas comparable authors in a control group didn't. To be more precise, Stretch lost 16 lb, Shrimp lost 14. Their counterparts across the way lost 1 and gained 1. Then if we looked simply at - post-test weights, using a straight *t*-test, the difference between the groups would be 15 lb. However, the variability in this difference, which goes into the denominator, includes all the differences among individuals, amounting to ±20 lb. By contrast, if we examine change scores, the numerator is still 15 lb, but the denominator includes the variability of the differences within the groups, which is ±1 or 2 pounds. So the net effect is a large gain in precision and a corresponding increase in statistical power.

This, then, is the basic idea that we pursue here. We begin by examining two measurements per person, but eventually we explore the situation where there are any number of measures, and they may be a result of more than one factor. Pretesting and posttesting are only one example of these within-subject, or **repeated-measures**, designs.

As we have seen, the main advantage of these strategies is the potential gain in statistical power. It is also possible to correct for baseline differences between groups, such as may occur if randomization were inadequate or intact groups were used. But we should point out that this is not the universal panacea it would appear from our contrived example, and we will eventually explore situations where you can lose, as well as gain, power and sensitivity.

Having explored the theoretical issues around the issues of excess *avoirdupois*,[2] perhaps we can proceed to an actual example. The simplest example of a repeated-measures design involves two measurements on a series of subjects, such as those weighing in before and after a round of dieting.

[2] *Showing off our Canadian bilingualism, this bizarre word is printed on many scales, but means, literally, "have some weight." We sure do.*

It goes like this. We all know that the closer you get to the equator, the hotter the food gets. It's a puzzlement until you apply basic physics to the issue. Spicy food makes your body hotter, which makes you sweat, which evaporates, which absorbs heat from your flesh, which cools you off.[3] If this is so, then there may be real benefit, in calorie loss, of a fiery hot curry diet. First, most folks can't eat it anyway. Second, if they do, then the fire in their bellies raises their body temperature, which in turn results in a net energy loss to the environment. Voilá! The fat literally burns off!

[3] *It's the same reason that mad dogs and Englishmen go out in the noonday sun and is also the origin of that classic ex-pat line "There's nothing like a nice cuppa tea (pronounced TAY) on a hot summer day."*

So enter Casper Casimir, the charming chap from Chittigong, with *Captain Casper Casimir's Choice Curried Calorie-Consuming Cuisine for Cold Canadian Consumers* (the C^{11} Diet). All the prospective clients weigh in. For the treatment, they consume, to the best of their ability, suicide-level vindaloos, curries, and Rogan Josh's, at which point they sweat the pounds off. They undergo a second weigh-in after a month. The data are given in Table 10–1. We have taken the liberty, in the right-hand column, of calculating the difference for each individual (*after* minus *before*). We have also calculated the mean and SD of the prediet and postdiet weights and also the weight differences, as shown at the bottom of the table. Note that the SD of preweights and postweights are quite large, about 25 kg, reflecting the large stable differences among *Homo sapiens*. However, the SD of the differences is much smaller, only 3.03 kg.

Now, if we follow the logic of statistics, our null hypothesis is that no loss in weight has occurred. In terms of the individual differences, this is equivalent to the null hypothesis that the true *difference* in the population is zero. Our best estimate of this difference is the calculated difference, 2.08. Moreover, the estimated SD of the differences is the calculated SD, 3.03. The statistical question is, What is the likelihood that a difference of 2.08 or greater could have occurred by chance in a sample of size 12 drawn from the population with a mean difference of 0 and an SD of 3.03?

The approach is to determine a signal-to-noise ratio, naturally. Here the signal is the observed difference (d), 2.08, and the noise is the SE of the difference, $3.03 \div \sqrt{12}$. So the test, called a **paired *t*-test**, is equal to:

$$t = \frac{d}{s/\sqrt{n}} \tag{10–1}$$

In this case, it equals $2.08 \div (3.03 \div \sqrt{12}) = 2.38$.

Now, the critical value of a one-tailed *t*-test with 11 *df* (12 data − 1 mean) at the .05 level is equal to 1.80. Casimir will undoubtedly proclaim to the world that the C^{11} diet is "scientifically proven" and cite papers to back up his claim. Of course, you recall Chapter 6 and are a little more suspicious of one-tailed tests.

For illustration, if we were intent on randomizing to two groups at all costs, we could have gone ahead with an independent sample *t*-test. For the sake of argument, assume that the pretest values were instead derived from a control group of 12 who were destined to pass up the benefits of the curry plan. If they must maintained their wicked ways, it is likely that they would be the same as the treatment group before the treatment began. We could then compare the treatment group after treatment to the control group with an independent sample test as we did in Chapter 8:

$$t = \frac{2.08}{\sqrt{(24.8^2 + 24.0^2) \div 12}} = 0.216 \tag{10–2}$$

Given all the previous discussion, you should not be surprised to see that this *t*-test is minuscule and doesn't warrant a peek at Table C in the appendix.

TABLE 10–1

Pretest and post-test weights of 12 Casimir subjects

Subject	Pretest	Post-test	Difference
1	65	62	−3
2	88	86	−2
3	125	118	−7
4	103	105	+2
5	90	91	+1
6	76	72	−4
7	85	81	−4
8	126	122	−4
9	97	95	−2
10	142	145	3
11	132	132	0
12	110	105	−5
Mean	103.3	101.2	−2.1
SD	24.0	24.8	3.03

Therein lies the power of repeated observations. In the situation where small differences resulting from treatments are superimposed on large, stable differences between individuals, it can't be beat.

So why do all these randomized trials, where folks are assigned to one group or another and measured at the end of the study? There are three reasons, all of which go against the simple paired observation design; one a design issue, one a logistic issue, and one a statistical issue. We'll take them in that order.

The design problem is that a simple pretest–post-test design does not control for a zillion other variables that might explain the observed differences. Maybe the local union went on strike and the study subjects had to cut back on the food bill. Maybe "20/20" came out with a new Baba Wawa piece on the beneficial effect of kiwi fruit for dieters.[4] All of these are alternative "treatments" that might have contributed to the observed weight loss. For these reasons, most textbooks on experimental design mention this design only to dismiss it out of hand.

The logistic problem is more complicated. In many situations a pretest is not possible or desirable. If the outcome is mortality rates, it makes little sense to measure alive/dead at the beginning of the study. If it is an educational intervention, it is often dangerous to measure achievement at the beginning because the pretest measurement may be very much a part of the intervention, telling students what you want them to learn as well as anything you teach to them. Or it may be far too costly to measure things at the beginning.

Finally, there is a statistical issue. If *no* large, stable between-individual differences exist, not only will you not gain ground with a paired comparison, but you could possibly *lose* statistical power. The reason is that the difference score involves two measurements, each with associated error or variability. Comparing groups on the basis of only post-treatment scores introduces error from (1) within-subject variation and (2) between-subject variation. Taking differences introduces within-subject variation *twice*. If within-subject variation exceeds between-subject variation, the latter test will have less power than has the former.

To illustrate this point a bit more, and also to confront the design issue, let's consider a slightly more elaborate design. As we indicated, the difficulty with the pre/post design is that any number of agents might have come into play between the first and second measurement, and we have no justification for taking all the credit. One obvious way around the issue is to go back to the classic randomized experiment: randomizing folks to get and not get our ministrations, and then measuring *both* groups before and after the treatment. Now the data might look like that in Table 10–2.

First of all, this is not exactly a classic randomized controlled trial; that would only measure weights after treatment and then compare treatment and control groups with an unpaired t-test. The calibrated eyeball indicates that such a test is not worth the trouble; the mean in the treatment group is 101.17 kg and in the control group is 105.16 kg. The difference amounts to 4 kg, but the SDs are about 25 kg in each group. Nonetheless, for completeness, we'll go ahead and do it.

$$t = \frac{101.17 - 105.17}{\sqrt{(24.83^2 + 26.19^2) \div 12}} = 0.384 \qquad \textbf{(10–3)}$$

However, an alternative approach that takes advantage of the difference measure is to simply ask whether the average weight **loss** in the treatment group is different from the average weight loss in the

[4] *Who can forget the great grapefruit diet? Seduce the population, make zillions of dollars off the suckers, take a mistress who then shoots you full of holes, and lose about 5 pounds instantly as the blood drains away. And you never gain the weight back!*

TABLE 10–2

Pretest and post-test weights of 12 Casimir subjects and 12 controls

	Experimental			Control		
Subject	Pretest	Post-test	Difference	Pretest	Post-test	Difference
1	65	62	−3	68	70	+2
2	88	86	−2	122	123	+1
3	125	118	−7	84	83	−1
4	103	105	+2	95	97	+2
5	90	91	+1	106	106	0
6	76	72	−4	71	72	+1
7	85	81	−4	87	86	−1
8	126	122	−4	147	152	+5
9	97	95	−2	129	131	+2
10	142	145	+3	136	138	+2
11	132	132	0	105	104	−1
12	110	105	−5	99	100	+1
Mean	103.3	101.2	−2.1	104.1	105.2	+1.1
SD	24.0	24.8	3.03	25.2	26.2	1.73

control group.[5] If we call the weight loss Δ, the null hypothesis comparing treatment (T) and control (C) groups is:

$$H_0 : \Delta_T = \Delta_C$$

(10–4)

[5] *Of such things, Nobel prizes are not made. Nor are we implying that this is something you might not have thought of yourself.*

Having framed the question this way, the obvious test is an unpaired t-test on the *difference* scores:

$$t = \frac{-2.1 - 1.1}{\sqrt{(3.03^2 + 1.73^2) \div 12}} = 3.145$$

(10–5)

This is significant at the .05 level ($t[22] = 2.07$). The test of significance for the difference score is considerably higher than is the t-test for the post-test scores, even though the absolute difference was smaller (3.2 instead of 4.0), because the between-subject SD (about 24 to 26) is much larger than the within-subject SD (1.7 to 3.5).

This conclusion will likely always be true for diets. However, it should be obvious that if we simply shuffle the postdiet weights around, so there is not a close link with pretest measures, this drastically increases the within-subject SD and reduces the test of significance without affecting the post-test comparison at all. There are many real-world places where this may arise. If you use a measure (such as subjective pain rating in arthritic patients) that has a large amount of within-subject variability over time, the use of paired observations can actually reduce power.

Another cost of the paired observation is in the df. Because the unit of analysis is the *pair*, instead of having $2N$ observations from a study (and $2N - 2$ *df*), we only have N pairs and $(N - 1)df$. This is an issue, however, only when the sample size is quite small, as the t-test changes dramatically with sample size only in small samples.

SAMPLE SIZE CALCULATION

Sample size calculations for paired t-tests are the essence of simplicity. We use the original sample size calculation introduced in Chapter 6:

$$n = \left[\frac{(z_\alpha + z_\beta)\,\sigma}{\delta}\right]^2$$

(10–6)

where δ is the hypothesized difference, σ is the SD of the difference, and z_α and z_β correspond to the chosen α and β levels. The only small fly in the ointment is that we must now estimate not only the treatment difference, but also the SD of the difference within subjects—which is almost never known in advance. But look on the bright side—more room for optimistic forecasts.

SUMMARY

The comparison of differences between treatment and control groups using an unpaired t-test on the difference scores (between initial and final observations, or between matched subjects) is the best of both worlds—almost. The basic strategy is to use pairs of observations to eliminate between-subject variance from the denominator of the test. The test is used in pre/post designs, and a variant of the strategy is useful in the more powerful pre/post, control group designs.

The advantage of the test exists as long as the subjects or pairs have systematic differences between them. If this is not the case, then the test can result in a loss, rather than a gain, in statistical power.

EXERCISES

1. As we discussed, at least three kinds of *t*-tests can be applied to data sets—unpaired *t*-tests, paired *t*-tests, and unpaired *t*-tests on difference scores. For the following designs, select the most appropriate.
   a. Scores on this exercise before and after reading Chapter 10.
   b. Crossover trial, with joint count of patients with rheumatoid arthritis, each of whom undergoes (1) 6 weeks of treatment with gold, and (2) 6 more weeks with fool's gold (iron pyrites). Order is randomized.
   c. School performance of only children, versus children with one brother or sister.
   d. School performance of younger versus older brother/sister in two-child families.
   e. School performance of older brother/sister in one-parent versus two-parent families.
   f. Average intelligence of older and younger siblings, reared apart and reared together.
2. You may recall that we did a *t*-test on hair restorers in Chapter 7. Let's return to the data, but add a piece of information: subjects were related. Subjects 1 and 6, 2 and 7, and so on are brothers. How does this change the analysis?

Drug		Placebo	
Subject	Hairs	Subject	Hairs
1	12	6	5
2	14	7	10
3	28	8	20
4	3	9	2
5	22	10	12
Mean	15.8		9.8
SD	8.59		6.21

How to Get the Computer to Do the Work for You

- From **Analyze**, choose **Compare Means** → **Paired-Samples *T* Test**
- Click on the two variables to be analyzed from the list on the left [e.g., the before and after scores], and click the arrow to move them into the box marked **Paired Variables**
- Click **OK**

Analysis of variance techniques are extended to include situations where there are repeated observations on each subject—called repeated-measures ANOVA. The methods amount to inclusion of the Subject as an explicit factor in the analysis.

CHAPTER THE ELEVENTH

Repeated-Measures ANOVA

SETTING THE SCENE

One of the greatest occupational risks facing academics is "myalgic gluteatitis" (ME), a searing pain in the derrière resulting from overexposure to so-called "pains in the butt" at committee meetings. Do any over-the-counter analgesics provide relief from this dreadful agony?

Although many folks think of the academic life as a carefree idyll in which you arrive late, spend long pub lunches in quasi-intellectual debate, and leave early, the life of the mind does have its inherent risks. In particular, one of the activities most feared by academics of all stripes is the dreaded committee meeting. It's not just that it's stupifyingly boring for the most part; it can also represent a real occupational health risk. Blood clots in the leg (DVTs) and hemorrhoids from the long hours sitting motionless in a straight-backed chair are the obvious health concerns, but a far more debilitating risk awaits every committee member. It seems that just about every academic committee has at least one member who unnecessarily drags the proceedings down to a snail's pace worrying about minutiae, and is apparently oblivious to the sighs and groans of the other members. You know the type—the one who corrects the spelling and punctuation errors in the minutes; the one who spends endless time playing with his electronic appointment book when it's time to book the next meeting; the one who can find a reason to argue every point, no matter how seemingly mundane. He's commonly called the one who "is a pain in the butt."

But would it not be more correct to say that he "*induces* a pain in the butt?" Putting a clinical hat on, he might appropriately be viewed as a carrier of a disease, the dreaded new syndrome "myalgic gluteatitis" (ME),[1] a terrible affliction characterized by a searing pain through the nether regions which persists long after the offensive individual has left and the meeting is over. Faced with endless academics begging for relief, we chose to mount the following study: after a meeting attended by one of the carriers, subjects were randomized to receive one of Entrophen, Tylenol, or Motrin to see whether any provided relief.

If this was a normal drug trial, subjects would be randomly assigned to receive one of the treatments, suitably blinded, and then induced to rate the degree of pain relief on some scale, something like the one below:

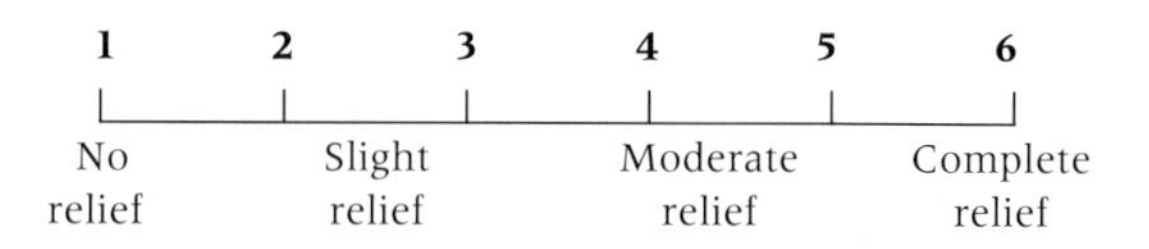

However, because of the nature of the syndrome, we are at liberty to do the study a little differently than the usual approach. Assuming that committees meet only every week or two, and that the pain subsides in hours, then we have a suitable **washout period**, so that we can use each subject again every 2 weeks.

So, instead of randomly assigning individuals to take one of the analgesics, we can actually ask each person to take all of the analgesics, in turn, separated by 2 weeks between meetings, and suitably blinded, of course. It's just a generalization of the paired *t*-test, in which each person was assessed twice and the difference between the two occasions calculated. Now we will deal with multiple assessments,[2] but the basic idea of each subject becoming his or her

[1] Literally inflammation of the muscles of the gluteas maximus. Not to be confused with myalgic encephalitis, which was the Latin name briefly adopted for Chronic Fatigue Syndrome. As near as we can tell, it has something to do with having a muscle in the head.

[2] I never was much good at arithmetic, even back in grade school. "Mr. Norman, count to 10." "OK, teacher. One, uh, two, uh, multiple."

own control holds true. This amounts to directly assessing the variance due to systematic differences between subjects, leaving less variance in the error term (if all goes well).

As a design aside, in order to ensure that order effects are not confounded with treatment effects, we would likely randomize the order of treatment so that ⅓ got Entrophen first, ⅓ got Tylenol first, and ⅓ got Motrin first. If we didn't do that, and all the people are in a long, natural healing course as they grow tolerant of the irritant and learn to tune him out, it might appear that the last medication worked the best. Down the road, we'll talk about how to analyze the data to see if there are order effects.

TABLE 11-1

Pain relief from various analgesics

Subject	Entrophen	Tylenol	Motrin	Average
1	5	3	2	3.33
2	5	4	3	4.00
3	5	6	5	5.33
4	6	4	2	4.00
5	6	6	6	6.00
6	4	2	1	2.33
7	4	4	3	3.67
8	4	5	5	4.67
9	4	2	2	2.67
10	5	3	1	3.00
MEANS	4.80	3.90	3.00	3.90

REPEATED-MEASURES ANOVA (ONE FACTOR)

The design we're talking about is called a **repeated-measures ANOVA** for, we hope, fairly obvious reasons. The data may look something like Table 11–1.

Conceptually, we are in the same situation as we were when we made the transition from an unpaired *t*-test to one-way ANOVA. In the first instance, we just want to determine whether there is any overall difference among the analgesics; it is of secondary importance to figure out whether 1 differs from 2, 2 from 3, and so on. The approach, just like the other ANOVAs we have encountered to date, is to examine the sources of variance. The important distinction in this design, however, is that there are repeated observations on each subject so that we can separate out subject variance from error variance. In the ordinary ANOVA designs, subjects are assigned at random (hopefully) to different groups, and any differences between subjects in the variable of interest ultimately ends up as error variance in the test of the effect of the grouping factors. Here, however, we can take the average of all the observations on each subject as a best guess at the true value of the variable for each subject. The subject variance is then calculated as the difference among these subject means, and the error variance is determined by the dispersion of individual values around each subject mean.

Looking at it this way, then, there are actually three sources of variance:

1. Differences among analgesics overall (at the bottom of the columns)
2. Differences among subjects in the average rating of improvement (right-hand column)
3. Error variance—the extent to which an individual value in a cell is not predictable from the margins

If we continue to look at it this way a bit longer, we see that the design is actually a two-way ANOVA, with the individual subjects as one factor and the pain reliever as a second factor. So, the cells are now defined by the factors Subject with 10 levels and Drug with 3 levels. There are 30 cells and 30 observations, so there is only one observation per cell.

TABLE 11-2

Observed and expected values (in square brackets) for first two subjects

Subject	Entrophen	Tylenol	Motrin	Average
1	5 [4.23]	3 [3.33]	2 [2.43]	3.33
2	5 [4.90]	4 [4.00]	3 [3.10]	4.00
MEANS	4.80	3.90	3.00	3.90

Let's plow ahead, using exactly the same approach as before.

Sum of Squares (Drug) $= 10\,[(4.8 - 3.9)^2 + (3.9 - 3.9)^2 + (3.0 - 3.9)^2] = 16.2$

Remember that since this effect doesn't include Subject, which has 10 levels, we have to multiply the whole ruddy thing by 10. Putting it another way, there are only three terms so we multiply by 10 to make it come out to 30.

Sum of Squares (Subject) $= 3\,[(3.33 - 3.90)^2 + (4.00 - 3.90)^2 + (5.33 - 3.90)^2 + \ldots + (3.00 - 3.90)^2] = 36.7$

Similarly, since this effect is missing the Drug, with 3 levels, we multiply the sum of squared differences by 3.

Now, to calculate the interaction term, it is necessary to estimate the expected values in each cell. We went through the logic before, and it results in an expected value for the first few cells as shown in Table 11–2.

So, the interaction sum of squares is:

Sum of Squares (Interaction) $= [(5 - 4.23)^2 + (3 - 3.33)^2 + (2 - 2.43)^2 + (5 - 4.90)^2 + (4 - 4.00)^2 + (3 - 3.10)^2 + \ldots + (1 - 2.10)^2] = 15.8$

No multiplier is necessary this time since there are already 30 terms.

TABLE 11–3 Analysis of variance summary table

Source	Sum of squares	*df*	Mean square	*F*	*p*
Drug	16.2	2	8.100	9.225	< .005
Subject	36.7	9	4.078		
Error (Subject × Drug)	15.8	18	0.878		
TOTALS	68.7	29			

TABLE 11–4 Analysis of variance summary table for an equivalent between-groups design

Source	Sum of squares	*df*	Mean square	*F*	*p*
Drug	16.2	2	8.100	4.167	< .05
Error	52.5	27	1.944		
TOTALS	68.7	29			

Estimation of degrees of freedom is just like before. For the Drug main effect, there are three data points and one grand mean, so $(3 - 1) = 2$ degrees of freedom. For the Subject main effect, there are 10 data points and one mean, so $(10 - 1) = 9$ *df*. And finally, for the interaction, there are $(10 - 1) \times (3 - 1) = 18$ *df*. This all totals to $18 + 9 + 2 = 29$ *df*, one less than the total number of data, so we must have got it right.[3]

[3]Note from long-suffering co-author: "It's about bloody time!"

We can now go the last step and create the Mean Squares and the ANOVA table, as shown in Table 11–3.

Our research hypothesis asks whether there is any significant difference in relief from the various pain relievers. This is simply captured in the main effect of Drug. The error term for this comparison is the Subject × Drug Mean Square, since this reflects the extent to which different subjects respond differently to the different drug types.[4] That is, just as in an ordinary ANOVA, differences between subjects is a measure of error in the estimate of the effect. The only difference is that, since we have multiple measures for each subject, this amounts to a Subject × Drug interaction.

[4]As we shall soon see, for more complex designs, the difference in response of different subjects, which shows up as a (Subject × Factor) interaction, ends up as the error term for the test of the effect.

The test of significance, then, is based on the ratio of Mean Square (Drug) to Mean Square (Subject × Drug), and equals 9.23, as shown in the ANOVA table (Table 11–3).This is significant at the .005 level.

One last run through these data, before we go on to bigger and better things. As we indicated at the start, we could have done the experiment in a different way by simply randomly assigning individuals to each of the three analgesic groups. That is, instead of measuring 10 people three times, we could have used 30 people and measured each only once. Let's pretend, for the moment, that the data from Table 11–1 came from this design. The appropriate analysis would be a simple one-way ANOVA, and the ANOVA table would look like Table 11–4.

Although the total Sum of Squares is the same, the variance due to subjects and the error variance are now lumped together, leading to a bigger error term for the test of the drug effect, which is now not as highly significant. The reason is that we are no longer able to extract variance due to systematic differences between subjects, since we only have one observation for each subject.

We have described repeated-measures ANOVA in its simplest form. It is a natural extension of the paired *t*-test, just as the one-way ANOVA is an extension of the unpaired *t*-test. In fact, some people call it a **one-way repeated-measures ANOVA.**

Another parallel holds. Like the paired *t*-test, repeated-measures ANOVA is able to take account of systematic differences between subjects, which usually leads to more powerful tests. Finally, like the one-way ANOVA, it is the simplest of the class of designs, and the basic method extends naturally to more complex ones.

GENERALIZATION TO INCLUDE OTHER TRIAL FACTORS

You may have noticed in passing that all the drugs we have used to date were brand names. Here's a chance to address the age-old question of brand name versus generic and, at the same time, learn a little more about the arcane delights of repeated-measures ANOVA. The strategy we used to go from one-way to two-way ANOVA, and to include a second factor in the design, can be applied to repeated observations, only this time we capture the unsuspecting academics for six meetings, three with brand names (Entrophen, Tylenol, Motrin) and three with the equivalent generics (ASA, acetaminophen, ibuprofen). The design is shown in Table 11–5; we have thrown in some additional data.

The first challenge is to work out all the possible lines in the ANOVA table. To start, there are now three factors: Subject (as before) with 10 levels, Drug (as before) with three levels, and now Brand/Generic with two levels. These are the three **main effects**. As we showed in Table 11–5, this means there are $3 \times 2 = 6$ observations on each subject. We now examine some interactions: Drug × Brand/Generic (whether the differences between drugs are the same for brand names and generics); and several interactions with subjects—Subject × Drug (whether some subjects respond differently to some drug types more than others), Subject × Brand/Generic (whether some subjects show different responses to brands and generics), and Subject × Drug × Brand/Generic (we won't even try to put this one into words). Again, generalizing from the first example, every Subject × Something interac-

TABLE 11–5

Two-factor repeated-measures design (analgesic type, brand/generic)

Subject	Brand			Generic		
	Entrophen	Tylenol	Motrin	ASA	Aceta-minophen	Ibuprofen
1	5	3	2	4	3	3
2	5	4	3	4	3	3
3	5	6	5	5	6	6
4	6	4	2	5	4	3
5	6	6	6	5	6	6
6	4	2	1	3	2	1
7	4	4	3	4	4	2
8	4	5	5	4	5	5
9	4	2	2	3	2	3
10	5	3	1	4	3	2
MEAN	4.8	3.9	3.0	3.9	3.8	3.4

tion is the error term to test the effect, since it amounts to differences in the effect for different people.

The Sums of Squares are computed using the general strategy of computing differences between individual means and the grand mean for main effects, and differences between individual cell means and their expected value from the marginals for interactions, in each case multiplying by some fudge factor corresponding to the number of levels of the remaining factors. So, for the main effect of Drug, there are three terms in the Sum of Squares, corresponding to the mean of ASA and Entrophen (4.35) subtracted from the grand mean (3.833), the mean of Tylenol and acetaminophen (3.85), and the mean of Motrin and ibuprofen (3.20), all squared up and multiplied by the famous fudge factor, which is now 2 × 10 = 20, to get a result of 60 terms. After the dust settles, we have Table 11–6.

TABLE 11–6

Analysis of variance summary table of data in Table 11–5

Source	Sum of squares	*df*	Mean square	*F*	*p*
Subject	76.33	9	8.48	6.52	.001
Drug	15.63	2	7.80	6.02	.01
Error 1 (Subject × Drug)	23.37	18	1.30		
Brand/ Generic	0.27	1	0.27	2.25	.168
Error 2 (Subject × B/G)	1.07	9	0.12		
Drug × B/G	3.03	2	1.52	10.37	.001
Error 3 (Subject × Drug × B/G)	2.63	18	0.15		

B/G = Brand/Generic.

As the table shows, we found an overall main effect of Drug ($F = 6.02$, $df = 2/18$, $p = .01$), but no overall effect of Brand/Generic ($F = 2.25$, $df = 1/9$, $p = .168$). However, we are a bit surprised to see a Drug × Brand/Generic interaction, indicating that the differences between drug classes are different for brands and generics. That shows up when we inspect the table a bit closer and see that, although ASA-based drugs are most effective, the effect is smaller for the generic (4.1 versus 4.8); similarly, ibuprofen is the least effective drug, but the generic is actually a bit more effective than the brand (3.4 versus 3.0). Such are the vagaries of science.[5]

If you want a name for this design, it's called a two-way repeated-measures ANOVA. Note that we are simply using the same basic strategy to encompass a design where the repeated observations arise from more than one factor.

Between-Subjects and Within-Subjects Factors

Now that we are used to the idea of including more than one repeated factor in the design, we can bring up the heavy artillery and contemplate a world where both repeated observations within each subject, and other factors that group subjects into classes, arise together in the same design. One way to think about the distinction is to class all the factors as either **within-subjects** or **between-subjects**. The two factors that we have encountered to date, Drug and Brand/Generic, both occur as repeated observations for each subject in the design, and are, therefore called within-subject factors.

A **within-subjects (trial) factor** is one in which all levels of the factor are present for each subject, that is, it results in repeated measures.

[5] *Or would be the vagaries if these were real data. Our lawyer advised us to add the disclaimer that these data are completely artificial.*

TABLE 11–7

Experimental design for inclusion of between- and within-subject factors

Group	Subject	Brand: Entrophen	Brand: Tylenol	Brand: Motrin	Generic: ASA	Generic: Acetaminophen	Generic: Ibuprofen
Research	1						
	2						
	3						
	.						
	10						
Admin	11						
	12						
	13						
	.						
	20						

What would a between-subject factor look like? Well, let's expand the design one last time. Perhaps we are concerned that ME is only prevalent for some kinds of committees. From personal experience, ME carriers seem to gravitate to administrative committees, which are inevitably burdened with routine decision-making necessary to keep the academic ship afloat. Carriers tend to avoid the more creative committees like research group meetings. Perhaps the severity of ME is lower in research meetings, or the effects are more transient so that relief comes more rapidly.

How would we put it to a test? We would use the same design as before, only this time we recruit 10 academics from an administrative committee and 10 *different* academics from a research group.[6,7] The design would now look like Table 11–7.

We have the same two within-subject factors as before. However, we now have an additional factor, Group (i.e., type of committee), which is different. It can only assume one value for each subject, that is, a particular person can only be on an administrative committee or a research committee, not both (see note 6), and subjects are grouped under each level of this factor. By extension, this is called a **between-subjects** or **grouping** factor.

> A **between-subjects** or **grouping** factor is one in which each subject is present at only one level of the factor, that is, subjects are grouped under a level of the factor.

As a matter of course, it is usually easier in these repeated-measures designs to put all the between-subjects factors on the left and all the within-subjects factors on the top. This guarantees that the innermost column on the left will be "Subject," and that each row corresponds to all the measurements on one subject—in this case, six.

Now, let's proceed to anticipate what the ANOVA table might look like. We won't do it, since, as yet, there are no data in the boxes.[8] To begin, there are four **main effects**: Group (Do folks on research committees get faster relief from ME pain than those on administrative committees?) and Subject (Do some people get more or less relief than others?) from the left column; and Drug (Do all drugs give the same relief?) and Brand/Generic (Do brand name drugs give the same relief as generic drugs?) from the top rows. There are also some two-way interactions—Group × Brand/Generic, Group × Drug, and as before Drug × Brand/Generic. There are also several interactions with Subject, which will find their way into the assorted error terms.

The one interaction with Subject that is missing is the Subject × Group term. Since each subject is on either an administrative or a research committee, not both, we can't estimate this interaction. This is just another way of saying that Group is a between-subjects factor.

A more general way of describing this idea is to speak of **crossed** and **nested** factors.

> Two factors are **crossed** if each level of one factor occurs at all levels of the other factor. A factor is **nested** in another factor if each level of the first factor occurs at only one level of the second factor.

What was that again? Well, it's a lot like between- and within-subjects factors. Since Group is a between-subjects factor, each subject can be in either the administrative category or the research category, not both; so, Subject is nested within Group. However, since both Group and Subject occur at all levels of Drug and Brand/Generic (each subject has an observation at all levels of Drug and all levels of Brand/Generic), we say that Subject and Group are crossed with Drug and Brand/Generic. Finally, for

[6] *To ensure that committee type is a true between-subjects factor, we must ensure that there are different folks in each group. We could, of course, try to find a situation in which the same people are on both an administrative and a research committee, but then it would be another within-subject factor.*

[7] *It's not essential to have the same number of people in each committee type; we did it just for convenience.*

[8] *This presents an ideal situation to cook the data as we see fit. We will.*

TABLE 11–8

Analysis of variance summary for three-factor ANOVA

Source	Sum of squares	*df*	Mean square	*F*	*p*
Group	25.34	1	25.34	3.94	.008
Error (Subject:Group)	51.45	8	6.43		
Brand/Generic	0.42	1	0.42	3.12	.110
Brand/Generic × Group	0.02	1	0.02	0.12	.730
Error (Subject:B/G × Group)	1.06	8	0.13		
Drug	14.40	2	7.20	5.35	.020
Drug × Group	0.40	2	0.20	0.20	.860
Error (Subject × Drug:Group)	21.53	16	1.34		
Brand/Generic × Drug	2.53	2	1.27	8.68	.003
Group × B/G × Drug	0.13	2	0.07	0.46	.640
Error (Subject:Group × B/G × Drug)	2.33	16	0.15		

B/G = Brand/Generic.

completeness, since we have acetylsalicylic acid as both Brand (Entrophen) and Generic, Drug and Brand/Generic are themselves crossed factors. Nested factors are often signified with a colon (:) in the ANOVA summary table, so we would write Subject nested within Group as S:Group.

So now the whole "kit and caboodle" appears as Table 11–8.

There are a few things to note. First, as we have alluded, there is not a single error term; there is a different error term for each main effect and its associated interaction. All the error terms amount to the associated interaction with subjects. Second, the ANOVA, although complicated, still obeys some of our fundamental rules: (1) the degrees of freedom add up to one less than the number of data points; (2) the Sums of Squares for individual terms could be summed to yield a Total Sum of Squares; (3) Mean Squares and *F*-ratios are calculated just as before, except that the correct error term must be used (by the computer of course); and (4) the degrees of freedom for numerator and denominator of the *F*-ratio must use the right degrees of freedom (but, again, the computer takes care of all this).

In the end, we are simply partitioning variance across multiple factors in order to (1) investigate the possible effects and interactions, and (2) reduce the corresponding error terms and, thereby, increase the power of the test. In particular, one explicit factor in all repeated-measures designs is Subject, so that any variance due to systematic differences between subjects can be removed and the power of other tests correspondingly increased.

We need not stop here of course. We are limited in the number of factors only by the number of degrees of freedom. In particular, at the outset, we indicated that as good researchers, we should vary the order in which subjects get the different drugs. There's an easy way to do this and to ensure that it all balances out in the end. It's called a **Latin Square** and it is performed as follows. Pick some order—for the sake of argument, Entrophen → Tylenol → Motrin. Now just move everything to the right and rotate the last one to the beginning: Motrin → Entrophen → Tylenol. Do it one last time: Tylenol → Motrin → Entrophen. This exhausts the possibilities. Now ensure that ⅓ of the subjects gets the first sequence, ⅓ the second, and ⅓ the last. If we do this, then we have created a second between-subjects factor. If there is an order effect, it will be apparent and will show up as an Order × Drug interaction.

Finally, let us take a minute to try to convince you that there really is a grand unity to the whole thing. We could have analyzed the Group effect differently. We could have said to ourselves, "Ah heck, at the end of the day, we've got a bunch of data on 20 folks in two groups. Let's just average it and do a *t*-test." The numerator of the *t*-test would involve the difference between the overall means for all the data in the administrative and research groups, the same data that are going into the Group main effect. The denominator of the *t*-test would involve differences between the subject means, the same data that went into the Subject main effect. The resulting *t* would be just the square root of the Group main effect, believe it or not.[9]

Other Applications of Repeated-Measures ANOVA

Repeated-measures ANOVA is a very useful strategy to look at the effects of interventions in situations in which the same person can receive multiple treatments. Any time that a washout period is feasible, then the repeated-measures ANOVA design acts as a generalized crossover study. It also can work with repeated observations at the same time, for example, patch tests for suntan lotions.

However, it has many other applications. Basically, it can be used any time there are repeated measurements on the same set of individuals.[10] One sit-

[9] *After all we've put you through in this chapter, the answer may well be "Not."*

[10] *We do, however, restrict this to repeated measurements of the same quantity—weight, serum sulphur, or whatever—usually at different points in time. Many studies do make repeated observations on subjects, but of different variables, such as grip strength, morning stiffness, joint count, sedimentation rate, and so on, for arthritic patients. For this situation, use multivariate statistics, covered in Chapter 12.*

uation in which this occurs is when there are sequential measurements over time, as subjects are followed up for days, weeks, or months. If we did a trial with two groups in which we made repeated observations, we would likely analyze the data with a two-factor repeated-measures ANOVA, with treatment/control as a between-subjects factor and time as a within-subjects factor.[11] As we will see in Chapter 17, this is one of several ways to approach the general issue of measuring change.

Another place where we use repeated-measures ANOVA a lot is in measurement studies. When you look at reliability or agreement, whether you use an intraclass correlation or Cohen's kappa, the starting point is a repeated-measures ANOVA. Interestingly, measurement folks, whose passion is differentiating among people, are really mainly interested in all the things we called "error."

Finally, repeated-measures designs are one special case of factorial designs and, like factorial designs, the number of possible designs is limited only by imagination and resources.

[11]Note that if a baseline measurement were made, this should not *be treated as another level of the time factor. More appropriately, this should be used as a* ***covariate*** *and the analysis should be a* ***repeated-measures analysis of covariance*** *(see Chapter 16).*

[12]Usually, that is, in psychology and education. Many clinicians, however, seem to think that this approach is a bit of esoterica foisted on them recently by zealous psychometricians. Ironically, nothing could be further from the truth. We found an entire chapter devoted to the intraclass correlation in Fisher's 1925 statistics book.

[13]One exception is a Time factor that consists of very different intervals (e.g., Day 1, Day 2, Day 10, Day 30). Then, the assumption of homoscedasticity may break down and, as we'll see in Chapter 12, it may be better to analyze the data using MANOVA.

REPEATED-MEASURES ANOVA AND RELIABILITY OF MEASUREMENT

Repeated-measures ANOVA has one other very useful application—to compute variance components for use in studies of agreement. For example, if we go back to Table 11–1, it is evident that some folks, such as Patient 6, get better relief across the board than others, such as Patient 10. If we turn things around and use the data as a measure of individual susceptibility to pain relief, then we can start to ask whether the measure can reliably distinguish between those who get a lot of relief and those who get a little.

Reliability is usually assessed by the **Reliability Coefficient**,[12] which is defined as the proportion of variance in the scores related to true variance between the objects of measurement—in this case, patients. The formal definition looks like:

$$Reliability = \frac{\sigma^2_{Subj}}{\sigma^2_{Subj} + \sigma^2_{Residual}}$$

Conveniently, we have the makings of a reliability coefficient in the ANOVA (Table 11–3). Subject variance is directly related to the Mean Square (Patient), and error variance is related to Mean Square (Patient × Drug). Because of the relationship between variances and mean squares, this can be transformed into a computational formula involving Mean Squares. Note that the Mean Square (Residual) is from the Patient × Drug term:

$$Reliability = \frac{Mean\ Square\ (Subj) - Mean\ Square\ (Res)}{Mean\ Square\ (Subj) + (k-1)\ Mean\ Square\ (Res)}$$

In the present example, then, the reliability is equal to:

$$Reliability = \frac{4.078 - 0.878}{4.078 + (2)0.878} = \frac{3.19}{7.13} = 0.54$$

This coefficient is called an **Intraclass Correlation Coefficient** (ICC), since all observations are from the same variable or class; this distinguishes it from the Pearson correlation that has observations from different variables. Fisher called this the "interclass correlation." It ranges from 0 (no systematic difference between subjects) to 1 (all the variance in scores is due to systematic differences between subjects).

ASSUMPTIONS AND LIMITATIONS OF COMPLEX ANOVA DESIGNS

Are there no costs incurred in this exercise? Of course there are. Nothing comes free, except to selected dictators, capitalists, warlords, and other unscrupulous types. First, just as the case for two-way ANOVA and all other parametric tests, there is the assumption that the data are at least interval level and are normally distributed. We also demand **homoscedasticity**—a lovely word meaning equal variances. However, we have discussed in Chapter 6 the extent to which the tests are robust to the violation of these assumptions and, as you recall (or we recall), the **Central Limit Theorem** indicates that for sample sizes over 10 to 20, the normality assumption is unnecessary.[13] Repeated-measures ANOVA, like all factorial ANOVA designs, imposes one additional constraint: the designs must be **balanced** or nearly so; this was discussed in Chapter 9. The good news is that, as long as the design is balanced, the ANOVA is robust with respect to assumptions about distributions.

Are there any more limitations? Indeed there are. There are two reasons why one must not continue to add factors into a design at random. First, unless these factors are designed into the study from the outset, they will likely result in imbalance, and we already indicated where that slippery slope leads. Second, there is a law of diminishing returns. Each factor you add, even if you have only two levels of the factor, costs one degree of freedom for the main effect and each interaction. If there are more than two levels, the dfs escalate. Unless the factor is accounting for useful variance, the paradoxical situation can arise that even though the factor carries away some of the Error Sum of Squares, the Error Mean Square term for other analyses actually increases, because the degrees of freedom have been reduced more than the Sum of Squares as a result of the addition of the factor. The upshot is that the mean square—which enters into the statistical test—actually goes up.

Nevertheless, despite the constraints imposed by the addition of more than one factor into a design, the power of analysis and interpretation obtained from factorial and repeated-measures ANOVA is often remarkable, and has added tremendously to the versatility of experimental research.

SAMPLE SIZE ESTIMATION

For all sorts of reasons, there is no exact formula to calculate sample size for two- or three-factor repeated-measures designs. If it is a single factor, and has only two levels, then the procedures outlined for the paired t-test in Chapter 10, which are the essence of simplicity, are appropriate. However, anything more complicated, forces us to estimate in advance (1) what might be the appropriate change **within subjects** and, then, (2) the approximate interaction **between subjects** and this effect. The last grant reviewer who went for such long shots jumped off a building in the Crash of '29.

The best strategy to survive the vagaries of reviewers is to take an approximate approach. Pick the one effect you really care about, which, hopefully, is a main effect with two levels, and use an approximate calculation based on the paired t-test. It still requires a bit of imagination to come up with the error term, but it's not impossible.

The only exception to this approach is, unfortunately, fairly common: when the effect of concern is a two-way interaction. Here, an even more sweeping approximation is needed. Again we convert this to a pairwise comparison, and then go back to the paired t-test.

SUMMARY

We have considered a number of extensions to the paired t-test, all described as repeated-measures designs. They amount to variations on factorial ANOVA methods, with Subjects as an explicit factor in the design.

EXERCISES

1. For the following designs, name the factor equivalent to "subjects," then name the between-subjects and within-subjects factors.

a. Thirty spondylitis patients are treated by chiropractors on a weekly basis for 12 weeks. After each treatment, range of motion of the SI joint is measured.

b. Twelve patients suffering from chronic headaches are treated by three different headache medications. At the onset of a headache, each patient selects either a red, white, or blue pill, which he or she selects by throwing a dart at a Union Jack on the basement wall. An hour later, the patient rates the pain on a 10-point scale. This continues until the patient has treated 6 headaches with each color of pill, for a total of 18 headaches per patients.

c. Twelve patients suffering from chronic headaches are treated by three different headache medications. Each patient is randomly assigned to be treated by red, white, or blue pills by the attending physician throwing a dart at a Stars and Stripes on the clinic wall. An hour after the onset of each headache, the patient rates the pain on a 10-point scale. This continues until the patient has treated six headaches.

d. Histologic slides of lymph gland biopsies are judged by pathologists on a 5-point scale for likelihood of cancer. There are 20 slides in total. Each slide is rated by 6 pathologists.

e. Histologic slides of lymph gland biopsies are judged by pathologists on a 5-point scale for likelihood of cancer. There are 20 slides in total. Each slide is rated by 6 pathologists, at 3 levels of experience—2 first-year residents, 2 final-year residents, and 2 pathologists.

f. Histologic slides of lymph gland biopsies are judged by pathologists on a 5-point scale for likelihood of cancer. There are 20 slides in total, all derived from patients with a minimum of 10 years follow-up. Half the slides were from proven normal patients, and the other 10 were from patients who eventually died of lymphoma (cancer of the lymph glands). Each slide is rated by 6 pathologists, at 3 levels of experience—2 first-year residents, 2 final-year residents, and 2 pathologists.

2. To compare 3 of the NSAIDs for the treatment of rheumatoid arthritis, 45 subjects were divided into 3 groups of 15 subjects each and given 1 of the drugs. They rated their degree of pain at the end of 10 days, using a 100-point scale. The results of the one-way ANOVA was: $F(2, 42) = 2.99$; $.05 < p < .10$. The investigator approaches you for some suggestions for what she might do to increase the likelihood of getting p below .05. Would you expect that each of the strategies listed MIGHT WORK or WOULDN'T WORK?

a. Increase the number of drugs from 3 to 5.

b. Increase the number of subjects from 15 to 25 per group.

c. Use a within-subject (e.g., crossover) design with the same number of subjects (45).

d. Use a simpler pain scale (Present/Absent) to increase agreement.

3. For the following designs and ANOVA tables, you get to fill in the blanks:

a. Seventeen Scottish lairds are assembled in the manor, plied with a "wee dram o' the malt" all night long, then asked to rate their state of euphoria (1) the night before and (2) the morning after.

Source	Sum of squares	*df*	Mean square	*F*
Laird (L)	320	16		
Night/Morn (NM)	42			
NM × L	160			

b. An entomologist (bug freak) counts the number of spikes on the legs of North American and South American horned cockroaches (*Stylopyga orientalis*, yet another Japanese import!) to see if they have different lineages. The bug freak has 20 bugs per group, and 6 legs per bug.

Source	Sum of squares	*df*	Mean square	*F*
North American/ South American (NS)	1,300			
Bug (B)	3,800			
Leg (L)	5,000			
L × NS	550			
L × NS × B	950			

c. Twenty medical students are observed and rated on five different patient workups. Each workup is observed by two staff clinicians.

Source	Sum of squares	*df*	Mean square	*F*
Students (S)	950	19		
Patient (P)	300	2		
P × S	190			
Observer (O)	120	1		
O × S	95			
P × O	34			
P × O × S	38			

How to Get the Computer to Do the Work for You

Because repeated-measures ANOVAs are somewhat more complex than straight factorial ANOVAs, we're going to break with tradition a bit and show the actual contents of the computer commands for the analyses we did in this chapter. We'll use the most complicated analysis, in which there are two groups (Administrative/Research); three drugs; and two types of each drug (Brand/Generic). When we entered the data, we added a variable called **Group**, which had the value 1 for the administrative group subjects and 2 for the research group subjects.

- From **Analyze**, choose **General Linear Model → GLM—Repeated Measures**
- Enter **Drug** <Tab> in the **Within-Subjects Factor Name**
- Enter **3** <Tab> in the **Number of Levels** box and press **Add**
- Enter **Brndgnrc** <Tab> in the **Within-Subjects Factor Name**
- Enter **2** <Tab> in the **Number of Levels** box and press **Add**
- Click the **Define** button
- Click **Entrophen, ASA, Tylenol, Acetaminophen,** and so on to replace the lines **__?__(1,1), __?__(1,2)**, until the six lines in **Within-Subject Variables** are filled
- Move **Group** to the **Between-Subjects Factor(s)** box
- Click **OK**

CHAPTER THE TWELFTH

Multivariate ANOVA (MANOVA)

Multivariate analysis of variance (MANOVA) is an extension of ANOVA that allows for two or more dependent variables to be analyzed simultaneously. This avoids the problem of inflation of the alpha level because of multiple testing, and takes into consideration the correlations among the dependent variables. It is also useful in analyzing repeated measures.

SETTING THE SCENE

The previous examples, which looked at condom brand and circumcision status, had to deal with a major problem: From whose perspective do we measure satisfaction—the man's or the woman's? It is possible that what may be better for men is worse for women, and vice versa, so that the results of a study may be different depending on the choice of the outcome measure. We can get around this problem, and others, by using techniques that allow us to consider more than one dependent variable at a time; one such technique is an extension of ANOVA, called multivariate *analysis of variance, or MANOVA.*

In keeping with the abbreviation of this statistical test (MANOVA), the problem we will grapple with (in a manner of speaking) is one that has bedeviled sex therapists and missionaries alike for generations—does it matter who's on top?[1] We'll start off easy, and have only two groups of subjects: a group that does "it" in the missionary position, and a second group in which the women are on top.[2] To avoid the problem of having to make an arbitrary decision regarding whose opinion we will assess, we'll ask both partners[3] to complete a satisfaction scale that ranges from 1 ("It would have been better if I had done it myself") to 100 ("The Earth has not stopped moving yet"). The first (clean) question that could be asked is, Why would obtaining two measures lead to an entire chapter in a book? Why not simply do two *t*-tests and be done with it? This question is even more cogent when we realize that the mathematics of univariate statistics are relatively straightforward, but multivariate tests—those that involve two or more dependent variables—require something called **matrix algebra**, which has its own arcane language.[4] The answer is that once we have more than one dependent variable (DV), a host of problems arises if we treat them as if they were unrelated variables.

The first problem is one of interpretation. If we did separate *t*-tests on the measures, and found that both were significant, would it mean that the groups differed on both variables? The answer is a very definite "Yes and No." The groups did, in fact, differ on the two outcomes, but, because the two variables are likely correlated with one another (if they weren't correlated, we wouldn't bother with multivariate analyses), we can't really conclude that they differed in two discrete areas. Let's use an extreme example to illustrate this point. If the intervention were a weight reduction program, and the outcomes were both weight and the body mass index, it would be ridiculous to become ecstatic about the fact that we found change in both variables. Of course both variables changed, because they're highly correlated with one another; so, if one changes, the other must, too. His and her satisfaction scores probably aren't as highly correlated, but since we are dealing with warm, sensitive, gender-neutral, whale-loving tree-huggers, it would be highly unusual for them to engage in practices that satisfy one person and are a turnoff for the other. MANOVA takes the correlations among the DVs into account, so the results aren't distorted by having redundancies among the variables.

[1] *If you're not sure how this question relates to the test, check the Glossary under "MANOVA."*

[2] *We will not use the term "in the superior position," because that will only cause havoc among those who want to generalize the term to other situations.*

[3] *We recognize that restricting this study to situations involving only two partners may impose limitations for some people, but it is necessary to keep the analysis simple.*

[4] *Don't be scared off. We don't understand it too well, either, so we'll avoid it as much as we can. You'll have to know some of the language, but since this is your first time, we'll be gentle.*

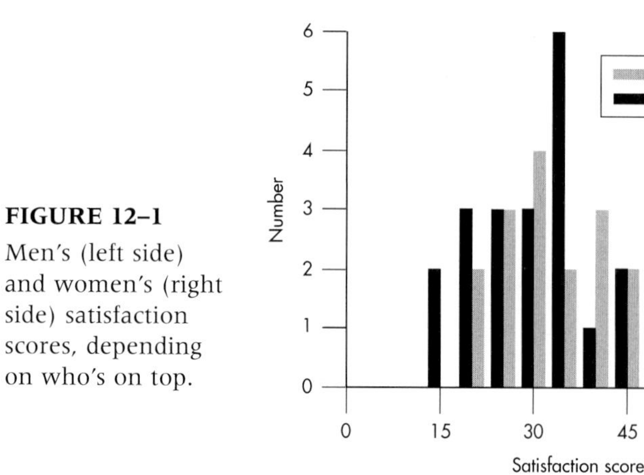

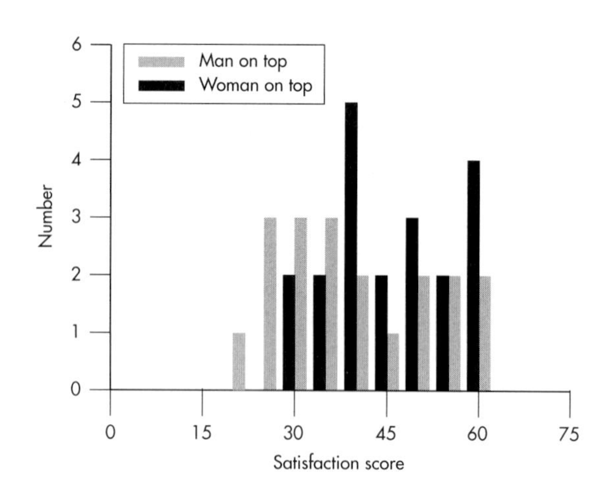

FIGURE 12–1
Men's (left side) and women's (right side) satisfaction scores, depending on who's on top.

TABLE 12–1
Results of t-tests between the groups for both variables

Rater	Who's on top	Mean	SD	t	df	p
Male	Man	34.0	10.3	0.73	38	.47
	Woman	31.8	8.73			
Female	Man	41.95	12.79	0.61	38	.55
	Woman	44.2	10.37			

The second problem is that of multiple testing. As we saw in Chapter 5, the probability of finding at least one outcome significant by chance increases according to the formula:

$$1 - (1 - \alpha)^N \tag{12-1}$$

where N is the number of tests we do. If we have two outcome variables (and, hence, have performed two tests), the probability that at least one will be significant by chance at a .05 level is:

$$1 - (1 - .05)^2 = .0975 \tag{12-2}$$

If we have five outcomes, the probability is over 22%; by the time we have 10 outcome variables, it would be slightly over 40%. We could try to control this inflation of the alpha level with a Bonferroni correction or some other technique, but, as we discussed in Chapter 8, these tend to be overly conservative. Also, neither the estimate of how many tests would be significant by chance, nor the corrections for this would take into account the correlations among the dependent variables, which often can be considerable.

A third problem is the converse of the second: rather than finding significance where there is none, we may overlook significant relationships that are present. First, we'll show that this *can* happen and then *why* it can happen. Let's start off in the usual way, by plotting the data to help us see what's going on; the results of the males' satisfaction questionnaires are shown on the left side of Figure 12–1, and those of the females' satisfaction on the right side of the figure. Using that most sensitive of tests, the calibrated eyeball, it doesn't look as if much of anything is happening (insofar as the data are concerned, at least);[5] the distributions of the two groups (positions) seem to overlap quite a bit for both variables.

The next step is to do a couple of t-tests on the data, and these are reported in Table 12–1. Again, the groups look to be fairly similar on the two variables; the means are relatively close together, differing by less than half a standard deviation, and only someone truly desperate for a publication would look twice at the significance levels of the t-tests. We could even go so far as to correlate each variable with the grouping variable, using the point-biserial correlation.[6] This doesn't help us much either; the correlation between the men's ratings and position is –0.12, and that for the women's is an equally unimpressive 0.10.

Now, though, we'll pull another statistical test out of our bag of tricks and analyze both dependent variables at the same time using a **multivariate analysis of variance** (MANOVA). We won't show you the results right now, but trust us that there is now a statistically significant difference between the groups.[7] So, what's going on? Why did we find significant results using a multivariate test when it didn't look as if there was anything happening when we used a whole series of univariate tests? The reason is that the *pattern* of the variables is different in the two groups. If you go back to Table 12–1, you can see that when men do the rating, they score higher for the male superior position than for the female superior position; but when women do the rating, the situation is reversed—both physically and psychometrically. This couldn't be seen when we looked at each variable separately. It's analogous to

[5] *Looking at the magnitudes of those evaluations, it doesn't look as if much is happening elsewhere, either.*

[6] *Which, as you no doubt remember, is simply a Pearson correlation of one continuous variable (each satisfaction scale) with a dichotomous one (who's on top).*

[7] *If you don't trust us, you can skip ahead a few pages to Table 12–3 and check for yourself. Now, aren't you sorry you doubted us?*

the advantage of a factorial design over separate one-way ANOVAs: with the former, we can examine interactions among the independent variables that wouldn't be apparent with individual tests; with MANOVA, we can look at interactions among the *dependent* variables in a way that is impossible with univariate tests.

So, the conclusions are clear: when you have more than one dependent variable, you're often ahead of the game if you use multivariate procedures. At the end of this chapter, we'll discuss some limitations to this approach, but for now let's accept the fact that multivariate statistics are the best way of analyzing multivariate data.

WHAT DO WE MEAN BY "MULTIVARIATE?"

We've used the term *multivariate* a few times, and we've even defined it—more than one dependent variable—but let's be more explicit (and confusing). If we went to the bother of checking in a dictionary, we would find that *multivariate* simply means "many variables." In Chapter 9, we looked at the interaction among condom brand and circumcision status on estimations of prowess. To most people, this may seem to be multivariate; after all, we're examining a total of three variables at once. However, statisticians aren't like most people. They have their own terms that bear only a passing resemblance to English (e.g., "heteroscedasticity" or "polychotomous"), or they use English words in their own unique and idiosyncratic way that the British would likely describe as "quaint." We already saw that there is nothing inherently normal about the "normal distribution," and we will see that "regression analysis" doesn't involving sucking your thumb or recalling a previous incarnation as a Druid princess. In the same way, statisticians have their own meaning for the word "multivariate." It refers, as we've said, to the analysis of two or more *dependent* variables (DVs) at the same time. In the example we just used, the type of prophylactic and whether or not the person was circumcised are *independent* variables (IVs), and there was only one DV, prowess. So, to a statistician, this would be a case of univariate statistics—factorial ANOVA—with two IVs and one DV. It would become a multivariate problem if, in addition to looking at prowess, we also measured the duration of the encounter.[8]

Doesn't this sound simple and straightforward? That's a sure sign that something will go awry. The fly in the ointment is that statisticians aren't consistent. Some of them would call multiple regression a multivariate technique, even though there is only one DV, and it is mathematically identical to ANOVA. Others prefer the term *multivariable* but, even here, some use the term to refer to many IVs and some just to indicate that many variables—dependent and independent—are involved. For most of us, however, multivariate means more than one DV, and that's the usage that we'll adopt.

t FOR TWO (AND MORE)

If we had only one dependent variable, the statistic we would use for this study would be the *t*-test, which starts with the null hypothesis:

$$H_0\colon \mu_1 = \mu_2 \tag{12–3}$$

that is, that the means of the two groups are equal. However, we have two dependent variables in the current example, so that each group has two means. In this case, then, the null hypothesis is:

$$H_0\colon \begin{pmatrix} \mu_{11} \\ \mu_{12} \end{pmatrix} = \begin{pmatrix} \mu_{21} \\ \mu_{22} \end{pmatrix} \tag{12–4}$$

where the first subscript after the μ indicates group membership (1 = Man on Top, 2 = Woman on Top), and the second shows the dependent variable (1 = Male Rater, 2 = Female Rater). What this indicates is that the list of means for Group 1 (the technical term for a list of variables like this is a **vector**) is equal to the vector of means for Group 2.[9] In other words, we are testing two null hypotheses simultaneously:

$$\mu_{11} = \mu_{21} \textit{ and } \mu_{12} = \mu_{22} \tag{12–5}$$

If our study involved a factorial design, such as looking at the effect of being at home versus on a romantic holiday, then we would have another null hypothesis for this new main effect of Setting, and a null hypothesis involving the interaction term of Position by Setting. This is just the same as what we do in the univariate case, except that we are dealing with two or more dependent variables.

When we plot the data for a *t*-test, we would have two distributions (hopefully normal curves) on the *X*-axis. The picture is similar but a bit more complicated with multivariate tests. For two variables, we would get an ellipse of points for each group, as in Figure 12–2. If we had three dependent variables, the swarm of points would look like a football (without the laces); four variables would produce a four-dimensional ellipsoid, and so on. These can be (relatively) easily described mathematically, but are somewhat difficult to draw until someone invents three-, four-, or five-dimensional graph paper.

Sticking with the analogy of the *t*-test, we compare the groups by examining how far apart the centers of the distributions are, where each center is represented by a single point, the group mean. In the multivariate case, we again compare the distance between the centers, but now the center of each ellipse is called the **centroid**.[10] It can be thought of as the overall mean of the group for each variable; in this case, in two-dimensional space (Male's rat-

[8] *A polite term for "it."*

[9] *You've just been introduced to your first term in matrix algebra, "vector." See, that was painless, wasn't it? There actually is a link between the use of the term vector in matrix algebra and in disease epidemiology, obscure though it may be: both are represented symbolically by arrows.*

[10] *That's your second term in matrix algebra.*

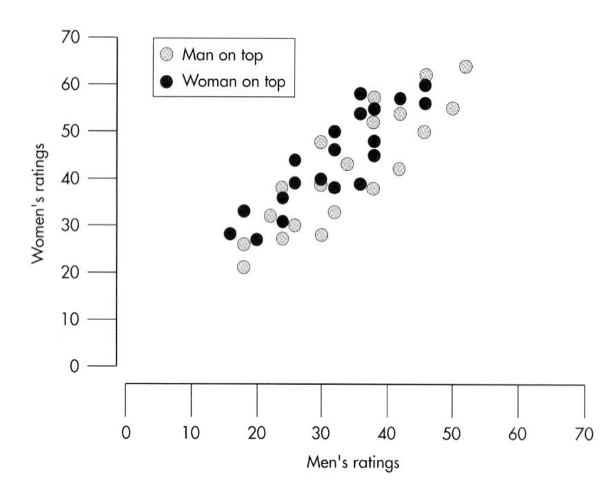

FIGURE 12–2
Scatterplot of the two DVs for both groups.

ing and Female's rating). If we had a third variable, we would have to think in three dimensions, and each centroid would consist of a vector of three numbers—the means of the three variables for that group. The logic of the statistical analysis is the same, however: the greater the distance, the more significant the results (all other things being equal).

LOOKING AT VARIANCE

Comparing the means in a *t*-test is necessary, but not sufficient; we also have to compare the differences between the means to the variances within the groups. Not surprisingly, the same applies in the multivariate case. However, there's an added level of complexity; in addition to the variances of each of the variables, we also have the **covariances** between the variables.[11] What this means is that instead of having just one number for the variance, we now have a *matrix* of variance and covariance terms. Not surprisingly, it's referred to as the **variance-covariance matrix** (VCV) and looks like:

$$VCV = \boldsymbol{\Sigma} = \begin{bmatrix} s_{11}^2 & s_{12}^2 \\ s_{21}^2 & s_{22}^2 \end{bmatrix} \quad (12\text{–}6)$$

First, note that we just stuck in a third term from matrix algebra, "matrix" itself, which is simply a rectangular array of numbers.[12] Second, the abbreviation for a variance-covariance matrix is either VCV or the symbol $\boldsymbol{\Sigma}$. That's right, it's the same symbol we use to indicate summation. You'd think that with so many Greek letters lying around that aren't being used in statistics, statisticians would have thought to use something different, but they didn't. Usually, the context makes the meaning clear; in addition, the symbol for a matrix is usually printed in **boldface**, to avoid confusion.[13]

In the VCV matrix, the terms along the main diagonal (s_{11}^2 and s_{22}^2, where the subscripts are the same) are the variances, and those off the diagonal (s_{12}^2 and s_{21}^2, where the subscripts are different) are the covariances. Just as with a correlation matrix, the VCV is symmetrical, so the value of s_{12}^2 is the same as that for s_{21}^2. Needless to say (but we'll say it anyway), there would be more terms in the VCV matrix if we had more dependent variables: for three variables, we would have three variances and three unique covariances; for four variables, four variances and six covariances, and so on.

Testing for equivalence of the VCV matrices means that the variance of X_1 is the same across all groups; the covariance between X_1 and X_2 is the same; and so forth for all of the variances and all of the covariances. The usual test for homogeneity of the VCV matrices is **Box's *M*** statistic. After some statistical hand waving over the matrices, the number is transformed into either an *F*-ratio or a χ^2. If $p > .05$, then it's safe to proceed, because the matrices do not differ significantly from each other. However, like many other tests for homogeneity, *M* is unduly sensitive to differences, especially when the sample size is large. The consequence is that a significant *M* statistic doesn't always mean that you have to either stop with the analysis or transform the data; unfortunately, there is nothing but our "feel" for the data to tell us whether the deviation from homogeneity is worth worrying about. Tabachnick and Fidell (1996) offer the following guidelines:

- If the sample sizes are equal, don't worry about *M*, because the significance tests are robust enough to handle any deviations from homogeneity
- If sample sizes are not equal, and the *p* associated with *M* is greater than .001, it's fairly safe to proceed
- If the cell sizes are unequal, and *p* is less than .001, interpret the results of the MANOVA very cautiously

The consequence of violating the assumption of homogeneity of the VCV matrices if the sample sizes are equal is a slight reduction in the power of the MANOVA. If the sample sizes vary considerably, then a significant *M* can indicate that the Type I error rate may be either inflated or deflated. Because this depends on which matrices are the most different, it is almost impossible to just look at the data and figure out which it will be.

FROM ANOVA TO MANOVA

The logic of MANOVA is very similar to that of ANOVA. With the one-way ANOVA, we partition the total variance (the Sum of Squares Total, or SS_{Total}) into that due to differences *between* the groups ($SS_{Between}$) and the error variance, which is the Sum of Squares *within* the groups (SS_{Within} or SS_{Error}). Then, the *F* test looks at the ratio of the explained variance (that due to the grouping factor) to the error (or unexplained) variance, after adjusting for the number of subjects and groups (the Mean Squares). With more complicated designs, such as factorial or repeated-measures ANOVAs, we "simply" split the SS_{Total} into more sources of variance, such as that due to each factor separately, that due to the interaction between the factors, and that due

[11] *A covariance is similar to a correlation between two variables, except that the original units of measurement are retained, rather than transforming the variables to standard scores first, as is done with correlations. It reflects the variance shared by the two variables.*

[12] *We didn't want to tell you at the time, but now that you've progressed this far, you can finally know: you've been dealing with matrices throughout the book. Any time you saw a table listing subjects down the side and variables across the top, you were looking at a data matrix.*

[13] *Unless, of course, you forget whether it's sums or matrices that are printed in boldface.*

to measurements over time; divide by the appropriate error term; and get more *F*-ratios.

We do exactly the same thing in MANOVA, except that we have more terms to worry about—the relationships between or among the DVs. This is equivalent to expanding the measurement of variance into a variance-covariance matrix when we looked at the assumption of homogeneity of variance. Similarly, we expand the Sum of Squares terms into a corresponding series of **Sum of Squares and Cross-Products** (SSCP) matrices: $SSCP_{Total}$, $SSCP_{Between}$, and $SSCP_{Within}$. That sounds somewhat formidable, but, actually, it is quite straightforward. A small data set, consisting of two variables (*X* and *Y*) and five subjects, is shown in Table 12–2. For each variable, the sum of squares is simply the sum of each value squared.[14] The cross-product is the first value of *X* multiplied by the first value of *Y*; these are then added up to form the sum of cross-products. The SSCP matrix for these number is therefore:

$$SSCP = \begin{bmatrix} 135 & 105 \\ 105 & 165 \end{bmatrix} \qquad (12\text{–}7)$$

where the off-diagonal cells are the same, since X • Y is the same as Y • X.

The *F* test is now just the ratio of the $SSCP_{Between}$ to the $SSCP_{Within}$, after the usual corrections for sample size, number of groups, and the number of DVs. As we'll see, however, we will run into the ubiquitous problem of multivariate statistics—a couple of other ways to look at the ratios. So, stay tuned.

TABLE 12–2

Calculating sums of squares and cross-products

	X	Y	X²	Y²	(X) (Y)
	3	9	9	81	27
	4	7	16	49	28
	5	5	25	25	25
	6	3	36	9	18
	7	1	49	1	7
TOTALS	25	25	135	165	105

FINALLY DOING IT

Sad to say, "it" in this case only means running the test (now that we've gotten the foreplay out of the way). The first question is, what test do we run? So far, we've been referring to the test as MANOVA, and, in fact, that's how we'll continue to refer to the test. If you look at older books, however, you'll see this test is also called Hotelling's T^2. Just as a *t*-test is an ANOVA for two groups (or, in the case of the paired *t*-test, two related variables), T^2 is a MANOVA for two groups or two variables. In the years B.C.,[15] it made sense to have a separate test for the two-group case, since there were some shortcuts that could simplify the calculations, which were all done by hand. Now that computers do all the work for us, the distinction isn't as important, and references to Hotelling's T^2 are increasingly rare, and nonexistent in the menus for some computer programs.

If we (finally) run a MANOVA on the satisfaction data, we would get an output similar to that shown in Table 12–3. As we mentioned, the multivariate equivalent of the test for homogeneity of variance is Box's *M*, which tests for equivalence of the variance-covariance matrices; this is the first thing we see in the output. The *p* level shows that the test is not significant, so we don't have any worries in this regard.

The next set of homogeneity tests belong to **Levene**;[16] they are univariate tests that look at each of the DVs separately. The usual rule of thumb is that if they are not significant, we can use the results from the MANOVA; if they are significant, we should stick with univariate tests. In our case, there are no significant differences in the error variances between groups, so it's safe to use the output labelled **Multivariate Tests.**

To make our job easier, we can skip the first four lines of the table, those dealing with the intercept. They simply tell us that something is going on, and that the data as a whole deviate from zero.[17] Finally, we arrive where we want to be—we have a multivariate test of the difference between the groups based on all of the DVs at once. But, as is all too common with multivariate statistics, we don't have just one test but, rather, four of them! Actually, things aren't quite as bad as they seem, especially when we have only two groups. As you can see, **Hotelling's trace**[18] and **Roy's largest root** have the same value; **Pillai's trace**[19] and **Wilks' lambda**[20] add up to 1.0; and with appropriate transformations, all of the tests end up with the same value of *F*. These relationships don't necessarily hold true when there are three or more groups, but let's start off easy.

Because many multivariate tests use Wilks' lambda (λ), that's where we'll start. As you no doubt remember, the *F*-ratio for the between-groups effect in an ANOVA is simply:

$$F = \frac{MS_{Between\ Groups}}{MS_{Within\ Groups}} \qquad (12\text{–}8)$$

Analogously, λ is:

$$\lambda = \frac{SS_{Within\ Groups}}{SS_{Total}} \qquad (12\text{–}9)$$

Notice that, for some reason no one except Wilks understands, λ is built upside down: the smaller the within-groups sum of squares (that is, the error), the smaller the value of λ. So, unlike almost every other

[14] *Makes sense, doesn't it?*

[15] *That's Before Computers; non-Christians prefer the term B.C.E., for Before Calculating Engines.*

[16] *Although he lets us use them.*

[17] *If this ever does come out as nonsignificant, we should question whether we should be in this research game at all or become neo-post-modern deconstructionists, so that no one will ever know that our hypotheses amount to nothing.*

[18] *Also called the Hotelling-Lawley trace, or* T.

[19] *Which is also called the Pillai-Bartlett trace, or* V.

[20] *A.k.a. Wilks' likelihood ratio, or* W. *(Ever get the feeling that everything is called something else in this game?)*

TABLE 12–3

Output from a MANOVA program

Box's M = 1.696	$F = .533$	df1 = 3	df2 = 259920	$p = .660$

Levene's test

	F	df 1	df 2	Sig
Male	0.932	1	38	.341
Female	1.205	1	38	.279

Multivariate tests

Effect		Value	F	df N	df D	Sig
Intercept	Pillai's trace	0.936	270.24	2	37	.001
	Wilks' lambda	0.064	270.24	2	37	.001
	Hotelling's trace	14.607	270.24	2	37	.001
	Roy's largest root	14.607	270.24	2	37	.001
Group	Pillai's trace	0.159	3.49	2	37	.040
	Wilks' lambda	0.841	3.49	2	37	.040
	Hotelling's trace	0.189	3.49	2	37	.040
	Roy's largest root	0.189	3.49	2	37	.040

Tests of between-subjects effects

		SS hyp	df	Mean square	F	Sig
Intercept	M	43296.4	1	43296.4	475.07	.001
	F	74218.225	1	74218.225	547.614	.001
Group	M	48.4	1	48.4	0.531	.471
	F	50.625	1	50.625	0.374	.545
Error	M	3463.2	38	91.137		
	F	5150.15	38	135.53		

statistical test we'll ever encounter, smaller values are better (more significant) than larger ones. What λ shows is the amount of variance *not* explained by the differences between the groups. In this case, about 6.4% of the variance is unexplained, which means that 93.6% *is* explained.[21] Not coincidently, .936 is also the value of Pillai's trace. In the two-group case, it's simply $(1 - \lambda)$, or the amount of variance that *is* explained.[22] Because most people use either Wilks' λ or Pillai's criterion, we won't bother with the other two. When there are more than two groups, Pillai's trace does *not* equal $(1 - \lambda)$, so you'll have to choose one of them on which to base your decision for significance (assuming they give different results). If you've done just a superb job in designing your study, ending up with equal and large sample sizes in each cell, and managing to keep the variances and covariances equivalent across the groups, then use Wilks' λ. However, if you're as human and fallible as we are, use Pillai's criterion, because it is more robust to violations of these assumptions (Olson, 1976), although slightly less powerful than the other tests (Stevens, 1986). In actual fact, however, the differences among all of the test statistics are minor except when your data are *really* bizarre.

The last part of Table 12–3 gives the univariate tests. Just a little bit of work with a calculator shows that they're exactly the same as the univariate tests in Table 12–1; square the values of t and you'll end up with the Fs. These tests are used in two ways. If the assumptions of homogeneity of variance and covariance aren't met, we would rely on these, rather than on the multivariate tests, to tell us if anything is significant. If we can use the results of the multivariate analysis, these tests tell us which variables are significantly different, in a univariate sense. Hence, they're analogous to post-hoc tests used fol-

[21] *If you didn't suspect it before, this should convince you that these are artificial data; if we did* any *study that accounted for 94% of the variance, our picture would show us holding Nobel prize medals, not a lousy cardboard maple leaf.*

[22] *Pillai's trace is also equivalent to* η^2 *(eta-squared), which is the usual measure of variance accounted for in ANOVAs.*

lowing a significant ANOVA. These results are somewhat unusual, in that the multivariate tests are significant but the univariate ones aren't, indicating that we have to compare the patterns of variables between the groups.

MORE THAN TWO GROUPS

If we had a third group,[23] the output would look very much the same as in Table 12–3. Naturally, the Sums of Squares and Mean Squares would be different, and perhaps the significance levels would also be different (depending on how much or little the participants enjoyed themselves), but the general format would be the same. The major difference in terms of interpretation is the Group effect; there will now be two degrees of freedom, reflecting the fact that there are three groups. If the Group effect is significant, then we have the same problem as with a run-of-the-mill, univariate ANOVA: figuring out which groups are significantly different from the others. Fortunately, the method is the same—post-hoc tests, such as Tukey's LSD or HSD. Most programs will do this for you, as long as you remember to choose this option.

DOING IT AGAIN: REPEATED-MEASURES MANOVA

What would be the MANOVA equivalent of a repeated-measures ANOVA? At first glance, it would seem to be two or more dependent variables, both of which are measured on two or more occasions and, in fact, this is one possible design (called a **doubly repeated MANOVA**). However, even if we have only one DV measured two or three times, it is often better to use a repeated-measures MANOVA than an ANOVA. There are two reasons for this. The first is that, while between-subjects designs are relatively robust with respect to heterogeneity of variance across groups, within-subjects designs are not, resulting in a higher probability of a Type I error than the α level would suggest (LaTour and Miniard, 1983). The second reason is that with ANOVA, there is an assumption of **sphericity**—that for each DV, the variances are equal across time, as are the correlations (that is, the correlation between the measures at Time 1 and Time 2 is the same as between Time 2 and Time 3 and is the same as between Time 1 and Time 3). Most data don't meet the criterion of sphericity, and this is especially true when the time points aren't equally spaced (such as measuring a person's serum rhubarb level immediately on discharge, then one week, two weeks, four weeks, and finally six months later). It's more likely that there's a higher correlation between the measures at Week 1 and Week 2 than between Week 1 and Month 6. Repeated-measures MANOVA, however, treats each time point as if it were a different variable, so the assumption of sphericity isn't required.

To keep things a bit simpler, we'll go back to using two groups, and modify our study a bit by having only one rater,[24] but we'll repeat the experiment

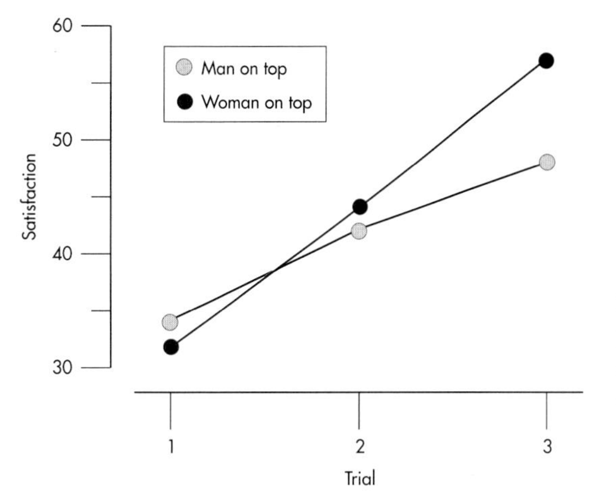

FIGURE 12–3 Repeated measures of satisfaction for the two groups.

three times, so we have three time points. The design, then, is a 2 (Who's on Top) by 3 (Trials) factorial. The results are shown graphically on Figure 12–3, and the computer output appears in Table 12–4.

The first part of the table gives the results for the between-subjects effect, which in this case is Group or Position. This is actually a univariate test, a one-way ANOVA, in which all of the trials are averaged. In this case, the F is not significant, showing that, overall, position doesn't make a difference.[25]

Next, we see the multivariate tests for the within-subjects factors and the interactions between the within-subjects and the between-subjects factors. In this example, we have only one within-subject factor (Trials) and one interaction term (Trials by Group). The same comments apply to this output as previously with regard to the meaning of the four different tests: they all yield the same F test when there are two groups, and stick with Pillai's trace or Wilks' lambda. In this example, there is an overall Trials effect and a significant interaction. If we go back to Figure 12–3, it appears that the ratings increase over time; and when women are on top, the ratings start somewhat lower but increase more than men's.[26]

Following this, we have univariate tests for the within-subject factor, used if Box's M shows us we should be concerned about the assumption of equality of the covariance matrices. The same general guidelines apply that we discussed before: don't worry about this if the sample sizes are equal or $p > .001$. If we do feel that it's more appropriate to use the univariate tests, then we first have to look at the test of sphericity, known as **Mauchly's *W***. For the purposes of significance testing, W is transformed into an approximate χ^2 test. If it is not significant (and you have a sufficient sample size), then we can proceed with abandon, using the appropriate lines in the succeeding tables. If it is significant (as it is here), then the numerator and denominator degrees of freedom for the F tests are "adjusted" by some

[23] *You can use your imagination for this, as long as it doesn't involve more than two people; that would require another dependent variable (and perhaps a larger bed). As a hint, a bed isn't* de rigueur.

[24] *You can determine for yourself whether the scores show it's a man or a woman doing the ratings.*

[25] *And if you believe that, we have some property in Florida we'd like to sell you.*

[26] *As before, we leave it to you to figure out the meaning of this.*

TABLE 12–4

Output from a repeated-measures MANOVA program

Tests of between-subjects effects

Source	Sum of squares	df	Mean square	*F*	Sig
Intercept	219564.08	1	219564.08	637.20	.001
Group	261.08	1	261.08	0.76	.390
Error	13093.85	38	344.58		

Multivariate tests

Effect		Value	*F*	df N	df D	Sig
Trial	Pillai's trace	0.924	224.68	2	37	.001
	Wilks' lambda	0.076	224.68	2	37	.001
	Hotelling's trace	12.145	224.68	2	37	.001
	Roy's largest root	12.145	224.68	2	37	.001
Trial × Group	Pillai's trace	0.575	25.00	2	37	.001
	Wilks' lambda	0.425	25.00	2	37	.001
	Hotelling's trace	1.351	25.00	2	37	.001
	Roy's largest root	1.351	25.00	2	37	.001

Mauchly's test of sphericity

					Epsilon		
Within-subjects effect	*W*	χ^2	df	Sig	Greenhouse-Geisser	Huynh-Feldt	Lower Bound
Trial	0.4	32.33	2	0	0.632	0.661	0.5

Box's M = 4.419	*F* = .673	df1 = 6	df2 = 10462	*p* = .671

Tests of within-subjects effects

Source		SS	df	MS	*F*	Sig
Trial	Sphericity assumed	7571.45	2	3785.73	259.16	.001
	Greenhouse-Geisser	7571.45	1.265	5986.69	259.16	.001
	Huynh-Feldt	7571.45	1.323	5724.38	259.16	.001
	Lower Bound	7571.45	1	7571.45	259.16	.001
Trial × Group	Sphericity assumed	612.35	2	306.18	20.96	.001
	Greenhouse-Geisser	612.35	1.265	484.18	20.96	.001
	Huynh-Feldt	612.35	1.323	462.97	20.96	.001
	Lower Bound	612.35	1	612.35	20.96	.001
Error (trial)	Sphericity assumed	1110.2	76	14.61		
	Greenhouse-Geisser	1110.2	48.06	23.10		
	Huynh-Feldt	1110.2	50.26	22.09		
	Lower Bound	1110.2	38	29.22		

value, which is referred to as ϵ (epsilon). As is so often the case, we have more adjustments than we can use. The most conservative one is the **Lower Bound**, which assumes the most extreme departure from sphericity.[27] The **Huynh-Feldt** adjustment is the least conservative, with the **Greenhouse-Geisser** adjustment falling in between. Most people use the Greenhouse-Geisser value, except when the sample size is small, in which case the Huynh-Feldt value is used.[28]

The caveat we added in the previous paragraph about *W* ("and you have a sufficient sample size") is one that holds true for every statistical test trying to prove a null hypothesis: the result may not be significant if the sample size is small, and there just isn't the power to reject the null of sphericity. The trouble is, nobody knows how much is enough,[29] so if your sample size is on the low side, you'd be safer to use one of the correction factors.

[27] It's called the Lower Bound because it is the lowest value that ε can have: 1/(k − 1), where k is the number of groups. The maximum value of ε is 1.0, indicating homogeneity.

[28] If the Lower Bound adjustment is rarely used, then why is it printed out? Probably because the programmer's brother-in-law devised it.

[29] At least in the area of statistics.

ROBUSTNESS

We mentioned earlier, when discussing tests of homogeneity of the VCV matrices, that MANOVA is relatively robust to violations here, especially if the group sizes are equal. With that same proviso of equal sample size, it is also relatively robust to deviations from multivariate normality. "Relatively" means two things: if the degrees of freedom associated with the univariate error terms are over 20, and if the deviation from normality is due to skewness, you can get away with almost anything. However, if the df_{Error} is less than 20, or if the non-normality is due to outliers, then it would be safer to trust the results of the univariate tests, *trim* the data to get rid of the outliers,[30] or use ranks instead of the raw data (discussed below). It's also a reminder (as if one were necessary) to plot your data before you do anything else to make sure you don't have outliers or any other pathologic conditions.

Is there anything we *can't* get away with if we have large and equal groups? There are two things we can't mess around with—random sampling and independence of the observations. As with most statistical techniques, MANOVA assumes that the data come from a random sample of the underlying population, and that the scores for one person are independent of those for all other people. If you violate either of these, the computer will still blithely crunch the numbers and give you an answer, but the results will bear little relationship to what's really going on.

POWER AND SAMPLE SIZE

As we have seen, the first test that's performed when we do a MANOVA is a test for homogeneity of the VCV matrices. For this test to run, we must have more subjects than variables in every cell (Tabachnick and Fidell, 1996). But, this is an absolute minimum requirement. There are tables for estimating sample size, based on the number of groups, variables, and power (Läuter, 1978), and some of them appear in Table N. These tables first appeared in a very obscure journal from what was once East Germany, so we're not sure if they are legitimate or part of a conspiracy to undermine capitalist society by having the West's researchers waste their time with under- or over-powered studies. In any case, we offer them for your use.[31] Going the other way, Table O (adapted from Stevens, 1980) gives the power for the two-group MANOVA (Hotelling's T^2) for different numbers of variables, sample sizes, and effect sizes.

For the most part, MANOVAs are less powerful than univariate tests. This can result in the anomalous situation that some or all of the univariate tests come out significant, but multivariate tests such as Pillai's trace are not significant.[32]

WHEN THINGS GO WRONG: DEALING WITH OUTLIERS

As we've mentioned, MANOVA can't handle data with outliers very well. If you're reluctant to trim the data and still want to use a multivariate procedure, an alternative exists for the one-way test. First, transform your raw data into ranks; if there are n_1 subjects in Group 1 and n_2 subjects in Group 2, then the ranks will range from 1 to $(n_1 + n_2)$ for each variable. Then, run MANOVA on the rank scores. Finally, multiply the Pillai trace (V) by $(N - 1)$, and check the results in a table of χ^2, with $df = p\ (k - 1)$, where p is the number of variables and k the number of groups (Zwick, 1985).

A CAUTIONARY NOTE

From all of the above, it may sound as if we should use MANOVA every time we have many dependent measures, and the more the better. Actually, reality dictates just the opposite. The assumption of homogeneity of the VCV matrices is much harder to meet than the assumption of homogeneity of variance; so, it's more likely that we're violating something when we do a MANOVA. Second, the results are often harder to interpret, because there are more things going on at the same time. Third, as we've seen, we may have less power with multivariate tests than with univariate ones. The best advice with regard to using MANOVA is offered by Tabachnick and Fidell (1983), who are usually strong advocates of multivariate procedures. They write:

> Because of the increase in complexity and ambiguity of results with MANOVA, one of the best overall recommendations is: Avoid it if you can. Ask yourself *why* you are including correlated DVs in the same analysis. Might there be some way of combining them or deleting some of them so that ANOVA can be performed? (p. 230)[33]

If the answer to their question is "No," then MANOVA is the way to go—but, don't throw everything into the pot. The outcome variables included in any one MANOVA should be related to each other on a *conceptual* level, and it's usually a mistake to have more than six or seven at the most in any one analysis.

WRAPPING UP

Taking Tabachnick and Fidell's advice to heart, we should try to design studies so that MANOVA isn't needed: we should rely on one outcome variable, or try to combine the outcome variables into a global measure. If this isn't possible, then MANOVA is the test to use. Analyzing all of the outcomes at once avoids many of the interpretive and statistical problems that would result from performing a number of separate *t*-tests or ANOVAs. We pay a penalty in terms of reduced power and more complicated results, but these are easier to overcome than those resulting from ignoring the correlations among the dependent variables.

[30] *"Trimming" is how you can get rid of data you don't want and still get published, as opposed to saying that you simply disregarded those values you didn't like.*

[31] *The authors (hereinafter referred to as the Parties of the First Part) do verily state, declare, and declaim, that all warranties, expressed or implied, assumed by users of these Tables (Parties of the Second Part) are hereby and forthwith null and void.*

[32] *Proving yet again (as if further proof were needed) that there ain't no such thing as a free lunch.*

[33] *Somewhat poignantly, they answer their own question in the next edition of their book: "In several years of working with students ... we have been royally unsuccessful at talking our students out of MANOVA, despite its disadvantage" (Tabachnick and Fidell, 1996, p. 376).*

EXERCISES

1. For the following designs, indicate whether a univariate or a multivariate ANOVA should be used.
   a. Scores on a quality of life scale are compared for three groups of patients: those with rheumatoid arthritis, osteoarthritis, and chronic fatigue syndrome.
   b. Each of these patient groups is divided into males and females as another factor.
   c. All of these groups are tested every 2 months for a year.
   d. The same design as 1.a, but now the eight subscales of the quality-of-life scale are analyzed separately.
   e. Same design as 1.b, but using the eight subscales.
   f. Same as 1.c, but with the subscales.

2. If the three groups did not differ with regard to their quality of life, and if the eight subscales were analyzed separately, what is the probability that at least one comparison will be significant at the .05 level by chance?

3. Based on the results of Box's *M*, you should:
   a. proceed with the analysis without any concern.
   b. proceed, but be somewhat concerned.
   c. stop right now.

Box's M = 22.745	F = 3.240	df1 = 6	df2 = 1193	p = .004

4. Based on the results of Levene's test, you should:
   a. use the results of the multivariate tests.
   b. use the results of the univariate tests.

Levene's test

	F	df N	df D	Sig
Variable A	.055	1	46	.815
Variable B	.155	1	46	.695
Variable C	.865	1	46	.357

5. Looking at the output from the multivariate tests:
   a. Is there anything going on?
   b. Does the variable SETTING have an effect?

Multivariate tests

Effect		Value	*F*	df N	df D	Sig
Intercept	Pillai's trace	.913	154.618	3	44	.000
	Wilk's lambda	.064	154.618	3	44	.000
	Hotelling's trace	10.542	154.618	3	44	.000
	Roy's largest root	10.542	154.618	3	44	.000
SETTING	Pillai's trace	.060	.941	3	44	.429
	Wilk's lambda	.940	.941	3	44	.429
	Hotelling's trace	.064	.941	3	44	.429
	Roy's largest root	.064	.941	3	44	.429

How to Get the Computer to Do the Work for You

- From **Analyze**, choose **General Linear Model → GLM—Multivariate**
- Click the variables you want from the list on the left, and click the arrow to move them into the box labeled **Dependent Variables**
- Choose the grouping factor(s) from the list on the left, and click the arrow to move them into the box labeled **Fixed Factor(s)**
- If you have more than two groups, click the **Post Hoc** button and select the ones you want [good choices would be **LSD, Tukey**, and **Tukey's-b**], and then click the **Continue** button
- Click the **Options** button and check the statistics you want displayed. The least you want is **Homogeneity Tests**; if you haven't analyzed the data previously, you will also want **Descriptive Statistics** and perhaps **Estimates of Effect Size**
- If you have a repeated-measures design, use the instructions at the end of Chapter 11

SECTION THE SECOND

C.R.A.P. DETECTORS

II–1. A cardiovascular researcher did yet another randomized clinical trial of a new antihypertensive agent. He randomized patients into three groups: (1) captopril, (2) methyldopa, and (3) placebo. After 6 weeks he measured their blood pressures and classified patients as normotensive (diastolic blood pressure < 90 mm Hg) or hypertensive (diastolic blood pressure > 90 mm Hg). He then analyzed the 3 × 2 table (Drug × Normal/Hypertensive) with the usual chi-square test. Would you?

It's studies like this which make statisticians go bald, from all the hair tearing. There are several problems, and we'll deal with them in stages.

First, and most important, *never* take a ratio variable such as blood pressure and categorize it into groups before analysis. You can do it afterward for ease of interpretation among those folks who see the world in two categories; but *never* categorize when you don't have to. The cost in sample size and power is typically a factor of 10 or so. A one-way ANOVA (Chapter 8) on the diastolic blood pressure (DBP) would be more appropriate.

> C.R.A.P. DETECTOR **II–1**
>
> **Never categorize data that start off as interval or ratio data unless the distributions are absolutely awful.**

Second, he likely measured DBP at the beginning of the study, and unless the inclusion criteria were incredibly tight such that every patient's initial blood pressure was about the same, stable, systematic differences probably exist among patients. So, a repeated-measures ANOVA (Chapter 11) using baseline DBP with drug as a between-subjects factor and time as a within-subjects factor and looking for an interaction would be more powerful still.

> C.R.A.P. DETECTOR **II–2**
>
> **Baseline measures can, and generally should, be incorporated into analysis with repeated-measures ANOVA.**

II–2. Another cardiovascular researcher wanted to investigate the effect of antihypertensive agents on quality of life.[1] He randomized patients to three groups that received captopril, methyldopa, and propranolol, respectively. After 24 weeks, he measured quality of life every way but Sunday with the following scales: (1) general well-being, (2) physical symptoms, (3) sexual dysfunction, (4) work performance, (5) sleep dysfunction, (6) cognitive function, (7) life satisfaction, and (8) social participation. He did *t*-tests comparing captopril to methyldopa to propranolol on all the measures. What would you do?

ANOVA methods are usually misused by not being used at all. A total of 24 *t*-tests are here, and 9 are significant. At the least, he should have done a one-way ANOVA (Chapter 8) to see if there was any difference among the three groups on each variable, then pursued any differences with post-hoc contrasts. Better still would be a MANOVA (Chapter 12).

C.R.A.P. DETECTOR II–3

ANOVA methods are usually abused when they're not used. Whenever you see multiple *t*-tests, suspect that ANOVA would be better.

[1] *This example is based on Croog et al. (1986). They did the analysis exactly right.*

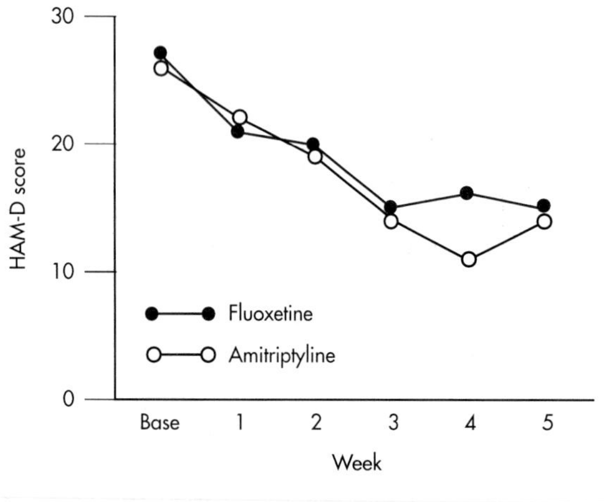

FIGURE II–1

HAM-D data over 5 weeks for 2 drug groups. (Modified from Feighner, JP [1985]. *Journal of Clinical Psychiatry*, **46**, 369–372.)

II–3. Feighner (1985) did a randomized control trial with a small sample of patients. He measured three outcomes: the HAM-D (a depression scale), the Raskin Depression Inventory, and the Covi Anxiety scale, at baseline and at weeks 1, 2, 3, 4, and 5. He reported that "the changes were statistically significant . . . in the fluoxetine group and for several of the efficacy measurements in the amitriptyline group." For the sake of interest, the data for the HAM-D are shown in Figure II–1. He also compared the treatment groups at the end of the study and found no significant difference between the two drugs. Would you analyze it this way?

We sure hope not. This one is so wrong, one wonders how it made it into print. Incidentally, only 16 of 44 patients actually completed the trial anyway, but we'll pretend they were all there. Here goes!

1. He analyzed the data from only week 0 and week 5 and totally ignored the data from weeks 1, 2, 3, 4. They should have used a repeated-measures ANOVA to look at all the data.

C.R.A.P. DETECTOR II–4

When data are taken on repeated occasions, use repeated-measures ANOVA, not a paired *t*-test.

2. He measured changes from baseline separately for the 2 drug groups, then compared the 2 groups at week 5. If the real interest is the new drug (fluoxetine), the separate analysis essentially ignores the control condition. The combined analysis at week 5, by contrast, ignores all the data gathered at baseline and along the way.

   If he had simply used an assessment at time 0 and time 5, the right analysis would be an unpaired *t*-test on the difference scores. Because he had multiple measures, he should have used repeated-measures ANOVA with one grouping factor (fluoxetine/amitriptyline) and one within-subjects factor (Time).

C.R.A.P. DETECTOR II–5

When you have a control group, you cannot analyze the results of the treatment and control groups separately.

SECTION

THE THIRD

REGRESSION AND CORRELATION

The previous section dealt with ANOVA methods, which are suitable when the independent variable is nominal categories and the dependent variable approximates an interval variable. However, there are many problems in which both independent and dependent variables are interval-level measurements. In these circumstances (with 1 independent variable) the appropriate method is called simple regression and is analogous to one-way ANOVA.

CHAPTER THE THIRTEENTH

Simple Regression and Correlation

SETTING THE SCENE

You notice that many of the Yuppie patients in your physiotherapy clinic appear to suffer from a peculiar form of costochondrotendonomalaciomyalagia patella (screwed-up knee), apparently brought on by the peculiar shift patterns of the BMW Series 17. You investigate this new syndrome further by developing an index of Yuppiness, the CHICC score, and attempting to relate it to range-of-motion (ROM) of the knee. But CHICC score and ROM are both continuous variables. You could categorize one or the other into High, Medium, and Low and do an ANOVA, but this would lose information. Are there better ways?

BASIC CONCEPTS OF REGRESSION ANALYSIS

The latest affliction keeping Beverly Hills and Palm Springs physiotherapists employed is a new disease of Yuppies. The accelerator and brake of the BMW Series 17 are placed in such a way that, if you try any fancy downshifting or upshifting, you are at risk of throwing your knee out—a condition that physiotherapists refer to as costochondrotendonomalaciomyalagia patella (Beemer Knee for short). The cause of the disease wasn't always that well known until an observant therapist in Sausalito noticed this new affliction among her better-heeled clients and decided to do a scientific investigation. She examined the relationship between the severity of the disease and some measure of the degree of Yuppiness of her clients. She could have simply considered whether they owned a series 17 BMW, but she decided to also pursue other sources of affluence. Measuring the extent of disease was simple—just get out the old protractor and measure range of motion (ROM).

But what about Yuppiness? After studying the literature on this phenomenon of the 1980s, she decided that Yuppiness could be captured by a **CHICC** score, defined as follows:

CARS—Number of European cars + Number of Off-road vehicles − Number of Hyundai Ponies, Chevettes, or minivans.
HEALTH—Number of memberships in tennis clubs, ski clubs, and fitness clubs.
INCOME—Total income in $10,000 units.
CUISINE—Total consumption of balsamic vinegar (litres) + number of types of mustard in refrigerator.
CLOTHES—Total of all Gucci, Lacoste, and Saint Laurent labels in closets.

CHICC and ROM are very nice variables; both have interval properties (actually, ROM is a true ratio variable). Thus we can go ahead and add or subtract, take means and SDs, and engage in all those arcane games which delight only statisticians. But the issue is, How do we test for a relationship between CHICC and ROM?

Let's begin with a graph. Suppose we enlisted all the suffering Yuppies in Palm Springs.[1] We find 20 of them, all claiming some degree of Beemer Knee, and measure CHICC score and ROM. The data might look like Figure 13–1. At first glance, it certainly seems that some relationship exists between CHICC and ROM—the higher the CHICC, the less

[1] *We would likely have to go outside Palm Springs. The "Y" in Yuppie stands for young, and everybody in Palm Springs is over 80, or looks it because of the desert sun. It's the only place on earth where they memorialize you in asphalt (Fred Waring Drive, Bob Hope Drive, Frank Sinatra Drive) before you are dead.*

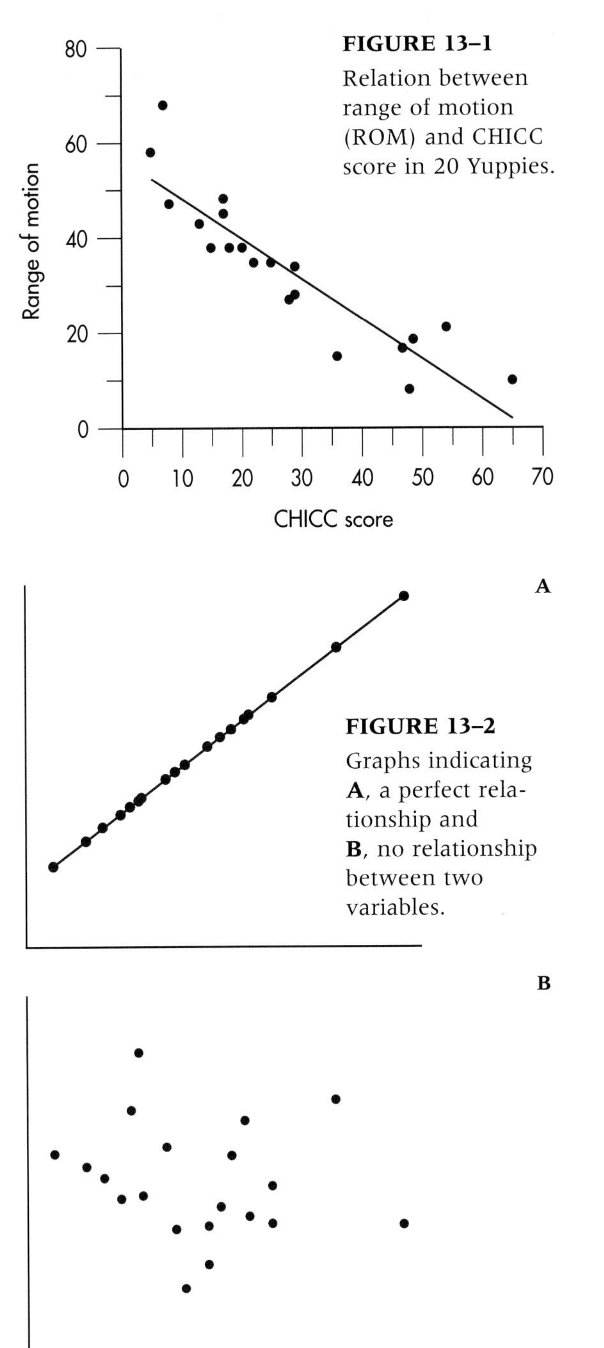

FIGURE 13–1
Relation between range of motion (ROM) and CHICC score in 20 Yuppies.

FIGURE 13–2
Graphs indicating **A**, a perfect relationship and **B**, no relationship between two variables.

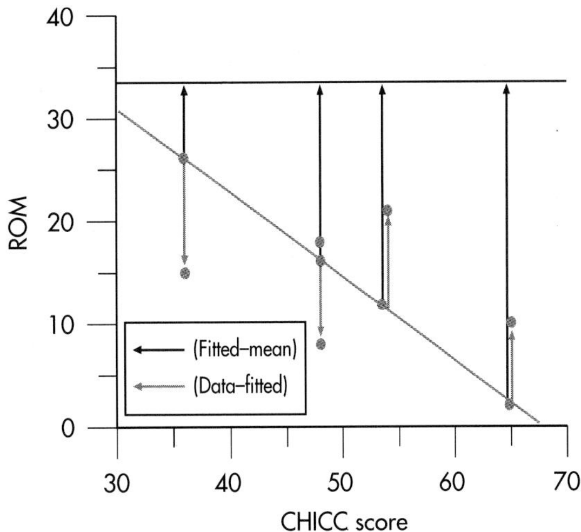

FIGURE 13–3
Relation between ROM and CHICC score (enlarged).

the ROM.[2] It also seems to follow a straight-line relationship—we can apparently capture all the relationship by drawing a straight line through the points.

Before we vault into the calculations, it might be worthwhile to speculate on the reasons why we all agree[3] on the existence of some relationship between the two variables. After all, the statistics, if done right, should concur with some of our intuitions. One way to consider the question is to go to extremes and see what conditions would lead us to the conclusion that (1) no relationship or (2) a perfect relationship exists.

Examine, if you will, Figure 13–2. Seemingly, the relationship depicted in the upper graph is as perfect as it gets. To the untrained eye (yours, not ours), Y is perfectly predictable from X—if you know one, you know the other. By contrast, even a sociologist would likely give up on the lower graph because of the lack of an apparent association between the two variables.[4]

Two reasons why we might infer a relationship between two variables are (1) the line relating the two is not horizontal (i.e., the slope is not zero). In fact, one might be driven to conclude that the stronger the relationship, the more the line differs from the horizontal. Unfortunately, although this captures the spirit of the game, it is not quite accurate. After all, we need only create a new ROM, measured in tenths of degrees rather than degrees, to make the slope go up by a factor of 10. (2) Perhaps less obviously, the closer the points fall to the fitted line, the stronger the relationship. That's why we concluded there was a perfect linear relationship on the top left of Figure 13–2. The straight-line relationship between CHICC and ROM explained all the variability in ROM.

Actually, both observations contain some of the essence of the relationship question. If we contrast the amount of variability captured in the departures of individual points from the fitted line with the amount of variability contained in the fitted function, then this is a relative measure of the strength of association of the two variables.

To elaborate a little more, consider Figure 13–3, where we have chosen to focus on the narrow window of CHICC scores between 30 and 70, which were extracted from the original data of Figure 13–1. Now the *signal* (there's that ugly word again!) is contained in the departure of the fitted data from the grand mean of 33.6. The *noise* is contained in the variability of the individual data about the corresponding fitted points.

[2] *Once again, we have broken with tradition. Most relationships are depicted so that more of one gives more of the other. We could have achieved this, of course, with some algebra, but we decided to make you do the work. Now the bad news—no wall mirror will save you; you have to stand on your head.*

[3] *One good reason is that the teacher says so. When we were students, this never held much appeal; strangely, now it does.*

[4] *Graphs such as the one on top are as rare as hen's teeth in biomedical research; the graph on the bottom is depressingly common.*

TABLE 13-1

CHICC scores and range of motion for 20 Palm Springs Yuppies

Subject	CHICC	ROM	Fitted ROM
1	5	58	52.6
2	8	47	50.0
3	13	43	45.8
4	15	38	44.1
5	22	35	38.2
6	20	38	39.9
7	17	45	42.4
8	29	34	32.3
9	17	48	42.4
10	25	35	35.7
11	28	27	33.2
12	36	15	26.4
13	48	8	16.3
14	65	10	2.0
15	48	18	16.3
16	29	28	32.3
17	18	38	41.6
18	7	68	50.9
19	54	21	11.3
20	47	17	17.2

If this is not starting to look familiar, then you must have slept through Section II.[5] We could apply the same, now almost reflex, approach of calculating a Sum of Squares (Signal) based on deviations of the fitted points from the grand mean and a Sum of Squares (Noise) based on deviations of individual data from the corresponding fitted points.

One mystery remains however, before we launch into the arcane delights of sum-of-squaring everything in sight. In several locations we have referred to the *fitted* line rather glibly, with no indication of how one fits such a line. Well, the moment of reckoning has arrived. For openers, you must search through the dark recesses of your mind to retrieve the formula for a straight line, namely:

$$Y = a + bX$$

where a is the *intercept*, the value of Y when X is equal to zero, and b is the *slope*, or the amount of change in Y for one unit of change in X.

Let's rewrite the equation to incorporate the variables of interest in the example and also change "a" and "b" to "b_0" and "b_1":

$$\widehat{ROM} = b_0 + b_1 \times CHICC$$

That funny-looking thing over *ROM* goes by the technical name of "hat," so we would say, "ROM hat equals . . ." It means that for any given value of CHICC, the equation yields an *estimate* of the ROM score, rather than the original value. So, a ^ over any variable signifies an estimate of it.

Still, the issue remains of how one goes about selecting the value of b_0 and b_1 to best fit the line to the data. The strategy used in this analysis is to adjust the values in such a way as to maximize the variance resulting from the fitted line, or, equivalently, to minimize the variance resulting from deviations from the fitted line. Now although it sounds like we are faced with the monumental task of trying some values, calculating the variances, diddling the values a bit and recalculating the values, and carrying on until an optimal solution comes about, it isn't at all that bad. The right answer can be determined analytically (in other words, as a solvable equation) with calculus.

Unfortunately, no one who has completed the second year of college ever uses calculus, including ourselves, so you will have to accept that the computer knows the way to beauty and wisdom, even if you don't.[6] For reasons that bear no allegiance to Freud, the method is called **regression analysis**[7] and the line of best fit is the **regression line**. A more descriptive and less obscure term is **least-squares analysis** because the goal is to create a line that results in the least square sum between fitted and actual data. Because the term doesn't sound obscure and scientific enough, no one uses it.

The *regression line* is the straight line passing through the data that minimizes the sum of the squared differences between the original data and the fitted points.

Now that that is out of the way, let's go back to the old routine and start to do some sums of squares. The first sum of squares results from the signal, or the difference between the fitted points and the horizontal line through the mean of X and Y.[8] In creating this equation, we call $\hat{Y}$ the fitted point on the line that corresponds to each of the original data; in other words, $\hat{Y}$ is the number that results from plugging the X value of each individual into the regression equation.

$$SS_{regression} = \sum(\hat{Y}_i - \bar{Y})^2 \quad (13\text{–}1)$$

This tells us how far the predicted values differ from the overall mean, analogous to the Sum of Squares (Between) in ANOVA.

The second sum of squares reflects the difference between the original data and the fitted line. This looks like:

$$SS_{residual} = \sum(Y_i - \hat{Y}_i)^2 \quad (13\text{–}2)$$

This is capturing the error between the estimate and the actual data, analogous to the Sum of Squares (Within) in ANOVA. It should be called the error sum of squares, or the within sum of squares, but it isn't—it's called the **Sum of Squares (Residual)**, expressing the variance that remains, or residual variance, after the regression is all over.

To make this just a little less abstract, we have actually listed the data used in making Figure 13–1 in Table 13–1. On the left side is the calculated CHICC score for each of the afflicted, in the middle

[5] *A not uncommon experience among readers of statistics books; however, we had hoped the dirty jokes would reduce the soporific effect of this one.*

[6] *The key to the solution resides in the magical words maximum and minimum. In calculus, to find a maximum or minimum of an equation, you take the derivative and set it equal to zero, then solve the equation, equivalent to setting the slope equal to zero. The quantity we want to maximize is the squared difference between the individual data and the corresponding fitted line. To get the best fit line, this sum is differentiated with respect to both* b_0 *and* b_1*, and the resulting expression is set equal to zero. This results in two equations in two unknowns, so we can solve the equations for the optimal values of the bs.*

[7] *The real reason it's called regression is that the technique is based on a study by Francis Galton called Regression Toward Mediocrity in Hereditary Stature. In today's language, tall people's children "regress" to the mean height of the population. (And one of the authors is delighted Galton discovered that persons of average height are mediocre; he always suspected it).*

is the corresponding ROM, and on the right is the fitted value of the ROM based on the analytic approach described above (i.e., plugging the CHICC score in the equation and estimating ROM).

As an example of the looks of these sums of squares, the Sum of Squares (Regression) has terms such as:

$$SS_{reg} = (52.6 - 33.6)^2 + (50.0 - 33.6)^2 + \ldots + (17.2 - 33.6)^2 = 3892 \quad (13\text{–}3)$$

and the Sum of Squares (Residual) has terms such as:

$$SS_{res} = (58 - 52.6)^2 + (47 - 50.0)^2 + \ldots + (17 - 17.2)^2 = 864 \quad (13\text{–}4)$$

To save you the anguish, we have worked out the Sum of Squares (Regression) and Sum of Squares (Residual) and have (inevitably) created an ANOVA table, or at least the first two columns of it (Table 13–2).

TABLE 13–2 ANOVA table for CHICC against ROM (step 1)

Source	Sum of squares	*df*	Mean square	*F*
Regression	3892			
Residual	864			

However, the remaining terms are a bit problematic. We can't count groups, so it is a little unclear how many df to put on each line. It's time for a little logic. The idea of *df* is the difference between the number of data values and the number of estimated *parameters*. The parameters were means up until now, but the same idea applies. We have two parameters in the problem, the slope and the intercept, so it would seem that the regression line should have 2 *df*. The residual should have $(n - 2 - 1)$ or 17, to give the usual total of $(n - 1)$, losing 1 for the grand mean. Almost, but not quite. One of the parameters is the intercept term, and this is completely equivalent to the grand mean, so only 1 *df* is associated with this regression, and $(n - 1 - 1)$ with the error term.

Now that we have this in hand, we can also go on to the calculation of the Mean Squares and, for that matter, can create an F test. So the table now looks like Table 13–3. The p-value associated with the F test, in a completely analogous manner, tells us whether the regression line is significantly different from the horizontal (i.e., whether a significant relationship exists between the CHICC score and ROM). In this case, yes.

TABLE 13–3 ANOVA table for CHICC against ROM (step 2)

Source	Sum of squares	*df*	Mean square	*F*
Regression	3892	1	3892.0	81.1
Residual	864	18	48.0	

THE COEFFICIENT OF DETERMINATION AND THE CORRELATION COEFFICIENT

All is well, and our Palm Springs physiotherapist now has a glimmer of hope concerning tenure. However, we have been insistent to the point of nagging that statistical significance says nothing about the magnitude of the effect. For some obscure reason, people who do regression analysis are more aware of this issue and spend more time and paper examining the size of effects than does the ANOVA crowd. One explanation may lie in the nature of the studies. Regression, particularly multiple regression, is often applied to existing data bases containing zillions of variables. Under these circumstances, significant associations are a dime a dozen, and their size matters a lot. By contrast, ANOVA is usually applied to experiments in which only a few variables are manipulated, the data were gathered prospectively at high cost, and the researchers are grateful for any significant result, no matter how small.

We have a simple way to determine the magnitude of the effect—simply look at the proportion of the variance explained by the regression. This number is called the **coefficient of determination** and usually written as R^2 for the case of simple regression. The formula is:

$$R^2 = \frac{SS_{reg}}{SS_{reg} + SS_{res}} \quad (13\text{–}5)$$

This expression is just the ratio of the signal (the sum of the squares of Y accounted for by X) to the signal plus noise, or the total sum of squares. Put another way, this is the proportion of variance in Y explained by X. For our example, this equals $3892 \div (3892 + 864)$, or 0.818. (If you examine the formula for eta^2 in Chapter 8, this is completely analogous.)

R^2, the *coefficient of determination*, expresses the proportion of variance in the dependent variable explained by the independent variable.

The square root of this quantity is a term familiar to all, long before you had any statistics course—it's the **correlation coefficient:**

$$r = \pm\sqrt{\frac{SS_{reg}}{SS_{reg} + SS_{res}}} \quad (13\text{–}6)$$

Note the little ± sign. Because the square of any number, positive or negative, is always positive, the converse also holds: the square root of a positive number[9] can be positive or negative. This is of some

[8] *The reason for examining differences from the horizontal line is clear if we project the data onto the Y-axis. The horizontal through the mean of the Ys is just the Grand Mean, in our old ANOVA notation, and we are calculating the analogue of the Sum of Squares (Between). Another way to think of it is—if no relationship between X and Y existed, then the best estimate of Y at each value of X is the mean value of Y. If we plotted this, we'd get a horizontal line, just as we've shown.*

[9] *Note that the coefficient of determination should not be less than zero because it is the ratio of two sums of squares. It can happen, when no relationship exists, to have an estimated sum of squares below zero; this is due simply to rounding error. Usually, it is then set equal to zero.*

value; we call the correlation positive if the slope of the line is positive (more of X gives more of Y) and negative if, such as in the present situation, the slope is negative. So the correlation is $-\sqrt{.818} = -.904$. One other fact, which may be helpful at times (e.g., looking up the significance of the correlation in Table G in the Appendix), is that the *df* of the correlation is the number of pairs -2.

The *correlation coefficient* is a number between -1 and $+1$ whose sign is the same as the slope of the line and whose magnitude is related to the degree of linear association between two variables.

We choose to remain consistent with the idea of expressing the correlation coefficient in terms of sums of squares to show how it relates to the familiar concepts of signal and noise. However, this is not the usual expression encountered in more hidebound stats texts. For completeness, we feel dutybound to enlightened you with the full messy formula:

$$r = \frac{\Sigma(X_i - \bar{X})(Y_i - \bar{Y})}{\sqrt{[\Sigma(X_i - \bar{X})^2]\,[\Sigma(Y_i - \bar{Y})^2]}} \tag{13–7}$$

Because we can write $(X_i - \bar{X})$ as x_i, and $(Y_i - \bar{Y})$ as y_i, this can also be written as:

$$r = \frac{\Sigma xy}{\sqrt{\Sigma x^2\ \Sigma y^2}} \tag{13–8}$$

However messy this looks, some components are recognizable. The denominator is simply made up of two sums of squares, one for X and one for Y. If we divide out by an N here and there, we would have a product of the variance of X and the variance of Y, all square-rooted.

The numerator is a bit different—it is a cross-product of X deviations and Y deviations from their respective means. Some clarification may come from taking two extreme cases. First, imagine that X and Y are really closely related, so that when X is large (or small) Y is large (or small)—they are *highly correlated*. In this case, every time you have a positive deviation of X from its mean, Y also deviates in a positive direction from its mean, so the term is $(+) \times (+) = +$. Conversely, small values of X and Y correspond to negative deviations from the mean, so this term ends up as $(-) \times (-) = +$. So if X and Y are highly correlated (positively), each pair contributes a positive quantity to this sum. Of course, if X and Y are negatively correlated, large values of X are associated with small values of Y, and vice versa. Each term therefore contributes a negative quantity to the sum.

Now imagine there is no relationship between X and Y. Now, each positive deviation of X from its mean would be equally likely to be paired with a positive and a negative deviation of Y. So the sum of the cross-products would likely end up close to zero, as the positive and negative terms cancel each other out. Thus this term expresses the extent that X and Y vary together, so it is called the **covariance** of X and Y, or **cov** (X,Y).

The *covariance* of X and Y is the product of the deviations of X and Y from their respective means.

The correlation coefficient, then, is the covariance of X and Y, standardized by dividing out by the respective SDs. So, yet another way of representing it is:

$$r = \frac{\text{cov}(X,Y)}{\sqrt{\text{var}(X) \times \text{var}(Y)}} \tag{13–9}$$

Incidentally, of historical importance, this version was derived by another one of the field's granddaddies, Karl Pearson. Hence it is often called the **Pearson Correlation Coefficient**. This name is used to distinguish it from several alternative forms, in particular the Intraclass Correlation. Its full name, used only at black-tie affairs, is the Pearson Product Moment Correlation Coefficient. Whatever it's called, it is always abbreviated *r*.

INTERPRETATION OF THE CORRELATION COEFFICIENT

Because the correlation coefficient is so ubiquitous in biomedical research, people have developed some cultural norms about what constitutes a reasonable value for the correlation. One starting point that is often forgotten is the relationship between the correlation coefficient and the proportion of variance we showed above—the square of the correlation coefficient gives the proportion of the variance in Y explained by X. So a correlation of .7, which is viewed favorably by most researchers, explains slightly less than half the variance; and a correlation of .3, which is statistically significant with a sample size of 40 or so (see Table G in the Appendix), accounts for about 10% of the variance.

Having said all that, the cultural norms now re-establish themselves. In some quarters, such as physiology and some epidemiology, any correlation below .7 is suspect. In other domains, a correlation of .15, which is statistically significant with a sample size of about 400, is viewed with delight. To maintain some sanity, we have demonstrated for you how correlations of different sizes actually appear.[10] In Figure 13–4, we have generated data sets corresponding to correlations of .3, .5, .7, and .9. Our calibrated eyeball says that, even at .9, a lot of scatter occurs about the line; conversely, .3 hardly merits any consideration.[11]

[10] *If people took Section I seriously, this demonstration would not be necessary. However they don't, so it is.*

[11] *We 'fess up. You don't have to track our own C-Vs very far back to find instances where we were waxing ecstatic in print about pretty low correlations. Those are the circles we move in.*

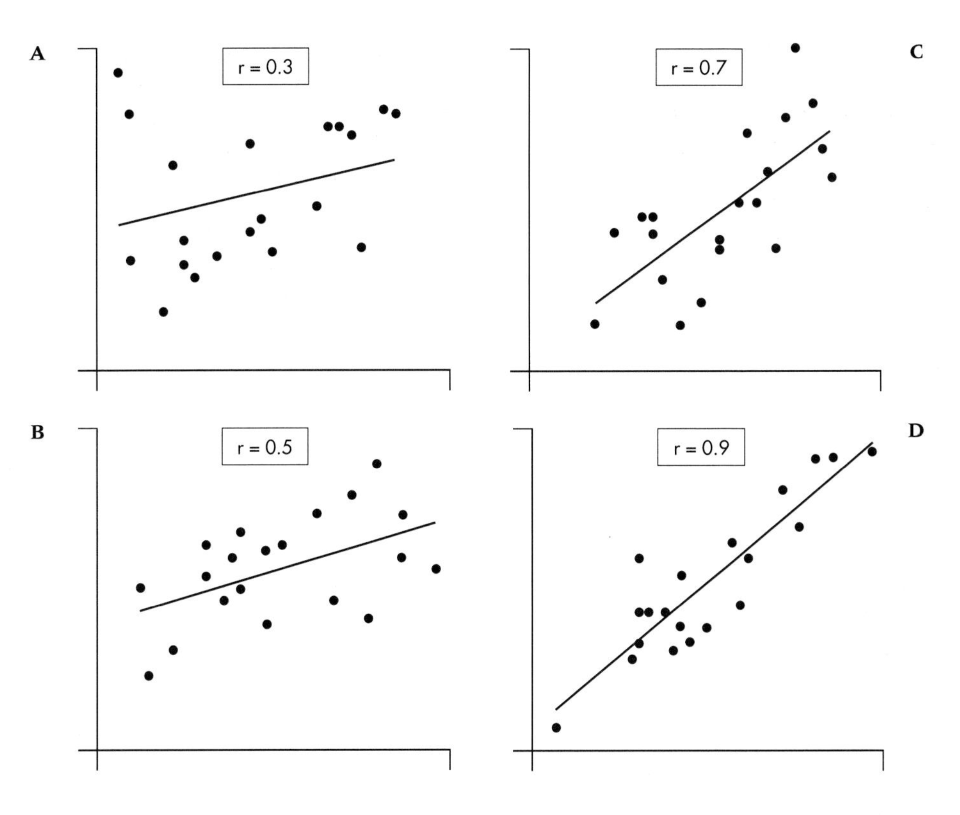

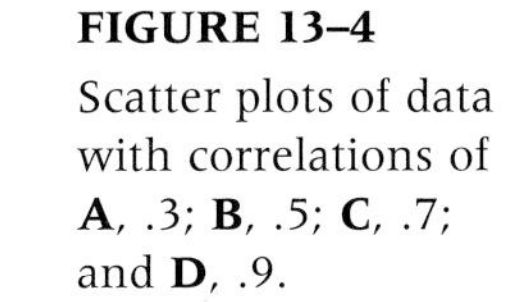

FIGURE 13–4
Scatter plots of data with correlations of **A**, .3; **B**, .5; **C**, .7; and **D**, .9.

Another way to put a meaningful interpretation on the correlation is to recognize that the coefficient is derived from the idea that *X* is partially explaining the variance in *Y*. Variances aren't too easy to think about, but SDs are—they simply represent the unexplained scatter. So a correlation of 0 means that the SD of *Y* about the line is just as big as it was when you started; a correlation of 1 reduces the scatter about the line to zero. What about the values in the middle—how much is the SD of *Y* reduced by a given correlation? We'll tell you in Table 13–4.

TABLE 13–4 Proportional reduction in standard deviation of *Y* for various values of the correlation

Correlation	SD $(Y\|X)$ ÷ SD (Y)*
.1	.995
.3	.95
.5	.87
.7	.71
.9	.43

*The expression (Y|X) is read as "*Y* given *X*," and means the new value of the standard deviation of *Y* after the *X* has been fitted.

What Table 13–4 demonstrates is that a correlation of .5 reduces the scatter in the *Y*s by about only 13%, and even a correlation of .9 still has an SD of the *Y*s that is 43% of the initial value! It should be evident that waxing ecstatic and closing the lab down for celebration because you found a significant correlation of .3 is really going from the sublime to the ridiculous.

There's a third way to get some feel for the magnitude of a correlation. If we take only those people who score above the median on *X*, where do they fall on *Y*? If the correlation between *X* and *Y* is 0, then the score on one variable doesn't affect the score on the other, so we'd expect that half of these people would be above the median on *Y*, and half would be below. Similarly, if the correlation were 1.0, then all of the people who are above the median on *X* would also be above the median on *Y*. Unfortunately, the relationship isn't linear between $r = 0$ and $r = 1$; the actual relationship is shown in Figure 13–5.

One last point about the interpretation of the correlation coefficient. If there is one guiding motto in statistics, it is this: CORRELATION DOES NOT EQUAL CAUSATION! Just because *X* and *Y* are correlated, and just because you can predict *Y* from *X*, and just because this correlation is significant at the .0001 level, *does not* mean that *X* causes *Y*. It is equally plausible that *Y* causes *X* or that both result from some other thing such as *Z*. If you compare country statistics, you find a correlation of about −.9 between the number of telephones per capita and the infant mortality rate. However much fun it is to speculate that the reason is because mums with phones can call their husbands or the taxis and get to the hospital faster, most people would recognize that the underlying cause of both is degree of development.

Simple as the idea is, it continues to amaze us how often it has been ignored, to the later embarrassment (we hope) of the investigators involved. For example, amidst all the hoopla about the dan-

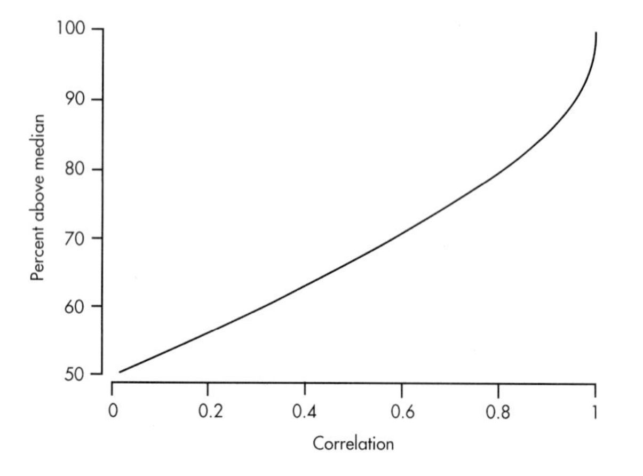

FIGURE 13–5

Percentage of people above the median on one variable who are above the median on the second variable, for various magnitudes of the correlation.

gers of hypercholesterolemia, one researcher found that *hypo*cholesterolemia was associated with a higher incidence of stomach cancer and warned about lowering your triglyceride levels too much. It turned out that he got it bass-ackwards—the cancer can produce hypocholesterolemia. Closer to home, several studies showed an association between an "ear crease" in the earlobe and heart disease. Lovely physiologic explanations have been made of the association—extra vascularization, an excess of androgens, etc. However, in the end, it turned out that both ear creases and coronary artery disease are strongly associated with obesity, and the latter is a known and much more plausible risk factor.

CORRELATIONS: CONFIDENCE INTERVALS AND SIGNIFICANCE TESTS

Because researchers spend so much time calculating correlation coefficients, often without examining the regression analysis on which they are based or even looking at the plots of the data, naturally, and perhaps unfortunately, someone has devised statistical tests of significance for the correlation. The first step is to determine the SE of the correlation coefficient, which happens to take a simple form:

$$SE_r = \sqrt{\frac{1 - r^2}{n - 2}}$$

(13–10)

In our example, this equals:

$$SE_r = \frac{\sqrt{1 - .904^2}}{\sqrt{20 - 2}} = .10$$

(13–11)

Note that this is independent of anything happening in the data—means or SDs. It is related only to the correlation coefficient itself and the sample size. The 95% CI around the sample correlation coefficient, then, is just 1.96 times this quantity.

Finally, we can use this estimate of the SE to devise a statistical test of the significance of the correlation coefficient. The coefficient, divided by its SE, is a t value with $(n - 2)$ df:

$$t = \frac{r}{\frac{\sqrt{1 - r^2}}{\sqrt{n - 2}}} = \frac{r\sqrt{n - 2}}{\sqrt{1 - r^2}}$$

(13–12)

which is equal to $.940 \div .1 = 9.04$.

For completeness, you may recall an earlier situation where we indicated that an F-value with 1 and n df was equal to a squared t value. This case is no exception; the equivalent F-value is 9.04^2, or 81.1, which is what emerged from our original ANOVA (Table 13–3).

SAMPLE SIZE ESTIMATION

Hypothesis Testing

In the previous chapters on ANOVA and the t-test, we determined the sample size required to determine if one mean was different from another. The situation is a little different for a correlation; we rarely test to see if two correlations are different. However, a more common situation, particularly among those of us prone to data-dredging, is to take a data base, correlate everything with everything, and then see what is significant to build a quick post-hoc ad-hoc theory. Of course, these situations are built on existing data bases, so sample size calculations are not an issue—you use what you got. However, the situation does arise when a theory predicts a correlation, and we need to know whether the data support the prediction (i.e., the correlation is significant). When designing such a study, it is reasonable to ask what sample size is necessary to detect a correlation of a particular magnitude.

The sample size calculation proceeds using the basic logic of Chapter 6—as do virtually all sample size calculations involving statistical inference. We construct the normal curve for the null hypothesis, the second normal curve for the alternative hypothesis, and then solve the two z equations for the critical value. However, one small wrinkle makes the sample size formula a little hairier, and it revealed itself in Equation 13–10 earlier. The good news is that the SEs of the distribution are dependent only on the magnitude of the correlation and the sample size, so we don't have to estimate (read "guess") the SE.[12] The bad news is that the dependence of the SE on the correlation itself means that the widths of the curves for the null and alternative hypotheses are different. The net result of some creative algebra is:

$$n = \left(\frac{z_\alpha + z_\beta\sqrt{1 - r^2}}{r}\right)^2 + 2$$

(13–13)

[12] *Those of us who have developed sample size fabrication (oops, estimation) to an art form regard this as a disadvantage because it reduces the researcher's* df.

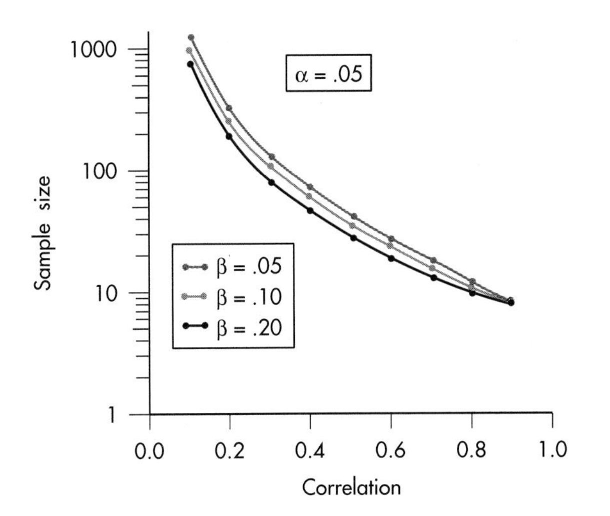

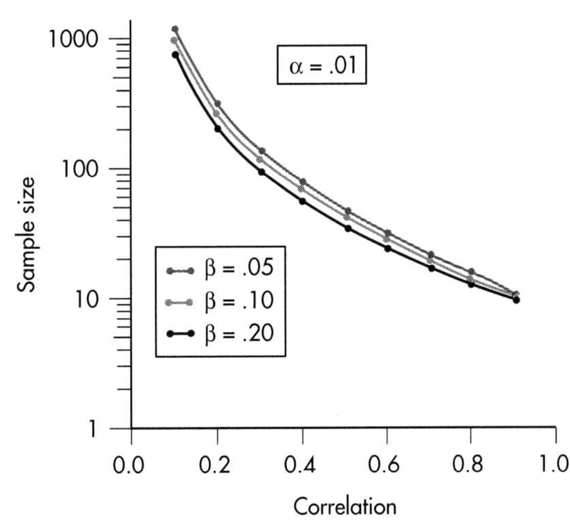

FIGURE 13–6
Sample size for correlation coefficients related to magnitude of the correlation and α and β level.

To avoid any anguish putting numbers into this equation, and also to reinforce the message that such calculations are approximate, we have put it all onto a graph (actually the two graphs in Figure 13–6).

To read these families of curves, first decide what the α level is going to be: .05 or .01; α = .05 puts you on the left graph, and α = .01 puts you on the right. Next, pick a β level from .05 to .20, which orients you on one of the three curves on each graph. The next guess is related to how big a correlation you want to declare as significant, which puts you somewhere on the *X*-axis. Finally, read off the approximate sample size on the *Y*-axis.

SUMMARY

Simple regression is a method devised to assess the relationship between a single interval level *independent* variable and an interval level *dependent* variable. The method involves fitting an optimal straight line based on minimizing the sum of squares of deviations from the line. The adequacy of fit can be expressed by partitioning the total variance into variance resulting from regression and residual variance. The proportion of variance resulting from the independent variable is expressed as a correlation coefficient, and significance tests are derived from these components of variance.

EXERCISES

1. Two studies are conducted to see if a relation exists between mathematics ability and income. Study 1 uses 100 males, ages 21 to 65, drawn from the local telephone book. Study 2 uses the same sample strategy but has a sample size of 800. What will be the difference between the studies in the following quantities?

	1 > 2	1 = 2	1 < 2	1 ? 2
Sum of Squares (Regression)	____	____	____	____
Sum of Squares (Residual)	____	____	____	____
Coefficient of determination	____	____	____	____
Correlation	____	____	____	____
Significance of the correlation	____	____	____	____
Slope	____	____	____	____
Intercept	____	____	____	____

2. Study 3 uses the sample size as 2, but the men are sampled from subscribers to Financial Times. Now what will happen to these estimates?

	3 > 2	3 = 2	3 < 2	3 ? 2
Sum of Squares (Regression)	____	____	____	____
Sum of Squares (Residual)	____	____	____	____
Coefficient of determination	____	____	____	____
Correlation	____	____	____	____
Significance of the correlation	____	____	____	____
Slope	____	____	____	____
Intercept	____	____	____	____

3. An analysis of the relationship between income and SNOB (Streiner-Norman Obnoxious Behavior) scores among 50 randomly selected men found a Pearson correlation coefficient of 0.45. Would the following design changes result in an INCREASE, DECREASE, or NO CHANGE to the correlation coefficient:

a. Increase the sample size to 200
b. Select only upper-echelon executives
c. Select only those whose SNOB scores are more than +2 SD above or less than −2 SD below the mean

How to Get the Computer to Do the Work for You

- From **Analyze**, choose **Correlate** → **Bivariate**
- Click the variables you want from the list on the left, and click the arrow to move them into the box labeled **Variables**
- Click **OK**

If you want to see a scatterplot, then:

- From **Graphs**, choose **Scatter**
- If it isn't already chosen, click on the box marked **Simple**, then select the **Define** button
- Select the variable you want on the *X*-axis from the list on the left, and click the **X Axis** arrow
- Do the same for the *Y*-axis
- Click **OK**

CHAPTER THE FOURTEENTH

Multiple Regression

In this chapter, we generalize the methods of regression analysis to cope with the situation where we have several independent variables that are all interval-level and one dependent variable.

SETTING THE SCENE

Having described (and published about) the new syndrome, Beemer Knee, and shown that it is indeed a result of a decadent lifestyle, you decide to explore further exactly what aspects of "lifestyle" are causing the problem. You want to look at all the variables in the CHICC score, both individually and together. How do you combine all these multiple measures into one regression analysis?

In the last chapter, our intrepid physiotherapist ventured into behavioral medicine by examining the relationship between Beemer Knee and a number of factors associated with the Yuppie lifestyle. It may have occurred to you that she was perhaps oversimplifying things by picking five variables and then ramming them all into a single total score.

You may recall that Yuppiness was codified by a **CHICC** score, defined as follows:

CARS—Number of European cars + Number of Off-road vehicles − Number of Hyundai Ponies, Chevettes, or minivans.

HEALTH—Number of memberships in tennis clubs, ski clubs, and fitness clubs.

INCOME—Total income in $10,000 units.

CUISINE—Total consumption of balsamic vinegar (litres) + number of types of mustard in the fridge.

CLOTHES—Total of all Gucci, Lacoste, and Saint Laurent labels in the closets.

Looking closer at the cause of the affliction, it seems at first blush that some of these variables may play a larger role in the disease than do others. CARS is an obvious prime candidate because the disease was first recognized among Beemer drivers and appeared to be related to fast shifting or heel-and-toe braking. HEALTH might aggravate the condition, despite the label, as a result of all the twisting and knee strain from tennis, squash, or skiing. CLOTHES might hurt too, if subjects are wearing skin-tight slacks too often, constricting the circulation in the lower extremities. But INCOME and CUISINE seem to be a bit of a stretch.

What is the effect of stuffing extra variables in the summary score? First, collecting, coding, and analyzing all these extra data costs more.[1] Second, beyond a certain point, they are likely contributing only noise to the prediction, reducing the sensitivity of the analysis. We want to keep track of the contribution made by individual variables while still allowing for the joint prediction of the dependent variable by all the variables (or, as we shall see, all the variables contributing significantly to the prediction). Although seemingly complex, the method is actually a conceptually straightforward extension of simple regression to the case of multiple variables. Not surprisingly, it goes by the name of **multiple regression**.

[1] *Although some researchers might view this as a good thing.*

> *Multiple regression* involves the linear relationship between one dependent variable and multiple (more than one) independent variables.

CALCULATIONS FOR MULTIPLE REGRESSION

The first step in multiple regression is to create a new regression equation that involves all the independent variables of interest. Ours would look like:

$$\hat{y} = b_0 + b_1\text{CARS} + b_2\text{HEALTH} + b_3\text{INCOME} + b_4\text{CLOTHES} + b_5\text{CUISINE}$$

TABLE 14–1 Analysis of variance of prediction of ROM from five independent variables

Source	Sum of squares	*df*	Mean square	*F*	*p*
Regression	4280	5	856.0	25.17	.001
Residual	476	14	34.0		
Total	4756	19			

This is just longer than what we had before, not fundamentally different. A reasonable next step would be to graph the data. However, no one has yet come up with six-dimensional graph paper, so we'll let that one pass for the moment. Nevertheless, we will presume, at least for now, that were we to graph the relationship between ROM and each of the independent variables individually, an approximately straight line would be the final result.

We can then proceed to stuff the whole lot into the computer and press the "multiple regression" button. Note that "the whole lot" consists of a series of 20 data points on this six-dimensional graph paper, one for each of the 20 Yuppies who were in the study. Each datum is in turn described by six values corresponding to ROM and the five independent variables. The computer now determines, just as before, the value of the *b*s corresponding to the *best fit line*, where "best" is defined as the combination of values that result in the minimum sum of squared deviations between fitted and raw data. The quantity that is being minimized is:[2]

[2] *From here on in, the independent variables are abbreviated to conserve paper; our bit for the "green revolution" and as compensation for the contribution of all our hot air to global warming.*

$$\sum[ROM_i - (b_0 + b_1\text{CA}_i + b_2\text{HE}_i + b_3\text{INC}_i + b_4\text{CL}_i + b_5\text{CU}_i)]^2 \tag{14–1}$$

We will call this sum, as before, the Sum of Squares (Residual) or SS_{res}.

Of course, two other Sums of Squares can be extracted from the data, Sum of Squares (Regression), or SS_{reg}, and Sum of Squares (Total), or SS_{tot}.

$$SS_{reg} = \sum[\overline{ROM} - (b_0 + b_1\text{CA}_i + b_2\text{HE}_i + b_3\text{INC}_i + b_4\text{CL}_i + b_5\text{CU}_i)]^2 \tag{14–2}$$

Although this equation looks a lot like SS_{res}, the fine print, particularly the bar across the top of ROM instead of the *i* below it, makes all the difference. SS_{res} is the difference between individual data, ROM_i and the fitted value; SS_{reg} is the different between the fitted data and the overall grand mean $\overline{\text{ROM}}$. Finally, SS_{tot} is the difference between raw data and the grand mean:

$$\text{SS}_{\text{tot}} = \sum[ROM_i - \overline{ROM}]^2 \tag{14–3}$$

And of course, we can put it all together, just as we did in the simple regression case, making an ANOVA table (Table 14–1).

Several differences are seen between the numbers in this table and the tables resulting from simple regression in the previous chapter. In fact, only the Total Sum of Squares (4756.0) and the *df* (19) are the same. How can such a little difference make such a big difference? Let's take things in turn and find out.

1. Sum of Squares—Although the Total Sum of Squares is the same as before, the Sum of Squares resulting from regression has actually gone up a little, from 3892 to 4280. This is actually understandable. In the simple regression case, we simply added up the five subscores to something we called CHICC. Here we are estimating the contribution of each variable separately so that the overall fit more directly reflects the predictive value of each variable. In turn, this improves the overall fit a little, thereby increasing the Sum of Squares (Regression) and reducing the Sum of Squares (Residual) by the same amount.
2. Degrees of Freedom—Now the *df* resulting from regression has gone from 1 to 5. This is also understandable. We have six estimated parameters, rather than two, as before; one goes into the intercept. The overall *df* is still 19, with 5 *df* corresponding to the coefficients for each variable. Then, because the overall *df* must still equal the number of data –1, the *df* for the residual drops to 14.
3. Mean Squares and *F*-ratio—Finally, the Mean Squares follow from the Sum of Squares and df. Because Sum of Squares (Regression) uses 5 *df*, the corresponding Mean Square has dropped by a factor of nearly four, even though the fit has improved. This then results in a lower *F*-ratio, now with 5 and 14 *df*, but it is still wildly significant.

Significant or not, this is one of many illustrations of the Protestant Work Ethic as applied to stats: "You don't get something for nothing." The cost of introducing the variables separately was to lose *df*, which could reduce the fit to a nonsignificant level while actually improving the fitted Sum of Squares. Introducing additional variables in regression, ANOVA, or anywhere else can actually cost power unless they are individually explaining an important amount of variance.

We can now go the last step and calculate a correlation coefficient:

$$R = \sqrt{\frac{\text{SS}_{\text{reg}}}{\text{SS}_{\text{reg}} + \text{SS}_{\text{err}}}} = \sqrt{\frac{4280}{4280 + 476}} = .95 \tag{14–4}$$

As you might have expected, this has gone up because the Sum of Squares (Regression) is larger. Note the capital *R*; this is called the **Multiple Cor-**

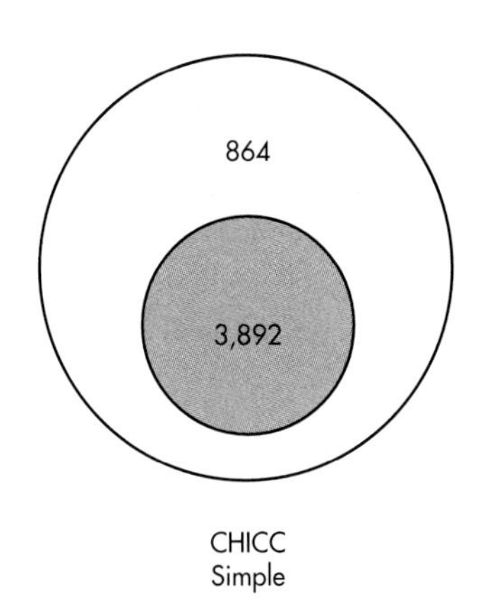

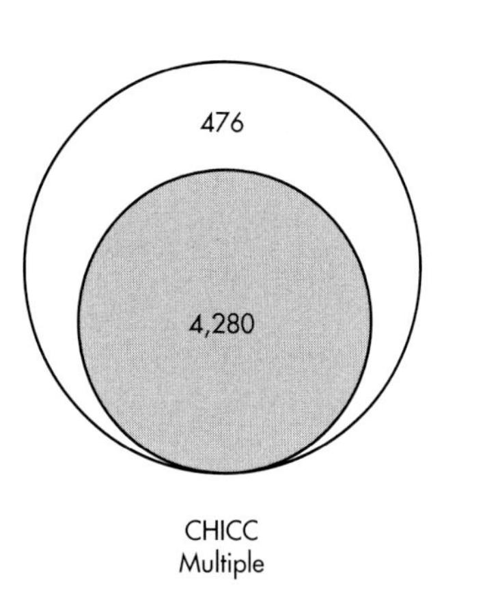

FIGURE 14–1

Proportion of variance (shaded) from simple regression of CHICC score and multiple regression of individual variables. The numbers represent the relevant Sum of Squares.

relation Coefficient to distinguish it from the simple correlation. But the interpretation is the same.

> The *Multiple Correlation Coefficient (R)* is derived from a multiple regression equation, and its square (R^2) indicates the proportion of the variance in the dependent variable explained by all the specified independent variables.

As always, a graphic display aids interpretation of the sums of squares. In Figure 14–1, we have shown the proportion of the Total Sum of Squares resulting from regression and residual. As we already know, a bit of difference exists, with the multiple regression taking a bit more of the pie.

So that's it so far. You might rightly ask what the big deal is because we have not done much else than improve the fit a little by estimating the coefficients singly, but at the significant cost of *df*. However, we have not, as yet, exploited the specific relationships among the variables.

TABLE 14–2

ANOVA of regression of individual variables

Source	Sum of squares	*df*	Mean square	*F*	*r*
Cars:					
Regression	3405.0	1	3405.0	45.4	.85
Residual	1351.0	18	75.0		
Health:					
Regression	1622.0	1	1622.0	9.31	−.58
Residual	3134.0	18	174.1		
Income:					
Regression	643.0	1	643.0	2.81	.36
Residual	4113.0	18	228.5		
Clothes:					
Regression	214.5	1	214.5	.85	.21
Residual	4541.5	18	252.3		
Cuisine:					
Regression	237.0	1	237.0	.95	.22
Residual	4519.0	18	251.0		

RELATIONSHIPS AMONG INDIVIDUAL VARIABLES

Let's backtrack some and take the variables one at a time, doing a simple regression, as discussed previously. If you permit a little poetic license, the individual ANOVAs (with the corresponding correlation coefficients) would look like Table 14–2. These data give us much more information about what is actually occurring than we had before. First, note that the total sum of squares is always 4756, as before. But CARS alone is most of the sum of squares and has the correspondingly highest simple correlation. This is as it should be; it was clinical observations about cars that got us into this mess in the first place. HEALTH comes next, but it has a negative simple correlation; presumably if you get enough exercise, your muscles can withstand the tremendous stresses associated with Beemer Knee. INCOME is next, and still significant; presumably you have to be rich to afford cars and everything else that goes with a yuppie lifestyle. Last, CUISINE and CLOTHES are not significant, so we can drop them from further consideration.

Although we confess to having rigged these data so that we wouldn't have to deal with all the complications down the road, the strategy of looking at simple correlations first and eliminating from consideration insignificant variables is not a bad one. The advantage is that, as we shall see, large numbers of variables demand large samples, so it's helpful to reduce variables early on. The disadvantage is that you can get fooled by simple correlations—in both directions.

At first blush, you might think that we can put these individual Sums of Squares all together to do a multiple regression. Not so, unfortunately. If we did, the Regression Sum of Squares caused by just the three significant variables would be:

$$SS_{reg} = (3405 + 1622 + 643) = 5670$$

Not only is this larger than the Sum of Squares (Regression) we already calculated, it is larger than the total Sum of Squares! How can this be?

Not too difficult, really. We must recognize that the three variables are not making an independent contribution to the prediction. The ability to own a Beemer and belong to exclusive tennis clubs are both related to income—the three variables are

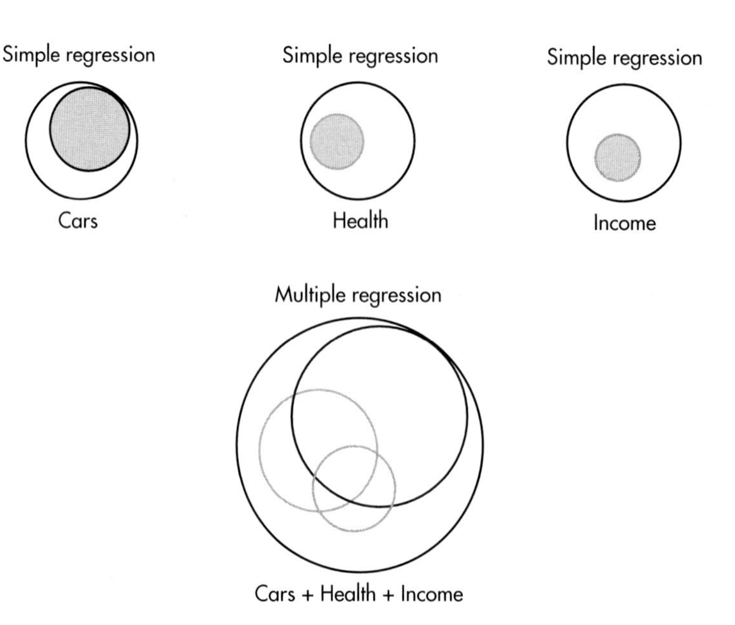

FIGURE 14–2
Proportion of variance from simple regression of Cars, Health, and Income, and multiple regression.

intercorrelated. This may suggest that income causes everything, but then *real* income may lead to a Rolls, and legroom is not an issue in the driver's seat of a Rolls.[3] We are not, in any case, concerned about causation, only correlation, and as we have taken pains to point out already, they are not synonymous. From our present perspective, the implication is that, once one variable is in the equation, adding another variable will account only for some portion of the variance that it would take up on its own.

[3] *This is not from personal experience, although if this book sells well, one day it may be.*

As a possibly clearer example, imagine predicting an infant's weight from three length measurements—head circumference, chest circumference, and length. Because all are measures of baby bigness, chances are that any one is pretty predictive of baby weight. But once any of them is in the regression equation, addition of a second and third measurement is unlikely to improve things that much.

We can also demonstrate this truth graphically. First, consider each variable alone and express the proportion of the variance as a proportion of the total area, as shown in Figure 14–2. Each variable occupies a proportion of the total area roughly proportional to its corresponding Sum of Squares (Regression). Note, however, what happens when we put them all together as in the lower picture. This begins to show quantitatively exactly why the Sum of Squares (Regression) for the combination of the three variables equals something considerably less than the sum of the three individual sums of squares. As you can see, the individual circles overlap considerably, so that if, for example, we introduced CARS into the equation first, incorporating HEALTH and INCOME adds only the small new moon-shaped crescents to the prediction. In Figure 14–3 we have added some numbers to the circles.

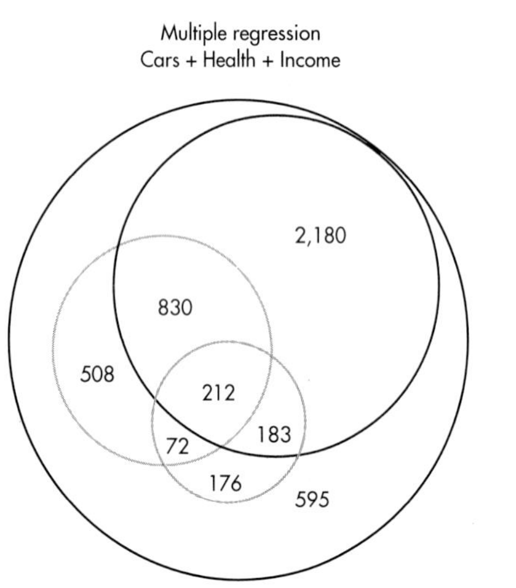

FIGURE 14–3
Proportion of variance from multiple regression with partial sums of squares.

We already know that the Sum of Squares (Regression) for CARS, HEALTH, and INCOME are 3405, 1622, and 643, respectively. But Figure 14–3 shows that the overall Sum of Squares (Regression), as a result of putting in all three variables, is only SS (Total) − SS (Err) = (4756 − 595) = 4161. (Alternatively, this equals the sum of all the individuals areas [2180 + 830 + 212 + 183 + 508 + 72 + 176] = 4161.) For thoroughness, the new multiple correlation, with just these three variables, is:

$$R = \sqrt{\frac{SS_{reg}}{SS_{reg} + SS_{err}}} = \sqrt{\frac{4161}{4161 + 595}} = .935 \tag{14–5}$$

PARTIAL *F* TESTS AND CORRELATIONS

Partial *F* Tests

We can now begin identifying the unique contributions of each variable and devising a test of statistical significance for each coefficient. The test of significance is based on the *unique* contribution of each variable after all other variables are in the equation. So, for the contribution of CARS, the unique variance is 2180; for HEALTH, it's 508; and for INCOME, it is 176. Now we devise a test for the significance; each contribution called, for fairly obvious reasons, a **partial *F* test**. Its formula is as follows:

$$\text{Partial } F = \frac{SS_{reg}(\text{variables in}) - SS_{reg}(\text{variable out})}{MS_{res}(\text{variables in})}$$

(14–6)

where **(variables in)** means all the variables that are in the model, and **(variable out)** means all of the variables in the model except the one being evaluated.

The *partial F test* is the test of the significance of an individual variable's contribution after all other variables are in the equation.

The numerator of this test is fairly obvious: the relevant Sum of Squares, divided by the number of *df*. Because we have only one coefficient, the numerator *df* is always equal to 1.

The denominator of the test is a bit more subtle. What we require is an estimate of the true error variance. As any of the Sums of Squares within the "regression" circles is actually variance that will be accounted for by one or another of the predictor variables, the best guess at the Residual Sum of Squares is the SS (Res) after all variables are in the equation, in this case equal to 595. In turn, the Mean Square is this divided by the residual *df*, now equal to (19 − 3) = 16. So, the denominator for all of the partial *F* tests is 595 ÷ 16 = 37.18, and the tests for each variable are in Table 14–3.

Partial and Semipartial Correlations

Another way to determine the unique contribution of a specific variable, after the contributions of the other variables have been taken in account, is to look at the **partial** and **semipartial** (or **part**) correlations. Let's say we have a dependent variable, *z*, and two predictor variables, *a* and *b*. The **partial correlation** between *a* and *z* is defined as:

$$r_{az.b} = \frac{r_{az} - r_{ab}r_{zb}}{\sqrt{1 - r_{ab}^2}\sqrt{1 - r_{zb}^2}}$$

(14–7)

The cryptic subscript, $r_{az.b}$, means the correlation between *a* and *z*, partialling out the effects of *b*. This statistic tells us the correlation between *a* and *z*, with the contribution of *b* removed from *both* of the other variables. However, for our purposes, we want the effect of *b* removed from *a*, but we don't want it removed from the dependent variable. For this, we need what is called the **semipartial correlation**, which is also known by the alias, the **part correlation**. Its numerator is exactly the same as for the partial correlation, but the denominator doesn't have the SD of the partialled scores for the dependent variable:

$$sr_{az.b} = \frac{r_{az} - r_{ab}r_{zb}}{\sqrt{1 - r_{zb}^2}}$$

(14–8)

Partial and semipartial correlations highlight one of the conceptual problems in multiple regression. On the one hand, the more useful variables we stick into the equation (with some rare exceptions), the better the SS_{reg}, *R*, and R^2, all of which are desirable. On the other hand (and there's always another hand), the more variables in the equation, the smaller the unique contribution of a particular variable (and the smaller its *t*-test). This is also desirable, because the contribution of any one variable is not usually independent of the contributions of others. We could, in fact, end up with the paradoxical situation that none of the predictors makes a significant unique contribution, but the overall *R* is "whoppingly" significant. We'll return to some of these pragmatic issues in a later section.

*b*s AND *β*s

As you may have noticed, we have been dealing with everything up to now by turning them into sums of squares. The advantage of this strategy is that all the sums of squares add and subtract, so we can draw pretty pictures showing what is going on. One disadvantage is that we have lost some information in the process. In particular, we have not actually talked about the *b* coefficients, which is where we began. A second disadvantage is that every other statistics book does things the other way around, and unfortunately this time the issue cannot be resolved by looking in a mirror. However, because we long passed the point where your pocket calculator would bail you out, we had better toe the line a little, so you can make some sense out of the computer printouts.

First, virtually all programs test the significance of each coefficient with a form of *t*-test. Generally, a table is created that lists each coefficient, called *b*, and its standard error, called *SE(b)*,[4] or something similar. It looks like Table 14–4. The ratio of these is then presented in the form of a *t*-test, and an associated level of significance is shown. This is not so mysterious because a *t*-test is a ratio of a value to its SE. Further, the *t*-test is simply the square root of

[4] *For some reason that surpasseth human understanding, some programs, such as Minitab, call this the SD.*

TABLE 14–3 Partial *F* tests for each variable

Variable	Numerator MS	Denominator MS	*F*	*p*
Cars	2180	37.18	58.63	<.0001
Health	508	37.18	13.66	<.0001
Income	176	37.18	4.73	<.05

TABLE 14-4

Coefficients and standard errors

Variable	b	$SE(b)$	β	t	p
Cars	.135	.0176	.354	7.65	<.0001
Health	.106	.0287	.245	3.70	<.0001
Income	.037	.0170	.155	2.17	<.05

the associated partial F-value, which we determined already in Equation 14–6.

The coefficient, b, also has some utility independent of the statistical test. If we go back to the beginning, we can put the prediction equation together by using these estimated coefficients. We might actually use the equation for prediction instead of publication. For our above example, the prediction equation from the CHICC variables could be used as a screening test to estimate the possibility of acquiring Beemer Knee.

The b coefficients can also be interpreted directly as the amount of change in Y resulting from a change of one unit in X. For example, if we did a regression analysis to predict the weight of a baby in kilograms from her height in centimeters and then found that the b coefficient was .025, it would mean that a change in height of 1 cm results in an average change of weight of .025 kg, or 25 g. Scaling this up a bit, a change of 50 cm results in an increase in weight of 1.25 kg.

Next in the printout comes a column labeled β. "Beta?" you ask. "Since when did we go from samples to populations?" Drat—an exception to the rule. This time, the magnitude of beta bears no resemblance to the corresponding b value, so it is clearly *not* something to do with samples and populations. Actually, a simple relationship is found between b and β, which looks like this:

$$\beta = b \times \frac{\sigma_x}{\sigma_y} \qquad \textbf{(14–9)}$$

In words, β is standardized by the ratio of the SDs of x and y. As a result, it is called a **standardized regression coefficient.** The idea is this: although the b coefficients are useful for constructing the regression equation, they are devilishly difficult to interpret relative to each. Going back to our babies, if weight is measured in grams and height in meters, the b coefficient is 10,000 times larger than if weight is measured in kilograms and height in centimeters, even though everything else stayed the same. So by converting all the variables to standard scores (which is what Equation 14–9 does), we can now directly compare the magnitude of the different βs to get some sense of which variables are contributing more or less to the regression equation.

STEPWISE REGRESSION

One additional wrinkle on multiple regression made possible by cheap computation is called **stepwise regression**. The idea is perfectly sensible—you enter the variables one at a time to see how much you are gaining with each variable. It has an obvious role to play if some or all of the variables are expensive or difficult to get. Thus economy is favored by reducing the number of variables to the point that little additional prediction is gained by bringing in additional variables. Unfortunately, like all good things, it can be easily abused. We'll get to that later.

Hierarchical Stepwise Regression

To elaborate, let's return to the CHICC example. We have already discovered that Cuisine and Clothes are not significantly related to ROM, either in combination with the other variables or alone. This latter criterion (significant simple correlation) is a useful starting point for stepwise regression because the more variables the computer has to choose from, the more possibility of chewing up *df* and creating unreproducible results.

Physiotherapy research is notoriously underfunded, so our physiotherapist has good reason to see if she can reduce the cost of data acquisition. She reasons as follows:

1. Information on the make of cars owned by a patient can likely be obtained from the Department of Motor Vehicles without much hassle about consent and ethics.[5]
2. She might be able to get income data from the Internal Revenue department, but she might have to fake being something legitimate, such as a credit card agency or a charity. This could get messy.
3. Data about health, the way she defined it, would be really hard to get without questionnaires or phone surveys.

So if she had her druthers, she would introduce the variables into the equation one at a time, starting with CARS, then INCOME, then HEALTH. This perfectly reasonable strategy of deciding on logical or logistical grounds *a priori* about the order of entry is called **hierarchical stepwise regression**. Because it requires some thought on the part of the researcher, it is rarely used.

Hierarchical stepwise regression introduces variables, either singly or in clusters, in an order assigned in advance by the researcher.

What we want to discover in pursuing this course is whether the introduction of an additional variable in the equation is (1) statistically significant, and (2) clinically important. Statistical significance inevitably comes down to some F test expressing the ratio of the additional variance explained by the new variable to the residual error variance. Clinical importance can be captured in the new multiple correlation coefficient, R^2, or, more precisely, the change in R^2 that results from introducing the new variable. This indicates how much additional variance was accounted for by the addition of the new variable.

[5] *There is no ethical behavior on the road.*

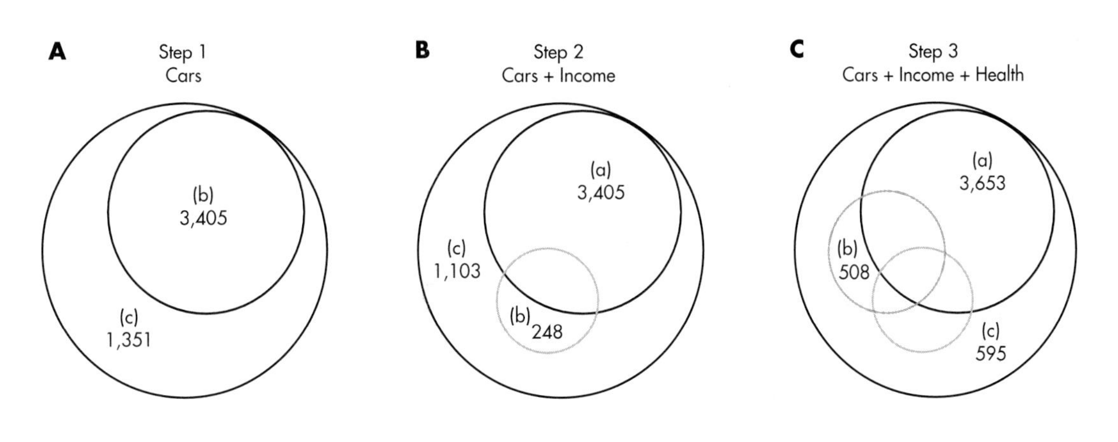

FIGURE 14–4
Proportion of variance from stepwise regression of Cars, Income, and Health. The sums of squares correspond to **a**, regression from the previous step, **b**, additional Sum of Squares from present step, and **c**, residual Sum of Squares.

All this stuff can be easily extracted from Figure 14–3. We have rearranged things slightly in Figure 14–4. Now we can see what happens every step of the way. In Step 1, we have one independent variable, CARS, and the results are exactly the same as the simple regression of CARS on ROM. The Sum of Squares (Regression) is 3405, with 1 *df*, and the Sum of Squares (Error) is 1351, with 18 *df*. The multiple R^2 is just the proportion of the Sum of Squares explained, 3405 ÷ 4756 = .716 as before, and the F test of significance is the Mean Square (Regression) ÷ Mean Square (Residual) = (3405 ÷ 1) ÷ (1351 ÷ 18) = 45.36.

Now we add INCOME. Because all the independent variables are interrelated, this adds only 248 to the Sum of Squares (Regression), for a total of 3653, with 2 *df*, leaving a Sum of Squares (Error) of 1103 with 17 *df*. Now the multiple R^2 is 3653 ÷ 4756 = .768, and the F test for the addition of this variable is (248 ÷ 1) ÷ (1103 ÷ 17) = 3.82. This is conventionally called the ***F*-to-enter** because it is associated with entering the variable in the equation. The alternative is the ***F*-to-remove**, which occurs in stepwise regression (discussed later). This score results from the computer's decision that, at the next step, the best thing it can do is remove a variable that was previously entered.

A subtle but important difference exists between this partial sum of squares and the partial sum of squares for INCOME, which we encountered previously. In ordinary multiple regression, the partials are always with *all* the other independent variables in the equation, so it equalled only 176. Here, it is the partial with just the *preceding* variables in previous steps in the equation; consequently, this partial sum of squares is a little larger.

Finally, we throw in HEALTH. This adds 508 to the Sum of Squares (Regression) to bring it to 4161, with 3 *df*. The Sum of Squares (Residual) is further reduced to 595, with 16 *df*. The multiple R^2 is now 4161 ÷ 4756 = .875, and the F test is (508 ÷ 1) ÷ (595 ÷ 16) = 13.66.

All of this is summarized in Table 14–5, where we have also calculated the change in R^2 resulting from adding each variable. Addition of INCOME accounted for only another 5% of the variance. Although this is not too bad (most researchers would likely be interested in variables that account for 2% to 3% of the variance), this time around it is not significant. How can this be? Recognize that both the numerator and denominator of the F test are contingent on what has gone before. The numerator carries variance in addition to that already explained by previous variables, and the denominator carries variance that is *not* explained by all the variables in the equation to this time. When we examined the partial F tests in Table 14–3, all three variables were in the equation. The additional Sum of Squares resulting from INCOME was 176 instead of 248 because both CARS and HEALTH were in the equation. However, the denominator (the Mean Square [Residual]) was reduced further from 1103 ÷ 17 = 64.9 to 37.18. The net effect was that the partial F test for introduction of INCOME was just significant in the previous analysis.

This illustrates both the strength and limitations of the stepwise technique. By considering the combination of variables, it is possible to examine the independent effect of each variable and use the method to eliminate variables that are adding little to the overall prediction. Unfortunately, therein also lies a weakness because the contribution of each variable can be considered only in combination with the particular set of other variables in the analysis.

As we shall see, these problems are amplified when we turn to the next method.

Ordinary Stepwise Regression

In this method, the researcher begins by turning over all responsibility for the logical relationship

TABLE 14–5
Stepwise regression analysis of ROM against Cars, Income, and Health

Variable	Multiple R^2	Change in R^2	F	p
Cars	.716	—	45.36	<.0001
Income	.768	.052	3.82	ns
Health	.875	.107	13.66	<.0001

ns = not significant.

among variables to the machine. Variables are selected by the machine in the order of their power to explain additional variance. The mathematics are the same as used in hierarchical regression described above, except that, at the end of every step, the computer calculates the best next step for *all* the variables that are not yet in the equation, then selects the next variable to enter based on a statistical criterion. The usual criterion is simply the largest value of the *F*-to-enter, determined as we did before.

The process carries on its merry way, entering additional variables with gay abandon, until ultimately the beast runs out of steam. "Out of steam" is also based on a statistical criterion, usually an *F*-to-enter that does not achieve significance.

Of course, we have yet one more wrinkle. It can happen (with all the interactions and interrelationships among the variables) that, once a whole bunch of variables are in the model, the best way to gain ground is to throw out a variable that went into the equation at an earlier stage but has now become redundant. The computer approaches this by determining not only what would happen if any of the variables *not* in the equation were *entered*, but also what would happen if any of the variables currently *in* the equation were *removed*. The calculation just creates another *F*-ratio, and if this *F*-to-remove is the largest, the next step in the process may well be to throw something out.

So what's the matter with letting the machinery do the work for you? Is it just a matter of the Protestant Work Ethic? Unfortunately not, as several authors have pointed out (e.g., Leigh, 1988; Scailfa and Games, 1987; Wilkinson, 1979). At the center of the problem is the stuff of statistics: random variation. Imagine we have 20 variables that we are anxious to stuff into a regression equation, but in fact *none* of the 20 are actually associated with the dependent variable (in the population). What is the chance of observing at least 1 significant *F*-to-enter at the .05 level? As we have done before in several other contexts, it is:

$$1 - (1 - .05)^{20} = 1 - 0.358 = .642$$

If we had 40 variables, the probability would be .87 that we would find something significant somewhere. So when we begin with a large number of variables and ask the computer to seek out the most significant predictor variables, inevitably, buried somewhere among all the "significant" variables are some that are present only because of a Type I error. In short, stepwise regression procedures to determine which of a large number of variables are significant predictors are useful primarily to determine which of the variables are *not* significant.

Stepwise regression procedures, using a statistical criterion for entry of variables, should therefore be regarded primarily as an exploratory strategy to investigate possible relationships to be verified on a second set of data. Naturally, very few researchers do it this way.

INTERACTIONS

One simple addition to the armamentarium of the regressive (oops, regression) analyst is the incorporation of *interaction* terms in the regression equation. We have already described the glories of systematic use of interactions in Chapter 9, and the logic rubs off here as well.

As an example, there are several decades of research into the relationship between life stress and health. A predominant view is that the effect of stress is related to the accrual of several stressful events, such as divorce, a child leaving home,[6] or a mortgage (Holmes, 1978). In turn, the model postulates that social supports can buffer or protect the individual from the vagaries of stress (Williams, Ware, and Donald, 1981). Imagine a study where we measured the number of stressful events and also the number of social relationships available, and now want to examine the relationship to doctor visits.[7] The theory is really saying that, in the presence of more stressful events, more social supports will reduce the number of visits; in the presence of less stress, social supports are unrelated to visits. In short, an *interaction* exists between stress, social supports, and visits.

How do we incorporate this interaction in the model? Nothing could be simpler—we create a new variable by multiplying the stress and support variables. So, the equation is:

$$\text{Visits} = b_0 + b_1\text{STRESS} + b_2\text{SOCSUP} + b_3(\text{STRESS} \times \text{SOCSUP})$$

Finally, we would likely test the theory using hierarchical regression, where we would do one analysis with only the main effects and then a second analysis with the interaction term also, to see whether the interaction added significant prediction.

DUMMY CODING

DO NOT SKIP THIS SECTION JUST BECAUSE YOU'RE SMART! We are not casting aspersions on the intelligence of people who code data; dummy coding refers to the way we deal with predictor variables that are measured on a nominal or ordinal scale. Multiple regression makes the assumption that the dependent variable is normally distributed, but it doesn't make any such assumption about the predictors. However, we'd run into problems if we wanted to add a variable that captures that other aspect of "Yuppiness"—type of dwelling. If we coded Own Home = 1, Condo with Doorman = 2, and Other = 3, we would have a nominal variable. That is, we can change the coding scheme and not lose or gain any information. This means that treating DWELLING as if it were a continuous variable would lead to bizarre and misleading results; the b and β wouldn't tell us how much Beemer Knee would increase as we moved from one type of dwelling to another, because we'd get completely different numbers with a different coding scheme.

[6] *In contrast to most parents, psychologists view this event as stressful.*

[7] *Actually, we don't have to imagine one, we did it (McFarlane et al., 1983).*

The solution is breaking the variable down into a number of *dummy variables*. The rule is that if the original variable has k categories, we will make k – 1 dummies. In this case, because there are three categories, we will create two dummy variables. So, who gets left out in the cold? Actually, nobody. One of the categories is selected as the reference, against which the other categories are compared. Mathematically, it really doesn't make much difference which one is chosen; usually it's selected on logical or theoretical grounds. For example, if we had coded income into three levels (for Beemer owners, that would be 1 = $150,000 *to* $174,999; 2 = $175,000 to $199,999; and 3 = $200,000 and up[8]), then it would make sense to have the paupers (Level 1) as the reference, and see to what degree greater income leads to more Beemer Knee. If there is no reason to choose one category over another (as is the case with DWELLING), it's good practice to select the category with the largest number of subjects, because it will have the lowest standard error. Following Hardy (1993), the three guidelines for selecting the reference level are:

- If you have an ordinal variable, choose either the upper or lower category.
- Use a well-defined category; don't use a residual one such as "Other."
- If you still have a choice, opt for the one with the largest sample size.

If we chose OWN HOME as the reference category, then the two dummy variables would be CONDO (with Doorman, naturally) and OTHER. Then, the coding scheme would look like Table 14–6; if a person lives in her own home, she would be coded 0 for both the CONDO and the OTHER dummy variables. A condo dwelling person would be coded 1 for CONDO and 0 for OTHER; and someone who lives with his mother would be coded 0 for CONDO and 1 for OTHER.[9]

Now our regression equation is:

$$\hat{y} = b_0 + b_1 CARS + b_2 HEALTH + b_3 INCOME + b_4 CLOTHES + b_5 CUISINE + b_6 CONDO + b_7 OTHER$$

where b_6 tells us the increase or decrease in Beemer Knee for people who live in condos as compared with those who own a home, and b_7 does the same for the OTHER category as compared to OWN HOME.

The computer output will tell us whether b_6 and b_7 are significant, that is, different from the reference category. One question remains: How do we know if CONDO and OTHER are significantly different from each other? There are two ways to find out. The first method is rerunning the analysis, choosing a different category to be the reference. The second method is to get out our calculators and use some of the output from the computer program, namely, the variances and covariances of the regression coefficients. With these in hand, we can calculate a t-test:

$$t = \frac{b_6 - b_7}{\sqrt{\sigma^2_{b_6} + \sigma^2_{b_7} - 2\,cov\,(b_6 b_7)}} \qquad \textbf{(14–10)}$$

where $cov\,(b_6\ b_7)$ means the covariance between the two *bs*.

WHERE THINGS CAN GO WRONG

By now, you're experienced enough in statistics to know that the issue is never, "Can things go wrong?" but, rather, "Where have things gone wrong?" However, before you can treat a problem, doctor, you have to know what's wrong. So, we're going to tell you what sorts of aches and pains a regression equation may come down with, and how to run all sorts of diagnostic tests to find out what the problem is.[10] There are two main types of problems: those involving the cases (the specific disorders are *discrepancy*, *leverage*, and *influence*[11]) and those involving the variables (mainly *multicollinearity*).

Discrepancy

The discrepancy, or *distance*, of a case is best thought of as the magnitude of its residual, that is, how much its predicted value of Y (Y-hat, or $\hat{Y}$) differs from the observed value. When the residual scores are standardized (as they usually are), then any value over 3.0 shows a subject whose residual is larger than that of 99% of the cases. The data for these *outliers* should be closely examined to see if there might have been a mistake made when recording or entering the data. If there isn't an error you can spot, then you have a tough decision to make—to keep the case, or to trim it (which is a fancy way of saying "toss it out"). We'll postpone this decision until we deal with "influence."

Leverage

Leverage has nothing to do with junk bonds or Wall Street shenanigans. It refers to how atypical the pattern of predictor scores is for a given case. For example, it may not be unusual for a person to have a large number of clothes with designer labels; nor would it be unusual for a person to belong to no health clubs; but, in our sample, it would be quite unusual for a person to have a closet full of such clothes *and* not belong to health clubs. So, leverage relates to combinations of the predictor variables that are atypical for the group as a whole. Note that

TABLE 14–6

Dummy coding for dwelling variable

Dwelling	Dummy variable: Condo	Dummy variable: Other
Own home	0	0
Condo	1	0
Other	0	1

[8] *Minus the yearly cost of maintaining their car, so we subtract $63,000 from each.*

[9] *There are other possible coding schemes (see Hardy, 1993), but this is the easiest.*

[10] *The cure is always the same—take two t-tests and call me in the morning.*

[11] *Not to be confused with the Washington law firm of the same name.*

the dependent variable is not considered, only the independent variables.

A case that has a high leverage score has the potential to affect the regression line, but it doesn't have to; we'll see in a moment under what circumstances it will affect the line. Leverage for case *i* is often abbreviated as h_i, and it can range in value from 0 to $(N - 1)/N$, with a mean of p/N (where *p* is the number of predictors). Ideally, all of the cases should have similar values of *h*, which are close to p/N. Cases where *h* is greater than $2p/N$ should be looked on with suspicion.

Influence

Now let's put things together:

$$\text{Influence} = \text{Distance} \times \text{Leverage}$$

This means that distance and leverage individually may not influence the coefficients of the regression equation, but the two together may do so. Cases that have a large leverage score but are close to the regression line, and those that are far from the regression line but have small leverage scores, won't have much influence; whereas cases that are high on both distance and leverage may exert a lot of influence. If you look at Figure 14–5, you'll see what we mean. Case A is relatively far from the line (i.e., it's discrepancy or distance is high), but the pattern of predictors (the leverage) is similar to that of the other cases. Case B has a high leverage score, but the discrepancy isn't too great. Case C has the largest influence score, because it is relatively high on both indices. In fact, Case A doesn't change the slope or intercept of the regression line at all, because it lies right on it (although it is quite distant from the other points). Case B, primarily because it is near the mean of *X*, has no effect on the slope, and lowers the value of the intercept just a bit. As you can see from the dashed regression line, Case C has an effect on (or "influences") both the slope and the intercept.

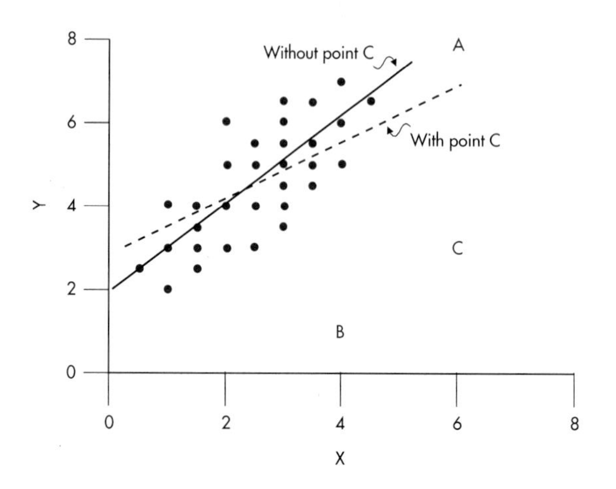

FIGURE 14–5
A depiction of cases with large values for discrepancy (Case A), leverage (Case B), and influence (Case C).

Influence is usually indicated in computer outputs as "Cook's distance," or CD. CD measures how much the residuals of all of the other cases would change when a given case is deleted. A value over 1.00 shows that the subject's scores are having an undue influence and it probably should be dropped.

Multicollinearity

Multiple regression is able to handle situations in which the predictors are correlated with each other to some degree. This is a good thing because, in most situations, it's unusual for the "independent variables" to be completely independent from one another. Problems arise, however, when we cross the threshold from the undefined state of "to some degree" to the equally undefined state of "a lot." We can often spot highly correlated variables just by looking at the correlation matrix. Multicollinearity refers to a more complex situation, in which the *multiple correlation* between one variable and a set of others is in the range of 0.90 or higher. This is much harder to spot, because the zero-order correlations among the variables may be relatively low, but the multiple correlation—which is what we're concerned about—could be high. For example, if we measured a person's height, weight, and body mass index (BMI), we'd find that they were all correlated with each other, but at an acceptable level. However, the multiple correlation of height and weight on the BMI would be well over 0.90.

Most computer programs check for this by calculating the squared multiple correlation (SMC, or R^2) between each variable and all of the others. Some programs try to confuse us by converting the SMC into an index called *tolerance*,[12] which is defined as $(1 - R^2)$. The better programs test the SMC or tolerance of each variable, and kick out any that exceed some criterion (usually SMC ≥ 0.99, or tolerance ≤ 0.01). We would follow Tabachnick and Fidell's (1996) advice to override this default and use a more stringent criterion (SMC ≥ 0.90 or tolerance ≤ 0.10).

[12]No, we don't know either why this term was chosen. Perhaps it refers to the tolerance of statisticians to use terms that are seemingly meaningless.

Yet another variant of the SMC that you'll run across is the *Variance Inflation Factor* (VIF), which is the reciprocal of tolerance. For variable *i*:

$$VIF_i = \frac{1}{(1 - R_i^2)} \qquad \textbf{(14–11)}$$

The main effect of multicollinearity is to inflate the values of the standard errors of the βs, which, in turn, drives them away from statistical significance. VIF is an index of how much the variance of the coefficients are inflated because of multicollinearity. If we stick with our criterion that any R^2 over 0.90 indicates trouble, then any VIF over 10 means the same thing. Another effect of multicollinearity is the seemingly bizarre situation that some of the β weights have values greater than 1.0, which, in fact, can be used as a diagnostic test for the presence of multicollinearity.

THE PRAGMATICS OF MULTIPLE REGRESSION

One real problem with multiple regression is that, as computers sprouted in every office, so did data bases, so now every damn fool with a lab coat has access to data bases galore. All successive admissions to the pediatric gerontology unit (both of them) are there—*in a data base*. Score assigned to the personal interview for every applicant to the nursing school for the past 20 years, the 5% who came here and the 95% who went elsewhere or vanished altogether, are there—*in a data base*. All the questionnaires, filled out by all the students, on all the courses, are there—*in a data base*. All the laboratory requisitions and routine tests ordered on the last 280,000 admissions to the hospital are there—*in a data base*.

A first level of response by any reasonable researcher to all this wealth of data and paucity of information should be, "Who cares?" But then, pressures to publish or perish being what they are, it seems that few can resist the opportunity to analyze them, usually without any previous good reason (i.e., hypothesis). Multiple regression is a natural for such nefarious tasks—all you need do is select a likely looking dependent variable (e.g., days to discharge from hospital, average class performance, undergraduate GPA—almost anything that seems a bit important), then press the button on the old "mult reg" machine, and stand back and watch the *F*-ratios fly. The last step is to examine all the significant coefficients (usually about 1 in 20), wax ecstatic about the theoretical reasons why a relationship might be so, and then inevitably recommend further research.

Given the potential for abuse, some checks and balances must exist to aid the unsuspecting reader of such tripe. Here are a few:

1. The number of data (patients, subjects, students) should be a minimum of 5 or 10 times the number of variables entered into the equation—not the number of variables that turn out to be significant, which is always small. Use the number you started with. This rule provides some assurance that the estimates are stable and not simply capitalizing on chance.
2. Inevitably, when folks are doing these types of post-hoc regressions, something significant will result. One handy way to see if it is any good is to simply square the multiple correlation. Any multiple regression worth its salt should account for about half the variance (i.e., a multiple R of about .7). Much less, and it's not saying much.
3. Similarly, to examine the contribution of an individual variable before you start inventing a new theory, look at the *change in* R^2. This should be at least a few percent, or the variable is of no consequence in the prediction, statistically significant or not.
4. Finally, look at the patterns in the regression equation. A gradual falloff should be seen in the prediction of each successive variable, so that variable 1 predicts, say, 20% to 30% of the variance, variable 2 an additional 10% to 15%, variable 3, 5% to 10% more and so on, up to 5 or 6 variables and a total R^2 of .6 to .8. If all the variance is soaked up by the first variable, little of interest is found in the multiple regression. Conversely, if things dribble on forever, with each variable adding a little, it is about like number 2 above—not much happening here.

So these are some ways to deal with the plethora of multiple regressions out there. They reappear at the end of this section as C.R.A.P. Detectors, but we place them here to provide some sense of perspective.

SAMPLE SIZE CALCULATIONS

For once, nothing could be simpler. No one could possibly work out ahead of time what a reasonable value for a particular regression coefficient might be, let alone its SE. About all that can be hoped for is that the values that eventually emerge are reasonably stable and somewhere near the truth. The best guarantee of this is simply that the number of data be considerably more than the number of variables. Thus the "sample size calculation" is the essence of simplicity:

Sample size = 5 (or 10) times the number of variables

If you, or the reviewers of your grant, don't believe us, try an authoritative source—Kleinbaum, Kupper, and Muller (1988).

SUMMARY

Multiple regression methods are the strategy of choice to deal with the common problem of predicting one dependent variable from several (or many) independent variables. Caution must be used in overinterpreting regression models based on relatively small samples, and stepwise regression procedures should be viewed with considerable suspicion (unless they are hierarchical).

EXERCISES

1. A researcher does a study to see if he can predict success in reflexology school (measured by the average number of skull bumps the student can detect on simulated plastic heads) using several admissions variables: age, GPA, and gender (M = 0, F = 1). He does a multiple regression analysis and determines the R^2s and *b*s. Comment on the results shown in the several displays below.

   a. Multiple $R = .15$
   $R^2 = .0225$
   $n = 17$

Variable	*b*	SE (*b*)	*t*	*p*
Age	.131	.044	2.97	.01
GPA	.034	.112	.303	ns
Gender	.003	.017	.176	ns

ns = not significant.

   b. Multiple $R = .15$
   $R^2 = .0225$
   $n = 1233$

Variable	*b*	SE (*b*)	*t*	*p*
Age	.131	.004	32.75	.0001
GPA	.034	.012	2.83	.01
Gender	.003	.0007	4.28	.001

   c. Multiple $R = .75$
   $R^2 = .5625$
   $n = 5$

Variable	*b*	SE (*b*)	*t*	*p*
Age	.561	.424	1.32	ns
GPA	.383	.312	1.22	ns
Gender	.137	.129	1.06	ns

ns = not significant.

2. In a study of high school depression, a sample of 800 children were selected at random from city high schools.

   A questionnaire was administered, including the categories (a) stress, (b) perceived comfort in social situations, (c) attitudes to parents, (d) social support from parents, (e) socioeconomic status, and (f) a standardized measure of depression.

   A regression analysis used the depression score as dependent variable. The multiple correlation was significant ($R^2 = .176$, $p < .001$), and all the individual variables entered the regression equation.

   What effect would the following strategies have on the listed measures?

	R^2	Significance of R^2	Beta
A. Increase sample size to 1600	_____	_____	_____
B. Select only kids from private schools	_____	_____	_____
C. Include family income as predictor	_____	_____	_____
D. Repeat study with kids who were depressed then had therapy	_____	_____	_____

How to Get the Computer to Do the Work for You

All forms of regression are run from the same dialog box; you simply choose different options to do the different types of analyses. For a straight multiple regression:

- From **Analyze**, choose **Regression** → **Linear**
- Click on the dependent variable from the list on the left, and click the arrow to move it into the box marked **Dependent**
- Click on the predictor variables from the list on the left, and click the arrow to move them into the box marked **Independent**
- Click **OK**

To do a hierarchical regression, click the **Next** button to the right of **Block 1 of 1** after you've chosen the variable(s) you want to enter first; select the variable(s) in the second block and click **Next**, and so on, until all the variables you want are selected.

For a stepwise solution, also click on the down-arrow in the **Method** box and choose **Stepwise**.

CHAPTER THE FIFTEENTH

Logistic Regression

Logistic regression is an extension of multiple regression methods for use where the dependent variable is dichotomous (e.g., dead/alive). The method determines the predicted probability of the outcome based on the combination of predictor variables.

SETTING THE SCENE

A dreaded disease of North African countries is "Somaliland Camelbite Fever," which is contracted, as the name implies, from intimate contact with camels. What combination of variables related to camel contact best predicts the likelihood of contracting the scourge?

We have had such a great time up to now collapsing some historical distinctions that we figure, "Why stop?" You may recall that at the beginning of this whole mess we made a big deal of the difference between categorical (nominal, ordinal) variables and continuous (interval, ratio) ones. The former use nonparametric statistics, which we will get into soon, and the latter use parametric statistics, which we just did.

One fairly advanced nonparametric statistic is called **logistic regression**. It is used when the dependent variable is dichotomous, for example, dead or alive, and the independent variables are usually continuous (but they don't have to be). You will notice that the boundaries are starting to smear. Although it's an advanced nonparametric method, once you get into it, it looks an awful lot like ordinary multiple regression.[1]

For illustration, let's acknowledge that many of the major scourges of mankind never reach the temperate shores of Europe and North America. One of the deadliest is Somaliland Camelbite Fever[2] (SCF), which results in an involvement of multiple systems. One early sign is the development of a hump in the middle of the back[3] (not to be confused with widow's hump), accompanied by water retention. The legs grow spindly, the breath grows more odoriferous, and eventually there are psychologic manifestations as the hapless victim becomes progressively more bad tempered and seeks solitude in sunny corners of sand boxes, where he crouches on all fours awaiting his demise.

One intrepid epidemiologist ventured forth to determine risk factors for the disease. Four potential variables were identified: (1) number of years spent herding camels (Years); (2) size of the herd (Herd); (3) family history of SCF (Fam), which is a dichotomous variable; and (4) a Buccal Coliform Count (BCC) from a mouth swab of the beasts (since it was thought that the disease is spread by bacteria residing in the camel's mouth—leading to the horrible odor).

Now if SCF were a continuous variable, the next step should be almost self-evident by now: construct a regression equation to predict SCF from a linear combination of Years, Herd, Fam, and BCC. The equation would then express the risk of coming down with SCF as a weighted sum of the four factors, and would look like:

$$z = b_0 + b_1 Year + b_2 Herd + b_3 Fam + b_4 BCC \qquad \textbf{(15–1)}$$

But probabilities don't go in a straight line forever; they are bounded by zero and one. So, it would be nice if we could transform things so that the expression ranges smoothly only between zero and one. One such transformation is the **logistic** transformation:[4]

$$y = Pr(SCF|z) = \frac{1}{1 + e^{-z}} \qquad \textbf{(15–2)}$$

[1] *So if you're rusty on multiple regression, you might want to have a fresh look at Chapter 14 before you proceed.*

[2] *First brought to the attention of modern medicine in* PDQ Statistics.

[3] *Two humps in Asia.*

[4] *On a historical note, for those so inclined, its name comes from the fact that it was originally used to model the* logistics *of animal populations. As numbers increased, less food was available to each member of the herd, leading to increased starvation. As the population dropped, more food could be consumed, resulting in an increase in numbers, and so on.*

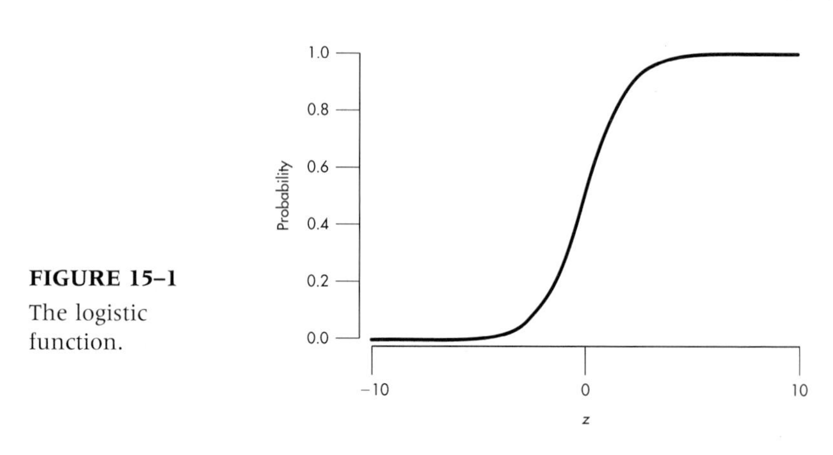

FIGURE 15–1
The logistic function.

TABLE 15–1
Data from the first 10 of 30 camel herders

Subject	Years of herding	Number in herd	Family history	Buccal count	SCF
1	3	22	No	300	No
2	5	3	No	350	No
3	25	344	Yes	446	Yes
4	14	28	Yes	121	No
5	2	77	No	45	No
6	16	34	No	233	Yes
7	28	66	Yes	654	Yes
8	19	100	Yes	277	No
9	13	87	Yes	321	Yes
10	33	45	Yes	335	Yes

A complicated little ditty. What it is saying in the first instance is that y is the probability of getting SCF for a given value of z, that is, for a given value of the regression equation. This function does some nice things. When $z = 0$, y is $1/(1 + e^0) = 1/(1 + 1) = 0.5$. When z goes to infinity (∞), it becomes $1/(1 + e^{-\infty}) = 1$. And when z goes to $-\infty$, it becomes $1/(1 + e^{\infty}) = 0$. So, it describes a smooth curve which approaches zero for large negative values of z, and goes to 1 when z is large and positive. A graph illustrating this is shown in Figure 15–1.

Until now, the shape of the curve is not self-evidently a good thing; now, we remind ourselves about the situation we're trying to model. The goal is to estimate something about the various possible risk factors for dreaded SCF. All the predictors are set up in an ascending way, so that if z is low, you don't have any risk factors, and the probability of contracting SCF should be low, which it is—that's the left-hand side of the curve. Conversely, if you have a lot of risk factors, you're off on the right-hand side, where the probability of getting SCF should approach one, which it does. If we had tried this estimate with multiple regression, it would have ended up as a straight-line relationship, so probabilities would be negative on the left side, and go to infinity on the right side—definitely not a good thing.

This ability to capture a plausible relationship between risk and probability is not the only nice feature of the logistic function, but we'll save some of the other surprises until later. For the moment, it's best to realize that the job is far from done, since we have this linear sum of our original values (which is the good news) hopelessly entangled in the middle of a complicated expression (which is the very bad news). Time to mess around a bit more. First, we'll rearrange things to get the linear expression all by itself:

$$\frac{(1 - y)}{y} = e^{-(b_0 + b_1 Years + b_2 Herd + b_3 Farm + b_4 BCC)} \tag{15–3}$$

Now, the next bit of sleight of hand. The way to get rid of an exponent is to take the logarithm, so we now get:

$$\log \frac{(1 - y)}{y} = -(b_0 + b_1 Years + b_2 Herd + b_3 Fam + b_4 BCC) \tag{15–4}$$

and $-\log (c/d) = \log (d/c)$, so the negative sign goes away:

$$\log \frac{y}{(1 - y)} = (b_0 + b_1 Years + b_2 Herd + b_3 Fam + b_4 BCC) \tag{15–5}$$

Son of a gun! We have managed to recapture a linear equation, so we can go ahead and analyze it as yet another regression problem. Here's another definition to add to the long list, by the way. The expression we just derived, $\log [y/(1-y)]$, is called the *logit function* of y. We'll talk later about the interpretation of this function, but first we have to figure out how the computer computes all this stuff.

To put some meat on the bones, let's conjure up some real data (real being a relative term, of course). We fly off to the Sahara,[5] rent the Land Rover, and wander from wadi to wadi and oasis to oasis, recruiting camel herders wherever we can find them, buying off each with a few handfuls of beads and bullets, administering a questionnaire, and swabbing their camels' mouths. After months of scouring the desert, we have found 30 herders. The pathologist does her bit and we have found that 18 herders had SCF and 12 didn't. The data for the first few are shown in Table 15–1.

[5]*Normally research assistants do the data gathering, but this rule is ignored when exotic travel and frequent flyer points are involved.*

Before we get to the output of the analysis, we should invest some time describing how the computer does its thing. One might think that, having converted the whole thing to a linear equation, we could just stuff the data into the regression program and let the beast go on its merry way, estimating the parameters and adjusting things so that the Sum of Squares (Residual) is at a minimum (see Chapter 14).

Unfortunately, things aren't quite that simple. The standard method of computing regression by minimizing the residual Sum of Squares, called the

method of least squares for (one hopes) fairly obvious reasons, can only be used when the underlying model is linear, and the dependent variable is approximately normally distributed. Although we have done our best to convert the present problem to a linear equation, this was at the cost of creating a dependent variable that is anything but linear. So, statisticians tell us that we cannot just crash ahead minimizing the error Sum of Squares. Instead we must use an alternative, computationally intensive,[6] method called **Maximum Likelihood Estimation** or **MLE.**

MAXIMUM LIKELIHOOD ESTIMATION PROCEDURES

To understand what MLE procedures are all about, it's necessary to first go back to basics. The whole point of logistic regression is to compute the estimated probability for each person, based on Equation 15–2. That estimate of the probability that the poor soul will have SCF is then compared with the actual probability, which is of a particularly simple form: 0 (the person don't got it) or 1 (the person does got it). Now, if the logistic regression is doing a good job, and the universe is being predictable, the estimated probabilities for those poor souls who actually have SCF will be high, and the probabilities for those who don't will be low. The **Likelihood Function** considers the fit between the estimated probabilities and the true state, computed as an overall probability across all cases.

Looking at things one case at a time, if p_i is the computed likelihood of disease given the set of predictor variables for person *i* who has the disease, then the probability of the overall function being correct associated with this person is just p_i. If person *j* doesn't have the disease, then the appropriate probability of a correct call on the part of the computer program is $(1 - p_j)$. So the overall probability of correct calls (i.e., the probability of observing the data set, given a particular value for each of the coefficients in the model) is just the product of all these *p*s and $(1 - p)$s:

$$L = \prod_i \hat{p}_i \prod_j (1 - \hat{p}_j) \tag{15–6}$$

where the *i*s include all the cases, and the *j*s include all the non-cases, and Π (Greek letter *pi*) is a symbol meaning "product".[7] This is the **Likelihood Function**, obtained by simply multiplying all the *p*s for the cases and $(1 - p)$s for the non-cases together to determine the probability that things could have turned out the way they did for a particular set of coefficients multiplying the predictors. It's actually conceptually similar to the standard approach of minimizing the residual sum of squares in ordinary regression. The residual sum of squares is the difference between the observed data and the fitted values estimated from the regression equation, except that we do the whole thing in terms of estimated probabilities and observed probabilities (which are either 0 or 1). And since the numbers of interest are probabilities, we multiply them together instead of adding them.

The rest of the calculation is the essence of simplicity (yeah, sure!). All the computer now has to do is try a bunch of numbers, compute the probability, fiddle them a bit to see if the probability increases, and keep going until the probability is maximum (the **Maximum Likelihood Estimation**). That's what happens conceptually, but not actually. Just as in ordinary regression, calculus comes to the rescue to find the maximum value of the function.[8] However, this analysis is so hairy that even after using calculus, it's still necessary for the computer to use iteration to get the solution, rather than calculate it from the equation.

Unconditional versus Conditional Maximum Likelihood Estimation

Unfortunately, that's still not the whole story. What we computed above is one kind of MLE, called **Unconditional MLE.** There's still another kind, called, not surprisingly, **Conditional MLE.** Basically, they differ in that unconditional MLE takes the computed probability at face value, and goes ahead and maximizes it. The conditional MLE compares the particular likelihood to all other possible configurations of the data. What actually gets maximized is the ratio of the likelihood function for the particular set of observations conditional on all other likelihood functions for all other possible configurations of the data.

What in the name of heaven do we mean by "all other configurations?" Assume we reorder the data so the first 13 are cases and the next 17 are controls. That's only one possibility; others include the first 12 are cases, the next one isn't, the one after that is, and the remaining 16 are not. That is, the other configurations are computed by keeping the predictors in a fixed order and reordering the 1s and 0s in the last column to consider all possible combinations, always keeping the totals at 13 and 17. For each one of these configurations the MLE probability is computed, and then they're all summed up.[9] Needless to say, the computation is *very* intensive.

All this would be of academic interest only, except that sometimes it really matters which one you pick. When the number of variables is large compared to the number of cases, then the conditional approach is preferred, and the unconditional MLE can give answers that are quite far off the mark. How large is large? How high is up? It seems that no one knows, but perhaps the "rule of 5"—5 cases per variable—is a safe bet.

SAMPLE CALCULATION

Now that we understand a bit about what the computer is doing, let's stand back and let it do its thing. We press the button on the SCF data, sit back, and

[6] *Which is why logistic regression is absent from many statistical packages.*

[7] *Note that we threw in a little hat over the ps; this is the mathematical way to say that these are estimates, not the truth.*

[8] *What really happens is that the process begins by differentiating the whole thing with respect to each parameter (calculus again), resulting in a set of* k *equations in* k *unknowns, the set of coefficients for each of the predictor variables, and a constant. See note 6 in Chapter 13.*

[9] *For your interest (who's still interested?), the number of possible combinations for our 30 herders is* 1.786×10^{35}*! If you want to see how we got there, look up the Fisher Exact Test.*

TABLE 15–2

B coefficients from logistic regression analysis of SCF data

Variable	*B*	SE	Wald	*df*	Sig	*Exp(B)*
Family history	3.36	1.48	5.13	1	.023	28.79
Herd	−.0073	.010	.547	1	.459	0.992
Years	.0086	.0238	.132	1	.716	1.008
BCC	.0128	.0061	4.42	1	.035	1.012
Constant	−1.45	2.044				

[10] *Although some folks might like to convince you that this came from epidemiology, it didn't—it came from horse racing. Imagine that the odds makers work out that the probability of Old Beetlebaum winning is 20%. The* ***odds*** *of him winning is 0.2/(1 − 0.2) = 0.25, or turning it around, the odds* ***against*** *Beetlebaum are 0.8/(1 − 0.8) = 4. So they say it's 4 to 1 against Beetlebaum in the 7th.*

watch the electrons fly, and, in due course, a few pages come spinning out of the printer. Some of it looks familiar; some does not. Let's deal with the familiar first. A recognizable table of *b* coefficients is one of the first things we see, and it looks like Table 15–2.

Just like ordinary multiple regression, this table shows the coefficients for all the independent variables and their associated standard errors (SEs). However, the column to the right of SE, which one might expect to be the *t*-test of significance for the parameter ($t = b/\text{SE}(b)$; see Chapter 13) is now labelled the **Wald** test, and it doesn't look a bit like the $b/\text{SE}(b)$ we expected.

As it turns out, this is familiar territory. The Wald statistic has two forms: in the first form, it is just the coefficient divided by the SE, and it is distributed approximately as a *z*-test (so Table A is appropriate). In the second form, the whole thing is squared, so it equals $[b/\text{SE}(b)]^2$. In the present printout, taken from SPSS, they chose to square it. So this part of the analysis, in which we have computed the individual coefficients, their SEs, and the appropriate tests of significance, is familiar territory indeed. As it turns out, Family History and BCC are significant at the .05 level; Herd and Years are not.

This is almost straightforward, except for the last column. What on earth is *Exp(B)*? So glad you asked. The short answer is that it stands for "exponential of B," which leaves most folks no further ahead. Remember, however, that we began with something that had all the *b* coefficients in a linear equation inside an exponential. Now we're working backward to that original equation.

Suppose the only variable which was predictive was Family History (Fam), which has only two values: 1 (present) or 0 (absent). We have already pointed out that the logistic function expresses the probability of getting SCF given certain values of the predictor variables. Focusing only on Fam now, the probability of SCF given a family history looks like:

$$\log\left[\frac{p_1(SCF)}{(1-p_1)(SCF)}\right] = b_0 + b_3\,FAM$$

(15–7)

If there is no family history then Fam = 0, and the formula is:

$$\log\left[\frac{p_0(SCF)}{(1-p_0)(SCF)}\right] = b_0$$

(15–8)

Now the ratio of $p/(1-p)$ is the **odds**[10] of SCF with Fam present or absent. The **odds ratio** is the ratio of the odds, naturally, and if you are good at diddling logs you can show that the log odds ratio is:

$$\log\left[\frac{p_1/(1-p_1)}{p_0/(1-p_0)}\right] = b_3$$

(15–9)

In words: for discrete predictor variables, the regression coefficient is equal to the log odds ratio of the event for the predictor present and absent. If we take the exponential of the *b* coefficient, we can get to the interpretation of the last column:

$$\frac{p_1/(1-p_1)}{p_0/(1-p_0)} = e^{b_3}$$

(15–10)

So, in the present example, the *b* coefficient for Fam is 3.36, and $e^{3.36}$ is 28.79; therefore, the relative odds of getting SCF with a positive family history is 28.79. That is, in this equation, a family history is a bad risk factor; the odds of getting SCF when you have a positive history is about 29 times what it would be otherwise. Perhaps you should stay away from your parents.

For continuous variables we can do the same calculation, but it requires a somewhat different interpretation. Now the e^b corresponds to the change in odds associated with a one unit change in the variable of interest. So looking at BCC, the other significant predictor, the relative odds is 1.013, meaning that a one-unit change in BCC increases the odds of contracting SCF by about 1%.

GOODNESS OF FIT AND OVERALL TESTS OF SIGNIFICANCE

That was the easy part; however, missing from the discussion so far is any overall test of fit—the equivalent of the ANOVA of regression and the R^2 we encountered in multiple regression. Since this is a categorical outcome variable, we won't be doing any ANOVAs on these data.

At one level, goodness of fit is just as easy to come by with logistic regression as with continuous data. After all, we began with a bunch of cases of SCF and non-cases. We could do the standard epidemiologic "shtick" and create a 2 × 2 table of observed versus predicted classifications. To do this, since the predicted value is a probability, not 0 or 1, we must establish some cutoff above which we'll call it a case—50% or .50 seems like a reasonable starting point.

If we do that for the present data set, the contingency table would look like Table 15–3.

The overall agreement is 26/30 or 86.67%. We could of course do *kappa*s or *phi*s on the thing (see Chapter 21)—but we won't.

While this table is a useful way to see how we're doing overall, it is not easily turned into a test of significance. We could just do the standard chi-squared on the 2 × 2 table, but this isn't really a measure of how well we're doing, because to create the table, every computed probability from 0 to .49999 was set equal to 0, and everything from .50000 to 1 was set equal to 1. So, the actual fit may well be somewhat better than it looks from the table.

In fact, the likelihood function, which we have already encountered, provides a direct test of the goodness of fit of the particular model, since it is a direct estimate of the likelihood that the particular set of 1s and 0s could have arisen from the computed probabilities derived from the logistic function. That is, the MLE *is* a probability—the probability that we could have obtained the particular set of observed data given the estimated probabilities computed from putting the optimal set of parameters into the equation for the logistic function (optimal in terms of maximizing the likelihood, that is). Now, the better the fit, the higher the probability that the particular pattern could have occurred, so we can then turn it around and use the MLE as a measure of goodness of fit.

It turns out that, although the programs could report this estimate, they don't—that would be too sensible. Instead, they do a transformation on the likelihood:

$$Goodness\ of\ Fit = -2 \log \hat{L} \quad (15\text{–}11)$$

The only reason to do this transformation is that some very clever statistician worked out that this value is approximately distributed as a chi-squared with degrees of freedom equal to the number of variables in the model.

In the present example, the value of the goodness of fit is 18.10. With 4 variables, there are 4 *df*, so the critical chi-squared, derived from Table F, is 14.86 for $p < .005$; so, it's highly significant.

There are a couple of tricks in the interpretation of this value. First, you have to recall some arcane high school math. For any number x less than one, the logarithm of x ($\log x$) is negative. Since the MLE is a probability (and, therefore, can never exceed 1.0), its log will always carry a negative sign, and further, the bigger the quantity (bigger negative that is), the smaller the probability, the worse the apparent fit. So, when we multiply the whole thing by "−2," it all stands on its head. Now, the bigger the quantity (this time in the usual positive sense), the lower probability, and the *worse* the fit. That is, just as you've gotten used to the idea that the bigger the test, the better, statisticians turn things "bass-ackwards" on you.

TABLE 15–3

Observed and predicted classification

Predicted	Observed		
	Non-case	Case	Total
Non-case	9	1	10
Case	3	17	20
TOTALS	12	18	30

A consequence of this inversion is that, like all multiple regression stuff, the more variables in the model, the *smaller* the goodness-of-fit measure. Tracing down this logic, as we introduce more variables, we expect that the fit will improve, so the ML probability will be higher (closer to 1), the log will be smaller (negative but closer to zero), and ($-2 \log L$) will be smaller. Of course, also like multiple regression, this is achieved at the cost of an increase in the degrees of freedom, so it's entirely possible that additional predictor variables will improve the goodness of fit (make it smaller), but it will be less significant, since the degrees of freedom for the chi-squared is going up faster than is the chi-squared.

STEPWISE LOGISTIC REGRESSION AND THE PARTIAL TEST

All of the above suggests a logical extension along the lines of the partial F test in multiple regression. We could fit a partial model, compute the goodness of fit, than add in another variable, recompute the goodness of fit, and subtract the two, creating a test of the last variable which is a chi-squared with one degree of freedom. In the present example, if we just include Fam, Herd, and Years on the principle that it's awfully difficult to persuade camels to "open wide" as we ram a swab down their throats, we find the goodness of fit is 25.88. So, the partial test of BCC is:

$$-2 \log \mathrm{L}\ (\text{full model}) - (-2 \log \mathrm{L}\ (\text{partial model})) = 25.89 - 18.10 = 7.79 \quad (15\text{–}12)$$

With $df = 1$, the critical chi-squared at .05 is 7.88, which is pretty close to the computed chi-squared.

One final wrinkle, then we'll stop. The p-value resulting from this stepwise analysis *should* be the same level of significance as that for the B coefficient of BCC in the full model. Regrettably, it isn't quite. Looking again at Table 15–2 the computed Wald statistic yields a p of .03, which is larger but still significant.

Why the discrepancy? As near as we can figure, it's because of the magical word "approximately," which has featured prominently in the discussion. For the discrepancy to disappear, we need a larger

[11] *Kleinbaum (1994), in the "To Read Further" section.*

sample size, but we have it on good authority[11] that no one seems to know how much larger it needs to be. Again, upon appealing to higher authorities than ourselves, we are led to believe that, when in doubt, we should proceed with the formal likelihood ratio test.

MORE COMPLEX DESIGNS

Now that we have this basic approach under our belts, we can extend the method just as we did with multiple regression. Want to include interactions? Fine. Just multiply Herd × Years to get an overall measure of exposure[12] and stuff it into the equation. Want to do a matched analysis, in which each person in the treatment group is matched to another individual in the control group? Now we have a situation like a paired *t*-test, in which we are effectively computing a difference within pairs. No big deal. Create a dummy variable for each of the pairs (so that, for example, pair 1 has a dummy variable associated with it, pair 2 has another, and so on). Of course, in this situation, where the number of variables is going up by leaps and bounds, the use of an unconditional ML estimator is a really bad idea, and you'll have to shell out for a new software package. And on and on—*ad infinitum, ad nauseam.*

[12] *If these were wolves, not camels, we would obviously call it a "pack-year" of exposure.*

SUMMARY

Logistic regression analysis is a powerful extension of multiple regression for use when the dependent variable is categorical (0 and 1). It works by fitting a logistic function to the 1s and 0s, estimating the fitted probability associated with each observed value. The test of significance of the model is based on adjusting the parameters of the model to maximize the likelihood of the observed data arising from the linear sum of the variables, the so-called **MLE** procedure. Tests of the individual parameters in the logistic equation proceed much like the tests for ordinary multiple regression.

SAMPLE SIZE

Regarding the question of how many subjects are necessary to run a logistic regression, there's a short answer and a long one. The short answer is, "Dunno"; the long answer is, "Neither does anyone else." We looked in many text books about multivariate statistics and logistic regression, and only one (Tabachnick and Fidell, 1996) even mentions the issue of sample size. Tabachnick and Fidell state that, "A number of problems may occur when there are too few cases relative to the number of predictor variables" (p. 579), and spend three paragraphs discussing the problems and possible solutions (such as collapsing categories). Unfortunately, they don't say what's meant by "too few cases," so, by default, we recommend the old and familiar standby, 10 subjects per variable, without the slightest evidence to back this up.

EXERCISES

1. For the following designs, indicate the appropriate statistical test.
   a. A researcher wants to examine factors relating to divorce. She identifies a cohort of couples who were married in 1990 through marriage records at the vital statistics office. She tracks them down and determines who is still married and who is not. She also inquires about other variables—income, number of children, husbands' parents married, wife's parents married.
   b. A surgeon wants to determine whether laparoscopic surgery results in improvements in quality of life. He administers a quality-of-life questionnaire to a group of patients who have had a cholecystectomy by (i) laparoscopic surgery or (ii) conventional surgery. The questionnaire is administered (i) before surgery, (ii) 3 days postoperatively, and (iii) 2 weeks postoperatively.
   c. What is the role of lifestyle and physiologic factors in heart disease? A researcher studies a group of post-myocardial infarction patients and a group of "normal" patients of similar age and sex. She measures cholesterol and blood pressure, activity level (in hours/week), and alcohol consumption in mL/week.
   d. At the time of writing, the latest cancer scare is that antiperspirants "cause" breast cancer. To test this, a researcher assembles a cohort of women with breast cancer and a control group without breast cancer and asks about the total use of antiperspirant deodorants (in stick-years).
   e. To determine correlates of clinical competence, an educator assembles a group of practicing internists, and gives them a series of 10 written clinical scenarios. The independent variables are (i) Year of graduation, (ii) Gender, and (iii) Academic standing (pass, honors, distinction). The dependent variable is the proportion of the cases diagnosed correctly.
2. A randomized trial of a new anticholesterol drug was analyzed with logistic regression (using cardiac death as an endpoint), including drug/placebo as a dummy variable, and a number of additional covariates. The output from a logistic regression reported that $\text{Exp}(b_1) = -2.0$. Given that the cardiac death rate in the control group was 10%, what is the relative risk of death from the treatment?

CHAPTER THE SIXTEENTH

Advanced Topics in Regression and ANOVA

This chapter reviews several advanced analytical strategies that bring together ANOVA and regression. Analysis of covariance (ANCOVA) combines continuous variables (covariates) and factors and is used for assessing treatment effects while controlling for baseline characteristics. Power series analysis is a regression including higher powers of the independent variable (e.g., quadratic, cubic, or quartic terms).

SETTING THE SCENE

You have been collecting data at your PMS (Pathetic Male Syndrome) Clinic for 15 years. Despite admonitions to the contrary, you just can't resist the temptation to analyze everything in sight with multiple regression. After graphing the data, three things are evident: (1) Pathos Quotient (PQ) increases linearly with belly size, (2) middle-aged males have the highest PQ, and (3) treatment with testosterone injections appears to have some effect on the PQ. Multiple regression tells us how to deal with straight-line relationships, ANOVA works on treatment groups, but how in the world will you deal with all this complexity?

By now we have given you the conceptual tools to master nearly every complexity of ANOVA and regression. However, we have left out one small detail—namely, how to put them together. It may not be self-evident why one should bother to try to merge two goods things. After all, it would seem that each is capable of handling a large class of complex problems. But reflect a moment on a simple twist to the designs we have encountered thus far.

The syndrome we investigate in this chapter, PMS, is commonly referred to as "mid-life crisis" or "male menopause" in its acute phase, but it has a more insidious onset than is implied by those terms. One sign is a gradual movement upward or downward in the belt line—after all, why else do elderly men buckle their pants somewhere around the nipple line or down around their knees? Another is the purchase of flamboyant hats to cover the shrinking number of hair follicles. The presence of satellite dishes in the backyard to receive dirty movies is a warning signal as well. We are now confident that you, as a health professional, will be able to recognize this new epidemic.

Certainly, there is some hope on the horizon, thanks to the miracles of modern medical science. As Bob Dole reminds us,[1] there is life after prostatectomy, and there are now pharmacologic means to keep more than our spirits up. Viagra, however, has only local effects down in the nether regions (albeit spectacular, by all accounts). To deal with the more systemic manifestations of masculine aging, we have to revert to more traditional therapy, specifically testosterone injections.

But how does one actually measure PMS? A simple diary, wherein the PM (pathetic male) counts the number of wistful sighs, the number of times he says to his significant other, "Not tonight dear, I have a backache," the number of unused notches (guess which side of the buckle) on his belt, the number of ounces of Greek Formula 18 consumed in a week, and the total dollar sum of subscriptions to various lewd or semi-lewd male magazines, makes a ratio variable (if not a rational one!), which we'll call the Pathos Quotient, or PQ.

As we indicated above, PMS is related to three other variables. Pathos Quotient increases linearly with belly size—that's a job for regression. On the other hand, if men are given male sex hormones, they seem to recover a bit. That is a comparison between two groups formed on the basis of a nominal variable, and it can be handled with a *t*-test or a one-way ANOVA. As far as the relationship with age goes, it sounds like a curve peaking at about 45 and falling off on both sides, which to those of mathematical inclination might suggest a quadratic term. (*Quadratic* means a term squared, *cubic* is a term cubed, and *quartic* is to the fourth power.) But how can we put it all together?

[1] *As part of his continuing efforts to commit political* hara-kiri.

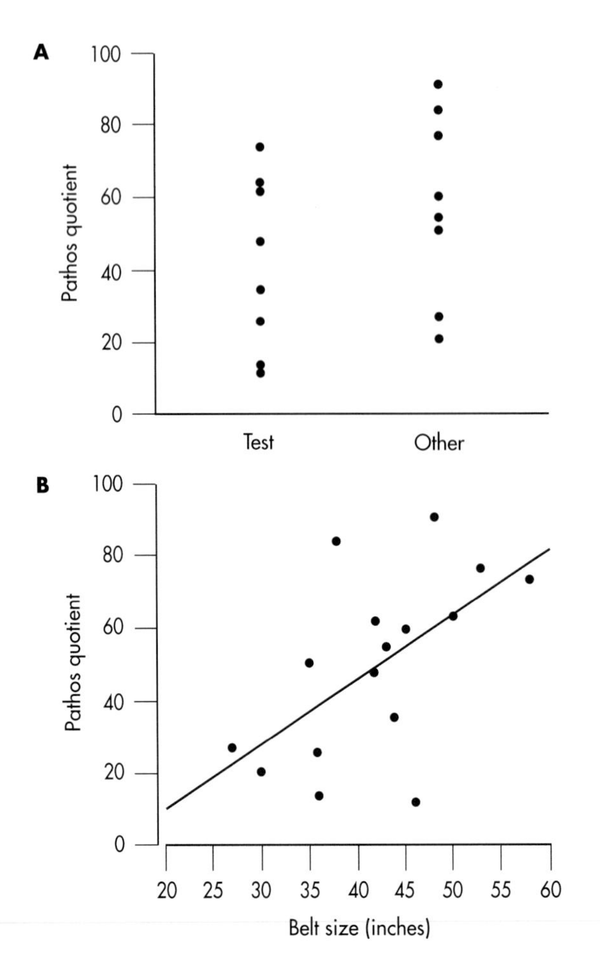

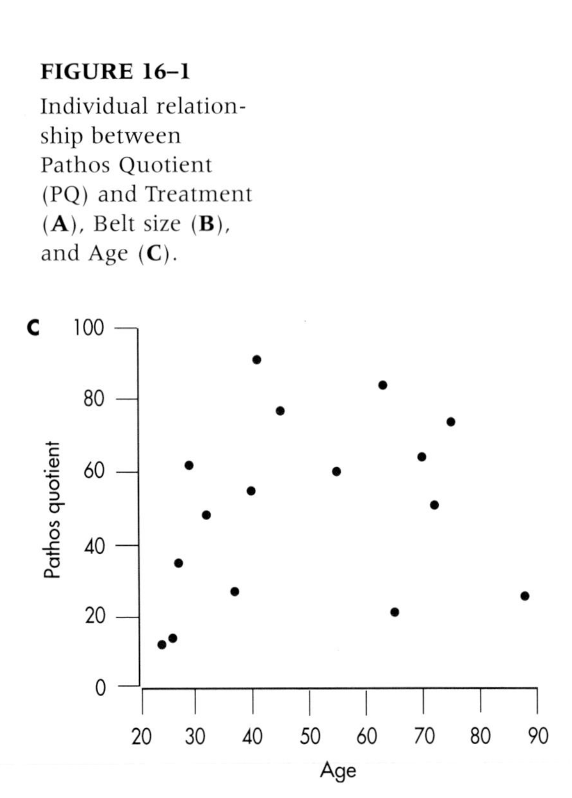

FIGURE 16–1 Individual relationship between Pathos Quotient (PQ) and Treatment (**A**), Belt size (**B**), and Age (**C**).

TABLE 16–1 Data for 16 PMS patients

Subject	Age	Belt size	Treatment	PQ
1	24	46	Testosterone	12
2	26	36	Testosterone	14
3	27	40	Other	27
4	88	44	Testosterone	35
5	32	36	Testosterone	26
6	29	30	Other	21
7	70	42	Testosterone	48
8	75	35	Other	51
9	37	42	Other	62
10	65	50	Testosterone	64
11	72	45	Testosterone	60
12	55	53	Other	77
13	45	48	Other	91
14	41	38	Other	84
15	63	43	Other	55
16	40	58	Testosterone	74

TABLE 16–2 ANOVA of regression for PMS against belt size

Source	Sum of squares	*df*	Mean square	*F*	*p*
Regression	2893.2	1	2893.2	6.11	.027
Residual	6629.0	14	473.5		
$R^2 = .303$					

Having gotten this far, we might like to see the appearance of these relationships on graphs. Figure 16–1 shows the PQ scores for 16 subjects in comparison to belt size, age, and treatment, based on the data of Table 16–1. It is evident from the graph that the data are pretty well linearly related to belt size. We could proceed to do a regression analysis on the data in the usual way. If we did, the ANOVA of the regression looks like Table 16–2, and the multiple of R^2 turns out to be .30, which is not all that great.

Looking at treatment, this is just a nominal variable with two levels, and the hormone group mean is a bit lower than the "Other" group. If we wanted to determine if there was any evidence of an effect of treatment, we could simply compare the two means with a *t*-test. For your convenience, we have done just that; the *t* value is 1.33, which is not significant.

Finally, we do have this slightly bizarre relationship with age, indicating that the mid-life crisis is a phenomenon to be reckoned with, and moreover, its effects seem to dwindle on into the 60s. It is anything but obvious how this should be analyzed, so we won't—yet. For the sake of learning, we'll leave age out of the picture altogether for now and simply deal with the other two variables—Belt Size (a ratio variable) and Treatment with testosterone/other (a nominal variable).

ANALYSIS OF COVARIANCE

Again, if you've learned your lessons well, you know by now that a first approach is to graph the data, and at least this time it really isn't too hard to put three variables on two dimensions. We simply use different points for the two groups, then plot the data against belt size again. Figure 16–2 shows the updated graph.

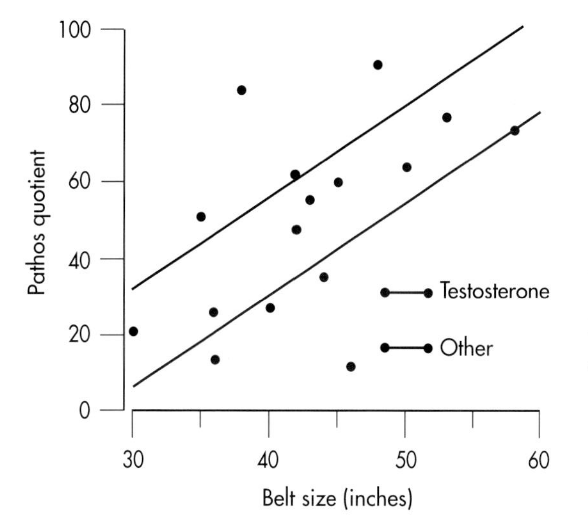

FIGURE 16–2 Relationship between PQ and Treatment and Belt size.

Now we have a slightly different picture than before. If we look back at the relationship to Belt Size, we can see that the data are actually pretty tightly clustered around *two* lines, one for Testosterone and one for Other. Some of the variability visible in the data in Figure 16–1B was a result of the treatment variable, as well as the belt size. Conversely, if we imagine projecting all the data onto the Y-axis, so that we have essentially two distributions of PQs, one for Testosterone and the other for Other, we recapture the picture of Figure 16–1A. And taking account of all of the variance from both sources, by determining two lines instead of one, we are able to reduce the scatter, or the error variance, around the fitted lines. This should result in a more powerful statistical test, both for analyzing the impact of belt size on PQ and also for determining if treatment has any effect.

Conceptually, we have the same situation as we had with multiple regression. We have two independent variables, Belt Size and Treatment, each of which is responsible for some of the variance in PQ. As a result, the residual variance, which results from other factors not in the study, is reduced. The effect of using both variables in the analysis is to reduce the error term in the corresponding test of significance, thereby increasing the sensitivity of the test.

The challenge is to figure out how to deal with both nominal and ratio independent variables. What we seem to need is a bit of ANOVA to handle the grouping factor and a dose of regression to deal with the continuous variable. Historically, the problem is dealt with by a method called **ANCOVA**, from **AN**alysis of **COVA**riance, once again using creative acronymizing to obscure what was going on. You may recall from Chapter 13, however, that the covariance was a product of X and Y differences that expressed the relationship between two interval-level variables, so this is a reasonable description of what might be the relationship to belt size. We then need some way to analyze the effect of the treatment variable, which amounts to looking at the difference between two groups, something we would naturally approach with an ANOVA, or a t-test, which is the same thing. Put it together and what have you got? Analysis of covariance.[2]

The time has now come to turn once again from words to pictures, employing what is now a familiar refrain—parceling out the total Sum of Squares in PQ into components resulting from Belt, Treatment, and error. To see how this comes about, refer to Figure 16–3, which is simply an enlargement of Figure 16–2 around the middle of the picture. We have also included the Grand Mean of all the PQs as a horizontal line, and we have thrown in a bunch of arrows (we'll get to those in a minute).

Three possible sources of variance are Treatment, Belt, and the ubiquitous error term. So far, so good, but how do they play out on the graph? Sum of Squares resulting from Treatment is related to the distance between the two parallel lines, so it expresses the treatment effect on PQ. The Sum of Squares resulting from Belt is the sum of all the squared vertical distances between the fitted points and their corresponding group mean, just as in a regression problem, except that the distances are measured to one or the other line. Sum of Squares (Error) is the distance between the original data points and the corresponding fitted data point. The better the fit between the two independent variables (Treatment and Belt), (1) the closer the data will fall to the fitted lines, and (2) the larger will be Sum of Squares (Belt) and Sum of Squares (Treatment) when compared with the Sum of Squares (Error).

Viewed this way, it's not such a difficult problem, showing once again that a picture is worth a few words. But we haven't actually started analyzing it numerically yet, so here we go. You will note that we have made a big deal of putting together both nominal and interval-level data, but in fact they both come down to sums of squared differences when we look at the variance components. In fact, we seem to be in the process of collapsing the distinction altogether between ANOVA and regression methods. After all, in the last two chapters we got used to the idea of ANOVAing a regression problem. Perhaps we can be forgiven if we now stand things on their heads and do a regression to an ANCOVA problem.[3]

Suppose we forget for a moment that these are a mixture of variables and just plow ahead stuffing them into a regression equation. It might look a bit like this:

$$PQ = b_0 + b_1 \times \text{Treatment} + b_2 \times \text{Belt}$$

[2] *Not bibbitty, bobbitty, boo—silly.*

[3] *And this is a conceptual headstand, to which no amount of mirrors will lend assistance.*

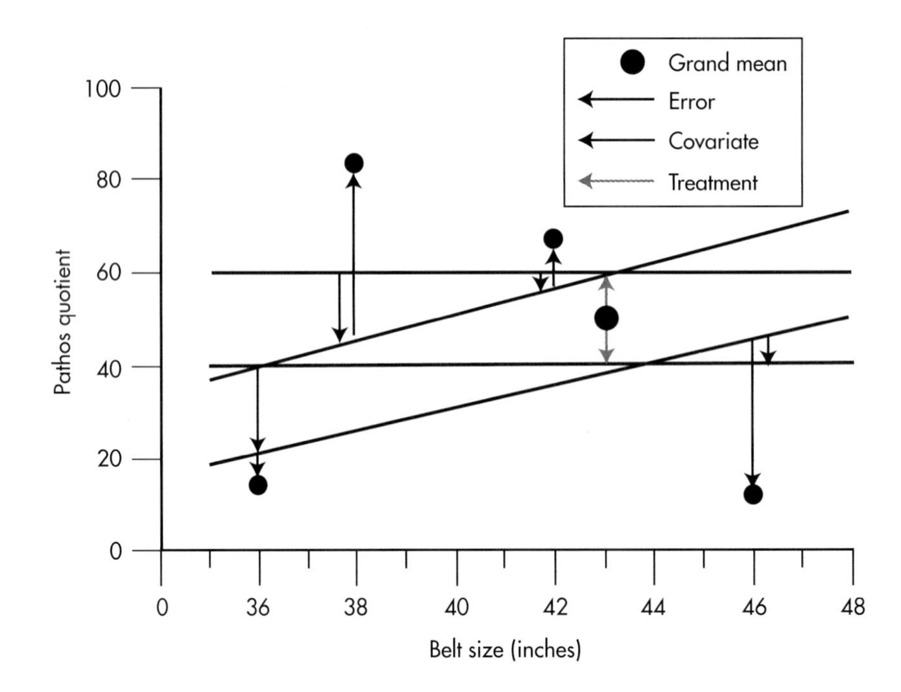

FIGURE 16–3 Relationship between PQ and Treatment and Belt size (expanded), indicating sums of squares.

That looks like a perfectly respectable regression equation. But we have only one little problem. When we put Belt into the equation it's pretty clear what belt size to use—32, 34, 36 ... 54 inches (or the metric equivalent). But what number do we use for Treatment? It's a nominal variable, so there is no particular relationship between any category and a corresponding number. Well ... suppose we try 0 for Other and 1 for Testosterone; what happens? Then the regression equation for the control group is:

$$PQ = b_0 + b_1 \times 0 + b_2 \times \text{Belt} = b_0 + b_2 \times \text{Belt}$$

and for the treatment group it is:

$$\begin{aligned} PQ &= b_0 + b_1 \times 1 + b_2 \times \text{Belt} \\ &= [b_0 + b_1] + b_2 \times \text{Belt} \end{aligned}$$

In other words, the choice of 0 and 1 for the Treatment variable creates two regression lines with the same slope, b_2, which differ only in the intercept. In the Testosterone group, the intercept is ($b_0 + b_1$); in the Other group it is just b_0. So b_1 is just the vertical distance between the two lines in the graph (i.e., the effect of treatment). That is just what we want. All that remains is to plow ahead just as with any other regression analysis and determine the value and statistical significance of the *b*s. In the course of doing so we have actually done what we set out to do: determine the variance attributable to each independent variable.[4]

In this case, the Sum of Squares resulting from regression, for the full model, is equal to:

$$\sum[(b_0 + b_1 \times \text{Treatment}_i + b_2 \times \text{Belt}_i) - \overline{PQ}]^2 \tag{16–1}$$

[4] *This is just creating dummy variables as we did in Chapter 14. As it turns out, ANCOVA is just a special case of multiple regression.*

Lest the algebra escapes you, this is just the difference between the fitted point at each value of PQ_i (the whole equation in the parentheses) and the overall mean of *PQ*, with all the individual differences squared and summed. So this is the sum of squares in *PQ* resulting from the combination of the independent variables.

The Sum of Squares (Residual) is equal to:

$$\sum[PQ_i - (b_0 + b_1 \times \text{Treatment}_i + b_2 \times \text{Belt}_i)]^2 \tag{16–2}$$

This takes the difference between the original data, PQ_i, and the fitted values (again, the stuff in the parentheses), all squared and added. So this represents the squared differences between the original data and the fitted points.

To test the significance of each independent variable, we must actually determine three regression equations: one with just Treatment in the equation, one with just Belt in the equation, and the last with both in the equation. This way we can determine the effect of each variable above and beyond the effect of the other variables. The ANOVAs for each of the models are in Table 16–3.

We then proceed to determine the individual contributions. For Belt, the additional Sum of Squares is (5274.5 − 1139.1) = 4135.4 with 1 *df*, and the residual term is 326.8. The *F* test for this variable is therefore 4135.4 ÷ 326.8 = 12.65, equivalent to a *t* of 3.56. We'll let you work out the equivalent test for Treatment. Suffice to say that it, too, is significant, with a *t* of 2.70, $p < .05$.

Actually, although we have structured the problem as a regression problem for continuity and simplicity, if the analysis were actually run as an

ANCOVA program, the contributions of each variable would be separately identified in the ANCOVA table (Table 16–4).

Note that a funny thing happened when both variables went in together. Because each variable accounted for some of the variance, independent of the other, the residual variance shrank, so the test of significance of both variables became highly significant. When each was tested individually, however, Treatment was not significant, and Belt was only marginally so. For those of you with a visual bent, the situation is illustrated in Figure 16–4.

Figure 16–4 nicely illustrates one potential gain in using ANCOVA designs: the apportioning of variance resulting from covariates such as Belt can actually increase the power of the statistical test of the grouping factor(s). Of course, this is true only insofar as the covariates account for some of the variance in the dependent variable. As with regression, it can work the other way, where adding variables decreases the power of the tests.

TABLE 16–3

ANOVAs of regressions for Testosterone/other, Belt size, and Both

Source	Sum of squares	*df*	Mean square	*F*	*p*
Treatment					
Regression	1139.1	1	1139.1	1.90	.189
Residual	8383.9	14	598.9		
R^2 = .120					
Belt size					
Regression	2893.7	1	2893.7	6.11	.027
Residual	6629.2	14	473.5		
R^2 = .303					
Treatment and Belt size					
Regression	5274.5	2	2637.2	8.07	.005
Residual	4248.5	13	326.8		
R^2 = .553					

TABLE 16–4

Summary ANCOVA table for Treatment and Belt size

Source	Sum of squares	*df*	Mean square	*F*	*p*
Covariate (Belt)	4135.4	1	4135.4	12.65	<.01
Treatment	2380.8	1	2380.8	7.29	<.01
Explained*	5274.5	2	2637.2	8.07	<.01
Residual	4248.5	13	326.8		

*Note that, in contrast to factorial ANOVA designs, here the sums of squares don't add up because there is an overlap in the explained variance. If you don't believe it yet, look at Figure 16–4.

ANCOVA for Adjusting for Baseline Differences

Actually, surprisingly few people are even aware of this potential gain in statistical power from using covariates. More frequently, ANCOVA is used in designs such as cohort studies where intact control groups are used and the two groups differ on one or more variables that are potentially related to the outcome or dependent variable.

As an example, consider the pitiless task of trying to drum some statistical concepts into the thick heads of a bunch of medical students.[5] In an attempt to engage their humorous side, one prof decides to try a different text this year—*Bare Essentials*, naturally. He gives this class the same exam as he gave out last year and finds that the mean score on the exam is 66.1% this year, whereas it was 73.5% last year. That's not funny for him or us.

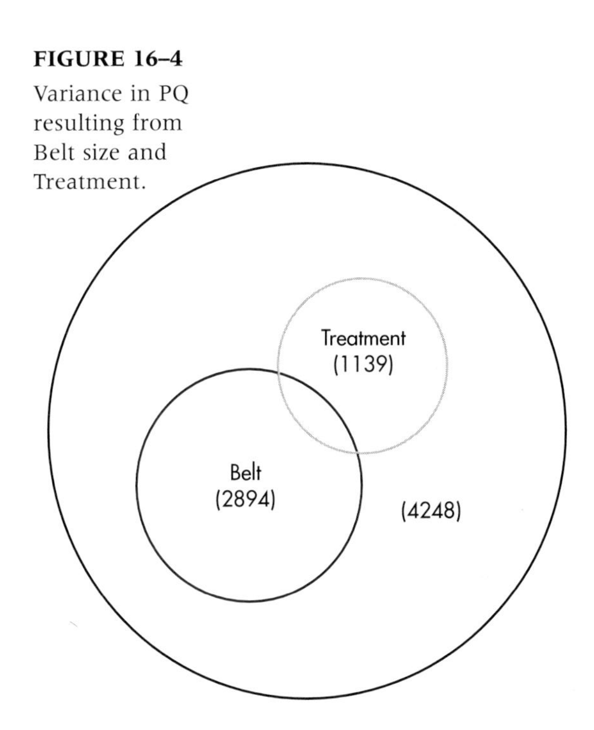

FIGURE 16–4
Variance in PQ resulting from Belt size and Treatment.

Do Norman and Streiner honor the money-back guarantee and forfeit their hard-earned cash? Not likely, for several reasons: (1) we're tight-fisted, (2) we already spent it recklessly on women,[6] and (3) we know the dangers of historical controls and other nonrandomized designs. A little detective work reveals the fact that the admissions committee has also been messing around and dropped the GPA standard, replacing it with interviews and other touchy-feely stuff. So one explanation is that this class has a slightly higher incidence of cerebromyopathy[7] than had the last. But what can we do about it? Clearly we need some independent measure of quantitative skills. Let's take the physics section of the Medical College Admissions Test (MCAT). If we plot MCAT physics scores and final grades for the two classes, we get Figure 16–5.

A different picture now emerges. It is clear that *Bare Essentials* delivered on the goods. The regression line for this year's class is consistently higher than

[5] *Frankly our sympathies go out to any medical or other students who are reading this book to survive a statistics course. In our view, it makes no more sense for an undergraduate student in health sciences heading for a clinical career to have to be able to do statistics than it does for an architect to be required to forge the I-beams in a building.*

[6] *Our wives.*

[7] *Muscle heads.*

TABLE 16–5

Mean scores (and SD) for the classes of 1990 and 1991 on MCAT physics and post-test

Test	Class 1990	Class 1991
MCAT	48.6 (10.1)	40.8 (8.2)
Post-test	73.0 (16.0)	66.1 (28.8)

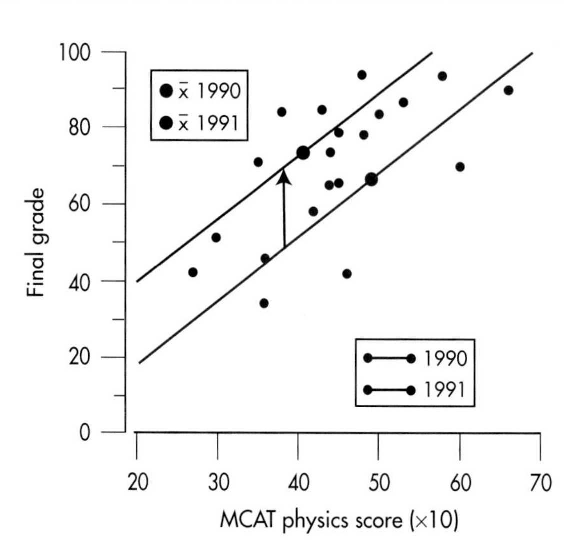

FIGURE 16–5 Relationship between MCAT physics and post-test statistics score for the classes of 1990 and 1991

last year's, by about 15%, as shown by the arrow. But what happened is that the admissions committee blew it (at least as far as stats mastery goes) by admitting a number of students with chronic cases of cerebromyopathy, so that they start off duller (i.e., to the left of the graph), and end up duller. But, relative to their starting point, they actually learn more from *Bare Essentials*, and we get to keep the dough.

We'll put some statistics into it (which is what we're here for), and the data for the two cohorts on MCAT and post-test are shown in Table 16–5. If we do a t-test on the post scores, the result is $t(18) = 0.82$, $p = .41$, which is a long way from significance and in the wrong direction anyway. Note that graphically, this is equivalent to projecting all the data onto the Y-axis and looking at the overlap of the two resulting distributions. However, if we bring up the heavy artillery and ANCOVA the whole thing, with MCAT as the covariate and 90/91 as the grouping factor, a whole new picture emerges.

First, the estimated effect of 90/91 (i.e., the *Bare Essentials* treatment effect) is now a super +19.75—the difference between the two lines. Further, the effect is highly significant ($t(18) = 3.60$, $p = .002$). So not only did we improve the sensitivity of the test in this analysis, we also corrected for the bias resulting from baseline differences, to the extent that the estimate of the treatment effect changed direction.

This then summarizes the potential gains resulting from using ANCOVA to account for baseline differences:

1. When randomization is not possible and differences between groups exist, ANCOVA can correct for the effect of these differences on the dependent variable.
2. Even when you have no reason to expect baseline differences, ANCOVA can improve the sensitivity of the statistical test by removing variance attributable to baseline variables.

Extension to More Complex Designs

So far we have considered only a simple design, with one grouping factor and one covariate, but there is no theoretical limit to the extension of the basic strategy to any number of covariates, or any number of grouping factors.[8] Any legitimate ANOVA design can be extended by the addition of covariates. We must, however, emphasize that this strategy is only useful to the extent that the covariates are related to the dependent variable and are unrelated to each other, as discussed earlier.

The approach is simple enough. You just continue to treat it as a regression problem, with a new b for each covariate and a new dummy variable for each grouping factor.[9] The analysis proceeds to estimate a Sum of Squares, Mean Square, and significance test for each variable, just as before.

Suppose, for example, that we were also interested in the effects of steroids on PQ. This amounts to another nominal variable in the design. As we discussed in Chapter 14, we must create a new coefficient to separately estimate the difference between steroid and placebo, and testosterone and placebo. In general, if there are k levels of a grouping factor, there will be $(k - 1)$ dummy variables created.[10]

Nonlinear Regression

Whatever happened to age and PMS? We established that there was a slightly bizarre relationship between age and the PQ score, so that it peaked at about 45, but we then went on to greener pastures. Time to recycle.

If you return to Figure 16–1C, you will note that the relationship of PQ with age adopted a peculiar form, somewhat akin to a 10-gallon hat. Actually, a flashlight reflector is a more appropriate, although not as glaringly obvious, analogy. If you ever survived one of those courses in plane and solid geometry, or optics,[11] you would immediately recognize this as the curve traced out by a function of the form $y = x^2$; take our word for it.

How can we crunch out the statistics of it all? Easy, by now—just create another regression equation containing both linear ($y = bx$) and quadratic ($y = cx^2$) terms. The equation would look like:

$$PQ = b_0 + b_1 \times Age + b_2 \times Age^2$$

Now to look at the contribution of each term, it is necessary to do three regression analyses: one containing only the term in Age, one containing only Age², and one containing both. The corre-

[8] *There are, of course, practical limits, both logistical and statistical. Like all regression problems, diminishing returns are the order of the day.*

[9] *See the section "Dummy Coding" in Chapter 14 if you forget what a dummy variable is—dummy!*

[10] *A more illuminating approach might have been to create a fourth group that gets both drugs. We would then likely want to examine whether there is an interaction—whether both drugs together are better than the simple additive effect. This is easily handled with yet another parameter, based on the product of the two dummy variables, so it is 1 if both are present, and 0 otherwise.*

[11] *Both are courses that have proven absolutely essential to your subsequent success in life—for those of you who are physicists or pure mathematicians.*

sponding ANOVA tables are shown in Table 16–6. Clearly Age alone does a fairly crummy job of fitting the data, and Age^2 is even worse, but the combination of the two is quite credible.

If we then proceed to determine the individual contributions, the additional Sum of Squares for Age^2 is (6831 − 1043) = 5788 with one degree of freedom, and the residual term is 207.6. So, the *F* test for this variable is 5788/207.6 = 27.8, equivalent to a *t* of 5.27, which is significant at the .0001 level. For Age, the *t*-value is 5.57, also significant at the .0001 level.

Clearly the addition of the quadratic term to the linear term resulted in a much better fit than either alone. Conceptually, the analysis proceeded in familiar fashion, simply creating a multiple regression equation in which there are terms for Age and Age^2. If the model still didn't look too great, and we saw from the figure some suggestion of additional curlicues, we could easily add in a cubic or quartic term.[12]

The general strategy of creating nonlinear terms in a regression equation is called **Power Series Regression** when it involves a series of terms with increasing powers of some variable (obviously). However, there are other times we might want to depart from a straight-line relationship.

One of the most common times occurs when things are either accruing, such as serum drug levels, or falling away, as in weight-loss programs (in theory, at least). In these cases, we would likely invoke negative exponential or logarithmic terms.

TABLE 16–6

Nonlinear regression with age and age^2

Age

Source	Sum of squares	*df*	Mean square	*F*	*p*
Regression	1043	1	1043.0	1.72	.210
Residual	8479	14	605.6		

$R^2 = 0.109$

Age^2

Source	Sum of squares	*df*	Mean square	*F*	*p*
Regression	412.6	1	412.60	0.634	ns
Residual	9110.0	14	650.7		

$R^2 = 0.043$

Age + Age^2

Source	Sum of squares	*df*	Mean square	*F*	*p*
Regression	6831	2	3416.0	16.49	.0001
Residual	2692	13	207.1		

$R^2 = 0.717$

ns = not significant.

General Linear Model

By now, it is apparent that we can recast just about every ANOVA problem as a regression problem by the appropriate choice of dummy variables. We can introduce interaction terms to cover all contingencies. We can also throw around nonlinear terms with gay abandon and again shove them into some regression form, then proceed with business as usual. In short, all these problems can be cast as a sum of terms, each with a coefficient to be estimated, something of the form:

$$Y_j = b_0 + \sum b_i f(X_j) \quad (16\text{–}3)$$

where $f(X)$ indicates some mathematical function of *X*—linear, quadratic, exponential, logarithmic, or whatever.

At some point in the not-too-distant past, the similarity between ANOVA, ANCOVA, and regression was noted,[13] and all of these analyses are now viewed as subsets of something called the **General Linear Model** (GLM). Nothing changes except the concept, but it's great to impress people at cocktail parties with.[14]

For a sense of closure, we'll finish off by completing the analysis of the PMS data according to the initial observations of the astute clinician. We'll fit a model of the form:

$$PQ = b_0 + b_1(Age) + b_2(Age)^2 + b_3(Belt) + b_4(Test/Other)$$

We have a nominal variable (Test/Other), a ratio variable (Belt), and a power series in Age. When the dust settles, the analysis appears as in Table 16–7.

So, as before, the addition of more terms resulted in an improved fit and a reduction of the error term. Note, however, that individually the only significant

TABLE 16–7

Analysis of Age, Age^2, Belt size, and Test/Other

Variable	*b* coeff.	SE	*t*	*p*
Test/Other	−13.50	7.65	1.76	.105
Age	4.71	1.32	3.56	.004
Age^2	−.0041	.012	3.33	.007
Belt size	1.241	.557	2.23	.048

ANOVA of the regression

Source	Sum of squares	*df*	Mean square	*F*	*p*
Regression	7755	4	1938.0	12.06	.001
Residual	1768	11	160.7		

[12] *We can't, however, add in a term in x^5, because we don't know how to express it in Latin. (From long-suffering coauthor: It's "quintic.")*

[13] *A similarity which we deliberately highlighted by treating ANCOVA*

with regression analysis, instead of drawing out the standard, and forgettable, formulas for it.

[14]*To* really *impress them, you don't say "I did a Gee Ell Emm yesterday at work." That would give the game away. You have to say "I did a Glim at work yesterday." That marks you as a real* cognoscento.

[15]*Actually one for each subject but the first one. The* **b** *coefficients for each subject express the difference between the mean of each subject (after the first one) and the first subject. The old "N – 1" routine strikes again.*

[16]*You have no idea how long it took to get data cooked right.*

[17]*Hardly a persuasive argument unless you designed them.*

[18]*Following on our previous discussion, a steroid preparation designed to kill off muscle tissue in the cerebral cortex.*

contributions came from the terms in Age and Belt Size, and there is no longer an effect of Treatment.

Although the last bit looks just like we changed the name to protect the guilty, it's actually a lot more profound than that. The fact that we can lump both nominal and interval-level variables into the same equation starts to collapse the distinction between the two historical classes—regression and ANOVA. Indeed, we can see that both are just special cases of the GLM. In principle, any of the designs we have discussed in the previous chapters can be put into the form of a linear equation, in which the interval-level predictors like Belt Size are entered directly, the nominal level predictors are accommodated through the use of dummy variables, and power terms and interactions can be thrown in with gay abandon. Even repeated-measures designs, in which *Subject* is an explicit factor, are handled by creating whole families of dummy variables—one for each subject.[15] Including interaction terms just amounts to more terms in the equation.

Although folks don't often conduct their analyses this way, since special purpose software to do ANOVA or regression is much simpler to use, there are two good reasons to know about GLM. First, it becomes a constant reminder that all of statistics—in fact, all of science—is really about explaining variance. Second, in some sticky situations, such as multifactor ANOVAs in which some subjects had the indecency to drop out leaving you with an unbalanced design (Oh, the shame of it!), GLM approaches must be used to avoid biased estimations.

Assumptions of ANCOVA

Unfortunately, ANCOVA comes with some costs, namely the usual raft of assumptions. Certainly one condition is that the lines are parallel. We neatly avoided this issue by cooking our data so that we always ended up with parallel lines.[16] The two reasons why the lines must be parallel are (1) because that's what the ANCOVA packages are designed for,[17] and (2) because that is the only way you can estimate a treatment effect. After all, if the lines are not parallel, that means that the effect of treatment (the distance between the lines) is different depending on where you are situated on the *X*-axis. So if somebody comes along and poses the question, "So, hotshot, how good is Corticomyostatin[18] anyway?" you would have to concede that it depends on how smart you are to begin with, as assessed by the MCAT score. And the last thing any clinician, pharmacist, or snake-oil salesman worth his fee wants to be caught saying is, "That all depends."

Actually, now that you have, through our guidance, achieved a sense of holistic serenity about the world of statistics, you may realize that this condition is not really too constraining. In the first place, many situations arise where there is no relationship between the treatment and the covariate. Patients may well respond about the same to a drug, regardless of the initial state of the disease (or they may not). Second, as we pointed out in Chapter 9, we rather like interactions because they can be informative, and this is just another example of an interaction.

In any case, the prudent and standard action to take is to always test for an interaction first, before proceeding with the ANCOVA. One way to do this is by performing a separate regression on each line, determining the slopes, and then testing whether the slopes are significantly different. If they are, then you don't proceed with the ANCOVA. Note that most computer programs automatically test for parallelism.

A better approach is to use a slightly more elaborate model, one that explicitly includes an interaction term. It involves an arcane and complex methodology called multiplication, where you multiply the treatment dummy variable and the covariate together, and then fit a new constant. Here's how. Recall that the model equation before was:

$$\text{PQ} = b_0 + b_1 \times \text{Treatment} + b_2 \times \text{Belt}$$

If we now add in an interaction term, the new equation looks like:

$$\text{PQ} = b_0 + b_1 \times \text{Treatment} + b_2 \times \text{Belt} + b_3 \times \text{Treatment} \times \text{Belt}$$

Now remember that the way we pulled this off was to use a dummy variable with values of 0 for Other or 1 for Testosterone. If we do the same stunt here, the equation for the control group, which is coded 0, is:

$$\text{PQ} = b_0 + b_2 \times \text{Belt}$$

And for the testosterone group, it is equal to:

$$\begin{aligned}\text{PQ} &= b_0 + b_1 + b_2 \times \text{Belt} + b_3 \times \text{Belt} \\ &= (b_0 + b_1) + (b_2 + b_3) \times \text{Belt}\end{aligned}$$

So the treatment effect is contained in the b_1 coefficient as before. The two slopes are estimated separately; for the control group the line has a slope of b_2, and for the treatment group the line's slope is $(b_2 + b_3)$. So any difference in the slopes shows up in the test of significance of the b_3 term, which is done as is any other regression coefficient.

Some other constraints on the selection of the covariates exist; they are more a matter of logic than of statistics.

1. *The covariate should be* related *to the dependent variable*. Because the whole game is to remove variance in the dependent variable attributed to the covariate, it should be almost self-evident (if we've been doing our jobs) why this is a good idea. But this condition does preclude the willy-nilly covarying of anything you can lay your hands on, such as age, gender, mar-

ital status, number of dogs, etc., most of which are virtually unrelated to everything. A good rule of thumb is that if the correlation between the covariate and the dependent variable is under 0.30, use ANOVA; if it's 0.30 or above, use ANCOVA.

2. *The covariate should be* unrelated *to the treatment variable*. This sounds a bit like what we were dealing with above, but it's not quite the same. Imagine, in our example, that our statistics teaching is so very good that it acts on general mathematical skills the way that teachers of yore insisted a Latin course would act on language skills[19] and that computer science teachers still insist BASIC will act on logic skills. If so, then we might expect that *Bare Essentials* would improve not just the post-test score but also the MCAT score. Now suppose further that we didn't dream up the idea of using MCAT as a covariate until we found the first conclusion from the *t*-test, and at that point we insisted that all the little dears had to take the MCAT as a condition of getting through the course.

   If all these supposes are so, then the treatment will change *both* the post-test and the MCAT score equally. The net result will be that the two groups will end up on the same regression line except that the treatment group will have moved up and to the right, reflecting improvement in both MCAT and post-test scores. Thus we would falsely conclude that treatment had no effect.

   For this reason, conservative statisticians demand that any covariates be measured before treatment. We are a little less severe; we'll accept that, for all its virtues, *Bare Essentials* is unlikely to influence height or religion,[20] so these could be measured anytime (although we're not sure why you would).

3. *If multiple covariates are used, these should be unrelated to each other.* It is straightforward to extend the strategy to include the analysis of multiple covariates—straightforward and usually dumb. The reason is already familiar (we hope). As you introduce additional variables, the law of diminishing returns rapidly takes hold so that each new variable accounts for relatively little additional variance but costs 1 df or more. The situations where gains can be had from more than one or two covariates are rare indeed.

SAMPLE SIZE

As you might have guessed, by the time we arrive at these complexities, any attempt to make an exact sample size calculation is akin to keeping an umbrella open in a tornado. There are therefore two strategies available:

1. Do as we did in Chapter 14. Add up all the independent variables (not forgetting to count dummy variables as appropriate), multiply by 10 (Kleinbaum, Kupper, and Muller, 1988), and that's the sample size.
2. Take the comparison you *really* care about and calculate a simple sample size for it. For example, in a two-group drug trial with a covariate, the comparison of real interest is drug/placebo. Use the formula for a *t*-test (Chapter 7), indicate that the use of a covariate will add statistical power, and stop. As another example, if you wanted to measure change with ANCOVA, you could use the formula for a paired *t*-test and again indicate that it is likely conservative.

SUMMARY

This chapter described several advanced methods of analysis based on regression analysis. Power series analysis and other nonlinear regressions are simply multiple regressions where coefficients are estimated for various functions of the X variable. ANCOVA methods combine continuous variables and grouping factors into a single regression equation, using dummy variables for the latter.

EXERCISES

1. In the following designs, identify the between-subject factors, within-subject factors, and covariates.
   a. A group of students are randomized to receive either (a) a wonderful, humorous, perceptive, brilliant, and witty new statistics book (this one, naturally) or (b) the same old boring, dull, inarticulate, condescending statistics book (any of the others) at the beginning of a stats course. The mark in their last undergraduate math course is recorded. At the end of the stats course, they complete a 60-item multiple choice test.
   b. Patients with chronic leg cramps are randomized to receive either calcium supplements or a placebo. After 6 weeks, they are asked to rate whether the pain has become better or worse and by how much (on a 100-mm Visual Analog Scale).
   c. The effect of transcutaneous electrical nerve stimulation (TENS) is assessed by physiotherapists. Each time patients with low back pain come in for treatment, they are given TENS at one of six different power levels assigned at random. Unbeknownst to the patient or therapist, a random device in the machine turns it on or off for a particular session. This continues until patients have completed 12 sessions—TENS/Placebo at 6 levels.

[19] *R.L. Thorndike conclusively disproved this one in 1904, but many of us were still taking Latin in the 1960s. So much for the influence of evidence.*

[20] *On the other hand, if L. Ron Hubbard can do it, why can't we?*

d. As in *C* above, but the sample is stratified on male/female.
e. Surgical performance, measured by the total time required to remove a gallstone, is predicted using the following variables: (a) Right-handed or left-handed, (b) Reaction time, (c) I.Q.

2. The "Dr. Fox Effect" demonstrates that a charming, witty speaker can suck everybody into believing his message. (That's where we get Presidents and Prime Ministers from, silly). To further explore this phenomenon, students received a series of seminars from a total of 12 speakers of varying ages. Six were dressed neatly and nattily (NN), and six were dressed soiled and shabbily (SS). The effect of dress and speaker age on student ratings were explored. As one final wrinkle, students were divided by gender, with 10 men and 10 women in the class.
   a. What variable corresponds to "Subjects"?
   b. What is the "Between-Subjects" factor? How many *df*?
   c. What is the covariate, and how many *df* does it have?
   d. How many repeated measures are there? What is the *df* associated with each?

How to Get the Computer to Do the Work for You

- To add covariates for both factorial and repeated-measures ANOVAs, there is just one more step: simply move the desired variable(s) from the list on the left to the box labeled **Covariates**
- To add covariates within a Regression approach, move the covariate(s) into the **Independent(s)** box
- Then press the **Next** button to the right of **Block 1 of 1**
- Enter the other predictor variables and again press **Next**
- Select **Forward** from the **Method** option
- For regression equations that include power terms and interactions, it is easiest to first create new variables using **Transform → Compute**

CHAPTER THE SEVENTEENTH

Measuring Change

There are a number of approaches to examining the change in a variable over time. The simplest involves difference scores, and is analyzed with a paired *t*-test. Analysis of covariance is more appropriate, and generally leads to a more powerful test. ANCOVA can also be generalized to the situation in which there are multiple occasions of measurement. Finally, Individual Growth Curve Analysis is unique in its ability to model individual change and deal with missing data.

SETTING THE SCENE

Over the years, people with arthritis have fallen prey to countless over-the-counter preparations, guaranteeing immediate relief from the pain, and charging outrageous prices considering that about 90% are just variants on aspirin. In this chapter, we examine one more of the family, Robert's Rectal Pills, which has the dubious advantage that its mode of administration is somewhat unusual, at least for an aspirin concoction. In the course of doing so, we explore a number of methods for assessing change (in this case, change in joint counts) ranging from the simple, obvious, and ubiquitous methods (difference scores and paired t*-tests) to more advanced and powerful methods, including ANCOVA and Individual Growth Curve Analysis.*

The measurement of change has received a lot of press recently (e.g., Norman, 1989, 1997). It has intuitive appeal to clinicians—something along the lines of, "We're in the business of making people better, so we should measure how much better we make them." It also seems like the proper thing to do statistically. In fact, in many situations, it is so self-evident that we do it almost without thinking about it. For example, if we were to do a study of some new diet plan by assigning everyone to a treatment or control group, no one in his or her right mind would just look at the weights after the treatment was over. Obviously, you would measure everyone before and after, find out how much each person gained or lost, then analyze these difference scores.[1] This step seems so intuitively correct that it couldn't possibly be wrong. Unfortunately, while it's not exactly wrong, it's not quite right either. In addition, it is limited to the simple case of only two measurements—a pretest and a post-test. In this chapter, we will begin with this simple case, then develop several other more powerful methods to assess change in more complex (and powerful) experimental situations.[2]

For illustration, consider briefly a disease like rheumatoid arthritis, which goes up and down with the weather, the time of day, the period of the moon, and the advancing perihelion of Mercury's orbit. Now we have a new wonder drug, called Robert's Rectal Pills (RRP).[3] Like most over-the-counter medications, it contains mostly ASA (acetylsalicylic acid) with a mere soupçon of baking soda.

To test it, we could bring a series of patients into the clinic and measure their joint counts or any of the dozen other things rheumatologists like to measure. For the moment, let's just stay with joint counts (JC).[4] Then, like all good experimenters, we randomize them into two groups. One group would get RRP, and another would get some other pill which looks like RRP, goes in the same place, but contains only the baking soda. We would wait a month, bring 'em all back, measure their new JC, do some analyses, and publish. To put some meat on the bones, the JCs for 20 patients (10 in the RRP group, 10 in the placebo group), before and after treatment, are shown in Table 17–1.

If that's all we did, then the straightforward analysis is, as we already indicated, an unpaired *t*-test on the difference scores for each subject in the treatment and control groups. We did that for the data in Table 17–1, and it turns out to be 1.99 which, with $df = 18$, is not significant at the .05 level.

[1] *And if you want to practice your statistical prowess, the right test is an* unpaired *t*-test *on the difference scores derived from the treatment and control groups.*

[2] *The reason it is wrong or at least suboptimal to use difference scores has to do with a phenomenon called* ***regression to the mean****, which we'll talk about in due course. If there is no measurement error, which excludes every study we've ever done or read about, then difference scores are perfectly OK.*

[3] *Carter and Dodd beat us to the better organs.*

TABLE 17–1

Data from a randomized trial of RRP versus placebo with a pretest and a 1 month post-test

Subject	Group	Pretest of JC (0 months)	Post-test of JC (1 month)	Difference
1	1	22	16	−6
2	1	24	17	−7
3	1	32	25	−7
4	1	24	21	−3
5	1	35	32	−3
6	1	27	22	−5
7	1	34	27	−7
8	1	15	13	−2
9	1	29	25	−4
10	1	25	21	−4
MEAN		26.7	21.9	−4.8
SD		6.1	7.8	1.78
11	2	32	31	−1
12	2	33	34	+1
13	2	42	34	−8
14	2	27	24	−3
15	2	22	24	+2
16	2	18	15	−3
17	2	16	13	−3
18	2	32	29	−3
19	2	25	19	−6
20	2	24	22	−2
MEAN		27.1	24.5	−2.6
SD		7.8	7.5	2.80

TABLE 17–2

ANOVA of pretest and post-test scores for the RCT of RRP

Source	Sum of squares	*df*	Mean square	*F*	*p*
Group	22.5	1	22.5	0.25	.63
Subject: Group	1617.4	18	89.85		
Time	136.9	1	136.9	44.8	<.001
Time × Group	12.1	1	12.1	3.96	.062
Error	55.0	18	3.05		

On the other hand, if you have been taking to heart all the more complicated stuff in the last few chapters, you might want to do a repeated-measures ANOVA, with one between-subjects factor (RRP group vs. Placebo group) and one within-subjects repeated measure (Pretest/Post-test). Since we expect the treatment group to get better over the time from pretest to post-test and the control group to stay the same, this amounts to a Pre-/Post-test × Group interaction. All these calculations are shown in Table 17–2 if you're interested.[5]

The mean scores for the RRP and control groups are about the same initially: 26.7 and 27.1. After a month, the JC of the RRP group has dropped some to 21.9, and the JC of the control group has also dropped a bit to 24.5. The Time × Group interaction resulted in an *F* test of 3.96 with 1 and 18 degrees of freedom, which has an associated *p*-value of .062. This is equivalent to the *t*-test of the difference scores, which came out to 1.99 (the square root of 3.96) with the same probability (.062), just as we had hoped. So the interaction term is equivalent to the unpaired *t*-test of the difference scores, and the *F* test of the interaction is just the square of the equivalent *t*-test. Once again, statistics reveals itself to be somewhat rational (at least some of the time).

[4] *If we really did measure a whole bunch of things, we should use Multivariate Analysis of Variance (MANOVA). See Chapter 12.*

[5] *They're shown in Table 17–2 even if you're not interested.*

REGRESSION TO THE MEAN AND ANCOVA

We keep dropping hints that this analysis of difference scores, while sensible, isn't quite right. What's so wrong about it? Well, one thing that's obviously wrong is that it wasn't significant, so we can't publish the results, and Robert may want his money back. Is that just the luck of the draw, or is there something happening in the data that we didn't pick up on?

Let's take a closer look at the data in Table 17–1. On the right side, we have displayed the difference in joint counts from pre- to post-treatment. A close inspection reveals that it seems that in both the treatment and control groups, the worse you are to begin with, the more you improve. Even in the placebo group, those with really bad joint counts initially seem to get quite a bit better after treatment. What an interesting situation. The drug and the placebo both do a world of good for severe arthritics, but don't help mild cases; in fact, they get worse.

What is really going on is a situation called **regression to the mean.** At issue is the reason people started off with severe or mild pain in the first place. Some folks in the severe group were just having a crummy day and felt much worse than they do on the average. A month later, on the average they got a bit better, closer to their normal state of disease, so they ended up with somewhat better scores. Conversely, some of the folks with initially "mild" disease were there because they were having a good day; a month later they had regressed closer to their normal state, and consequently appeared to get worse from treatment. If we were to proceed with a difference score on everybody, the high scorers would end up with negative differences as they moved closer to the mean, and the low scorers would have positive scores as they did the same. This occurred because of measurement error on the two occasions. This is more obvious if we plot the post-test scores against the pretest scores, as shown in Figure 17–1.

If we examine the two trend lines carefully, what we see is that, although both pass close to the origin, a pretest score of around 50 is associated with a post-test score of around 40. This means that, although everyone overall, in both the treatment and control groups, is getting a bit better (lower scores), the worse you are initially, the better you get. Now if we create a difference score by subtracting the baseline state from the post-test state, we will, in general, be subtracting some of this measurement error as well as the true score. The further we get from the mean, the worse will be the overcorrection for the baseline score.

Graphically, when we take the difference score, this amounts to fitting the data to a regression line through the origin with a slope of 1, and, as the figure shows, the data are better fitted by a line with an intercept above the origin and a slope less than one. Algebraically, the difference score is modelled by:

$$y - x = \delta; \text{ i.e., } y = x + \delta \tag{17–1}$$

A better model would be:

$$y - \alpha x = \delta; \text{ i.e., } y = \alpha x + \delta \tag{17–2}$$

where α is fitted by the model and is, in general, less than one. So, simply subtracting the initial state from the final state overcorrects for the initial state, and introduces additional error variance directly related to the error of measurement.

The solution is to do an Analysis of Covariance (ANCOVA), with final score as the dependent variable and initial state as the covariate. The ANCOVA fits the optimal line to the scores, in the regression sense of minimizing error, and avoids overcorrection. We have done this in Table 17–3. Now we find that the effect of treatment results in an F test of 4.36, which, with 1 and 17 df, is significant at the .05 level.

The use of ANCOVA in the design more appropriately corrected for baseline differences, leaving a smaller error term and a significant result. Under fairly normal circumstances, the gains from using ANCOVA instead of difference scores will be small, although, occasionally, there can be gains of a factor of two or more in power. Of course, if there is less measurement error, there is less possibility of regression to the mean, and less to gain from the use of ANCOVA.

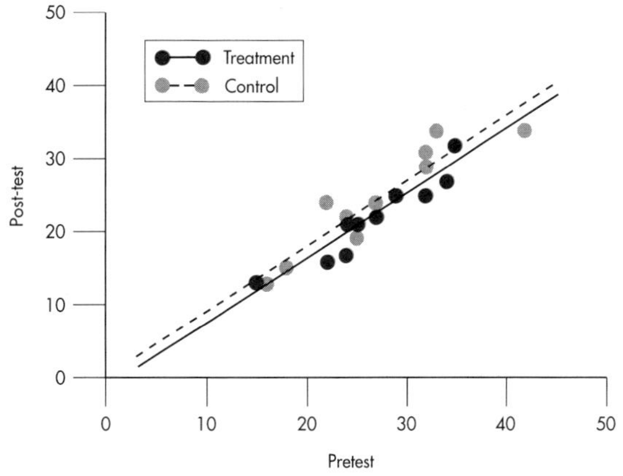

FIGURE 17–1 Relationship between pretest and post-test scores.

TABLE 17–3 Analysis of covariance of pretest and post-test scores for the RCT of RRP

Source	Sum of squares	df	Mean square	F	p
Group	25.2	1	25.2	4.36	.05
Pretest (covariate)	691.1	1	691.1	119.57	<.001
Error	98.3	17	5.78		

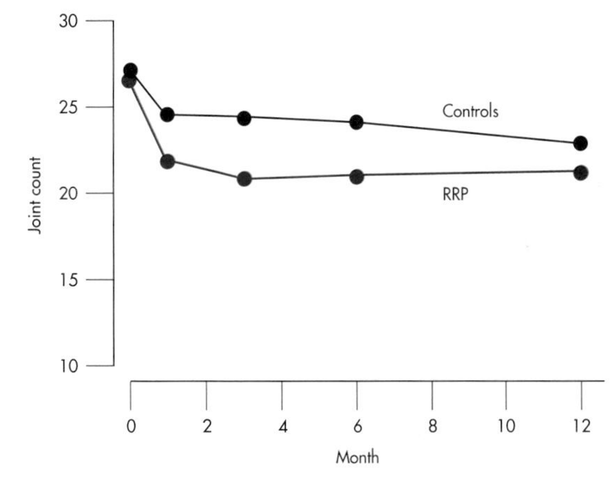

FIGURE 17–2 Joint count for the study of RRP (from Table 17–4).

MULTIPLE FOLLOW-UP OBSERVATIONS: ANCOVA WITH CONTRASTS

That's fine as far as it goes. However, it is rarely the case that people with chronic diseases have only one follow-up visit. They come back again and again, seemingly forever and ever. It seems a shame to ignore all these observations just because you can't do a t-test on them. Of course, one approach is to pick one time interval, either by design (e.g., specifying in advance that you'll look at the 12-month follow-up) or by snooping (e.g., looking at all the differences and picking the one time period when the treatment seems to have had the biggest impact). The former is inefficient; the latter is fraudulent, although we've seen both done, with great regularity.

What's so bad about the first strategy? Two things. First, it's throwing away half the data you gathered, which is something that statisticians really don't like to do. Second, and more fundamentally, it's treating change in a very simplistic manner, kind of like a quantum change—first you're in this state, later you're in that state.[6] There's a ton of information about *how* patients are getting from one state to the other that is lost in the simple look at just the first and last measurement.

Let's take a closer look. Examine Table 17–4 in which we've thrown in some more follow-up data. We've graphed the means in Figure 17–2. As you can see, the response to the drug is pretty immediate—people in the treated group have less pain after 1 month, and more or less stay at a decreased level of pain. Those in the control group get a little better, and again, stay a little better. Of course this is only one possibility (and perhaps an unlikely one); in a

[6] *As if there's something simplistic about quantum mechanics.*

TABLE 17–4

Data from an RCT of RRP versus placebo with post-tests at 1, 3, 6, and 12 months

Subject	Group	Pretest 0 mo	Post-test 1 mo	3 mo	6 mo	12 mo
1	1	22	16	17	13	14
2	1	24	17	18	19	22
3	1	32	25	24	26	22
4	1	24	21	19	20	23
5	1	35	32	29	31	27
6	1	27	22	21	19	17
7	1	34	27	27	24	22
8	1	15	13	11	12	14
9	1	29	25	22	21	19
10	1	25	21	20	23	24
MEAN		26.7	21.9	20.8	20.8	20.4
SD		6.1	5.6	5.1	6.1	5.3
11	2	32	31	29	27	28
12	2	33	34	33	30	35
13	2	42	34	35	32	28
14	2	27	24	24	26	25
15	2	22	24	23	19	19
16	2	18	15	16	14	16
17	2	16	13	14	14	11
18	2	32	29	28	31	29
19	2	25	19	20	24	17
20	2	24	22	21	24	21
MEAN		27.1	24.5	24.3	24.1	22.9
SD		7.8	7.5	6.9	6.5	7.3

TABLE 17–5

ANOVA and ANCOVA of scores with multiple follow-ups at 1, 3, 6, and 12 months for the RCT of RRP

ANOVA results

Source	Sum of squares	*df*	Mean square	*F*	*p*
Group	151.3	1	151.3	0.967	.33
Subject: Group	2815.9	18	156.44		
Time	339.3	4	84.8	20.1	<.001
Time × Group	30.3	4	7.56	1.79	.14
Error	304.0	72	4.22		

ANCOVA results

Source	Sum of squares	*df*	Mean square	*F*	*p*
Group	141.5	1	141.5	8.00	.012
Pretest (covariate)	1848.8	1	1848.8	104.5	<.001
Subject: Group	300.9	17	17.7	44.8	<.001
Time	11.13	3	3.71	0.97	.41
Time × Group	3.55	3	1.18	0.31	.82
Time × Pretest	20.21	3	6.74	1.76	.17
Error	195.1	51	3.82		

while we'll examine some other possibilities. For the moment, let's think about how we can analyze the data and be true to the pattern of change.

Two obvious approaches come to mind, namely the two we just did: an ANOVA with five repeated measures, and an ANCOVA, using the pretest as the covariate and post-tests at 1, 3, 6, and 12 months as the repeated observations. If we do an ANOVA, we might still look for a Time × Group interaction, since we are expecting that there will be no difference at Time 0, but significant differences at 1, 3, 6, and 12 months. On the other hand, the data from the follow-up times really look like a main effect of treatment, not an interaction.

If we do an ANCOVA, things are much clearer. The baseline data are handled differently as a covariate; we would expect that the effect of treatment will simply end up as a main effect. The results of both analyses are shown in Table 17–5.

Now that's interesting! Despite the fact that we have four times as many observations of the treatment effect as before, the ANOVA now shows no overall significant effect in the main effect of Group or the Time × Group interaction, where we had an almost significant interaction before. By contrast, in the ANCOVA, the main effect of Group is now significant at the .012 level. What is going on here?

The explanation lies in a close second look at Figure 17–2. When we do the ANOVA, the overall main effect of the treatment is washed out by the pretest values, which, since they occurred before the treatment took effect, are close together. Conversely, the Time × Group interaction amounts to an expectation of different differences between Treatment and Control groups at different times, and this effect only happens when you contrast the pretest values with the post-test values (which was fine when there was only one post-test value). In effect, the differences between treatments is now smeared out over the main effect and the interaction, neither of which are appropriate tests of the observed data. By contrast, the ANCOVA gives the pretest data special status and does not try to incorporate them into an overall test of treatment. Instead, it simply focuses on the relatively constant difference between treatment and control groups over the four post-test times, and appropriately captures this in the main effect of treatment.

In the situation where there are multiple follow-up observations, the right analysis is therefore an ANCOVA, with pretest as the covariate and the later observations as repeated dependent measures. Is this only the case when the treatment effect is relatively constant over time? As it turns out, no. But to see this, we have to go one further step into the analysis and also generate some new data.

MULTIPLE, TIME-DEPENDENT OBSERVATIONS

As we indicated, the situation in which the treatment acts almost instantly and does not change over time is likely as rare as hen's teeth. A more common scenario is one in which the treatment effect is slow to build and then has a gradually diminishing effect. One example of such a relationship is shown in the data of Table 17–6, where we have added some constants to the post-test observations to make the relationship over time somewhat more complex.

It is apparent that the treatment group shows continual improvement over time but with a law of diminishing returns, and the control group, as usual, just rumbles along. This might be more obvious in the graph of the data shown in Figure 17–3. Here we see clearly that the relation between Time and JC in the treatment group is kind of nonlinear—the sort of thing that might require a $(\text{Time})^2$ term as well as a linear term in Time. However, the control group data is still a straight horizontal line.

How do we put all this into the pot? First, we have to explicitly account for Time, since the straight ANOVA treats each X value as just nominal data—the results are the same regardless of the order in which we put the columns of data. We have somehow to tell the analysis that it's dealing with data at 0, 1, 3, 6, and 12 months. Second, it is apparent that we have to build in some kind of power series (remember Chapter 16?) in order to capture the curvilinear change over time. Finally, we might even expect some interactions, indicating that the treated group has linear and quadratic terms but that the control group doesn't.

Believe it or not, all this happens almost at the push of a button.[7] It's called **orthogonal decomposition**. When this button is pushed, the computer decomposes the Sum of Squares due to Time and to the Time × Group interactions into linear, quadratic, and higher order terms (one less term in the power series than the number of time points). The results are shown in Table 17–7.

Now, the first six lines of this horrendous mess should look familiar. They're completely analogous to the sources of variance we found before when we did an ANCOVA on the pretest and post-test scores. The numbers are different, of course, because we cooked the data some to yield a more complex relationship to time. In the next nine lines things get more interesting. By asking for an orthogonal decomposition, we told the computer to pay more attention to the time axis and fit the data over time to a power series regression, so we can test whether the relationship is linear, quadratic, or cubic, and so on. What emerges is an overall linear term ($F = 17.0$, $df = 1/18$, $p < .001$) showing that there is an overall trend downwards, taking both lines into account. There is also an interaction with Group ($F = 5.41$, $df = 1/18$, $p = .03$), which signifies that the slopes of the two lines differ. Further down, there is a quadratic component interacting with Group ($F = 5.01$, $df = 1/18$, $p = .04$) showing that the line for the treated group has some curvature to it (this is not explicit in the interaction but, rather, an observation from the graph). Note that if we add up all these components, we get the three lines above which express the Time main effect and the Time × Group interaction. That is, the sum of the linear, quadratic, and cubic effects of Time (139.6 + 11.9 + 15.5) just equals 167.1, the main effect of Time. The interactions and error terms also sum to the Total Sum of Squares for the respective terms. We have decomposed the effects related to Time into linear, quadratic, and cubic terms which are **orthogonal**—they sum to the original. It's the same idea that we encountered when we did orthogonal planned comparisons as an adjunct to the one-way ANOVA—

TABLE 17–6

Data from an RCT of RRP versus placebo with post-tests at 1, 3, 6, and 12 months, and linear and nonlinear changes over time

Subject	Group	Pretest	Post-test			
		0 mo	1 mo	3 mo	6 mo	12 mo
1	1	22	16	16	9.5	9
2	1	24	17	17	15.5	17
3	1	32	25	20	22.5	17
4	1	24	21	7	16.5	18
5	1	35	32	27	27.5	22
6	1	27	22	19	15.5	12
7	1	34	27	25	20.5	17
8	1	15	13	9	8.5	9
9	1	29	25	20	17.5	14
10	1	25	21	18	19.5	19
MEAN		26.7	21.9	17.8	17.3	15.4
SD		6.1	7.8	6.2	5.7	4.3
11	2	32	31	29	27	28
12	2	33	34	33	30	35
13	2	42	34	35	32	28
14	2	27	24	24	26	25
15	2	22	24	23	19	19
16	2	18	15	16	14	16
17	2	16	13	14	14	11
18	2	32	29	28	31	29
19	2	25	19	20	24	17
20	2	24	22	21	24	21
MEAN		27.1	24.5	24.3	24.1	22.9
SD		7.8	7.5	6.9	6.5	7.3

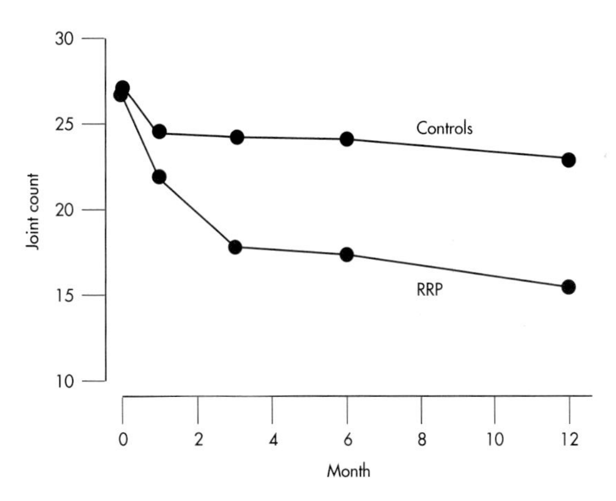

FIGURE 17–3

Joint count for the study of RRP (from Table 17–6).

[7] *In some software. The old standby, BMDP, does it in both the 2V and 5V subroutines. SPSS, as far as we can tell, will do an orthogonal decomposition, but demands equal spacing on the* X-*axis.*

TABLE 17–7

ANCOVA of follow-up scores with orthogonal decomposition for the RCT of RRP

Source	Sum of squares	df	Mean square	F	p
Group	599.9	1	599.9	36.5	<.001
Pretest (covariate)	2266.1	1	2266.1	137.7	<.001
Subject: Group	279.7	17	16.5		
Time	167.1	3	55.7	9.29	<.001
Time × Group	71.9	3	24.0	4.00	.012
Error	323.9	54	6.0		
Linear (T)	139.6	1	139.6	17.0	<.001
Linear × Group	44.2	1	44.2	5.41	.03
Error	147.2	18	8.17		
Quadratic (T^2)	11.9	1	11.9	3.06	.10
Quadratic × Group	19.5	1	19.5	5.01	.04
Error	70.0	18	3.9		
Cubic (T^3)	15.5	1	15.5	2.62	.12
Cubic × Group	8.2	1	8.2	1.39	.25
Error	106.7	18	5.92		

decomposing the Total Sum of Squares into a series of contrasts that all sum back to the original.

All this is quite neat (at least we think so), and all it requires is multiple observations over time and no missing data. Regrettably, while the multiple observations over time is easy enough to come by, persuading a bunch of patients to come back faithfully at exactly the appointed intervals, thereby missing the loving grandchild's birthday party or the free trip to Las Vegas, is as hard as Hades. This brings us to the last, and perhaps the most magical, of all the rabbits we'll pull out of this hat.

MULTIPLE OBSERVATIONS, MISSING DATA, AND INDIVIDUAL GROWTH CURVES

Up to now, we have been trapped into an ANOVA approach that is insistent to the point of obsession about complete data sets. Lose one datum and you lose the case, so it's not hard to see how all the best laid plans of mice and men gang aft a'glae.[8] Now that we are in the mood of fitting the equivalent of a regression line to the time course of events, why not fit a regression line to each person individually, then put them all together? That is, let folks make return appointments when the mood strikes them (within limits of course), fit a power series to each person's data, extract the coefficients of the fitted line (the βs), and then do some statistical shenanigans on the coefficients, instead of on the original data.

That is the basic approach to **individual growth curve** (IGC) analysis, which really puts the whole game of measuring change on a different footing altogether. What it amounts to is a two-step analysis. At step one we conduct a regression analysis for the data from each person. Thus, the repeated observations on each person, *i*, are modelled by his/her own regression equation, of the form:[9]

$$y_i(t) = \pi_0 i + \pi_{1i} t + \pi_{2i} t^2 + \pi_{3i} t^3 + \ldots + \pi_{ki} t^k \quad \textbf{(17–3)}$$

where k at a maximum is one fewer than the number of observations (time points). So, instead of dealing with the original k observations, we now use the k π coefficients derived from the regression fit of the person. Thus, in this step, there are a total of i regression equations—one for each person—and each person contributes one estimate for each of the regression parameters, π_{ik}. For example, in the data set of Table 17–6, there are 20 individuals, so $i = 20$. In turn, we took measures at 1, 3, 6, and 12 months, so $k = 4$, and we could fit up to a cubic term in time (one parameter is the intercept). (To keep things sane, from here on, we'll restrict the power series to quadratic terms, since the cubic terms were not significant in the last example anyway.)

The neat thing about the approach is that the time of follow-up for each patient need not be uniform. As long as we have enough data points, we can just fit the growth curve to that individual's data. By the same token, there need not be the same number of data for each person, since these are all just used to estimate the growth parameters. Of course, we can only estimate as many parameters (including the intercept) as we have data points for the individual.

The next step is to treat these coefficients as a new set of dependent variables so we now have a new set of k regression equations—one for each of the parameters. This second set of equations looks like:

$$\pi_{ik} = \beta_{0i} + \beta_{1i} X_{1i} + \beta_{2i} X_{2i} + \ldots + \beta_{mj} X_{mj} \quad \textbf{(17–4)}$$

Here we have a new set of m independent variables for each one of the growth parameters, so we have a total of $m \times k$ coefficients. Again, from our previous example, we will have i from 1 to 20, and we will have three X variables—the pretest score, the Group (treatment or control), and a new intercept, for each of the π coefficients. So, there will be three equations (constant, linear, and quadratic), which, in our example, look like:

$$\pi_{ik} = \beta_{0i} + \beta_{1i} Pretest_i + \beta_{2i} Group_i \quad \textbf{(17–5)}$$

[8] *This is no time to start spouting witticisms in a foreign tongue. But it's not all that foreign—it's Gaelic. The quote belongs to Robbie Burns. And please write to him, not us, about the sexist connotation. By the way, for the Yanks in our midst, "gang aft a'glae" is Gaelic for SNAFU.*

[9] *We use the symbol π (Greek letter pi) here for two reasons: (1) because that's what everyone else does, and (2) because we will shortly use β elsewhere. Note that, in this context, π doesn't equal 3.1416 and is different from how we used Π in Chapter 15 to indicate the product of a series of numbers. Thus is statistics kept precise and unambiguous.*

Thus, we can determine the relationship between the baseline score and the treatment and (a) the mean score for the individual overall, π_{0i}; (b) the linear coefficient or slope for the individual, π_{1i}; and (c) the quadratic growth term for the individual, π_{2i}.

The analysis is hierarchical, consisting of two levels. First, an initial within-subject regression analysis, which estimates the coefficients in the power series, π_{jk}, for each individual j and the various powers from 0 to k. There are as many regression equations as there are subjects. The output of this analysis is a set of regression coefficients, each expressed as a random variable with its own mean and standard deviation. At this point, we can estimate both the overall change over time of the variable, expressed as the magnitude of the various regression coefficients, and the individual variation in this change, expressed in terms of the standard deviation of the coefficients.

Second, these variables are then treated as a new set of data and predicted using a new set of linear equations, with terms for each of the predictors (in this case, Pretest and Group). This between-subject analysis results in $k \times m$ coefficients, which tells us how much each of the predictors contributes to the growth curves.

This is the overall conceptual framework behind individual growth curve analysis. The specific statistical method is called **Hierarchical Linear Modeling**, or HLM. Unfortunately, the method requires special software (Bryk and Raudenbush, 1987). However, all is not lost. A simplified one-step approach is available in one standard software package (BMDP). We will describe this approach and present results from the analysis.

In this simplified method, the analysis is conducted in one step, with a separate regression analysis for each subject. The individual parameter estimates are then combined to yield a distribution for each parameter, with its associated mean, standard deviation, standard error, and test of significance:

$$\pi_{ik} = \beta_{0i} + \beta_{1i}Pretest_i + \beta_{2i}\text{Group} + \beta_{3i}t + \beta_{4i}t^2 + \beta_{5i}Group \times t + \beta_{6i}Group \times t^2 \qquad (17\text{–}6)$$

While this looks like just another regression equation, the key distinction is that, instead of each subject furnishing a single data point for one overall regression, a set of regression coefficients is estimated for each individual, and these are then combined. Within the program, all the regression analyses are conducted simultaneously using Maximum Likelihood Estimation methods (see Chapter 15 for a description).

The results of the analysis of the data are shown in Table 17–8, along with the p-values from the previous repeated-measures ANCOVA.

As one might hope, the results are very similar to those obtained from the ANCOVA with one exception—there is no longer a main effect of treatment. The difference lies in the model. With the ANCOVA, the main effect of treatment is computed from the mean values in the treatment and control groups, averaged across all times; with this analysis, the difference between treatment and control reflects the difference in the intercepts (the values of the fitted curves at $t = 0$). Since the data tend to converge on the left side of the graph, the difference in the intercepts is smaller than the main effect. Nonetheless, we would draw a similar conclusion, namely, that the treatment results in a significant reduction in joint count over time.

TABLE 17–8 Individual growth curve analysis of follow-up scores for the RCT of RRP

Parameter	Estimate	SE	z-score	p	p from ANCOVA
Group	−.710	.655	−1.08	.27	<.001
Pretest	.856	.051	16.67	<.001	<.001
Time	−.723	.220	−3.32	.001	<.001
Group × Time	−.705	.220	−3.21	.001	.032
Time^2	−.031	.017	1.85	.06	.10
Group × Time^2	.039	.016	2.36	.02	.04

This analysis, however, doesn't really reflect the power of the method to deal with growth at the individual level, regardless of when the measurement is taken. To simulate this circumstance, we randomly deleted one follow-up observation from 14 of the 20 cases, leaving a total of 66 observations for analysis, instead of 80. If we had attempted an ANCOVA it would not have run, since we would be left with only 6 cases of the original 20. However, the IGC analysis ran happily and resulted in exactly the same conclusions as with the full data set—significant effects of pretest, Time, Group × Time, and Group × Time^2.

To summarize, IGC analysis is a powerful method for dealing with individual change, both conceptually and analytically. It is conceptually appealing as it treats growth as a continuous and individual process of change, and in the full HLM approach, it creates statistics which capture this individualized process. It is analytically useful because it can deal easily with the common situation of variable follow-up times and variable numbers of follow-up observations, across individuals.

While this sounds like the universal panacea of measuring change, it's not quite that omnipotent. It is important to keep in mind the reason some individuals may have more or fewer observations. If the number and timing of the observations might be related to the outcome (for example, sicker patients tend to die and, therefore, miss their follow-up visits), then no statistical "jimcrackery" will lead to unbiased conclusions. As usual, it also assumes that the variables (and the growth parameters) are normally distributed and that errors are independent. Because the methods are quite new, there is little known about the robustness of the approach to small violations of assumptions.

WRAPPING UP

We have explored a number of approaches to analyzing change. The simplest and most commonly used methods—difference scores and repeated-measures ANOVA—are less than optimal, although they may yield results which are approximately correct. ANCOVA methods, using the pretest or baseline measure as a covariate, are preferred and yield optimal and unbiased results. Finally, in the common situation of follow-up observations varying in number and timing, IGC analysis provides an attractive alternative.

EXERCISES

1. The bane of all statistical tests is measurement error. Suppose you did a study looking at the ability of new Viagro to regenerate the hair on male scalps. You do a before/after study with a sample of 12 guys with thinning hair, before and after 2 weeks of using Viagro. Being the compulsive sort you are and desperate for something—anything—to prevent baldness, you count every single hair on their heads. Although they are thinning, the counts are still in the millions and, so, are highly reproducible from beginning to end. Regrettably, with a paired t-test, the difference is not quite significant, $p = .063$. If you proceeded to use more advanced tests, particularly repeated-measures ANOVA and ANCOVA, what might be the result?
   a. $p < .05$ for both ANOVA and ANCOVA
   b. $p = .06$ for ANOVA, $p < .05$ for ANCOVA
   c. $p = .06$ for both ANOVA and ANCOVA
2. A common practice in analyzing clinical trials is to measure patients at baseline and at follow-up visits at regular intervals until the declared end of the trial. Frequently, the analysis is then conducted on the baseline and end-of-trial measures. Imagine a trial of a new antipsychotic drug, Loonix, involving measures of psychotic symptoms at baseline, 3, 6, 9, and 12 months (the declared end of the trial). The investigators report that there was a significant drop in psychotic symptoms in the treatment group (paired $t = -2.51$, $p < .05$), but the symptoms in the control group actually increased slightly (paired $t = +0.46$, n.s.)
   a. Is this analysis right or wrong?
   b. If the analysis was repeated, which of the following would be most appropriate? And what would be the likely result?
      i. Unpaired t-test on the difference scores from 0 to 12 months
      ii. Repeated-measures ANOVA on the scores at 0 and 12 months
      iii. ANCOVA on the scores at 3, 6, 9, and 12 months with time 0 as covariate

How to Get the Computer to Do the Work for You

For most of this chapter, the statistical methods have been encountered before. We have shown you how to do paired t-tests, repeated-measures ANOVA, and ANCOVA. Refer to the relevant chapters for advice if you need a refresher.

Individual Growth Curve analysis is, regrettably, more difficult. SPSS cannot do the full HLM analysis, nor to our knowledge does any other widely available software. There are some dedicated software packages around, such as HLM (www.ssicentral.com/hlm), but it tends to be expensive (over $400 US at the time this was written).

The simplified methods which we used were conducted using BMDP5V (BioMeDical Programs), a general software package that was formerly widely used (along with SPSS and SAS) but has since been bought out by SPSS and is difficult and expensive to find.

CHAPTER THE EIGHTEENTH

Principal Components and Factor Analysis

Fooling Around with Factors

Factor analysis looks at the pattern of relationships among variables and tries to explain that pattern in terms of a smaller number of underlying hypothetical factors.

SETTING THE SCENE

You have been appointed Dean of Admissions at the Mesmer School of Health Care and Tonsorial Trades. Your contract stipulates that you will receive a bonus of $100,000 each year that the graduation rate exceeds 75%. Only after signing the contract do you find that the success rate for the last 5 years has averaged only 23.7%. You decide that the only way to increase this abysmal figure is to impose tighter admissions criteria, and you meet with the faculty to draw up a list of the desired attributes of successful students. They arrive at three: (1) the eyes of an eagle, (2) the hands of a woman, and (3) the soul of a Byzantine usurer. You devise a test battery for applicants, with five tests in each area, just to be sure you've covered the areas well. Unfortunately, the test battery takes 32.6 hours to administer, and you're still not sure that all of the tests in each area are tapping the right skills. Is there any way you can (1) make sure you're measuring these three areas and (2) eliminate tests that are either redundant or measuring something entirely different?

As usual, we wouldn't be asking these questions unless the answers were "yes." The techniques we cover in this chapter to solve the Dean's dilemma are **principal components analysis** (PCA) and **factor analysis** (FA).[1] They differ from techniques we discussed earlier in one important way: no distinction is made between independent and dependent variables; all are treated equally and are based on one group of subjects. That is, the goal of these techniques is to examine the *structure* of the relationship among the variables, not to see how they relate to other variables, such as group membership or a set of dependent variables. For this reason, some people have referred to these techniques as "untargeted."[2]

To jump ahead of the story a bit, our beleaguered Dean will use these two procedures to: (1) explore the relationship among the variables, (2) see if the pattern of results can be explained by a smaller number of underlying constructs (sometimes called **latent variables** or **factors**), (3) test some hypotheses about the data, and (4) reduce the number of variables to a more manageable size.[3] In the dark, distant past, around 0 BC,[4] PCA and FA were used for quite different purposes. However, the distinction between them has gradually disappeared, and now PCA is used almost exclusively as simply the first step in FA. We'll keep using both terms because they're still around, and we'll indicate where one technique ends and the other begins.

So, let's get back to the Dean's dilemma. After searching the literature for appropriate tests to use, he comes up with the 15 listed in the box on page 164, which he administers to the 200 applicants over a 3-day period.

[1] *That's FA, not SFA, which means something else entirely.*

[2] *"Some people" means we forgot who, and we can't find the reference.*

[3] *There are better ways to test hypotheses using factor analysis, and we'll discuss them in the chapter on Structural Equation Modeling.*

[4] *That's "Before Computers."*

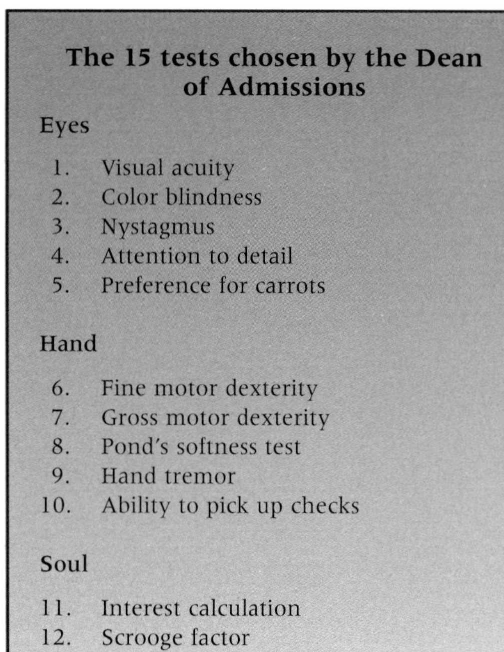
The 15 tests chosen by the Dean of Admissions

Eyes

1. Visual acuity
2. Color blindness
3. Nystagmus
4. Attention to detail
5. Preference for carrots

Hand

6. Fine motor dexterity
7. Gross motor dexterity
8. Pond's softness test
9. Hand tremor
10. Ability to pick up checks

Soul

11. Interest calculation
12. Scrooge factor
13. Dunning ability test
14. Overcharging index
15. Double billing

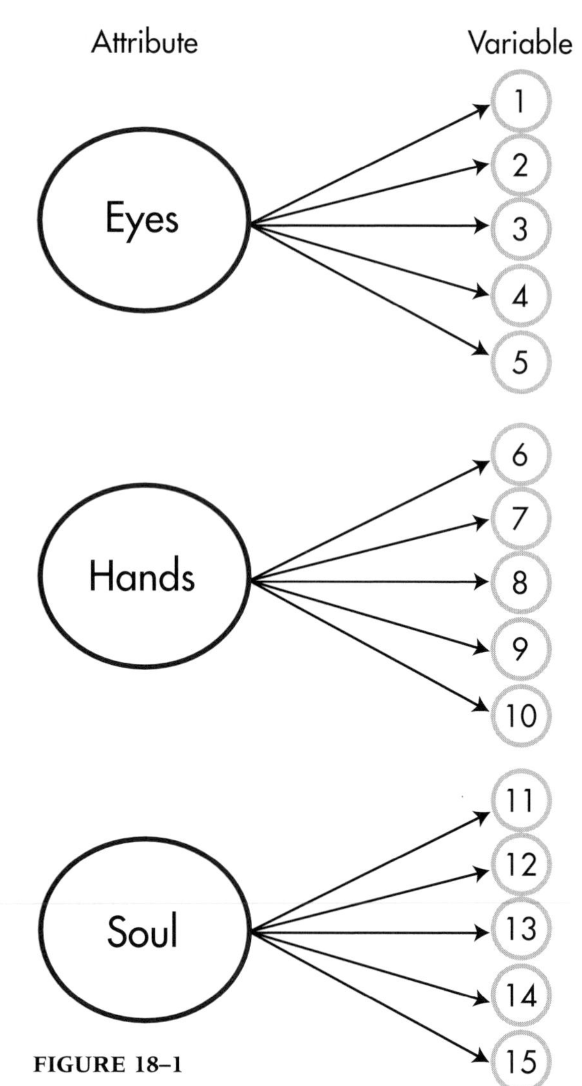

FIGURE 18–1
Three attributes (Eyes, Hands, and Soul), each measured by five tests.

WHAT ARE "FACTORS?"

What he hopes to find is shown in Figure 18–1: three different attributes, labeled in the large circles on the left and each tapped by five of the tests. Let's talk about the attributes for a moment. Strictly speaking, they don't really exist. You can't see or measure "Soul of a Byzantine Usurer" directly; you *infer* its presence from behaviors that are supposedly based on it. We expect (based on our theory of what Byzantine usurers are like) that people who have more of this attribute would charge higher interest rates, act more "Scrooge-like," overcharge more, and so on, than would people who have less of the attribute. To give another example, we can't see intelligence;[5] what we see and measure are various manifestations of intelligence. If our theory of intelligence is correct, people who have more of it should have a larger vocabulary, know more facts, work out puzzles faster, and complete more school than do people with less of it. What we measure are the purported consequences of the attribute, and we say that the common thread that makes them all correlate with each other is the underlying attribute itself.

[5] *There are some professors who maintain that it is impossible to see in their students because it isn't there. However, that is patently a base canard when applied to students who read this book.*

In psychological jargon, we call these attributes **hypothetical constructs;** in statistics, they are called **factors** or **latent variables**. One purpose of PCA and FA is to determine if numerous measures (these could be paper-and-pencil tests, individual items on the tests themselves, physical characteristics, or whatever) can be explained on the basis of a smaller number of these factors. In this example, the Dean wants to know if applicants' performance on these 15 tests can be explained by the 3 underlying factors; he will use these techniques to *confirm* his hypothesis (although, as we said, it's better to use *confirmatory* factor analysis to do this, which we'll explain in Chapter 19). In other situations, we may not know beforehand how many factors (if any) there are, and the object in doing the statistics is to determine this number. This is referred to as the *exploratory* use of PCA and FA.

Actually, Figure 18–1 oversimplifies the relationship between factors and variables quite a bit. If variables 1 through 5 were determined solely by the Eye of an Eagle factor, they would all yield identical results. The correlations among them would all be 1.00, and only one would need to be measured. In fact, the value of each variable is determined by *two* points (ignoring any measurement error): (1) the degree to which it is correlated with the factor (represented by the arrow coming from the large circles); and (2) its *unique* contribution—what variable 1 measures that variables 2 through 5 do not, and so

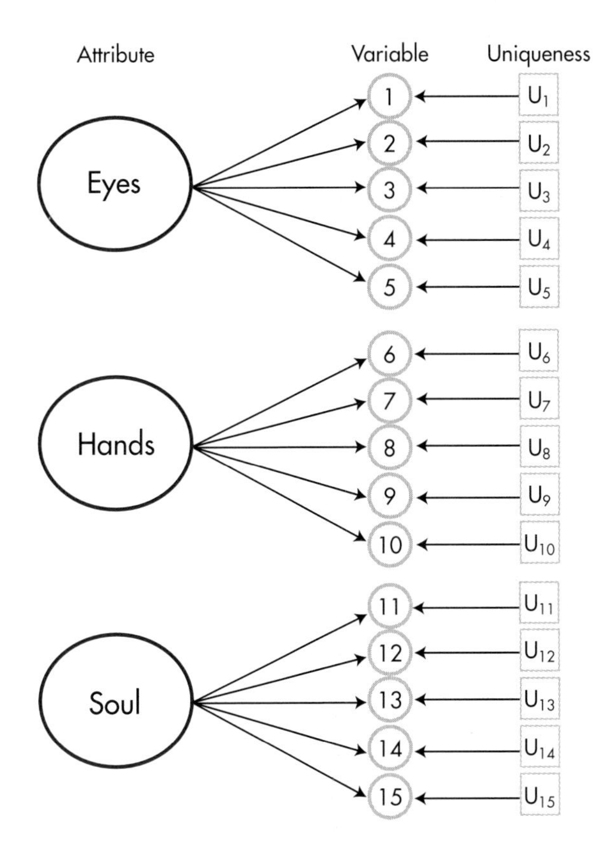

FIGURE 18–2

Adding the unique component of each variable to Figure 18–1.

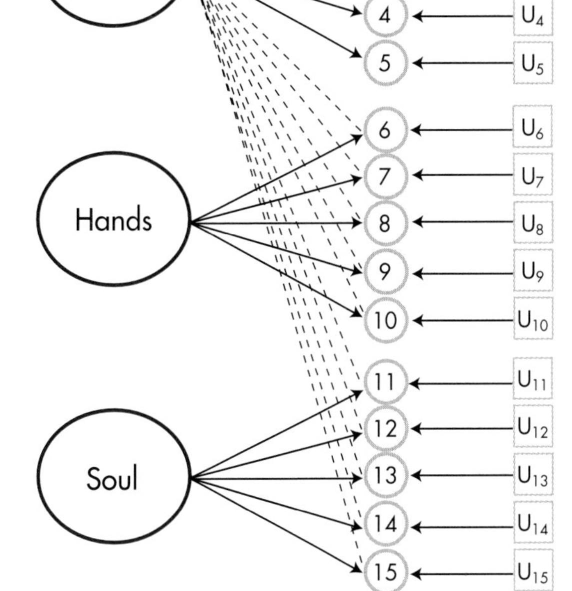

FIGURE 18–3

A more accurate picture, with each factor contributing to each variable.

on (shown by the arrow from the boxes labeled *U* in Figure 18–2). We can show this somewhat more complicated, but accurate, picture in Figure 18–2.

What exactly is meant by "uniqueness?"[6] We can best define it in terms of its converse, **communality**. The communality of a variable can be approximated by its multiple correlation, R^2, with all of the other variables; that is, how much it has in common with them and can be predicted by them. The *uniqueness* for variable 1 is then simply $(1 - R_1{}^2)$; that portion of variable 1 that cannot be predicted by (i.e., is unrelated to) the remaining variables.

Before we go on, let's complicate the picture just a bit more. Figure 18–2 assumes that factor 1 plays a role only for variables 1 through 5, factor 2 for 6 through 10, and factor 3 for 11 through 15. In reality, each of the factors influences all of the variables to some degree, as in Figure 18–3. We've added signs of these influences only for the contribution of the first factor on the other 10 variables. Factors 2 and 3 exert a similar influence on the variables, but putting in the lines would have complicated the picture too much. What we hope to find is that the influence of the factors represented by the dashed lines is small when compared with that of the solid lines.

HOW IT IS DONE

The Correlation Matrix

As we mentioned a bit earlier, the first few steps in FA, for historical reasons, go by the name of PCA. We begin with a correlation matrix. On a technical note, we start with a correlation matrix mainly because, in our fields, the variables are each measured with very different units, so we convert all of them to standard scores. If the variables all used a similar metric (such as when we factor analyze items on a test, each using a 0-to-7 scale), it would be better to begin with a variance-covariance matrix.

If life were good to us, we'd probably not need to go any further than a correlation matrix; we'd find that all of the variables that measure one factor correlate very strongly with each other and do not correlate with the measures of the other attributes (i.e., the picture in Figure 18–2). However, this is almost never the case. The correlations within a factor are rarely much above .85, and the measures are almost always correlated with "unrelated" ones to some degree (more like Figure 18–3). Thus we are left looking for patterns in a matrix of $[n \times (n - 1) \div 2]$ unique correlations; in our case, $(15 \times 14) \div 2$, or 105 (not counting the 1.00s along the main diagonal), as

[6] *What exactly is meant by any of this? However, that's a question we'd best leave for the philosophers.*

TABLE 18–1

Correlation matrix of the 15 tests

	Acuity	Color	Nystagmus	Detail	Carrots	Fine dexterity	Gross dexterity
Acuity	1.000						
Color	.318	1.000					
Nystagmus	.403	.317	1.000				
Detail	.468	.230	.305	1.000			
Carrots	.512	.423	.523	.412	1.000		
Fine dexterity	.321	.285	.247	.227	.213	1.000	
Gross dexterity	.335	.234	.268	.327	.275	.622	1.000
Softness	.304	.157	.223	.335	.301	.656	.722
Tremor	.332	.157	.382	.391	.219	.578	.527
Check	.326	.195	.184	.325	.032	.423	.414
Interest	.116	.057	−.075	.099	.105	.311	.203
Scrooge	.308	.150	.091	.110	.212	.344	.353
Dunning	.314	.145	.140	.160	.155	.215	.095
Overcharge	.489	.239	.321	.327	.222	.344	.309
Billing	.258	.301	.132	.217	.200	.303	.296

shown in Table 18–1. Needless to say, trying to make sense of this just by eye is close to impossible.

Before going on to the next step, it's worthwhile to do a few "diagnostic checks" on this correlation matrix. The reason is that computers are incredibly dumb animals. If no underlying factorial structure existed, resulting in the correlation matrix consisting of purely random numbers between −.30 and +.30 (i.e., pretty close to 0), with 1.00s on the main diagonal (because a variable is always perfectly correlated with itself), the computer would still grind away merrily, churning out reams of paper, full of numbers and graphs, signifying nothing. The extreme example of this is an **identity matrix**, which has 1.00s along the main diagonal and zeros for all the off-diagonal terms. So several tests, formal and otherwise, have been developed to ensure that something is around to factor analyze.

Some of the most useful "tests" do not involve any statistics at all, other than counting. Tabachnick and Fidell (1989)[7] recommend nothing more sophisticated than an eyeball check of the correlation matrix; if you have only a few correlations higher than .30, save your paper and stop right there.

[7] *Their book is filled with uncommonly good wisdom and should be on the shelf of anyone doing advanced stats.*

A slightly more stringent test is to look at a matrix of the **partial correlations**. This "test" is based on the fact that, if the variables do indeed correlate with each other because of an underlying factor structure, then the correlation between any two variables should be small *after partialing out the effects of the other variables*. Some computer programs, such as BMDP, print out the partial correlation matrix. Others, such as SPSS/PC, give you its first cousin (on its mother's side), an **antiimage correlation matrix**. This is nothing more than a partial correlation matrix with the signs of the off-diagonal elements reversed—for some reason that surpasseth human understanding. In either case, they're interpreted in the opposite way as is the correlation matrix; a large number of *high* partial correlations indicates you shouldn't proceed.

A related diagnostic test involves looking at the communalities. Because they are the squared *multiple* correlations, as opposed to *partial* correlations, they should be above .60 or so, reflecting the fact that the variables are related to each other to some degree. You have to be careful interpreting the communalities in SPSS/PC. The first time it prints them out, they may (depending on other options we'll discuss later) all be 1.00. Later in the output, there will be another column of them, with values ranging from 0.0 to 1.0; this is the column to look at.

Among the formal statistical tests, one of the oldest is the **Bartlett Test of Sphericity.** Without going into the details of how it's calculated, it yields a chi-squared statistic. If its value is small, and the associated p level is over .05, then the correlation matrix doesn't differ significantly from an identity matrix and you should stop right there. However, Tabachnick and Fidell (1989) state that the Bartlett test is "notoriously sensitive," especially with large sample sizes, so even if it is statistically significant, it doesn't mean that you can safely proceed. Consequently, Barlett's test is a one-sided test: if it says you shouldn't go on to the Principal Components stage, don't; but if it says you can go on, it ain't necessarily so.

Another test is the **Kaiser-Meyer-Olkin Measure of Sampling Adequacy** (usually referred to by its nickname, MSA), which is based on the squared partial correlations. In the SPSS/PC computer package, the MSA value for each variable is printed along the main diagonal of the antiimage correlation matrix, and a summary value is also given. This allows you to check the overall adequacy of the matrix and also see which individual variables may not be pulling their full statistical weight. If the value is in the .60s, Kaiser describes

Softness	Tremor	Checks	Interest	Scrooge	Dunning	Overcharge	Billing
1.000							
.619	1.000						
.385	.432	1.000					
.246	.285	.370	1.000				
.232	.300	.380	.484	1.000			
.181	.271	.413	.585	.428	1.000		
.345	.395	.480	.408	.535	.512	1.000	
.313	.287	.553	.412	.621	.555	.642	1.000

the measure as "mediocre." Those values in the .50s are "miserable" and lower ones are "unacceptable"; you should proceed with the next step accordingly. Similarly you can consider eliminating variables that show poor sampling adequacy.

Extracting the Factors

Assuming that all has gone well in the previous steps, we now go on to extracting the factors, a procedure only slightly less painful than extracting teeth. The purpose of this is to come up with a series of linear combinations of the variables to define each factor. For factor 1, this would look something like:

$$F_1 = w_{11}X_1 + w_{12}X_2 + \ldots + w_{1k}X_k \quad (18\text{–}1)$$

where the X terms are the k (in this case, 15) variables and the *w*s are weights. These w terms have two subscripts; the first shows that they go with factor 1, and the second indicates with which variable they're associated. The reason is that, if we have 15 variables, we will end up with 15 factors and therefore 15 equations in the form of the one above. For example, the second factor would look like:

$$F_2 = w_{21}X_1 + w_{22}X_2 + \ldots + w_{2k}X_k \quad (18\text{–}2)$$

Now, this may seem like a tremendous amount of effort was expended to get absolutely nowhere. If we began with 15 variables and ended up with 15 factors, what have we gained? Actually, quite a bit. The *w*s for the first factor are chosen so that they express the largest amount of variance in the sample. The *w*s in the second factor are derived to meet two criteria: (1) the second factor is uncorrelated with the first, and (2) it expresses the largest amount of variance left over after the first factor is considered. The *w*s in all the remaining factors are calculated in the same way, with each factor uncorrelated with and explaining less variance than the previous ones. So, if a factorial structure is present in the data, most of the variance may be explained on the basis of only the first few factors.

Again returning to our example, the Dean hopes that the first 3 factors are responsible for most of the variance among the variables and that the remaining 12 factors will be relatively "weak" (i.e., he won't lose too much information if he ignores them). The actual results are given in Table 18–2. For the moment, ignore the column headed "Eigenvalue" (we get back to this cryptic word a bit later) and look at the last one, "Cumulative percent." Notice that the first factor accounts for 37.4% of the variance, the first two for over 50%, and the first five for almost 75% of the variance of the original data. So he actually may end up with what he's looking for.

What we've just described is the essence of PCA. What it tries to do, then, is explain the variance among a bunch of variables in terms of uncorrelated (the statistical term is **orthogonal**) underlying factors or latent variables. The way it's used now is to try to *reduce* the number of factors as much as possible so as to get a more parsimonious explanation of what's going on. In fact, though, PCA is only one way of determining the factors. BMDP has four different methods, and SPSS/PC has seven.

Principal Components Analysis and Factor Analysis

So, given this plethora of techniques, which one do we use—PCA or one of the others? To answer this, we have to delve a bit more[8] into the difference between PCA and FA. PCA tries to account for *all* of the variance of the variables, and if all of the factors (or *components*, yet another jargon term for the same thing) are retained, no information is lost or gained.

[8] *And we promise it will be just a bit more.*

TABLE 18–2

The 15 factors

Factor	Eigenvalue	Percent of variance	Cumulative percent
1	5.6025	37.4	37.4
2	2.0252	13.5	50.9
3	1.5511	10.3	61.2
4	.8968	6.0	67.2
5	.7888	5.3	72.4
6	.7113	4.7	77.2
7	.6826	4.6	81.7
8	.5345	3.6	85.3
9	.4514	3.0	88.3
10	.4423	2.9	91.2
11	.3598	2.4	93.6
12	.3158	2.1	95.7
13	.3015	2.0	97.8
14	.1785	1.2	98.9
15	.1580	1.1	100.0

We can use all of the factors to recapture what the original data were, with perfect accuracy. In FA, however, we are interested only in the variance that a variable shares with the other variables (the variable's *communality*). So, in PCA, we start with 1.0s in the main diagonal of the correlation matrix, since that reflects the total variance for each variable. With FA, we start by placing an initial estimate of the communalities (the squared multiple correlations between the variable and the other variables) in the main diagonal, and the technique then partitions each variable's variance into two parts: the *common* variance, which is associated with the factor or latent variable, and the *unique* variance for that specific variable. FA can't determine what portion of the unique variance is *reliable* variance (actually due to what that variable taps that the others don't) and what is actually *error* variance; both are lumped together as the unique variance. This also means that the total amount of variance analyzed by PCA is equal to the number of variables, but (because the squared multiple correlations are always less than 1.0) the total amount of variance analyzed by FA is somewhat lower.

So, when do we use what? If we're trying to find the optimal weights for the variables to combine them into a single measure, then PCA is the way to go. We would do this, for example, if we were concerned that we had too many variables to analyze and wanted to combine some of them into a single index. Instead of five scales tapping different aspects of adjustment, for instance, we would use the weights to come up with one number, thus reducing the number of variables by 80%. We would also use PCA if we wanted to account for the maximum amount of variance in the data with the smallest number of mutually independent underlying factors. On the other hand, if we're trying to create a new scale by eliminating variables (or items) that aren't associated with other ones or don't load on any factor, then "common" FA is the method of choice. This is the procedure that test developers would use.

In the rest of this chapter, we'll use the terms FA and PCA interchangeably; you put in whichever name you want, depending on what you're doing, since the methods we will describe apply to both.

On Keeping and Discarding Factors

A few paragraphs back, we mentioned that one of the purposes of the factor extraction phase was to reduce the number of factors, so that only a few "strong" ones remain. But first we have to resolve what we mean by "strong," and what criteria we apply. As with the previous phase (factor extraction) and the next one (factor rotation), the problem isn't a lack of answers, but rather a surfeit of them.

At the same time, the number of factors to retain is one of the most important decisions a factor analyst[9] must make. If too many or too few factors are kept, the results from later steps may be distorted to a marked degree.

The criterion that is still the most commonly used is called the **eigenvalue one** test, or the **Kaiser criterion**, after the person who popularized it.[10] It is the default (although, as we'll see, not necessarily the best) option in most computer packages. We should, in all fairness, describe what is meant by an eigenvalue. Without going into the intricacies of matrix algebra, an eigenvalue can be thought of as an index of variance. In PCA, each factor yields an eigenvalue, which is the amount of the total variance explained by that factor. We said previously that the *w*s are chosen so that the first factor expresses the largest amount of variance. It was another way of saying that the first factor has the largest eigenvalue, the second factor has the second largest eigenvalue, and so on. So why use the criterion of 1.0 for the eigenvalue?

The reason is that the first step in PCA is to transform all of the variables to *z*-scores so that each has a mean of 0 and a variance of 1. This means that the total amount of variance is equal to the number of variables; if you have 15 variables, then the total variance within the (*z*-transformed) data matrix is 15. If we add up the eigenvalues of the 15 factors that come out of the PCA (or any other factor extraction method), they will sum to—that's right, class, 15.[11] So you can think of a factor with an eigenvalue of less than 1.0 as accounting for less variance than is generated by one variable. Obviously then, dear reader,[12] we gain nothing by keeping factors with eigenvalues under 1.0 and are further ahead (in terms of explaining the variance with fewer latent variables) if we keep only those with eigenvalues over 1.0; hence, the eigenvalue one criterion.

This test has two problems. The first is that it's somewhat arbitrary: a factor with an eigenvalue of 1.01 is retained, whereas one with a value of .99 is rejected. This ignores the fact that eigenvalues, like any other parameter in statistics, are measured with some degree of error. On replication, these numbers will likely change to some degree, leading to a dif-

[9] *In psychiatric circles, it is said that one cannot become a factor analyst until one's self has been factor analyzed.*

[10] *That's Henry F. Kaiser, not Kaiser Wilhelm.*

[11] *If you don't believe us, add up the 15 numbers in the "Eigenvalue" column of Table 18–2. See, we told you so!*

[12] *A phrase much beloved by Albert Einstein, used when he was about to hit you with something that would take 6 months to figure out.*

ferent solution. The second problem is that the Kaiser criterion often results in too many factors (factors that may not appear if we were to replicate the study) when more than about 50 variables exist and in too few factors when fewer than 20 variables are considered (Horn and Engstrom, 1979).

The Lawley test tries to get around the first problem by looking at the significance of the factors. Unfortunately, it's quite sensitive to the sample size and usually results in too many factors being kept when the sample size is large enough to meet the minimal criteria (about which, more later). Consequently, we don't see it around much any more.

A somewhat better test is **Cattell's Scree Test**. This is another one of those very powerful statistical tests that rely on nothing more than your eyeball.[13] We start off by plotting the eigenvalues for each of the 15 factors, as in Figure 18–4 (actually, we don't have to do it; most computer packages do it for us at no extra charge). In many cases (but by no means all), there's a sharp break in the curve between the point where it's descending and where it levels off; that is, where the slope of the curve changes from negative to close to zero.[14] The last "real" factor is the one before the scree (the relatively flat portion of the curve) begins. If several breaks are in the descending line, usually the first one is chosen, but this can be modified by two considerations. First, we usually want to have at least three factors. Second, the scree may start after the second or third break. We see this in Figure 18–4; there is a break after the second factor, but it looks like the scree starts after the third factor, so we'll keep the first three. In this example, the number of factors retained with the Kaiser criterion and with the scree test is the same.

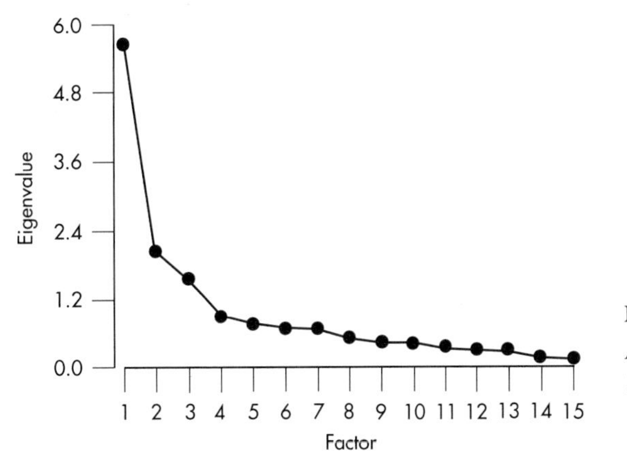

FIGURE 18–4
A scree plot for the 15 factors.

The fact that no statistical test exists for the scree test poses a bit of a problem for computer programs, which love to deal with numbers. Almost all programs use the eigenvalue one criterion as a default when they go on to the next steps of factor analysis. If you do a scree plot and decide you won't keep all the factors that have eigenvalues over 1.0, you have to run the FA in two steps: once to produce the scree plot, and again for you to override the eigenvalue criterion. You can usually do this by specifying either the minimum eigenvalue (equal to the value of the smallest one you want to retain) or the actual number of factors to keep. It's a pain in the royal derrière to have to do it in two steps, but it can be done.

Be aware, however, that the simplicity of the scree plot can be deceiving. When Cattell drew up his rules for interpreting them, he was very explicit about the scales on the axes. When computer programs draw scree plots, however, the ratio of the scale of the X-axis to that of the Y-axis varies from one output to the next, depending on how many factors there are. Therefore, what looks like a clean break using Cattell's criteria may seem smooth on the computer output, and vice versa. In fact, we found that even experienced factor analysts could not agree among themselves about how many factors were present (Streiner, 1998). So, don't use either the scree plot or the eigenvalue one criterion as absolute truth. Look on them as suggestions, and then have the computer rerun the analyses with that number of factors and also with more and fewer factors, and use the solution that makes the most sense.

The Matrix of Factor Loadings

After we've extracted the factors and decided on how many to retain, the computer gives us a table (like Table 18–3) that is variously called the **Factor Matrix**, the **Factor Loading Matrix**, or the **Factor Structure Matrix**. Just to confuse things even more, it can also be called the **Factor Pattern Matrix**. As long as we keep the factors orthogonal to each other, the factor structure matrix and the factor pattern matrix are identical. When we relax this restriction (a topic we'll discuss a bit later), the two matrices become different.

Table 18–3 tells us the correlation between each variable and the various factors. In statistical jargon, we speak of the variables *loading on* the factors. So "Visual Acuity" loads .627 on factor 1 (i.e., correlates .627 with the first factor), .285 on factor 2, and .347 on factor 3. As with other correlations, a higher (absolute) value means a closer relationship between the factor and the variable. In this case, then, "Visual Acuity" is most closely associated with the first factor.

A couple of interesting and informative points about factor loadings. First, they are *standardized regression coefficients* (β weights), which we first ran across in multiple regression. In factor analysis, the DV is the original variable itself and the factors are the IVs. As long as the factors are orthogonal, these regression coefficients are identical to correlation coefficients. (The reason is that, if the factors are uncorrelated, i.e., orthogonal, then the β weights are not dependent on one another.) This becomes important later, when we see what happens when we relax the requirement of orthogonality.

Second, the *communality* of a variable, which we approximated with R^2 previously, can now be

[13] *For this reason, Kaiser (1970) refers to this technique as "root staring" (because in matrix algebra, an eigenvalue is called a root of the matrix). Could this be an example of professional jealousy?*

[14] *In geology, "scree" is the rubble that accumulates at the foot of a hill; here it's the junk after the strong factors. In what other stats book can you also get a basic grounding in geology at the same time?*

TABLE 18–3

Unrotated factor loading matrix

	Factor 1	Factor 2	Factor 3
Acuity	.62684	.28525	.34653
Color	.42569	.28367	.35186
Nystagmus	.46046	.51771	.30305
Detail	.52711	.35772	.14661
Carrots	.48746	.49096	.43285
Fine dexterity	.68819	.12381	–.44882
Gross dexterity	.67587	.24838	–.46550
Softness	.68436	.22289	–.51248
Tremor	.69554	.16367	–.35142
Checks	.66956	–.24746	–.12809
Interest	.50922	–.57080	–.03457
Scrooge	.62186	–.42615	.09860
Dunning	.56544	–.50324	.28899
Overcharge	.73480	–.24955	.23411
Billing	.68151	–.43433	.19142

derived exactly. For each variable, it is the *sum of the squared factor loadings* across the factors that we've kept. Looking at Table 18–3, it would be $(.62684)^2 + (.28525)^2 + (.34653)^2 = .594$ for ACUITY. We usually use the abbreviation h^2 for the communality, and therefore the uniqueness is written as $(1 - h^2)$.

At this point, we still don't know what the factors mean. The first factor is simply the one that accounts for most of the variance; it does *not* necessarily reflect the first factor we want to find (such as the Eyes of an Eagle), or the variables higher up on the list. However, we'll postpone our discussion of interpretation until after we've discussed factor rotation below.

Rotating the Factors

Why rotate at all? Up to now, what we've done wouldn't arouse strong emotions among most statisticians.[15] We've simply transformed a number of variables into factors. The only subjective element was in selecting the number of factors to retain. However, if we asked for the factor matrix to include all of the factors, rather than just those over some criterion, we could go back and forth between factors and variables without losing any information at all. It is the next step, factor rotation, that really gets the dander up among some (unenlightened) statistical folks. The reason is that we have, literally, an infinite number of ways we can rotate the factors. Which rotation we decide to use (assuming we don't merely accept the program's default options without question) is totally a matter of choice on the analyst's part.

[15]To the extent to which strong emotions can be aroused in statisticians (which is why we refer to statisticians as they, rather than as us).

So, if factor rotation is still somewhat controversial, why do we do it? Unlike other acts that arouse strong passions, we can't explain it simply on the basis of the fact that it's fun. To us true believers, factor rotation serves some useful functions. The primary one is to help us understand what (if anything) is going on with the factors.

To simplify interpretation of the factors, the factor loading matrix should satisfy four conditions:

1. The variance should be fairly evenly distributed across the factors.
2. Each variable should load on only one factor.
3. The factor loadings should be close to 1.0 or 0.0.
4. The factors should be unipolar (all the strong variables have the same sign).

Let's see how well the factor loading matrix in Table 18–3 meets these criteria.

1. **Distribution of variance.** If we go back to Table 18–2, we can add up the eigenvalues of the first three factors. Their sum, 9.1788, shows the amount of variance explained by them (which is 61.2% of the total variance of 15). Of this amount, the first factor accounts for (5.6025 ÷ 9.1788), or 61.0%, the second factor for (2.0252 ÷ 9.1788), or 22.1%, and the third factor for the remaining 16.9%. So, the first factor contains a disproportionate share of the total variance explained by the three factors. We can also see this in the fact that all of the variables load strongly on this factor (Table 18–3): 12 of the 15 have loadings over .50 on factor 1, and only 2 variables (NYSTAGMUS and CARROTS) load higher on another factor than they do on factor 1. This situation is extremely common and is found because consistency tends to occur in people across various measures. What factor 1 often picks up is this "general factor," which only rarely tells us something we didn't already know.
2. **Factorial complexity.** Whenever a variable loads strongly on two or more factors, we call it **factorially complex.** In Table 18–3, NYSTAGMUS loads strongly on factors 1 and 2, CARROTS loads on all 3 factors to comparable degrees, and so on. Factorial complexity makes it more difficult to interpret the role of the variable. INTEREST is explained by both factor 1 and factor 2 and, conversely, the explanation of these factors must take CARROTS into account. It would make life much easier if we could understand the factors on the basis of mutually exclusive sets of variables.
3. **Magnitude of the loadings.** This is really a consequence of the second criterion. If a variable loads strongly on one factor, then its loadings on the other factors will be close to 0. The reason is that the sum of the squares of the loadings across factors (the variable's *communality*, you'll remember) remains constant when we rotate; so as some loadings go up, others have to go down.
4. **Unipolar factors.** If some loadings were positive and others negative, then a high score on the factor would indicate more of some variables, whereas a low score would indicate more of other variables. Again, in the inter-

est of interpretive ease, we'd like the factor to be unipolar; that is, a higher score on the factor means more of the latent variable, and a lower score simply means less of it. This occurs when all of the factor loadings have the same sign.

From a mathematical viewpoint, nothing is wrong with most of the variance being in one factor, or with factorial complexity, or with loadings in the middle range, or with bipolar factors. However, it is easiest to interpret the results of a factor analysis if we can meet these criteria and aim for *structural simplicity*. This is what rotating the factors tries to do. Unfortunately, no one's found a way to optimize all of these criteria at once. A rotation that spreads the variance equally across the factors may not necessarily reduce factorial complexity; and one that reduces complexity may not produce unipolar factors. Needless to say, this has resulted in a profusion of rotation techniques, each one designed to give priority to a different criterion, and all of which yield somewhat different results.[16] The one that's used most is called **varimax**, and that's what we'll go with first.

A simple example. Let's see how rotating the factors can help meet the four criteria and grant us our wish for simplicity. However, because it's hard to draw 3-dimensional patterns (1 dimension for each factor), we'll start off by forcing the PCA to give us only two factors. We can then generalize the procedure to three or more factors, although we won't be able to visualize the results as readily.

By asking for two factors, our factor loading table will have just two columns. Let's plot each variable using the loading on Factor 1 as the *X* coordinate and the loading on the second factor as the *Y* coordinate. What we'll get is Figure 18–5, where we can see problems with all of the criteria: (1) all of the variables show some degree of loading on Factor 1; (2) most of the variables are in the middle portions of the quadrants, showing that they are loading on both of the factors; (3) the factor loadings all seem to fall between .4 and .8 on Factor 1, and most of them are between .2 and .6 (absolute values) on Factor 2; and (4) Factor 1 is unipolar, but Factor 2 is definitely bipolar.

Now, keeping the axes orthogonal (at right angles) to each other, let's rotate them (Figure 18–6). The new axes are labeled Factor 1′ and Factor 2′. The only problem is that if we continue to rotate the axes clockwise until Factor 1′ is horizontal, all of the Factor 2′ coordinates will be negative; again, not a statistical problem, but it makes interpretation a bit harder. We can correct this little annoyance simply by reversing all of the signs of the Factor 2′ factor loadings, which is quite kosher, mathematically speaking. We end up with Figure 18–7.

How do our criteria fare in this picture? (1) A group of variables are showing a high loading on Factor 2 but not on Factor 1, demonstrating that not all of the variables are loading on the first factor any

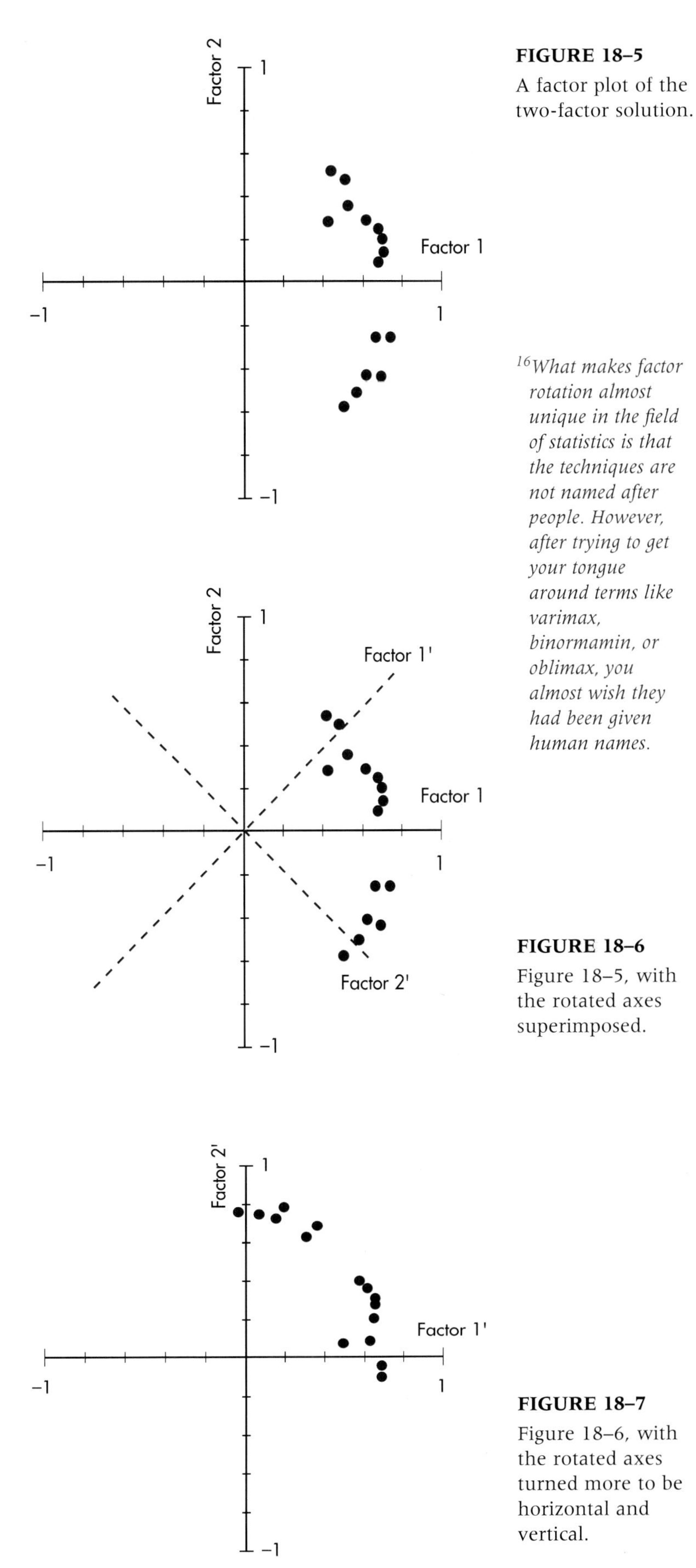

FIGURE 18–5 A factor plot of the two-factor solution.

FIGURE 18–6 Figure 18–5, with the rotated axes superimposed.

FIGURE 18–7 Figure 18–6, with the rotated axes turned more to be horizontal and vertical.

[16] *What makes factor rotation almost unique in the field of statistics is that the techniques are not named after people. However, after trying to get your tongue around terms like varimax, binormamin, or oblimax, you almost wish they had been given human names.*

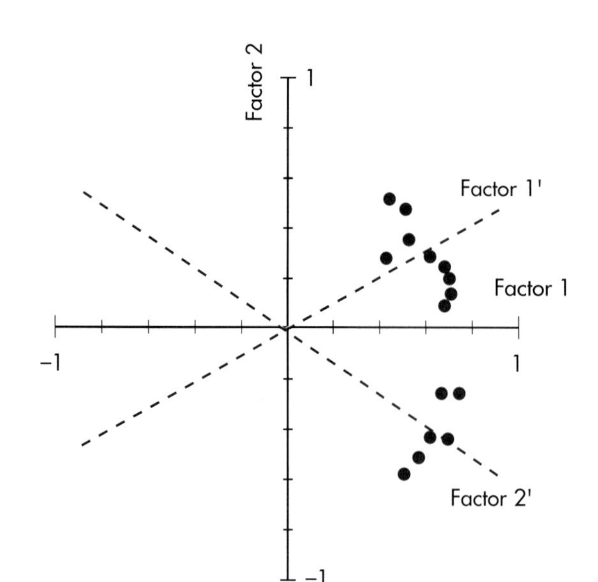

FIGURE 18–8
An oblique rotation to the factor plot in Figure 18–5.

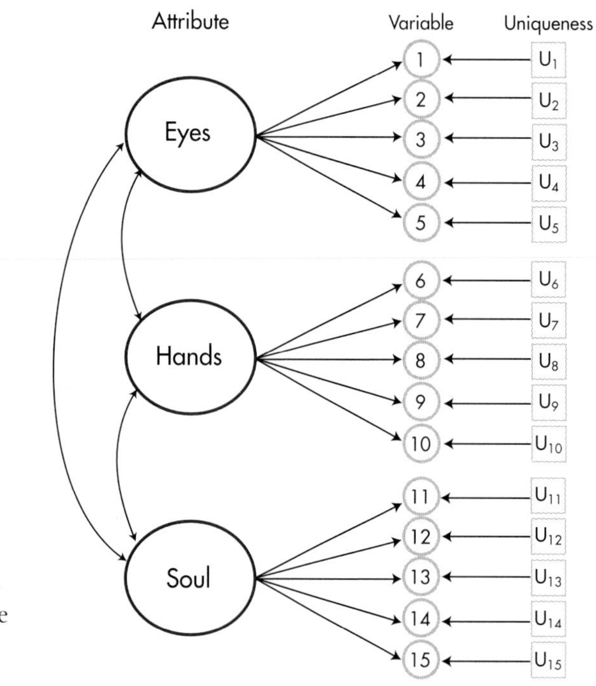

FIGURE 18–9
In an oblique rotation, the factors are correlated with each other.

more. (2) The variables seem to be closer to the axes than to the middle of the quadrant, indicating reduced factorial complexity. (3) Each variable is closer to the top on one factor and closer to the origin for the other factor, showing that the loadings are nearer to 1.0 or 0.0. (4) All of the variables are in or very near to the first quadrant. This means that all of the signs are positive (or those loadings which are negative are very small), resulting in unipolar factors. When we have more than 2 factors, we can plot all possible pairs of them. However, if we had as few as 5 factors, we'd have 10 graphs to wade through; 10 factors would result in 45 graphs, and so on.

Orthogonal versus oblique rotations. Before returning to our original problem, let's use this two-factor solution to illustrate one more point. You'll remember that earlier, we said, "...keeping the axes orthogonal (at right angles) to each other, let's rotate them" (Norman and Streiner, personal communication). However, the factors don't *have* to be orthogonal. In fact, having some degree of correlation among the factors is probably a better reflection of reality than having strictly independent ones. So, although it's easier to think of Hands of a Woman as being a completely separate attribute from Eyes of an Eagle, it's likely more accurate to think of them as being correlated to some degree.

When we rotated the axes in Figure 18–6, we were still left with some of the variables being near the middle of the quadrant. Because the angle between the axes was fixed at 90 degrees, there was little we could do. But, relaxing the condition that the factors have to be orthogonal, we can draw each axis closer to the middle of each group of variables, as in Figure 18–8. We call this an **oblique** rotation.

The advantage is that oblique solutions often lead to greater structural simplicity (using the criteria we listed before) than do orthogonal rotations. The tradeoff is that we now have to contend with the factors being correlated with each other to varying degrees. Instead of the relatively simple description of Figure 18–2, where the value of each variable is determined only by its "own" factor and its unique component, we have a more complicated situation (Figure 18–9).

In this case, to understand what Factor 1 is measuring, we not only have to look at the variables that have a high loading, but we also have to consider any correlation between Factor 1 and the others. The correlation among the factors leads to another issue, which we briefly mentioned earlier. As long as the factors were uncorrelated, each variable's regression coefficients for the factors were the same as the correlations between the variable and the factors; that is, the loadings could be interpreted either as simple correlations or as β weights. However, once we introduce some correlation between the factors, this equivalence doesn't hold any more. The factor *structure* matrix still consists of the loadings defined as partial regression coefficients, but now the factor *pattern* matrix holds the simple correlations between the variables and the factors. The higher the correlation among the factors, the greater the difference between these two matrices.

So, even though oblique rotations may mirror reality more closely than do orthogonal ones, most people prefer the latter. The reason is that orthogonal rotations have a number of desirable qualities. Because the factors are uncorrelated with each other (that's the mathematical meaning of "orthogonal"), any score derived from one factor will correlate 0 with scores derived from the other factors. This is a useful property if the results of a PCA or an FA are to be further analyzed with another statistical test. Also, as we've said, the interpretation of the factors is far easier if they are all independent from one another.

Back to the Dean. Before we leave the topic of rotations, let's just see how our three-factor solution fared with a varimax rotation. We'll skip the graphing stage because, in the absence of 3-dimensional graph paper, we would have to look at three factor plots for the unrotated solution (Factor 1 vs. 2; 1 vs. 3; and 2 vs. 3) and an equal number after the rotation. Instead, we'll focus on the factor matrix. The unrotated matrix was given in Table 18–3; the rotated one is in Table 18–4.

Before rotation, these three factors accounted for 61.2% of the total variance; this doesn't change. What does change is the distribution of the variance across factors. If you recall, of the variance that is explained, Factor 1 was responsible for 61.0%, Factor 2 for 22.1%, and Factor 3 for 16.9%. After rotation, these numbers become 37.0%, 33.2%, and 29.8%; obviously a much more equitable division. This is also reflected in the fact that now only five variables load strongly on Factor 1; previously, the majority of them did.

The other criteria did just as well. If we plot the absolute magnitudes of the unrotated factor loadings, as we did in the left side of Figure 18–10, we see that most of them fall between .3 and .7. The right side shows the same thing for the rotated loadings; the graph is much more bimodal, with relatively few values in the middle range. So we seemingly have succeeded in driving the loadings closer to 0.0 or 1.0. Also, in the unrotated solution, 12 of the 45 loadings were negative; in the rotated one, only 3 are, and they are relatively small. Last, only one variable, CHECKS, shows any degree of factorial complexity. The conclusion, then, is that rotating the axes got us a lot closer to structural simplicity.

TABLE 18–4 Rotated factor loading matrix

	Factor 1	Factor 2	Factor 3
Acuity	.26607	.18827	.69867
Color	.14288	.06270	.60095
Nystagmus	−.02079	.18035	.73411
Detail	.10065	.29793	.57307
Carrots	.04798	.09161	.80952
Fine dexterity	.22729	.78554	.14715
Gross dexterity	.12321	.82256	.20833
Softness	.13579	.85669	.16826
Tremor	.22645	.72647	.23456
Checks	.56992	.43687	.10139
Interest	.73234	.18696	−.12260
Scrooge	.72852	.19103	.10392
Dunning	.79744	−.00664	.14328
Overcharge	.70180	.20167	.35187
Billing	.79480	.15467	.18470

INTERPRETING THE FACTORS

Now that we've got the factors, what do we do with them? The first step is to determine which variables load on each factor. To do this, we have to figure out which loadings are significant and which can be safely ignored. We know a couple of ways of doing this. One way is to adopt some minimum value, such as .30 or .40. The problem is that any number we choose is completely arbitrary and doesn't take the sample size into account; a loading of .38 may be meaningful if we had 1,000 subjects, but it may represent only a chance fluctuation from 0 with 30 subjects.

A better method would be to retain only those loadings which are statistically significant. We can do this by looking up the critical value in a table for the correlation (see Table G in the appendix). But which value to use? Stevens (1986) recommends (1) using the 1% level of significance rather than the 5% because of the number of tests that will be done,

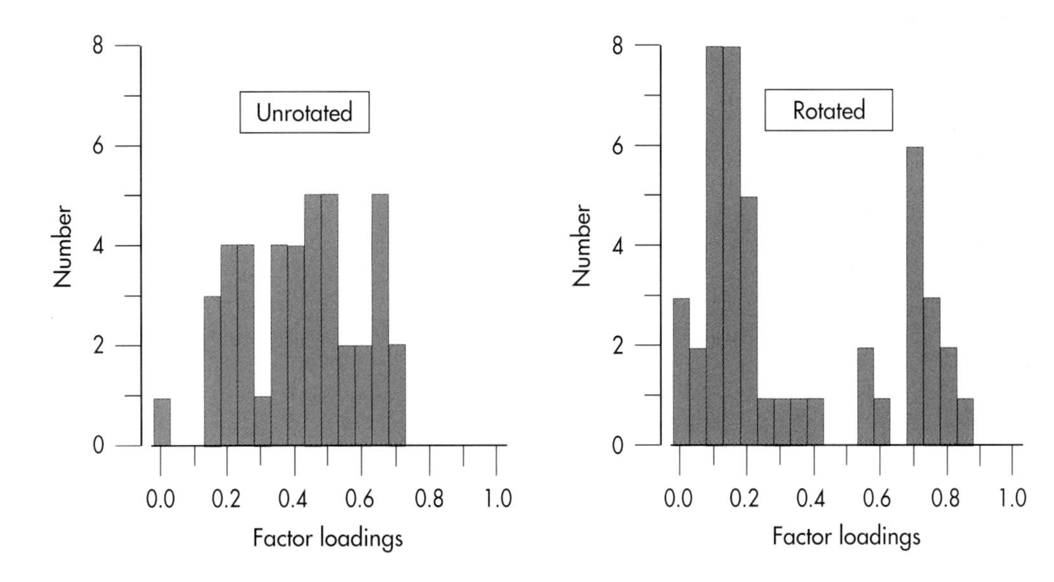

FIGURE 18–10 Plot of the factor loadings for the unrotated and rotated solutions.

TABLE 18–5

Matrix of significant factor loading

	Factor 1	Factor 2	Factor 3
Acuity			.699
Color			.601
Nystagmus			.734
Detail			.573
Carrots			.810
Fine dexterity		.786	
Gross dexterity		.823	
Softness		.857	
Tremor		.726	
Checks	.570	.437	
Interest	.732		
Scrooge	.729		
Dunning	.797		
Overcharge	.702		
Billing	.795		

and then (2) doubling that value because the SEs of factor loadings are up to twice those of ordinary correlations. When the sample size is over 100 (and we'll soon see why it had better be), a good approximation to use would be:

$$CV = \frac{5.152}{\sqrt{N - 2}} \tag{18–3}$$

Where did these numbers come from? When $N > 100$, the normal curve is a good approximation for the correlation distribution, and 2.576 marks off the 1% level of significance. Following Stevens, we double this (hence, 5.152) and then multiply by the SE for a correlation, which is $[1 \div \sqrt{(N - 2)}]$ and voila! So, if you want to use the 5% level, use 3.920 in the numerator.

Let's use this for our data. Because we had 200 most unwilling people taking the test,[17] we would get:

$$CV = \frac{5.152}{\sqrt{198}} = 0.366 \tag{18–4}$$

If we now go back to Table 18–4 and eliminate all loadings lower than this (and round down to three decimals to make the numbers easier to read), we get Table 18–5. Suddenly the light shines; it looks like we've pulled some degree of order out of chaos.[18] Factor 1 consists of six variables: CHECKS, INTEREST, SCROOGE, DUNNING, OVERCHARGE, and BILLING. This looks very much like the postulated Soul factor, with the addition of the CHECKS variable (a point we'll return to soon). Similarly, Factor 2 corresponds to the Hands attribute, and Factor 3 to Eyes.

However, there's one fly in the ointment. The CHECKS variable is both factorially complex (loading on Factors 1 and 2), and its highest loading is on the "wrong" factor. So, what do we do with it? We have three options:

1. We can throw that test out of the battery because it isn't tapping what we thought it would. If there are enough variables remaining in the factors (a minimum of three), this may be a sensible option. We would also toss out variables that didn't load well on any factor. This would be the case when the variable is quite complex, loading on a number of factors, or when it loads on some factor we didn't retain.
2. We can keep the variable in both factors. However, if our aim is to achieve simplicity and end up with uncorrelated factors, this wouldn't be a good choice.
3. If the variable is one we devised (e.g., an item on a test we're writing, or an entire test we're developing), we could rewrite it. The downside of this is that we would have to repeat the whole study with a new group of subjects to see if the revised variable is better than the original. However, if we're developing a scale, and one factor has relatively few items, this may be our only alternative.[19] In our example, because the Dean will have a new batch of 200 consenting adults next year, this option is feasible.

Table 18–5 can also help the Dean in another way. If he wants to make the test battery shorter, he can eliminate those tests with the lowest factor loadings. Needless to say, the reduced battery will not predict the factors as well, so yet another tradeoff has to be made.

Before we waltz away, though, we should make two last checks of the factors. A factor should consist of at least three variables (Tabachnick and Fidell say you can get away with two, but we feel that's low). Any factor that contains fewer should be discarded. Second, it's wise to go back to the original correlation matrix and see if the variables in the factor are indeed correlated with each other. Although it's unusual, situations can arise in which they're not, and again that factor should be thrown away.

USING THE FACTORS

In many cases, the steps we've gone through are as far as researchers want to go. They've used PCA and FA to either explore the data or confirm some hypotheses about them, and also to eliminate variables that were either not too helpful or factorially complex. However, we can use these procedures in another way: to reduce the number of variables. We may want to do this for a few reasons.

First, subject-to-variable ratios that are too low for some multivariable procedures may still be okay in FA (see below). So we can use PCA and FA to change a large number of variables into a smaller number of factors, which we can then analyze with linear regression or something else. Second, it may be easier for us to understand what a pattern of (say)

[17] *This was just an editorial comment; their state of mind does not affect the sample size.*

[18] *Doubters would say we've created chaos out of order, but what do those old sticks-in-the-mud know?*

[19] *Tooting our own horn a bit, see Streiner and Norman (1995) for more details on scale construction.*

three factors means rather than trying to juggle 15 scores in our mind all at once. What we would like to do, then, is to come up with one number for each factor. In our example, each person would have 3 scores, rather than 15, which, in essence, increases the subject/variable ratio by 5.

We mentioned earlier that the factor loadings are partial regression weights. So why not simply use them like a regression equation? The reason is that they were derived to predict the value of the *variable from the factors*. What we want to do is just the opposite, to predict the *factor from the variables*. So, if we want, we can command the computer to give us a **factor score coefficient matrix**, such as the one in Table 18–6. Each column is a regression equation, with the predicted factor score as the DV and the variables as the IVs.

TABLE 18–6

Factor score coefficient matrix

	Factor 1	Factor 2	Factor 3
Acuity	.02083	−.06561	.27844
Color	−.00017	−.09003	.26219
Nystagmus	−.08966	−.03297	.31804
Detail	−.04866	.03030	.21378
Carrots	−.05618	−.09673	.36291
Fine dexterity	−.04008	.30845	−.07620
Gross dexterity	−.08967	.33097	−.04558
Softness	−.08689	.35157	−.07099
Tremor	−.03831	.26677	−.02529
Checks	.14474	.10321	−.06656
Interest	.26002	−.00042	−.14349
Scrooge	.24091	−.03500	−.03712
Dunning	.29316	−.14363	.00856
Overcharge	.21052	−.06666	.08001
Billing	.26535	−.07488	.00191

So, the three-factor scores for subject 1 would be found by plugging her 15 *standardized* scores into the equations, which would then read:

$$FS_1 = (.02083)\text{ACUITY} - (.00017)\text{COLOR} + \cdots + (.26535)\text{BILLING}$$

$$FS_2 = (-.06561)\text{ACUITY} - (.09003)\text{COLOR} + \cdots - (.07488)\text{BILLING}$$

$$FS_3 = (.27844)\text{ACUITY} + (.26219)\text{COLOR} + \cdots + (.00191)\text{BILLING}$$

Most computer programs can calculate the factor scores for us and then save them in a file, making the job of transferring the results to another program much easier.

It almost goes without saying that if we have one way to do things in FA, a couple of other ways are lurking around just to complicate our lives. Computing factor scores is no exception. All of them yield scores with a mean of 0. Where they differ is in (1) the variance of the scores, and (2) the correlation among the factor scores. Although the factors are uncorrelated (assuming we've stopped at PCA or used an orthogonal rotation in FA), the factor scores themselves may be, depending on which technique we use.

However, we'll mention one more fact about factor scores that may actually simplify your life. When more than 10 variables are loading on a factor, you can probably forget about the equations entirely. If you use unit weights, set each significant loading equal to 1.00 (or −1.00 if it's negative) and the nonsignificant ones to 0.00; then all you have to do is add up the (standardized) scores—forget about multiplying them by the coefficients. The reason is that with more than 10 IVs, the β weights don't improve the predictive ability of the equation to any degree that's worth worrying about (Cohen, 1990; Wainer, 1976). Actually, this is most true when the variables are totally uncorrelated with each other: the greater the magnitude of the correlation, the greater the possible loss in efficiency when using these unit weights.

TYPES OF DATA TO USE

Most of the methods used for exploratory factor analysis are fairly robust against deviations from normality, and don't require multivariate normality (Floyd and Widaman, 1995). This means that five- or seven-point scales, often used in attitude and personality assessment, are fair game to be factor analyzed. But, it doesn't mean that anything goes; we do have *some* standards, after all.[20] The major no-no is dichotomous data: Yes/No, True/False, Present/Absent, 0/1, and other variables of that ilk. Despite this injunction, you'll find examples of this type of variable being factor analyzed in many journals. Comrey (1978) pointed out several problems that can arise with dichotomous data.[21] First, if about half of the people respond True on one variable, but 95% answer True on another variable, then the maximum correlation between these two variables is about ±0.23. Second, if 99 people say False to two items, and 1 person says True, then the correlation between the items will be 1.00. However, if this one person then changes her mind and also answers False, the correlation suddenly becomes 0.00. So, correlations with dichotomous data are often unstable and will be either artificially limited or grossly inflated, depending on the situation. Just as bad as dichotomous items are nominal variables. Because their coding is arbitrary, the numbers don't really mean anything; they're just names. If we change the coding scheme, the correlations among the variables will change drastically, so any "pattern" among the variables is artifactual.

A different problem arises when we have *ipsative* data, which is easier to define by starting off with an example. Let's say there is an item on a personality test that reads, "Sometimes I am sure that others can tell what I am thinking and they are bored."[22] If the person answers True, he gets scored on the Depression scale, while a False answer means that he gets scored on the Grandiosity scale (we'll ignore the

[20] *As low as they may appear, at times.*

[21] *Interestingly, he pointed them out only after he ran out of dichotomous item scales on the MMPI to factor analyze and publish.*

[22] *As far as we know, there isn't such an item, but there should be.*

other problem—that this is a dichotomous item). This means that we've artificially built in a negative correlation between these two scales—if you get a score on one, you can't get it on the other. These scales would be called ipsative. The sum of the rows or columns of an ipsative correlation matrix must be zero, and the expected correlations among the variables, assuming that they're independent, is $-1/(k - 1)$, where k is the number of scales (Dunlap and Cornwell, 1994). In this case, we have built in a correlation between the scales, merely because of the scoring scheme, and the correlation may not reflect a relationship between the scales at a theoretic level.

A third problem[23] arises when one variable is a combination of others in the same data matrix being analyzed. For example, a Total variable (such as the Full Scale IQ or the Verbal and Performance IQs on Wechsler IQ tests) could be the sum of two or more other variables. If this is the case, the computer program should have a major myocardial infarct and die on the spot. The only way around this problem is not to include variables that are sums or products of other variables in the same data matrix.

Finally, if you're developing a scale, most good texts[24] would recommend that about half of the items be scored in one direction and half in the other direction, to minimize various response biases. Before you factor analyze, change the scoring system so that high scores on all of the items reflect the same thing!

SAMPLE SIZE

In factor analysis, there are no power tables to tell exactly how many subjects to use. What we *do* have are firmly held beliefs[25] and some Monte Carlo simulations. Those simulations tell us that the sample size depends on a host of factors (pardon the pun), such as the number of items on each factor, the average strength of the factor loadings, the communalities of the items, and, for all we know, the astrological sign under which you were born. Very often, people who do these studies are emphatic that there are no fixed rules regarding the subject-to-variable ratio. However, in terms of planning a study, this advice fits the usual definition of an epidemiologist—he tells you something that is absolutely correct, that you already know, and that is totally useless. If we knew the magnitudes of all of these parameters, we wouldn't have to do the @#$%& study to begin with! What we are left with, then, is a rule-of-thumb guideline that we've been told we shouldn't have: we must have an absolute *minimum* of five subjects per variable (some people would go as low as three, but they're taking a big risk), with the proviso that we have at least 100 subjects.

Gorsuch (1983), one of the grand-daddies of FA, and the person who proposed these guidelines, said that this should suffice *only if* the communalities are high and there are many variables for each factor. If you don't meet these two conditions, then you should probably at least double the subject/variable ratio, as well as the total number of subjects analyzed. We dare say that if these rules are followed, the number of factor analyses performed each year will drop by about 70%, resulting in much joy among readers of journals and much consternation within the paper manufacturing business.

[23] *Or is it the fourth? We've lost track.*

[24] *To choose just one, completely at random, there's Streiner and Norman (1995)—yet again.*

[25] *Often argued with as much vehemence as two theologians debating whether angels can get dandruff (and with about the same degree of data to back them up).*

EXERCISES

In an attempt to gain immortality by attaching his name to a questionnaire, one of the authors develops a test for budding social workers called the Streiner Knowledge of Relationships, Empathy, and Warmth scale (the SKREW). He starts off with 12 items, which he hopes will tap these three areas, and administers them to a validation sample of 63 already blooming SW types. The rotated factor loading matrix is shown in the table.

1. The subject-to-variable ratio is:
   a. Acceptable, since it's 5:1.
   b. Acceptable, since they're only social workers.
   c. Too low; there should be at least 100 subjects.
   d. Too low; the ratio should be 10:1.
2. Using the Kaiser criterion, how many factors are there?
3. What proportion of the variance is accounted for by the retained factors?
4. Are there any items you would drop? Why (or why not, as the case may be)?

Rotated factor loadings of the 12 items

Item	Factor 1	Factor 2	Factor 3	Factor 4
1	.53	.33	.27	−.04
2	.14	.64	.12	.03
3	.30	.05	.57	.14
4	.15	.44	.23	.20
5	.02	.04	.64	.28
6	.07	.61	.14	.10
7	.78	.12	.08	−.16
8	.80	.15	.17	−.05
9	−.06	−.03	.22	.61
10	.13	.03	.57	.22
11	.10	.26	.11	.23
12	.11	.43	.09	.42
EIGENVALUES	1.709	1.380	1.323	0.828

5. a. What is the **communality** for Item 1?
   b. What is its **uniqueness**?
   c. What does this mean?
   d. Do you really care?
   e. Should you?

How to Get the Computer to Do the Work for You

- From **Analyze**, choose **Data Reduction → Factor**
- Click on the variables to be analyzed from the list on the left, and click the arrow to move them into the box marked **Variables**
- Click the **Descriptives** button
- In the **Statistics** box, **Initial solution** is the default; keep it
- In the **Correlation Matrix** box, choose **Coefficients** [if you haven't already run a correlation matrix], **Antiimage** [to get the Measure of Sampling Adequacy], and **KMO and Bartlett's test of sphericity**
- Click **Continue**
- Click the **Extraction** button
- In the **Method** box, keep the default of **Principal components** to run PCA, or change it to **Principal axis factoring** to run a common FA
- In the **Analyze** box, change the default from **Correlation matrix** to **Covariance matrix**
- In the **Display** box, keep **Unrotated factor solution** and add **Scree plot**
- In the **Extract** box, keep the default [Eigenvalues over 1 times the mean eigenvalue] or enter the number of factors you want to retain
- Click **Continue**
- Click the **Rotation** button
- The default for **Method** is **None**. Click **Varimax** for an orthogonal rotation, or **Oblimin** for an oblique one
- Click **Continue**
- Click the **Options** button
- In the **Coefficient Display Format** box, choose **Sorted by size**
- Click **Continue**
- Click **OK**

Path analysis is an extension of multiple regression, allowing us to look at the relationships among many "dependent" and "independent" variables at the same time. Structural equation modeling (SEM) takes this one step further, permitting the same type of analysis with latent variables. One form of SEM is confirmatory factor analysis.

CHAPTER THE NINETEENTH

Path Analysis and Structural Equation Modeling

SETTING THE SCENE

Although science has made tremendous advances in predicting which stars will become black holes instead of just flickering out, in explaining the movement of the continents over the face of the planet, and in eradicating many sources of disease, one area of knowledge has still eluded them: delineating what accounts for success in cheerleading. Dr. Yeigh Teeme has tried to solve this problem by gathering a lot of data (some say too much) about men and women who were and were not successful in this demanding task. He is bothered, however, by the fact that some variables may affect performance directly (e.g., the ability to jump high while smiling), while other variables may act indirectly (e.g., a "winning personality" may lead the coach to overvalue the person's ability). Compounding his problems, some variables are measured directly (like height and jumping ability), and others are derived from scores on two or more tests. How can Dr. Teeme tease these effects apart, take into account both measured and inferred variables, and figure out which ones are important?

PATH ANALYSIS

The two techniques we will discuss in this chapter—path analysis and structural equation modeling (SEM)—are extensions of procedures we have discussed earlier; namely, multiple regression and exploratory factor analysis.[1] So, if you feel a bit shaky about them, you may want to review those chapters first. Let's start off easy by using statistical tests we have already encountered and seeing what their limitations are and how we can get around them. We'll assume that success in cheerleading (the dependent variable) has been measured on a scale which goes from 1 ("Performance guaranteed to cause fans to root for the opposing team") to 10 ("Performance results in terminal happiness among fans"). We'll also assume that this scale, called the Cheer Leader Activity Profile/Teacher Rating of Athletic Performance or CLAP/TRAP, is normally distributed and has all of the other good qualities we would want. If Dr. Teeme wanted to predict CLAP/TRAP scores from the person's height (in inches), jumping ability (in inches), and intelligence (in IQ points),[2] he could run a multiple regression equation, which we discussed in Chapter 13; it would look something like:

$$\text{CLAP/TRAP} = b_0 + b_1 \text{ Height} + b_2 \text{ Jumping} + b_3 \text{ Intelligence}$$

A diagram of what this equation does is shown in Figure 19–1. For reasons which will become clear soon, we've drawn a box around each of the variables and a straight arrow[3] leading from each of the predictor variables to the dependent variable (DV). This implies that we are assuming that each of the variables acts directly on the DV. But, a little reflection may lead us to feel that the story is a bit more complicated than this. A person's height may act directly on the coach's evaluation, but it may also influence jumping ability. So, a more accurate picture may be the one shown in Figure 19–2. One question we may want to ask is, which model is a more accurate reflection of reality?

[1] *In fact, path analysis is "merely" a subset of SEM, so we've just halved what you have to learn. But, because it's easier to understand, we'll start off with path analysis.*

[2] *In the interests of political correctitude, we won't comment on the direction of the correlation between IQ and cheerleading, only its magnitude. But, have you ever stopped to wonder why so many cheerleaders are blond?*

[3] *That means a line that doesn't curve, with an arrowhead at one end; it doesn't refer to a well-scrubbed Boy Scout.*

We'll begin by just looking at how the variables are correlated with each other, to get a feel for what's going on; the correlation matrix is shown in Table 19–1. As we would suspect, CLAP/TRAP has a strong positive relationship to Height and Jumping Ability, and is strongly and negatively related to IQ;[4] Height and Jumping are positively correlated with each other and negatively related to IQ.[5]

Interpreting the Numbers

If we now ran a multiple regression based on the model in Figure 19–1, we would get (among many other things) three standardized regression weights (betas, or βs), one for each of the predictors. In this case, β_{Height} is 0.548, $\beta_{Jumping}$ is 0.199, and β_{IQ} is −0.245. While the relative magnitudes of the β weights and their signs parallel those of the correlations, the relationship between these two sets of parameters isn't immediately obvious. The problem is that the model in Figure 19–1 shows only part of the picture; it doesn't take into account the correlations among the predictor variables themselves. To introduce a notation we'll use a lot in this chapter, we can show that the variables are correlated with each other by joining the boxes with curved, double-headed arrows.[6] In Figure 19–3, we've added the arrows and some numbers. The numbers near the curved arrows are the zero-order correlations between the predictor variables, and those above the single-headed arrows are the correlations between the predictors and the DV. Below the arrows, and in parentheses, are the β weights. Now, believe it or not, we can see the relationship between the correlations and the β weights.

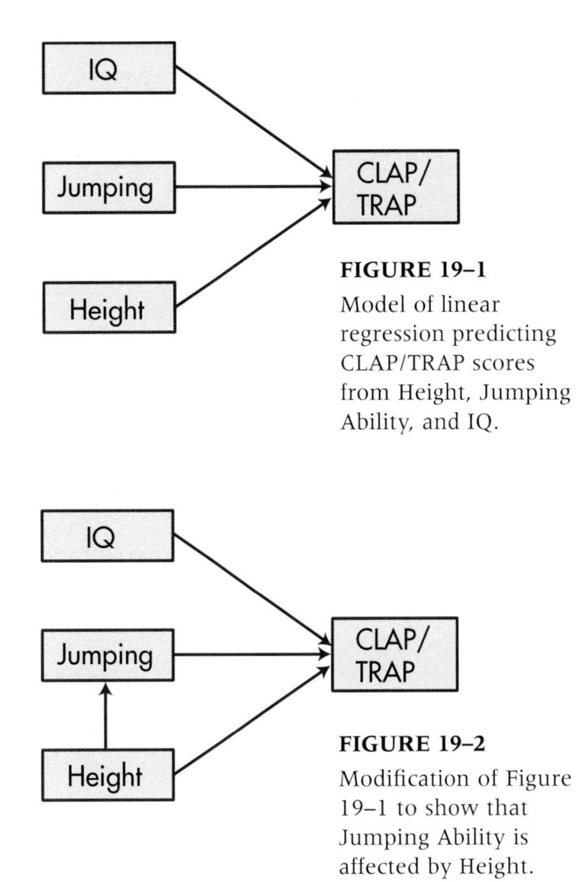

FIGURE 19–1
Model of linear regression predicting CLAP/TRAP scores from Height, Jumping Ability, and IQ.

FIGURE 19–2
Modification of Figure 19–1 to show that Jumping Ability is affected by Height.

[4] *OK, so we lied; we have commented on the relationship between IQ and cheerleading.*

[5] *The negative relationship between Height and IQ should give you some clue as to the author of this chapter. As a hint, Norman is 6'3"; Streiner is 5'8".*

[6] *That means a line with an arrowhead at each end; it's not the American Indians' equivalent of missiles with multiple warheads.*

TABLE 19–1
Correlations among the variables in predicting CLAP/TRAP scores

	CLAP/TRAP	Height	IQ	Jumping
CLAP/TRAP	1.000	.807	−.608	.677
Height		1.000	−.505	.678
IQ			1.000	−.433
Jumping				1.000

Let's start by looking at the effect of IQ on CLAP/TRAP. Its β weight is −0.245. Also, IQ exerts an indirect effect on CLAP/TRAP through its correlations with Jumping Ability and Height. Through Jumping Ability, the magnitude of the effect of IQ is the correlation between the two predictors (r = −0.433) times the effect of Jumping on CLAP/TRAP, which is its β weight of 0.199; and so $-0.433 \times 0.199 = -0.086$. Similarly, the indirect effect of IQ through Height is -0.505×0.548, or −0.277. Adding up these three terms gives us:

$$(-0.245) + (-0.086) + (-0.277) = -0.608$$

which is the correlation between IQ and CLAP/TRAP. Doing the same thing for Jumping Ability shows a direct effect of 0.199; an indirect effect through IQ of -0.433×-0.245 (which is 0.106), and an indirect effect through Height of 0.678×0.548, or 0.372. The sum of these three terms is 0.677, which, in this case,[7] is the zero-order correlation between Jumping and CLAP/TRAP.[8] Conceptually, then, r is the sum of both the direct and indirect effects between the two variables.

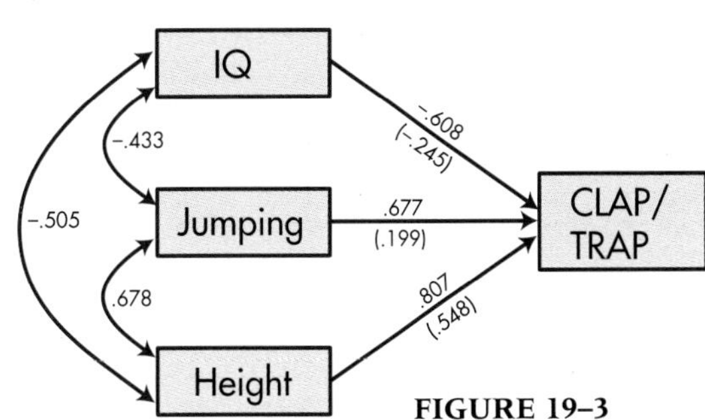

FIGURE 19–3
Figure 19–1 with the correlations among the variables and the β weights in parentheses added.

[7] *We'll see in a bit that this is true in this model, but it doesn't necessarily hold in others. We can actually exploit this to tell us how "good" each model is.*

[8] *To test your understanding of these concepts (but mainly to save the batteries in our calculator), we'll leave the calculations for the direct and indirect effects of Height to you.*

More formally, we can denote the correlation between pairs of predictors with the usual symbol for a correlation, r. Then, the correlation between

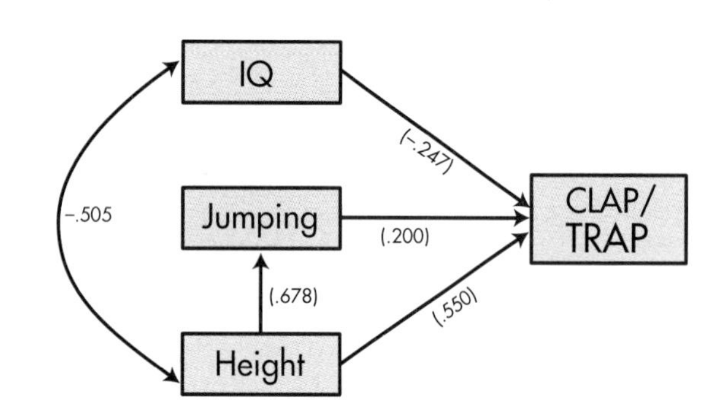

FIGURE 19–4
Figure 19–2 with path coefficients added.

TABLE 19–2

Original correlations in the upper triangle and reproduced correlations in the lower triangle

	CLAP/TRAP	Height	IQ	Jumping
CLAP/TRAP	1.000	.807	−.608	.677
Height	.810	1.000	−.505	.678
IQ	−.593	−.505	1.000	−.433
Jumping	.573	.678	−.342	1.000

IQ and Jumping is $r_{\text{IQ-Jump}}$, between Height and Jumping is $r_{\text{Height-Jump}}$, and between IQ and Height is $r_{\text{IQ-Height}}$. The path between each of the predictors and the dependent variable is a β weight, so that between Jumping and CLAP/TRAP is β_{Jump}, between IQ and CLAP/TRAP is β_{IQ}, and between Height and CLAP/TRAP is β_{Height}. Therefore:

$$r_{\text{IQ-CLAP/TRAP}} = \beta_{\text{IQ}} + (r_{\text{IQ-Height}} \times \beta_{\text{Height}}) + (r_{\text{IQ-Jump}} \times \beta_{\text{Jump}})$$

$$r_{\text{Height-CLAP/TRAP}} = \beta_{\text{Height}} + (r_{\text{IQ-Height}} \times \beta_{\text{IQ}}) + (r_{\text{Height-Jump}} \times \beta_{\text{Jump}})$$

$$r_{\text{Jump-CLAP/TRAP}} = \beta_{\text{Jump}} + (r_{\text{IQ-Jump}} \times \beta_{\text{IQ}}) + (r_{\text{Height-Jump}} \times \beta_{\text{Height}})$$

What we have just done is to **decompose** the correlations for the predictor variables into their **direct** and **indirect effects** on the DV. Now we are in a better position to see what adding the arrow in Figure 19–2 does to our model. The main effect is that it imposes directionality on the indirect effects; we are saying that Height can affect Jumping Ability (which makes sense), but that Jumping Ability doesn't affect Height (which wouldn't make sense[9]). In other words, we've delineated the **paths** through which the variables exert their effects; not surprisingly, Figure 19–2 is called a **path diagram**, and you have just been introduced to the basic elements of **path analysis**.

In Figure 19–4, we've added the "path coefficients" to the second model (the one in Figure 19–2). The direct effects are changed only slightly (although they are changed), but now the indirect effects are quite different. If we follow the arrows leading out of each predictor variable, we see that Jumping has a direct effect of 0.200 and, strange as it may seem, an indirect effect through Height of 0.678 × 0.550 (= 0.373), for a total of 0.573.[10] For Height, the direct effect of 0.550 is augmented by its indirect effect through Jumping Ability of 0.678 × 0.200, or 0.136, and its indirect effect through IQ of −0.505 × −0.247 =0.125, so that its total effect is 0.810. IQ is even less straightforward; it has a direct effect (−0.247), an indirect effect through Height to CLAP/TRAP (−0.505 × 0.550 = −0.278), and a *very* indirect effect through Height to Jumping to CLAP/TRAP (−0.505 × 0.678 × 0.200 = −0.068), for an overall effect of −0.593. If you were paying attention, you will have noticed that in this example, these numbers do *not* add up to equal the correlations in Table 19–2. We'll show you why they don't when we discuss how we can tell which models are better than others.

Finding Your Way through the Paths

When we decomposed the correlations in Figures 19–3 and 19–4, there were some paths we did not travel down and others that we took that seemed somewhat bizarre. For example, in Figure 19–3, we traced the indirect contribution of IQ → Jumping → CLAP/TRAP and IQ → Height → CLAP/TRAP; we did not have a path IQ → Jumping → Height → CLAP/TRAP. Similarly, in Figure 19–4, we had Height → Jumping → CLAP/TRAP, and, strangely enough, Jumping → Height → CLAP/TRAP, even though the arrow itself actually points from Height to Jumping. Back in 1934, Sewall Wright (the granddaddy of path analysis) laid out the rules of the road:

- For any single path you can go through a given variable only once.
- Once you've gone forward along a path using one arrow, you can't go back on the path using a different arrow.
- You can't go through a double-headed curved arrow more than one time.

Kenny (1979) added a fourth rule:

- You can't enter a variable on one arrowhead and leave it on another arrowhead.

Kenny's rule is the reason why, in Figure 19–3, we did not trace out the path IQ → Jumping → Height → CLAP/TRAP or IQ → Height → Jumping → CLAP/TRAP. These paths enter one variable on an arrowhead from IQ, and would then leave on an arrowhead to get to the other variable—maybe not a felony offense but definitely a misdemeanor with respect to the rules.

[9] *Unless, of course, you jumped from a high height and flattened yourself—but that would result in a negative correlation between the variables.*

[10] *We'll see the reason for this counter-intuitive path in the next section.*

Bizarre as it may seem, the path in Figure 19–4 starting at Jumping and then going through Height to CLAP/TRAP doesn't violate any of these rules. The rule prohibits only paths that go forward and then backward; this path goes backward and then forward. This path is meaningless in terms of our knowledge of biology (the technical term for this is a **spurious** effect), but it is legitimate insofar as decomposition of the correlation is concerned. It exists because Jumping and CLAP/TRAP have a common cause—Height.

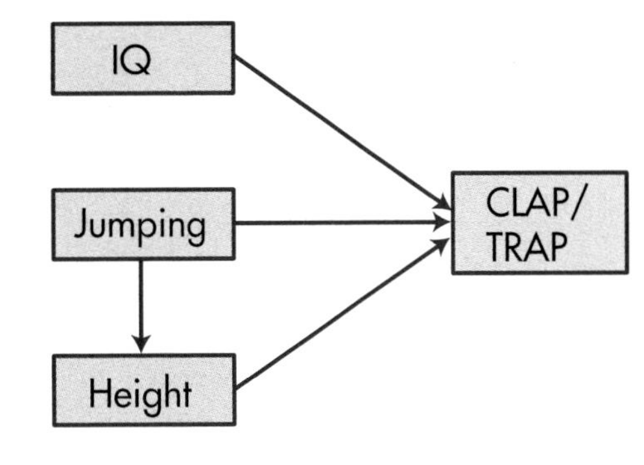

FIGURE 19–5
Figure 19–2 with Jumping Ability affecting Height.

Path Analysis and Causality

Until recently, path analysis was called "causal modeling."[11] Because we can specify paths by which we think one variable affects another, the hope was that we could make statements about causality, even if the data were actually correlational. In fact, if we specified a model that didn't make much sense, such as having the path go from IQ to Jumping Ability, we would know from the statistics that something was amiss; there shouldn't be a path between IQ and Jumping Ability. Does this mean that we actually can take correlational data and have them tell us about causation? Despite the hopes of the early developers of this technique, the answer is "No." If we had altered the model by having the path go from Jumping Ability to Height, rather than from Height to Jumping Ability, as in Figure 19–5, all of the statistics would be the same. That is, this technique can tell us whether or not there should be a path between two variables, but it cannot tell us which way the arrow should point; this can only be supplied by our theory, the study design, or knowledge of the literature. For example, if we find a relationship between gender and spatial or verbal ability, it's fairly safe to assume that gender (or factors correlated with gender, such as educational experiences or brain structure) leads to differences in ability. No matter how much we change people's abilities, their gender will remain the same.[12] Similarly, if we modify an educational program at Time 1, and find that the experimental, but not the control, group improved their skills at Time 2, the causality is apparent from the design of the study.

Path analysis can also disprove a "causal" model we may postulate or fail to disprove it. However, just as failing to disprove the null hypothesis does not mean that we have proven it; failing to reject a causal model isn't the same as showing that it's correct. The bottom line is that determining true causation from correlations is still akin to getting gold from lead.

Endogenous and Exogenous Variables

The models we've discussed so far are relatively simple ones. To show what more path analysis can do, let's look at some models it can handle and, at the same time, introduce you to some of the arcane vocabulary. When we looked at Figure 19–1, we referred to the variables on the left as the predictors and CLAP/TRAP as the dependent variable. Now that we have a new statistical technique, we have new terms for these variables.[13] In path analysis (and, as we'll see, in SEM in general), the variables on the left are referred to as **exogenous variables**.

> **Exogenous variables** have arrows emerging from them and none pointing to them.

This means that exogenous variables influence or affect other variables, but whatever influences them is not included in the model. For example, a person's height may be influenced by genetics and diet, but our model will ignore these factors.

What we had called the dependent variable, CLAP/TRAP, is an **endogenous variable** in SEM terms, in that it has arrows pointing toward it. What about Jumping Ability in Figure 19–2? It has an arrow pointing toward it as well as one emerging from it. As long as there is at least one arrow (a path) pointing toward a variable, it is called endogenous.

> Any variable that has at least one arrow pointing toward it is an **endogenous variable**.

This illustrates why the terms "independent," "predictor," and "dependent" variable can be confusing in path analysis. In Figure 19–4, Jumping is a dependent variable in relation to Height but a predictor in terms of its relationship to CLAP/TRAP. This further shows one of the major strengths of path analysis as compared to multiple regression; regression cannot easily deal with the situation in which a variable is both an IV and a DV.

Types of Path Models

Figure 19–6 shows a number of different path models, some of which we have already encountered. Those on the left are **direct** models, in that the exogenous variables influence the endogenous ones without any intermediary steps; in other words, the endogenous variables have arrows pointing toward them and none pointing away from them. In Part A of the figure (called an **independent** path model), the two exogenous variables affect an endogenous one, and the exogenous variables are not correlated

[11] Or "modelling;" the spelling doesn't affect the results.

[12] Although their attraction to the opposite gender may change.

[13] Actually, this isn't as capricious as it may first appear; for reasons we'll see shortly, keeping the original terms can lead to some confusion.

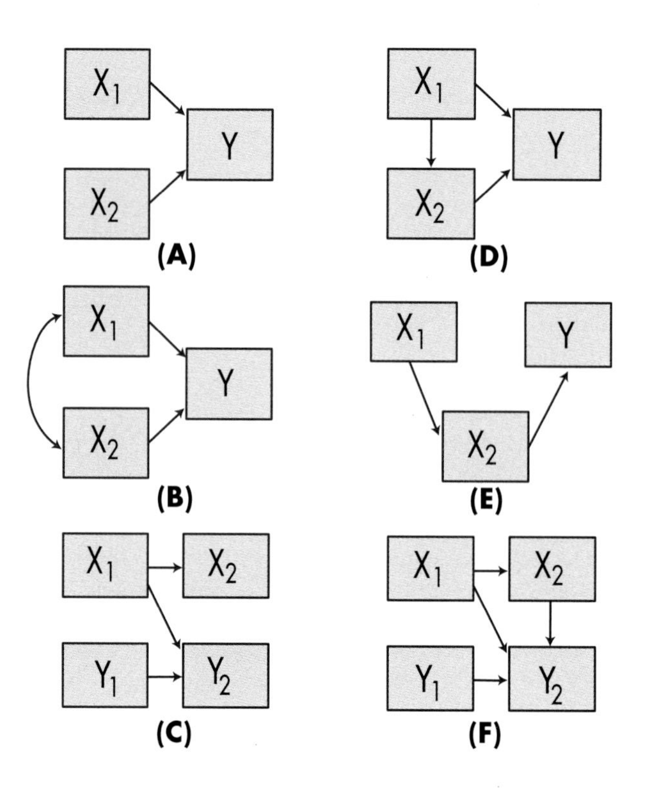

FIGURE 19–6
Different path models.

with each other; hence the name "independent." This situation isn't too common in research with humans, since, as Meehl said in his famous sixth law (1990), "everything correlates to some extent with everything else" (p. 204).[14] Part B of the figure, a **correlated** path model, is more common; it's equivalent to a multiple regression, although we often have more than two predictors, and drawing in all those arcs between pairs of variables can really make the picture look messy. The picture it portrays is what we usually deal with, in that we assume the predictors are correlated with one another to various degrees. In Part C, there are two exogenous and two endogenous variables. The interesting thing about this model is that X_1 and X_2 can be different variables, or they can be the same variable measured at two different times; the same applies to Y_1 and Y_2. For example, in one study we did (McFarlane et al, 1983), we wanted to see if stressful events in a person's life led to more illness. X_1 and X_2 were the amount of stress at Time 1 and 6 months later at Time 2, and Y_1 and Y_2 the number of illnesses at these two times. This model allowed us to take into account the fact that the best predictor of future behavior is past behavior (the path between Y_1 and Y_2), and then look at the added effect of stress at Time 1 on illness at Time 2.[15] The two endogenous variables don't have to be the same ones as the exogenous ones, nor do the two sets of variables have to be measured at different times, but they can be. Such is the beauty of path analysis.

The major difference between the diagrams on the left of Figure 19–6 and those on the right are that the latter have endogenous variables with arrows pointing both toward them as well as away from them. These are referred to as **indirect** or **mediated** models, because the variables with both types of arrows mediate the effect of the variable pointing to them on the variables to which they point. In Part D of Figure 19–6, variables X_1 and X_2 both influence Y. However, X_2 also mediates the effect of X_1 on Y. In our example, Jumping (X_2) affects CLAP/TRAP directly and also mediates the effects of Height (X_1). That is, we would expect that if two people can jump equally high, but one person is 6′3″ and the other "only" 5′8″,[16] we would expect the second person to get more "credit," since he's exerting himself more. In Part E of the figure, variable X_1 does not affect the endogenous variable directly, but only through its influence on X_2. For example, the height of one's father (X_1) affects a person's CLAP/TRAP score (Y) only by its effects on the cheerleader's height (X_2), which in turn affects Y.

The final model we'll show (there are infinitely many others), Part F of Figure 19–6, is the same as Part C, except that we've added a path between X_2 and Y_2. What this does is to turn X_2 (stress at Time 2) into a mediated variable. Now we are saying that the number of illnesses measured at Time 2 is affected by three things: illness at Time 1, stress at Time 1, and stress at Time 2. Furthermore, we are saying that stress at Time 1 works in two ways on illness behavior at Time 2: directly (sort of a delayed reaction to stress), and by affecting stress at Time 2, which, in turn, affects illness behavior immediately. The magnitude of the path coefficients tells us how strong each effect is.

[14] *No one really remembers his first five laws.*

[15] *A more complete model would add a path between X_2 and Y_2, but this is a different class of model, as we'll see in a moment.*

[16] *To choose two heights, not quite at random (see note 5).*

Getting Disturbed

The models in Figure 19–6 look fairly complete and, in Part F, relatively complex. But, in fact, something is missing. We're going to add some more terms, not to disturb you, but to disturb the endogenous variables. A more accurate picture (and a necessary one, from the standpoint of the statistical programs which analyze path models) is to attach **disturbance terms** to each endogenous variable. These are usually denoted by the letter *D* (for Disturbance), *E* (for Error), a small circle with an arrow pointing toward the variable, or a circle with one of the letters inside, as in Figure 19–7. This is similar to what we have in multiple regression, where every equation has an error term tacked on the end that captures the measurement error associated with the dependent variable. In path analysis (and in SEM in general), the disturbance term has a broader meaning; in addition to measurement error, it also reflects all of the other factors that affect the endogenous variable which aren't in our model, either because we couldn't measure them (e.g., genetic factors) or we were too dumb to think of them at the time (e.g., how much the cheerleader's parents bribed the coach for their kid to get a good performance score).

If we were to draw complete diagrams for the examples we've discussed, then Figure 19–3 would have a disturbance term attached to CLAP/TRAP. In Figure 19–4, there would be two disturbance terms: one associated with CLAP/TRAP and one for Jumping, since it has become an endogenous variable with the addition of the path from Height. So why didn't we draw them? Since every endogenous variable *must* have a disturbance term, they are superfluous to those of us who are "au courant" with path analysis. We *have* to draw them in when we use the computer programs, and we *may* put them in diagrams for the sake of completeness, but they're optional at other times.

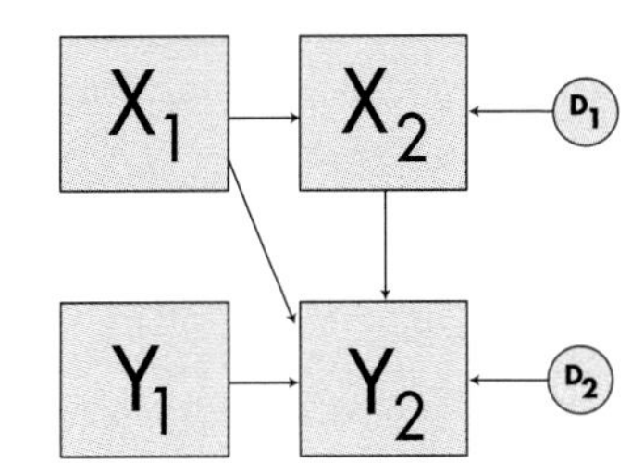

FIGURE 19–7
Addition of disturbance terms to Figure 19–6F.

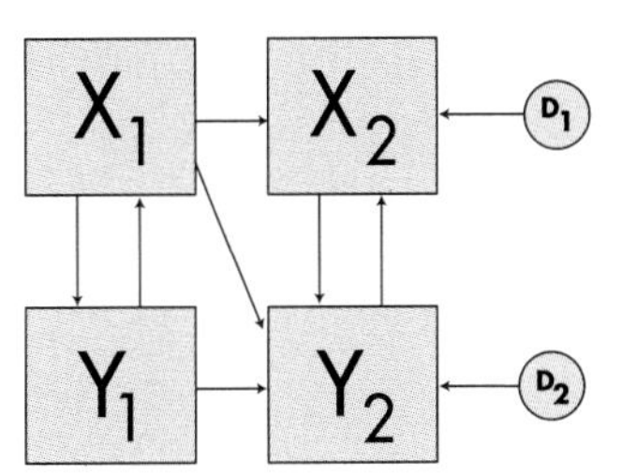

FIGURE 19–8
A nonrecursive path model.

Recursive and Nonrecursive Models

Although it may not be apparent at first, all of the models in Figure 19–6 are similar in two important ways. First, all of the paths are unidirectional;[17] that is, they go from "cause" to "effect" (using those terms very loosely, without really *meaning* causation). Second, if we had drawn the disturbance terms in the models, they would all be assumed to be independent from one another; that is, we wouldn't draw curved arrows between the disturbances. For some reason these models are referred to as **recursive** models.

Let's return to the example we used in Figure 19–6F and modify it a bit more. In Figure 19–8, we have added a path between X_1 and Y_1 and between X_2 and Y_2 to indicate that stress could affect illness concurrently. It could be just as logical that the relationship goes the other way and even more logical that the relationship was reciprocal: that stress at a given time affects illness and also that illness affects stress, so that Figure 19–8 may be a more accurate portrayal of what is actually occurring. A path diagram with this feedback loop is called a **nonrecursive** model.[18]

Note that there is a difference between connecting two variables with a curved, double-headed arrow, and joining them with two straight arrows going in opposite directions. The former means that we "merely" expect the two variables to covary, such as when we use two similar measures of the same thing. Covariance *may* be the result of one variable affecting the other, but it may also be due to the effect of some underlying factor affecting both of them. A feedback loop, on the other hand, explicitly states that each variable directly affects the other—stress leads to illness, and illness leads to stress.

Analysis and interpretation of nonrecursive models is much more difficult than with recursive ones, and far beyond the scope of one chapter in a book. If you ever do need to analyze a nonrecursive model, we suggest two things: (1) pour yourself a stiff drink and think again, and (2) if you still want to do it, read one of the books listed under "To Read Further."

K.I.S.S.[19]

At first glance, it may seem as if the safest strategy to use is to draw paths connecting everything with everything else, and to let the path coefficients tell you what's going on. This would be a bad idea for two reasons. First, as we will emphasize over and over again in this chapter (and indeed in much of the book), model building should be based on theory and knowledge. One disadvantage of the more sophisticated statistical techniques is that they are *too*

[17] *We don't count the curved arrows, as they reflect correlations among the variables, not paths between them.*

[18] *If the use of the terms "recursive" and "non-recursive" seems counter-intuitive and confusing, join the club. Membership isn't exclusive; it consists of everyone who has ever encountered the terms (with the exception of the dyslexic sadist who coined them).*

[19] *For the two people in the world who don't know what this means, it stands for Keep It Simple, Stupid.*

powerful in some regard; they capitalize on chance variance and, if you blithely and blindly use a computer program to replace your brain cells, you can be led wondrously astray, albeit with low p levels. The second reason is that there are mathematical limits to how many paths you can have in any one diagram.

The number of parameters (i.e., statistical effects such as path coefficients, variances, covariances, and disturbances) you can analyze is determined by the number of observations. "Observations" here is a function of the number of variables and is *not* related to the number of subjects. That is, a given path model has the same number of observations, whether the study used 10 subjects or 10,000 subjects. If there are k variables, then:

$$\textit{Number of Observations} = \frac{k \times (k + 1)}{2} \tag{19–1}$$

because there are $[k \times (k - 1)/2]$ covariances among the variables, plus k variances. In Figure 19–3, we have four observed variables, so we have 10 observations. This means that we can examine *at most* 10 parameters. Now the problem is, How many parameters do we actually have in this model? Another way of asking this question is, What don't we know that we should know?

To answer this, we have to elaborate a bit more on the purpose of path analysis and, more generally, of structural equation modeling. We are trying to find out what affects the endogenous variables; that is, how the exogenous variables work together (the curved paths, which represent correlations or covariances), and which of those paths (the straight arrows) are important. This is determined by the variances of the variables. If we had a perfect model, then knowing these variances would allow us to predict the person-to-person variation in the endogenous variables. Because the endogenous variables are, by definition, "caused" or influenced by the other variables, they are not free to vary or covary on their own but only in response to exogenous variables. Consequently, we are not interested in estimating the variances of endogenous variables but only those variances for variables which *can* vary. Note that the disturbance term attached to an endogenous variable is, in SEM thinking, free to vary and thus influence the endogenous variable—so it's one of those things we have to estimate.

Where does this leave us with respect to Figure 19–3? It's obvious we want to estimate the 3 path coefficients, so they're on the list of parameters to estimate. Also, we don't know the variances of the exogenous variables or the covariances among them, so that adds another 6 parameters (3 variances + 3 covariances). Finally, we have the variance of the disturbance term itself (which isn't in the figure, but is implied), meaning that there are 10 parameters to be estimated.

Now let's do the same for Figure 19–4. There are still 4 observed variables, so the limit of 10 parameters remains, but there are a different number of arrows. We have 4 path coefficients (the 3 from the variables to CLAP/TRAP, and 1 from Height to Jumping); the covariance between IQ and Height; the variances of 2 exogenous variables (IQ and Height); and 2 disturbance terms (those for Jumping and CLAP/TRAP), resulting in 9 parameters to be estimated.

If you've followed this so far, a number of questions should arise: Why didn't we count the variance of CLAP/TRAP in Figure 19–3 as one of the parameters to be estimated? Why did we count the variance of Jumping among the parameters in Figure 19–3, but not in Figure 19–4? Why can't we have more than 10 parameters? Does it make any difference that we had 10 parameters to estimate in Figure 19–3 and 9 in Figure 19–4? Is there intelligent life on Earth? If you've *really* been paying attention, you will have noticed that we already answered the first 2 questions. Stay tuned, and the remaining questions will be answered (except perhaps the last question, which has baffled scientists and philosophers for centuries).

To reiterate what we've said, we didn't count the variance of CLAP/TRAP in either model, or that of Jumping in Figure 19–4, for the same reason: they're endogenous variables and thus not free to vary and covary on their own. Since the goal of SEM is to explain the variances of variables and the covariances between pairs of variables that *can* vary, we aren't interested in the parameters of variables that are determined by outside influences. Hence, endogenous variables don't enter into the count.

Why can't we examine more parameters than observations? The analogy is having an equation with more unknown terms than data. For example, if we had a simple equation:

$$a = b + c$$

and we know that $a = 5$, then what are the values of b and c? The problem is that there are an infinite number of possible answers: $b = 0$ and $c = 5$; $b = 1$ and $c = 4$; $b = -19.32$ and $c = 24.32$; and so forth. We say that the model is **undefined** (or **under-identified**), in that there isn't a unique solution. If we had another observation (e.g., $b = -3$), then we can determine that c has to be 8; the model is now **defined** (the technical term is **just-identified**). If there were as many observations as parameters (i.e., we knew ahead of time that $a = 5$, $b = -3$ and $c = 8$), then the model would be correct, but there would be nothing left to estimate (this is referred to as being **over-identified**). This is the situation in Figure 19–3, where there are 10 observations and 10 parameters. In the next section, we'll see the implication of this.

As Good as it Gets: Goodness-of-Fit Indicators

How do we know if the model we've postulated is a good one? In path analysis, there are three things we look for: the significance of the individual paths; the **reproduced** (or **implied**) correlations; and the model as a whole. Of the three, the significance of the paths is the easiest to evaluate. The path coefficients are parameters and, as is the case with all parameters, they are estimated with some degree of error. Again like other parameters, the significance is dependent on the ratio of the parameter to its standard error of estimate. In this case, we end up with a z-statistic:

$$z = \frac{\textit{Estimate of Path Coefficient}}{\textit{Standard Error of Estimate}} \qquad \textbf{(19–2)}$$

and (yet again like other z-tests), if it is 1.96 or greater, it is significant at the .05 level, using a two-tailed test.[20]

A second criterion concerns the **reproduced** (or **implied**) correlation matrix. You'll remember that when we reproduced the correlations from the path coefficients in Figure 19–3, we were able to perfectly duplicate the correlations by tracing all the paths. However, when we tried to reproduce the correlations in Figure 19–4, we found differences between the actual and implied correlations. In fact, if we look at the reproduced correlations in the lower half of Table 19–2, some of them differ quite a bit from the original correlations, which are in the upper half of the matrix. This tell us two things: (1) the model in Figure 19–3 fits the data better than the model in Figure 19–4; and (2) the model in Figure 19–3 fits the data *too* well, in that there is no discrepancy at all between the original and the reproduced correlations. The reason is apparent if we compare the number of observations in Figure 19–3 with the number of parameters; there are 10 of each. Like the case in which we are told ahead of time that $a = b + c$ and that $a = 5$, $b = -3$, and $c = 8$, there's nothing left to estimate, and there is a perfect fit between the model and the data. Life is usually much more interesting when we have fewer parameters than observations; we can then compare how close different models come to estimating the data.

When we look at the model as a whole, the major statistic we use is called the **goodness-of-fit** χ^2 (χ^2_{GoF}), which we've met before within the context of logistic regression. In most statistical tests, bigger is better[21]—the larger the value of a t-test, F test, r, or whatever, the happier we are. This situation is reversed in the case of the χ^2_{GoF}, where we want the result to be as *small* as possible. Why this sudden change of heart?

Let's go back over the logic of χ^2 tests. Basically, they all reduce to the difference between what we *observe* and what would be *expected*, with "expected" meaning the values we would find if the null hypothesis were true. Thus, the larger the discrepancy between the two sets of numbers—observed and expected—the larger the χ^2. Because we usually want our results (the observed values) to be different from the null, we want the value of χ^2 to be large. However, when we use χ^2 to test for goodness of fit, we are not testing our observed findings against the null hypothesis, but rather against some hypothesized model, and we want our results to be congruent with it. So, the *less* the results deviate from the model, the better. In the case of χ^2_{GoF}, we hope to find a nonsignificant result, indicating that our results match the model.

The degrees of freedom associated with the χ^2_{GoF} is the difference between the number of observations and the number of parameters. In the case of the model in Figure 19–4, we have 10 observations and 9 parameters to estimate; hence, $df = 1$. In this example, χ^2_{GoF} is 2.044, which has a p level of .153. Because the χ^2_{GoF} is not statistically significant, we have no reason to reject the model. Just to reinforce what we said earlier, the χ^2_{GoF} for the model in Figure 19–5, where the arrow between Jumping and Height went the "wrong" way, is also 2.044; so this statistic doesn't tell us about the "causality" of the relationship between the variables. For the sake of completeness, we also discussed a different model, one which was so ridiculous that we were ashamed even to draw it. This was a model where there was a path between IQ and Jumping Ability. The χ^2_{GoF} for this model (which also has $df = 1$) is 42.525, which is highly significant, and means that there is a very large discrepancy between the model and the data.

In Figure 19–3, we cannot calculate a χ^2_{GoF}, because the number of parameters is equal to the number of observations, meaning that $df = 0$. This further illustrates that when a model is **fully determined** (a fancy way of saying that the number of parameters and observations is the same), the data fit the model perfectly.

So, if we find that χ^2_{GoF} is not significant, does this prove that our model is the correct one? Unfortunately, the answer is a resounding "No." As we just showed, the χ^2_{GoF} statistics associated with Figure 19–2 and Figure 19–5 are identical, even though one model makes sense and the other is patently ridiculous.[22] Also, the χ^2_{GoF} is dependent on the sample size, since this influences the standard errors of the estimates. If the sample size is low (say, under 100), even models that deviate quite a bit from the data may not result in statistically significant χ^2s. Conversely, if the sample size is very large (over 200 or so), it is almost impossible to find a model that *doesn't* deviate from the data to some degree. Furthermore, there is no guarantee that there is yet another, untested, model that may fit the data even better. Thus, χ^2_{GoF} can tell us if we're on the right track, but it cannot provide definite proof that we've arrived.

[20] *The standard error is estimated by the computer using methods that are beyond the scope of this chapter.*

[21] *The statistical term for this is the "Dolly Parton" effect. For those of you under the age of 30, it was known as the "Pamela Anderson" effect, until her return visit to the plastic surgeon.*

[22] *Although, to be fair, the fact that a model is asinine has never been a roadblock to its adoption. How many people believe in alien abductions (often in spaceships piloted by Elvis), in curing cancer by eating apricot pits, or that the next government will be more honest than the previous one?*

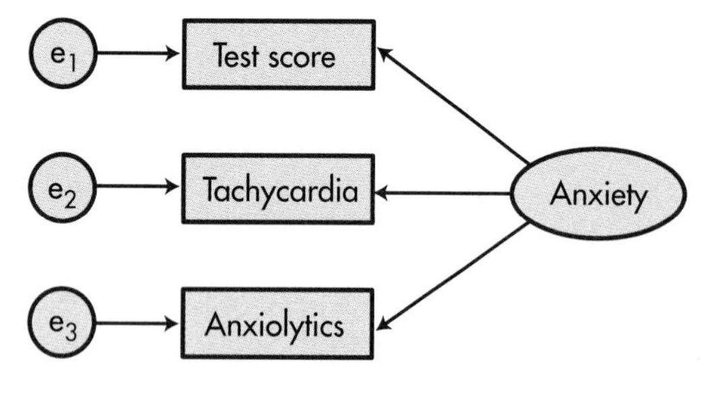

FIGURE 19–9 Relationship between three measured variables and a latent variable.

What We Assume

Path analysis makes certain assumptions, as do all statistical tests. Many of the assumptions are the same as for multiple regression, which isn't surprising, given that the two techniques are closely related. The first assumption is that the exogenous variables are measured without error. This is patently impossible in most (if not all) research, but it serves to remind us that we should try to use instruments that are as reliable as possible. When this rule is violated (as it always is), it results in underestimates of the effects of mediator variables (that is, the indirect paths) and overestimates of the effects of the direct paths (Baron and Kenny, 1986). The second assumption, which is much harder to detect, is that all important variables are included in the model. It's not a good idea to include variables that aren't important, but these usually become obvious from their weak path coefficients. When crucial variables are left out, though, the model fit may be poor or yield spurious results. No statistical test can tell us which variables have been omitted; only our theory can guide us in this regard. Third, multiple regression and path analysis assume that the variables are **additive**. If there are interactions among the variables, an appropriate interaction term should be built into the model (Klem, 1995). Finally, both techniques can handle a moderate degree of correlation among the predictor variables (**multicollinearity**), but the parameter estimates become unreliable if the correlations are high (Klem, 1995; Streiner, 1994).

A Word about Sample Size

The *df* associated with the χ^2_{GoF} test is the difference between the number of observations and the number of parameters. So where does the sample size come into play? The sample size affects the significance of the parameter estimates—the path coefficients, the variances, and the covariances. In all cases, the significance is the ratio of the parameter to the standard error, and the standard error, as we all know, is dependent on the square root of N. Having said that, how many subjects do we need? Unfortunately, there is no simple relationship between the number of parameters to be estimated and the sample size. A *very* rough rule of thumb is that there should be at least 10 subjects per parameter (some authors argue for 20), as long as there are at least 200 subjects. Yes, Virginia, path analysis (and SEM in general) is *extremely* greedy when it comes to sample sizes.

STRUCTURAL EQUATION MODELING

The major limitation with path analysis is that our drawings are restricted to circles and boxes, which become extremely monotonous. Wouldn't it be nice if we could add some variety, such as ovals? Of course the answer is "Yes"; why else would we even ask the question? This isn't as ludicrous as it first appears. In the drawing conventions of SEM, circles represent error or disturbance terms, and boxes are drawn to show **measured** variables—that is, ones we observe directly on a physical scale (e.g., Height) or a paper-and-pencil scale (e.g., CLAP/TRAP). In the chapter on factor analysis, however, we were introduced to another type of variable: **latent** variables (which we referred to in that chapter as **factors**). Very briefly, latent variables aren't measured directly; rather, they are inferred as the "glue" that ties together two or more observed (i.e., directly measured) variables. For example, if a person gets a high score on some paper-and-pencil test of anxiety (a measured variable), shows an increased heart rate in enclosed spaces (another measured variable), and uses anxiolytic medications (yet a third observed variable), we would say that these are all manifestations of the unseen factor (or latent variable) of "anxiety."

The primary difference between path analysis and SEM is that the former can look at relationships only among measured variables, whereas SEM can examine both measured and latent variables. Because we represent latent variables with ovals, there is the added advantage that our diagrams can now become more varied and sexy. Figure 19–9 illustrates the example we just used in SEM terms. Notice the direction of the arrows: they point *from* the latent variable *to* the measured ones, which may at first glance seem backwards. In fact, this reflects our conceptualization of latent variables—that they are the underlying causes of what we are measuring directly. Thus, the latent variable (or trait or factor) of anxiety is what accounts for the person's score on the test, his or her tachycardia, and is what leads to the use of pills. Also, since the three measured variables are endogenous, they have error or disturbance terms associated with them.

If Figure 19–9 looks suspiciously like the pictures we drew when we were discussing factor analysis, it's not a coincidence. Exploratory factor analysis (EFA—the kind we discussed in Chapter 18) is now seen as a subset of SEM. We don't gain very much, however, if we were to use the programs and techniques of SEM to do EFA. In fact, because SEM is a *model testing* or *confirmatory* technique, rather than an exploratory one, the computer output is less helpful than from programs specifically designed to do EFA (if that's what we want to do). Despite this, we'll start off with EFA to show you how the concepts we covered in path analysis apply in this relatively simple case. This will put us in a better position to understand how SEM works in more general cases.[23]

SEM and Factor Analysis

In addition to the variables of Height, Jumping Ability, and IQ, Dr. Teeme also hypothesized that, since the first part of the word "cheerleading" is "cheer," success in this endeavor also depends on the candidate's personality. However, unable to find a questionnaire that measures Cheer directly, he had to resort to using three other tests that he felt collectively measured the same thing: one tapping extroversion (the Seller of Used Cars Scale, or SUCS); another measuring positive outlook on life (the Mary Poppins Inventory, or MPI); and a third focusing on denial of negative feelings (the We're OK Scale, or WOKS). The correlations among these scales are shown in Table 19–3. As he suspected, the correlations were moderate, but positive.

TABLE 19–3 Correlations among three tests to measure "Cheer"

	SUCS	MPI	WOKS
SUCS	1.000	.699	.608
MPI		1.000	.577
WOKS			1.000

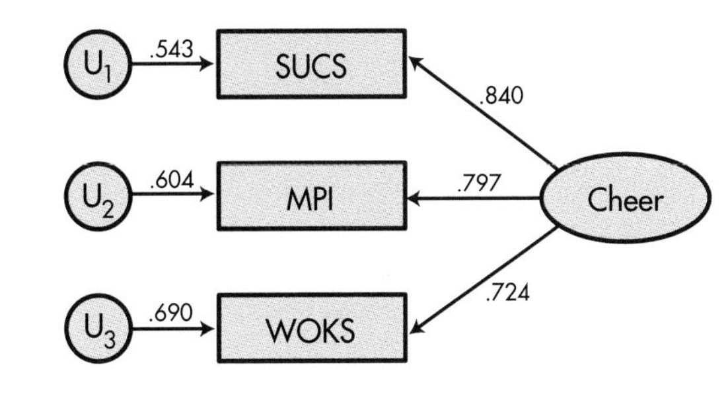

FIGURE 19–10 Factor loadings and uniqueness for the factor "Cheer" and its three measured variables.

If we now did an EFA, using a least-squares method of extracting the factor, we would find that the factor loadings were 0.840 for SUCS, 0.797 for MPI, and 0.724 for WOKS. A drawing of this using the SEM conventions is shown as Figure 19–10. There are a few points to note. First, the disturbance or error term in EFA is usually labelled with the letter U (which stands for "uniqueness"); the terminology is different, but the concept is the same.[24] Second, for each variable, the square of the factor loading (which is equivalent to a path coefficient) plus the square of the uniqueness equals 1.00 (e.g., $0.840^2 + 0.543^2 = 1.00$). In English, we've divided up the variance of the variable into two components: that explained by the factor (or latent variable); and that which is not explained by it (the uniqueness, error, or disturbance). Finally, the product of any two factor loadings is equal to the correlation between the variables. For example, the factor loadings for MPI and WOKS are 0.797 and 0.724, respectively; their product (0.797×0.724) is 0.577, or their correlation.

Now, if we ran this as if it were a structural equation model, we would find exactly the same thing! So, why make such a big deal about the difference between EFA and *confirmatory* factor analysis (CFA—the SEM approach to factor analysis)? Leaving the math aside for the moment,[25] the major difference is a conceptual one. In traditional, run-of-the-mill EFA, we are in essence saying, "I don't know how these variables are related to one another. Let me just throw them into the pot and see what comes out." That's why it's called "exploratory," folks. In fact, when we don't know what's going on, it can be a very powerful tool that can help us understand the interrelationships in our data. The downside is that we can end up with a set of factors that look good statistically but don't make a whole lot of sense from a clinical or scientific perspective; that is, it may not be at all obvious why the variables group together the way they do.

Confirmatory factor analysis, as is the case with all variants of SEM, is a *model testing* technique, rather than a *model building* one.[26] So, while we still get pages and pages of output (reams, if we're not careful what we ask for), the action is really at the end, where we see how well the model actually fits the data. This is a point that we've already mentioned and, as promised, we'll keep emphasizing: changes to the model to make it fit the data better should be predicated on our theoretical understanding of the phenomenon we're studying, and not based on moving arrows around to get the best goodness-of-fit index. We'll say a bit more about CFA later in the chapter.

Before we move on to discuss the steps in SEM, some other points about the advantages of using latent variables (in addition to the esthetic one of having ovals in our drawings) need to be discussed. In our example of the three scales to measure "Cheer," we said that none measures it exactly but that the three together more or less capture what we want; in other words, we're making a "super-scale" out of three scales. We can accomplish the same goal using other techniques, but the route would be more roundabout. We would have to run an EFA on these variables (and any other sets of variables we wanted to combine), use the output from this analysis to calculate a factor score for each person, and then use that factor score in the next stage of the analysis. With SEM, all of this can be accomplished in one step.

A second advantage arises from measurement theory. Any time we measure something, for example, whether it's blood pressure or pain, some error is involved. The error can arise from a variety of sources: the person's blood pressure may change from one moment to the next; the manometer may

[23] *Notice we said "general" and not "more difficult." That's an old didactic trick we learned to make things more palatable.*

[24] *The reason for the difference is purely historical; EFA was developed many years before SEM, and the term was chosen to denote the unique contribution of that variable: the variance it doesn't share with the other variables (see Chapter 18).*

[25] *Actually, we're going to leave it aside as long as we can.*

[26] *And you wondered why it is called* confirmatory *factor analysis?*

not be perfectly calibrated; and the observer may round up or down to the nearest 5 mm or make a mistake recording the number. When paper-and-pencil or observer-completed tests are involved, even more sources of error exist, such as biases in responding or lapses in concentration. These errors result in a measure that is less than reliable. A problem then arises when we correlate two or more variables: the observed correlation is lower than what we would find if the tests had a reliability of 1.0. Thus, we will almost always underestimate the relationships among the variables and, if the reliabilities are "attenuated" very much, we may erroneously conclude that there is no association when, in fact, one exists (i.e., we would be committing a Type II error). The solution is to **disattenuate** the reliabilities and figure out what the correlation would be if both tests were totally reliable.[27] One major advantage of SEM is that this is done as part of the process.

[Note to the reader: You can skip this paragraph if you're not interested in the application of SEM to scale development and are quite happy to remain ignorant of it. This should apply to about 93% of you.] If we had only one scale to tap some construct (that is, we would be dealing with a measured variable rather than a latent variable defined by two or more measured ones), we are sometimes further ahead if we randomly split the scale in half and then construct a latent variable defined by these two "subscales." We can then calculate the reliability[28] and, based on this, the disattenuated correlation.[29]

Now that we have a bit of background, let's turn to the steps involved in developing structural equation models. We'll follow the lead of Schumacker and Lomax (1996) and break the process down into five steps:[30]

- Model specification
- Identification
- Estimation
- Testing fit
- Respecification

Model Specification

To a dress designer, **model specification** may mean, "I want a woman who's 5′8″ tall, weighs 93 pounds, and has a perfect 20-20-20 figure." To many men, it means stating which supercharged engine to put into a sports car, and whether to get a hard or a rag top (to attract the model of the first definition). Here, however, we are referring to something far more mundane: explicitly stating the theoretical model that you want to test. We have already discussed much of this when we were looking at path analysis. The same concepts apply, but now we will broaden them to include both measured variables (which can be analyzed with path analysis) and latent variables (which cannot be). For example, all of the diagrams in Figure 19–6 can be drawn replacing the measured variables with latent ones. In addition, we can "mix and match," using both latent and measured variables in the same diagram, as long as we don't do obviously ridiculous things, such as having two latent variables define a measured one.

Let's use these concepts to fully develop a model of success in cheerleading, which is shown in Figure 19–11. We have a latent variable, Athletic Ability, which is measured by the three observed variables IQ, Height, and Jumping; and a second latent variable, Cheer, which is measured by our scales WOKS, SUCS, and MPI. Based on what we said ear-

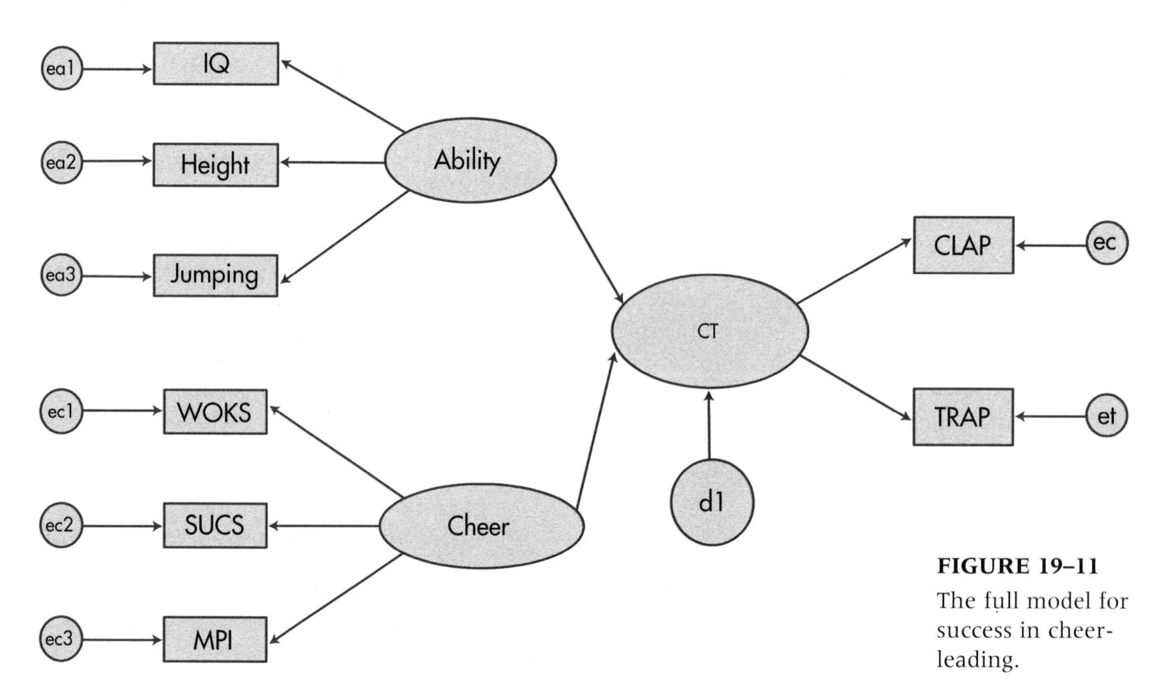

FIGURE 19–11
The full model for success in cheerleading.

[27] *If you want more details about how to do this, we can't recommend a better book than* Health Measurement Scales: A Practical Guide to Their Development and Use. *Rest assured that the fact that we wrote it (Streiner and Norman, 1995) in no way prejudiced our opinion.*

[28] *This would be equivalent to the split-half reliability, for those who are interested (Streiner and Norman, 1995). (Actually, it's the split-half reliability for those who aren't interested, too.) On a technical note, the reliability is corrected only for those sources of error that are captured in the model, usually due to items. If the model does not include the same scale administered on two occasions, then it cannot correct for test-retest unreliability, for example. For more details, see DeShon (1998).*

[29] *Some measurement theoreticians do not like using disattenuated reliabilities in scale development because they are overly optimistic regarding the utility of the tests. But, in this context, they give a more accurate estimate of the relationships among the variables, which is our primary interest.*

[30] *As you can see, this is much less difficult than recovering from*

lier,[31] it would make sense to randomly split CLAP/TRAP into two parallel forms (CLAP and TRAP) and have these define the latent variable, CT. Because CT is now an endogenous variable (it has arrows coming toward it from Cheer and Athletic Ability), we have to give it a disturbance term, which we've called d1.

This step of model building is called the **measurement model**, because we are specifying how the latent variables are measured. Some of the questions we could ask at this stage are as follows: (1) How well do the observed variables really measure the latent variable? (2) Are some observed variables better indices of the latent variable than are other variables? (3) How reliable is each observed variable? (We will look at these issues in more depth and actually analyze this model when we discuss some worked-out examples of SEM and CFA.)

Identification

When we were discussing path analysis, we said that the number of parameters can never be larger than the number of observations, and, ideally, they should be less. However, it is often the case that once we start adding up all the variances and covariances in our pretty picture, their number exceeds the limit of $[k \times (k + 1)]/2$. The solution is to put constraints on some of the parameters. In the example we used previously, we were able to solve the equation, $5 = b + c$, by specifying beforehand that $b = -3$; that is, we made b a **fixed** parameter. We could also have solved the equation if we had said that $b = c$, in which case both b and c would have to be $2\frac{1}{2}$. In this case, b and c are referred to as **constrained** parameters. So, we have three types of parameters in structural models: (1) **free** parameters, which can assume any value and are estimated by the structural equation model; (2) **fixed** parameters, which are assigned a specific value ahead of time; and (3) **constrained** parameters, which are unknown (as are free ones) but are limited to be the same as the value(s) of one or more other parameters. The joy[32] of SEM is figuring out how many parameters to leave as free and whether the remaining ones should be fixed or constrained.

Perhaps the easiest way of constraining parameters[33] is simply not to draw a path. We said that MPI loaded on Cheer and (implicitly, by not having a path) did not load on Athletic Ability. That's another way of saying that the parameter for the path from MPI to Athletic Ability is fixed to have a value of 0. Making the model as simple as possible is often the best way of avoiding "identification problems." A second method involves the latent variables. For each unobserved variable with one single-headed arrow emerging from it (such as the error terms), we must fix either the path coefficient or the variance of the error itself to some arbitrary, non-zero value; otherwise, we'd be trying to estimate b and c at the same time. But, because we don't actually measure (or observe) the error term, it doesn't make sense to assign a value to it, which leaves us fixing the path coefficient. The easiest thing to do is to give it the value 1, and this is what's usually done. In any case, assigning a different value won't affect the overall fit of the model, just the estimate of the error variance.[34]

When a latent variable has two or more single-headed arrows coming from it (as is the case with CLAP/TRAP, Cheer, and Athletic Ability), one of them must be set to be equal to 1 (i.e., be a **fixed** parameter). It doesn't really matter at all which one is chosen, but it's best to use the variable with the highest reliability, if this is known. Changing which measured variable has a coefficient of 1 will alter the unstandardized regressions of all the variables related to that latent variable (because the unstandardized regressions are relative to the fixed variable), but it does not affect the standardized regression weights. Notice that we didn't draw curved arrows between the exogenous variables. This is because we assume that the exogenous variables are correlated with themselves, and most programs, such as AMOS (Arbuckle, 1997), build this in automatically.

Finally, as we mentioned previously, there may be times when we believe that the error terms of two or more variables are identical (plus or minus some variation), and these become **constrained** parameters. In programs like AMOS, we indicate this by assigning the same name to the errors.

You might think that if you've gone to the trouble of counting up the number of observations, simplifying the model, and constraining some of the parameters, you will avoid identification problems. Such thinking is naive, and reflects a trust in the inherent fairness of the world that is rarely, if ever, justified. The steps we've just discussed are necessary but not sufficient. Identification problems will almost surely arise if you have a nonrecursive model (i.e., one that has reciprocal relationships), which is yet another reason to avoid them whenever possible. You can also run into difficulty if the **rank** of the matrix[35] you're analyzing is less than the number of variables. This can occur if one variable can be predicted from one or more other variables, meaning that the row and column in a correlation matrix representing that variable is not unique. For instance, the Verbal IQ score of some intelligence tests is derived by adding up the scores of a number of subscales (six on the Wechsler family of tests). If you include the six subscale scores as well as the Verbal IQ score in your data matrix, then you will have problems, since one variable can be predicted by the sum of the others; that is, you have seven variables (the six subscales plus the Verbal IQ) but only six unique ones, resulting in a matrix whose rank is less than the number of variables. Even high correlations among variables (e.g., between height and weight) may produce problems.

Often, there's no easy way to tell ahead of time if you're going to have problems with under-

alcoholism, gambling, drug abuse, "co-dependency," or any other such problems, which require 12 steps. We can also forgo the weekly meetings.

[31] *At least we pay attention to what we've said.*

[32] *A term to be taken with a large grain of salt.*

[33] *The terms "constraining parameters" or "constraints" can refer to making them either fixed or constrained; confusing, ain't it?*

[34] *If you want a more detailed explanation of the effect of choosing different values for the path coefficient, see Arbuckle (1997).*

[35] *Roughly speaking, the rank of a matrix is the number of unique rows and columns.*

identification of the model. You just do the best you can in setting up the model, pray hard, and hope the program runs. If the output says that the model needs more constraints, then you have to go back to your theory and determine if there is any justification in, for example, constraining two variables to have equal variances. You may suspect they may covary if they are the same scale measured at two different times, or two different scales tapping the same construct. If this doesn't help, you may want to consider a different career.

Estimation

Now that you've specified the model, the time has come to estimate all of those parameters. The easy work is all of that matrix algebra—inverting matrices, transforming matrices, pre- and post-multiplying matrices, and so forth. Because it's easy, the computer does it for us. What's left for us is the hard stuff—deciding the best method to use. This requires brain cells, a commodity that is in short supply inside the computer.[36] The fact that a number of different techniques exists should act as a warning, one that we've encountered in other contexts. If there were one approach that was clearly superior, then the law of the statistical jungle[37] would dictate that it would survive, and all of the other techniques would exist only as historical footnotes. The continued survival of all of them, and the constant introduction of new ones, indicates that no clearly superior solution exists; hence, the unfortunate need to think about what we're doing.

The **unweighted least squares** (ULS) method of estimating the parameters has the distinct advantage that it does not make any assumptions about the underlying distribution of the variables. However, it is scale-dependent; that is, if one or more of the indices is transformed to a different scale, the estimates of the parameters will change. In most of the work we do, the scales are totally arbitrary; even height can be measured either in inches or centimeters (or centimetres, if you live in the U.K. or one of the colonial backwaters[38]). This means that the same study done in Canada and the U.S. may come up with different results, simply because of different measurement scales used. This becomes more of a problem when we use paper-and-pencil tests which don't have meaningful scales. This is an unfortunate property, which is one of the reasons ULS is rarely used. **Weighted least squares** (WLS) is also distribution free and does not require multivariate normality, but it requires a very large sample size (usually three times the number of subjects whom you can enroll) to work well.

Many programs default to the maximum likelihood (ML) method of estimating the parameters. This works well as long as the variables are multivariate normal, and consist of interval or ratio data. But if the data are extremely skewed or ordinally scaled, then the results of the ML solution are suspect.

So, which one do you use? If you have access to the SEM program called LISREL8, and its "front-end" program, PRELIS2, then use those. They will calculate the correct matrix, based on the types of variables you have. If you use one of the other programs (e.g., EQS, PROC CALIS, or AMOS), then it may be worthwhile to run the model with a few different types of estimators. If all of the results are consistent, then you can be relatively confident about what you've found; if they differ, you'll have to go back and take a much closer look at the data, to see if you have non-normality, skewness, or some other problem. Then, either try to fix it (e.g., with transformations) or choose a method that best meets the type of data you have.

Testing the Fit

In the previous section, on estimation procedures, we lamented the fact that there were so many different approaches. While that is undoubtedly true, the problem pales into insignificance in comparison to the plethora of statistics used to estimate goodness of fit. We already mentioned the χ^2_{GoF} in the context of path analysis, and that test is also used in SEM. It has the distinct advantage that, unlike all of the other GoF indices, it has a test of significance associated with it. One rule of thumb for a good fit is that χ^2_{GoF} is not significant and that χ^2_{GoF}/df should be less than two. Unfortunately, as we mentioned when we were discussing path analysis, it is very sensitive to sample size and departures from multivariate normality.

Most of the other indices we will discuss[39] are scaled to take values between 0 (no fit) and 1 (perfect fit), although what is deemed a "good fit" is often arbitrary. Usually, a value of 0.90 is the minimum accepted value, but as we just mentioned, there are no probabilities associated with these tests. Let's go over some of the more common and useful ones to show the types of indices available, without trying to be exhaustive (and exhausting).

One class of statistics is called **comparative fit indices**, because they test the model against some other model. The most widely used index (although not necessarily the best) is the **Normed Fit Index** (NFI; Bentler and Bonett, 1980), which tests if the model is different from the null hypothesis that all of the variables are independent from one another (in statistical jargon, that the covariances are all zero).[40] It takes the form of:

$$NFI = \frac{\chi^2_{null} - \chi^2_{model}}{\chi^2_{null}} \qquad \textbf{(19–3)}$$

One disadvantage of the NFI is that it is sensitive to the sample size; models with small values of N may have NFIs < .90, even if they fit well. The **Normed Fit Index 2** (NFI2, also called the **Incremental Fit Index** or IFI) tries to compensate for this by incorporating the degrees of freedom:

[36] *And almost as rare outside it.*

[37] *Although somehow the thought of statisticians fighting it out, "red in tooth and claw" (yes, statisticians can quote Tennyson), appears somewhat oxymoronic to us.*

[38] *The authors are allowed to say this, because we live in one of them.*

[39] *We will only mention a few of them; if we listed them all, this book would be thicker than the large print, illustrated version of* War and Peace.

[40] *As Arbuckle (1997) says, the NFI and other indices that compare your model against the null "encourage you to reflect on the fact that, no matter how badly your model fits, things could always be worse" (p. 563).*

$$NFI2 = \frac{\chi^2_{null} - \chi^2_{model}}{\chi^2_{null} - df_{model}}$$

(19–4)

Other indices resemble R^2, in that they attempt to determine the proportion of variance in the covariance matrix accounted for by the model. One such index is the **Goodness-of-Fit Index** (GFI); fortunately for you, its formula involves a lot of matrix algebra, so we won't bother to show it.[41] The **Adjusted GFI** (AGFI) is analogous to the adjusted R^2, in that you're penalized for having a lot of parameters and a small sample size. The AGFI is a **parsimony** fit index, because you're rewarded for having a parsimonious model. There are a number of variants of this, all of which decrease the value of AGI proportionally to the number of parameters you have. Another widely used index is **Akaike's Information Criterion** (AIC) which is unusual in that smaller (closer to 0) is better:[42]

$$AIC = \chi^2_{model} - 2df_{model}$$

(19–5)

A slight variant of it is called the **Consistent AIC**, or CAIC:

$$CAIC = \chi^2_{model} - (log_e N + 1)df_{model}$$

(19–6)

Because nobody knows what a "good" value of AIC or CAIC should be, these indices are used most often to *compare* models—to choose between two different models of the same data. The one with the smaller AIC or CAIC is "better"; however, there isn't any statistical test for either the indices or the difference between two AICs or two CAICs, so we can't say if one model is statistically better or insignificantly better than the other.

Life is easy when all of the fit indices tell us the same thing. What do we do when they disagree? The most usual situation occurs when we get high values (i.e., over .90) for GFI or NFI2, but the χ^2_{GoF} is significant. Unfortunately, here we have to use some judgment. Can the significant χ^2_{GoF} be due to "too much" power? If so, it's probably better to trust one of the other indices. If there are roughly 10 subjects per parameter, then we should look at a number of indices. If they all indicate a "good fit," then go with that. But, if the χ^2_{GoF} is significant, and the other indices disagree with one another, then we have a fit that's marginal, at best, and whether or not to publish depends on your desperation level for another article.

Respecification

Respecification is a fancy term for playing with the model to achieve a better fit with the data. If you've been paying attention, you should realize that the statistical tests play a secondary role in this; the primary role should be your understanding of the area, based on theoretical and empirical considerations. Keep chanting this mantra to yourself as you read this section.

The major reason that a model doesn't fit is that you haven't included some key variables—the ones that are really important. Unfortunately, there are no statistical tests that can help us in this regard. There are no computer packages that will give you a Bronx cheer and say, "Fool,[43] you forgot to include the person's weight." All of your purported colleagues will be only too happy to perform this function, but only when it's too late, and you're presenting the findings at an international conference.

The statistical tests that do exist are for the opposite type of mis-specification errors: those due to variables that don't belong in the model or have paths connecting them to the "wrong" endogenous or exogenous variables. The easiest way to detect these is to look at the parameters. First, they should have the expected sign. If a parameter is positive and the theory states it should be negative, then something is dreadfully wrong with your model. The next step is to look at the significance of the parameters. As we've said, all parameters have standard errors associated with them, and the ratio of the parameter to its standard error forms a *t*- or *z*-test. If the test is not significant, then that parameter should likely be set equal to 0. Of course, this assumes that you have a sufficient sample size, so that you're not committing a Type II error.

All of the major SEM programs, such as LISREL, CALIS (part of SAS), AMOS, and EQS, can examine the effects of freeing parameters that you've fixed (the **Lagrange multiplier test**) and dropping parameters from the model (the **Wald statistic**). *These statistical tests should be used with the greatest caution.* Whether to follow their advice *must* be based on theory and previous research; otherwise, you may end up with a model that fits the current data very well but makes little sense and may not be replicable.

Now that we've given you the basics, let's run through a couple of examples.

A Confirmatory Factor Analysis

Let's assume that we have seven measured variables that we postulate reflect two latent variables: *a1* through *a4* are associated with the latent variable *f1*, and *b1* through *b3* with latent variable *f2*. We also think that the two latent variables may be correlated with each other. We start by drawing a diagram of our model (if we're using a program such as AMOS or EQS), which is shown in Figure 19–12. The program is relatively smart,[44] so it automatically fixed the parameters from all of the error terms to the measured variables to be 1. For each of the endogenous variables, it also set the path parameter for one measured variable to be 1. We didn't like the choice the program made, so we over-rode it and selected the variable in each set that, based on previous

[41] *If, for some reason you* really, really *want to see what it looks like, check out Tabachnick and Fidell (1996).*

[42] *To be consistent with our nomenclature, we would call this the Twiggy criterion (also known as the "Kate Moss criterion" by the younger set).*

[43] *We use this term only because our editor won't let us say "Schmuck" in such a family-oriented book.*

[44] *With the emphasis on the term "relatively."*

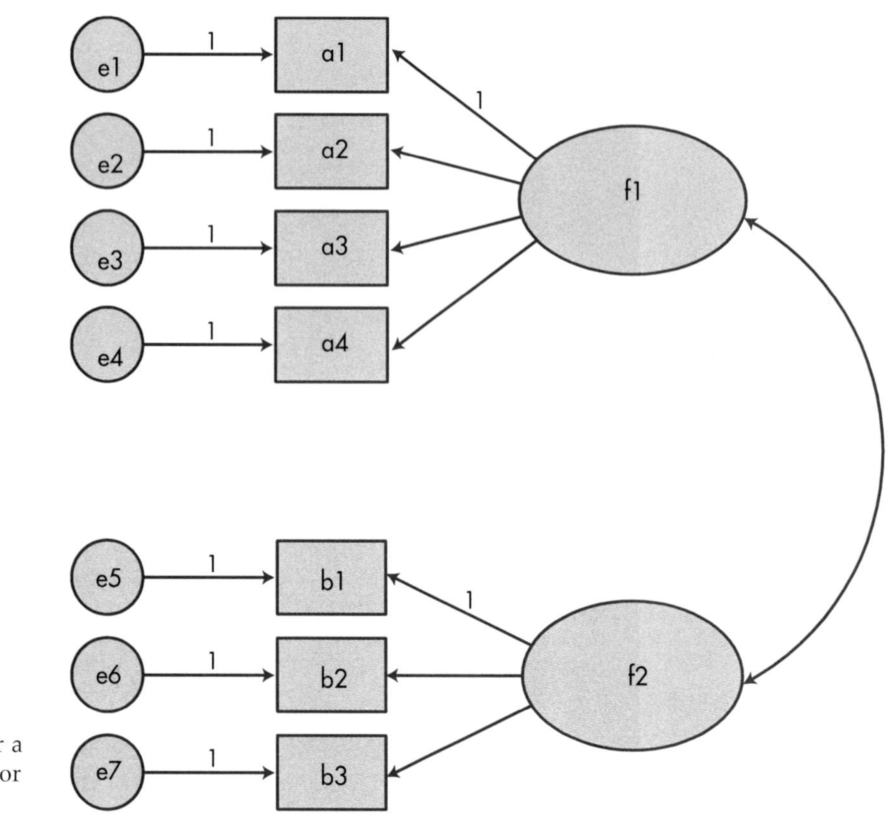

FIGURE 19–12
Input diagram for a confirmatory factor analysis.

research, has the highest reliability. After we push the right button,[45] we get the diagram shown in Figure 19–13 and reams of output, which we've summarized in Table 19–4.

First, what do all those little numbers in Figure 19–13 mean? The ones over the arrows should be familiar; they're the path coefficients or standardized regression weights (either term will do), which are equivalent to the factor loadings in EFA. The numbers over the rectangles are the squared multiple correlations, which are equivalent to the communality estimates in EFA. There are two other things this figure tells us. First, we goofed when it comes to variable *a4*; it doesn't really seem to be caused by the latent variable *f1*. Note that in contrast to EFA, we aren't told if it loads more on factor *f2* or it doesn't load on either factor; yet again, that's because we're testing a model, not trying to develop one. The second fact is that factors *f1* and *f2* probably aren't correlated; the correlation coefficient is only 0.06.

Now let's turn to the printed output in Table 19–4 and see what else we learn. First, there are 28 "sample moments"; in other words (i.e., English), 28 observations. This is based on the fact that there are seven measured variables, so there are $(7 \times 8) / 2 = 28$ observations. Our model specifies 15 parameters to be estimated: five regression weights (two others aren't estimated because we fixed them to be 1); the covariance between *f1* and *f2*; and the variances of the seven error (or disturbance) terms and two endogenous variables.[46] There are 13 degrees of freedom, which is the difference between the number of observations and the number of parameters. The next block of output tells us that the χ^2_{GoF} is 16.392, which, based on 13 degrees of freedom, has a p level of .229. So, despite the fact that variable *a4* doesn't work too well, the model as a whole fits the data quite well.

We then see the unstandardized and standardized regression weights. The unstandardized weights for *a1* and *b1* are 1.00, which is encouraging, because we set them to be equal to 1. The other five weights have standard errors associated with them, and the ratio of the weight to the SE is the **Critical Ratio** (CR) which is interpreted as a z-test. All of them are significant (at or over 1.96) except for *a4*, further confirming that it isn't correctly specified. Similarly, the covariance (and hence the correlation) between *f1* and *f2* is low and has a CR of only 0.551; that is, the two latent variables or factors aren't correlated. The next sets of numbers show the variances we're estimating and the squared multiple correlations (which are also given in the figure).

Because we asked for them, we also get the **Modification Indices** (MI), which tell us how much the model could be improved if we specified additional paths. The largest one, which is listed under Regres-

[45] *Unfortunately, you have to read a 350-page manual to figure out which button that is.*

[46] *Note that we estimate the* variances *of the error terms but not their* path coefficients; *those we constrained to be 1, since we can't estimate both at the same time.*

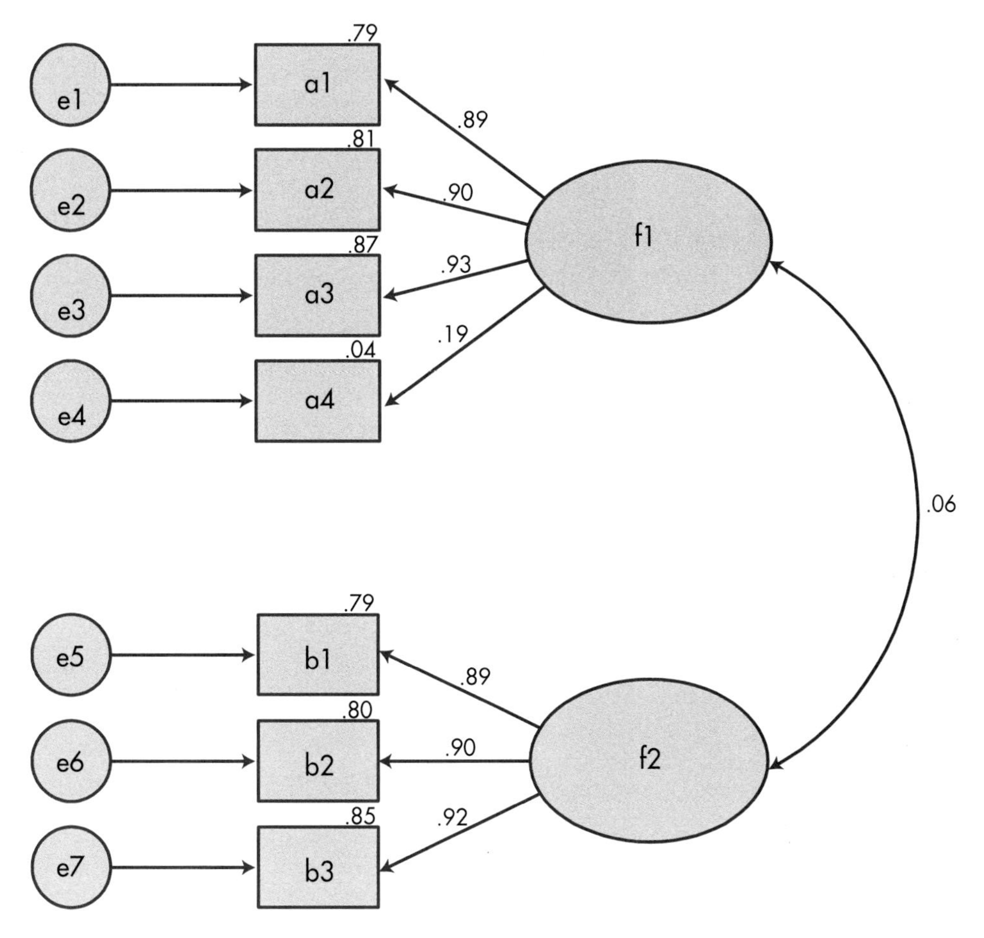

FIGURE 19–13 Output diagram based on Figure 19–12.

sion Weights, is $a3 \leftarrow b2$. That means that if we drew a path from *b2* to *a3*, our fit would improve. In fact, the path coefficient between the two is –0.123,[47] and the χ^2_{GoF} (based now on $df = 12$, because we specified another path) drops to 10.028.[48] But, because there is no theoretical rationale for this path (or for the other proposed modifications), we'll just ignore them.

Finally, we've given just a few of the myriad other GoF indices. The **saturated model** represents perfection (as many parameters as observations, meaning there's nothing more to estimate), and the **independence model** is the opposite (assuming nothing correlates with anything). Fortunately, our model is close to perfection; all of the indices are over 0.90. Our model would fit even better if we dropped variable *a4*. Yet again, this should be dictated by theory. If we believe that the variable is a substantive one, and that the nonsignificant path coefficient may be due to sampling error or a small sample size, we would keep it; otherwise, into the trash can it goes.

Comparing Two Factor Analyses

We're sometimes in a position where we want to compare two factor structures; for example, are the results for patients and controls or for men and women alike[49] and, if not, how do they differ? This *can* be done with EFA, but it is difficult, and the methods of comparing factor structures leave much to be desired. However, it's relatively easy to do it with CFA.[50] If we don't have any hypotheses beforehand regarding the factor structure, we can start by running an EFA with one group and then use the results to fix the parameter estimates in a CFA for the second group. Conversely, if we do have some idea of what the structure should look like, we can specify it for both groups and see where it fits and doesn't fit for each.

As an example, we'll stay with the problem presented in Figure 19–13 and assume we drew another sample, one in which *a4* actually does load on *f1* but we don't know this beforehand. Instead of fixing just one of the paths from the latent variables to the measured ones, we'll put in the unstandardized

[47] *Don't look for this in the output; it isn't there. We reran the model with this path included, just for your benefit, but we haven't shown the output.*

[48] *See note 47.*

[49] *We know the answer to the second question is "No"; men are from Mars and women from Venus, leaving open the question of who's on Earth (not to be confused with who's on first)?*

[50] *See note 44.*

TABLE 19–4

Selected output for the confirmatory factor analysis

Number of sample moments = 28
Number of distinct parameters to be estimated = 15
Degrees of freedom = 13

Chi-squared = 16.392
Degrees of freedom = 13
Probability = .229

Regression weights	Estimate	SE	CR
a1←f1	1.000	—	—
a2←f1	1.029	0.078	13.166
a3←f1	1.021	0.073	13.940
a4←f1	0.199	0.107	1.857
b1←f2	1.000		
b2←f2	1.100	0.086	12.779
b3←f2	1.091	0.082	13.317

Standardized regression weights	Estimate
a1←f1	0.888
a2←f1	0.901
a3←f1	0.934
a4←f1	0.190
b1←f2	0.888
b2←f2	0.897
b3←f2	0.920

Covariances	Estimate	SE	CR
f1↔f2	0.044	0.080	0.551

Correlations	Estimate
f1↔f2	0.059

Variances	Estimate	SE	CR
f1	0.824	0.148	5.560
f2	0.665	0.120	5.537
e1	0.220	0.044	5.050
e2	0.202	0.043	4.683
e3	0.127	0.036	3.487
e4	0.866	0.123	7.018
e5	0.178	0.037	4.854
e6	0.196	0.043	4.614
e7	0.143	0.038	3.809

Squared multiple correlations	Estimate
a1	0.789
a2	0.812
a3	0.871
a4	0.036
b1	0.789
b2	0.804
b3	0.847

Continued

regression weights from the first sample and again, based on our previous results, state that the covariance between *f1* and *f2* is 0. The output will be very similar to that in Table 19–4, with a few notable exceptions.

First, the number of parameters to be estimated drops from 15 to 9, because we've fixed an additional 6 parameters: 5 paths from the latent variables plus the covariance. Second, there will not be standard errors and critical ratios given for these 6 parameters, since we are not estimating them. The χ^2_{GoF}, now based on $df = 19$, is a whopping 144.47, indicating that the model doesn't fit the data worth a plugged nickel. If we look at the modification indices, the largest ones involve *a4* and *e4* in various forms, such as suggesting that we include covariance terms between *e4* and *f1*, or paths between *a4* and the three other variables associated with *f1*. None of these makes sense theoretically, but they all point to a mis-specification involving *a4*. We would again return to our theory and hypothesize that the path coefficient, which we fixed at 0.20 to be congruent with the results from the first sample, is wrong and perhaps should be closer to 0.80. Alternatively, we can set it free and see what the program does with it. Note (yet again) that our use of the modification indices is tempered by our knowledge and theory.

If the model actually fits the data, then we can conclude that the factor loadings that we found for one group would fit the second group, too. In this case, we can up the ante and make the comparison more stringent: Are the variances of the error terms similar across samples? This type of analysis is very useful for determining the equivalence of questionnaires in different groups of subjects.

A Full SEM Model

Now let's return to the complete model of success in cheerleading, shown in Figure 19–11, and add what we've learned. First, we have to fix all of the paths leading to the various disturbance terms to 1, and fix one path from each latent variable to be 1. Second, because CLAP and TRAP are random halves of the same test, it is logical to assume that their variances are similar. We indicate the fact that we've constrained these terms by giving the variances the same name. The results of all of this fixing and constraining are shown in Figure 19–14; the terms *vct* over the disturbance terms for CLAP and TRAP tell the program that these variances should be the same. This diagram now forms the input to the program, which should run as long as the variable names in the rectangles correspond to the variable names in our data file.

The output from the program is shown in Figure 19–15. To begin, the χ^2_{GoF} is 38.225 which, based on 19 degrees of freedom, is highly significant ($p = .006$). The other GoF indices are equivocal: GFI and NFI are both just slightly above the cutoff point of 0.90, while AGFI is only 0.831. All of this leads us to believe that the model could stand quite a bit of improvement, but where? Let's start with the measurement aspect of the model—how well are we measuring the latent variables of Athletic Ability, Cheer, and CLAP/TRAP?

The answer seems to be, not too badly, thank you, but perhaps we can do better with Ability and Cheer. If we look at the Modification Indices, most of them don't make too much sense from the perspective of our theory, but one bears a closer look—the suggestion of adding a covariance between *ea2* and *ea3*. Because Cheer as a whole seems to add little to the picture, let's leave it aside for now and rerun the model adding *e2* ↔ *e3*. Gratifyingly, the χ^2_{GoF} (df = 18) drops to 22.557, which has an associated p-value of .208. Because this model is a subset of the original one,[51] the difference between their respective χ^2s is itself distributed as a χ^2. So, if we subtract the χ^2s and the dfs, we get χ^2 (1) = 15.668, meaning that there was a significant improvement in the goodness of fit. This is also reflected in an increase in the path

TABLE 19–4 Continued

Covariances	MI	Par change
e2↔e6	5.408	0.064
e3↔f2	4.173	−0.081

Regression weights	MI	Par change
a2←b2	5.603	0.125
a3←f2	4.157	−0.122
a3←b2	5.749	−0.133

Model	GFI	AGFI	NFI
Your model	0.957	0.908	0.968
Saturated model	1.000		1.000
Independence model	0.454		0.000

SE = standard error; CR = critical ratio; MI = modification index; GFI = Goodness-of-Fit Index; AGFI = Adjusted GFI; NFI = Normed Fit Index.

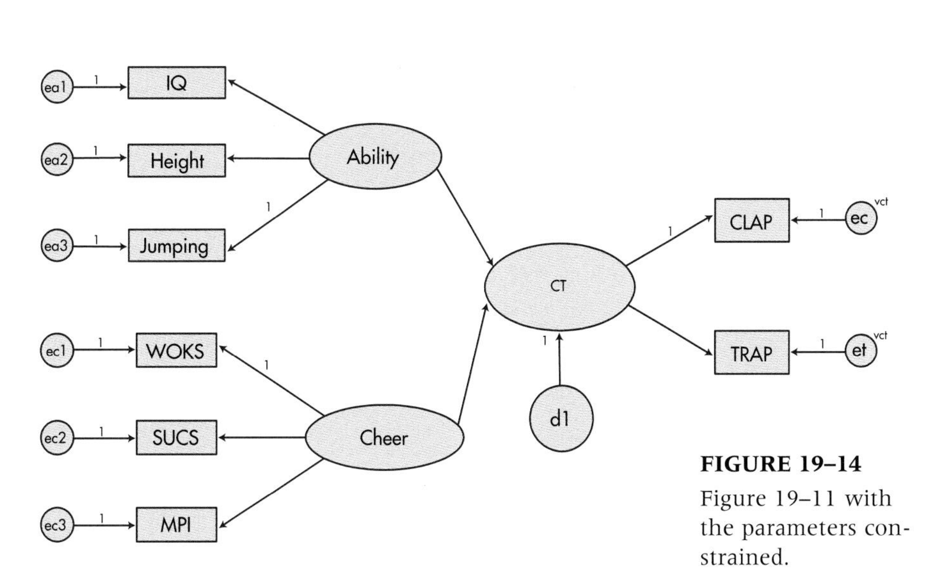

FIGURE 19–14
Figure 19–11 with the parameters constrained.

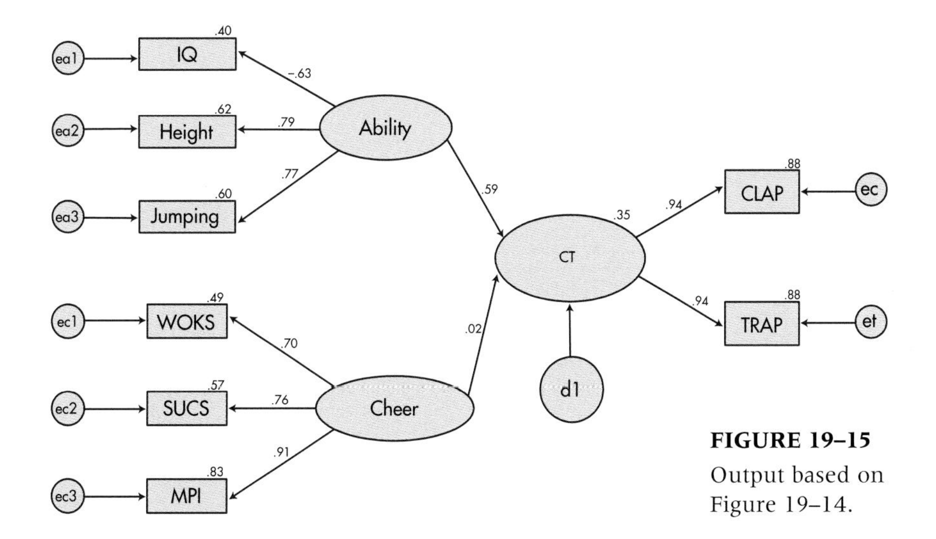

FIGURE 19–15
Output based on Figure 19–14.

[51] *Model X is a subset of Model Y (or is* ***nested*** *within Model Y) if all of the variables in Model X are also in Model Y. But there are some variables in Model Y that don't appear in Model X; for example, Model Y has variables A, B, and C and Model X has only variables A and C.*

coefficient from Ability to CLAP/TRAP, from 0.59 to 0.73; a drop in the coefficient from Cheer (from 0.02 to 0.01); and the fact that the other fit indices are in an acceptable range.

Finally, because Cheer doesn't help,[52] we can have a simpler model if we just drop it. Although the change in the χ^2_{GoF} isn't significant this time around, all of the parsimony-adjusted GoF indices increase. Also, from a research perspective, it means that we don't have to administer these three tests to all people, which at least makes us more cheerful.

[52] *Perhaps reinforcing Woody Allen's comment, "More than any other time in history, mankind faces a crossroads. One path leads to despair and utter hopelessness. The other to total extinction. Let us pray we have the wisdom to choose correctly."*

SUMMARY

Structural equation modeling is a relatively new and very powerful statistical method that can be seen as a general technique that includes, in its simpler forms, multiple regression, canonical correlation, path analysis, confirmatory factor analysis, and other tests. Its major advantages are that it can combine measured and latent variables in the same model; it can easily handle multiple predictor and outcome variables and complicated models; and it can accommodate mediating variables as well as variables measured over time. However, despite its early name, "causal analysis," it can determine causality only if the design of the study is appropriate; the statistic itself cannot assess causality from cross-sectional data.

EXERCISES

If the correlation between two predictor variables, A and B, is 0.50; the β weight between A and the outcome is 0.20; and the β weight between B and the outcome is 0.30; then:

1. What is the correlation between A and the outcome?
2. What is the correlation between B and the outcome?
3. Which variable(s) is (are) endogenous and which is (are) exogenous?
4. Which variable(s) should have a disturbance term?
5. How many observations are there?
6. If χ^2_{GoF} is 8.32, on $df = 2$, $p < .02$, does the model fit the data or not?

SECTION THE THIRD

C.R.A.P. DETECTORS

III–1. In an attempt to examine the relationship among height, IQ, and later success, Dr. Charlie Darvon, the noted pharmacopsychoanthropologist, analyzed data from the graduating class at Slippery State U. He administered an IQ test to all the graduates and measured their height. He then waited 10 years, following their progress in their respective careers, and measured their socioeconomic status on the Blishen scale (a ratio-level scale of measurement). To analyze the data, he classified the graduates as being in the top, middle, and bottom third of the class on height and IQ, then did a *t*-test on the two extreme groups. The *t*-test for IQ was significant ($t = 2.53$, $p < .05$), but the *t*-test for height was not. Would you approach things any differently?

Of course you would; that's why the question is here. It has several problems. The most obvious is that he couldn't resist the most common sin of biomedical researchers—he took perfectly respectable ratio-level variables, height and IQ, and collapsed them into three levels, thereby throwing away a pile of information. This is an absolute no-no! The solution is to retain the original data and use methods such as regression analysis, which deal with continuous data.

C.R.A.P. DETECTOR III–1

Never take data that are continuous and interval or ratio and classify them into categories *before* analysis.

Second, he threw out the middle group. This has two effects. The most obvious is that he has lost a third of his sample, affecting power. Third, by using extreme groups, he has biased the effect of the independent variables; thus the estimate of the effect, and the corresponding test of significance, can no longer be interpreted.

C.R.A.P. DETECTOR III–2

Analyses based on extreme groups are biased and lead to a potential loss of sample size and power. Use all the data.

Finally, he chose to analyze the two independent variables separately. More appropriate would be a joint analysis using multiple regression with two independent variables (Height and IQ) and one dependent variable.

C.R.A.P. DETECTOR III–3

All the independent variables should be analyzed together using ANOVA or regression methods.

III–2. Return to Question II–3 at the end of Section 2. Just to remind you, Feighner (1985) did an RCT with a small sample of patients, looking at fluoxetine versus amitriptyline. He measured three outcomes: the HAM-D, the Raskin Depression Inventory, and the Covi Anxiety scale, at baseline and at weeks 1, 2, 3, 4, and 5. He reported that "the changes were statistically significant...in the fluoxetine group and for several of the efficacy measurements in the amitriptyline group." He also compared the treatment groups at the end of the study and found no significant difference between the two drugs. We will pretend there was only one dependent variable. In Section 2, we suggested a repeated-measures ANOVA. With your new knowledge, would you do it any differently?

But of course. The baseline measure is not just one of six measures taken over the course of the study, and repeated-measures ANOVA treats it like a difference score. A better approach would be to treat the Time 0 measure as a covariate, then do a repeated-measures ANCOVA with Time

(5 levels) as the repeated measure and Drug (2 levels) as a grouping between-subject factor. Or use individual growth curves.

C.R.A.P. DETECTOR III–4

Baseline measures should be handled as a covariate, using ANCOVA methods.

III–3. A sociologist is investigating discrimination in employment practices of the local school board in Sexsex County. She studies all 27 teachers in the system and investigates the following variables: Age, Gender, Height, Religion (Christian, Jewish, Muslim, Hindu, Other), Handedness (right, left, ambi), and Degree (Bachelor, Master, Ph.D.). The dependent variable is income. She finds that the combination of variables has a multiple correlation of 0.37; that Gender enters the regression equation third, after Age and Degree; and Gender explains 15% of the variance. Do you believe her?

We hope not. There are several problems.

1. She has not just violated, she has crucified the old "rule of 10." Counting dummy variables, there are 11 independent variables in her regression equation and 27 subjects. Nothing coming out of this analysis is believable.

C.R.A.P. DETECTOR III–5

Watch the old "rule of 10" in all multiple regression analyses done on existing data bases.

2. A multiple correlation of .37, expressing $.37^2 = 13.6\%$ of the variance, is singularly unimpressive. Again, this leads towards discounting the study.

C.R.A.P. DETECTOR III–6

Square the multiple correlation. If it's not explaining about 30% or more of the variance, it's not a very impressive analysis.

3. Finally, there is a computation error. She claims that Gender enters the equation third and explains 15% of the variance, yet all the variables are additive and are good together for only 13.6%. More reason to reject.

C.R.A.P. DETECTOR III–7

Proportions of variances add, so add them. And don't get too excited over any variable that doesn't explain more than 5% of the variance.

III–4. Cohn (1982) used data from an animal study of cancer resulting from formaldehyde exposure to extrapolate the risk to humans. The rats were exposed to 2, 7, and 15 parts per million (ppm) of formaldehyde. In the 15-ppm group, about half of the rats developed nasal cancer. In the 7-ppm group, 2 of the 240 rats got cancer. In the 2-ppm group, none developed it. A multi-stage, multi-hit model (basically, a nonlinear regression) was fitted to these data and extrapolated to the excess exposure in homes containing urea formaldehyde foam insulation (UFFI), which releases gaseous formaldehyde into the air (.049 ppm vs .034 ppm in non-UFFI homes). The best estimate of risk was zero; however, the upper 95% confidence limit yielded an additional (attributable) risk from UFFI of 51 parts per million. The results are shown in Figure III–1, where each variable is shown as a logarithmic scale. Would you buy a home with UFFI in it?

There are two problems with the study. The minor one is that he committed a little fraud by using the upper 95% confidence estimate for his published estimates. Remember that his best estimate of the risk was zero; and the upper 95% CI **has to** be greater than zero. The major problem is that he assumed he could extrapolate downwards from 15 to .015 (.049 – .034), two orders of magnitude. Regardless of the sophistication of the model, no regression analysis should be extrapolated much beyond the original data—no model is good enough.

Unfortunately, environmental and occupational health folks have institutionalized this dangerous practice. That's why we have a new carcinogen every week. Nearly anything, in large enough doses, will cause cancer in susceptible rodents. And once the little beasties have it, then you draw your line down to minimal exposure and show that people will get it, too.

This also explains why some predictions go seriously awry. Anyone old enough will remember that in the 1960s, predictions were that the high birth rate would cause us to have standing-room-only on the planet by the year 2000.[1] In a similar vein, Binzel (1990) said that, at the rate that the estimates of Pluto's mass were decreasing, the planet would disappear entirely in 1980. The best comment, though, was made by Mark Twain, in *Life on the Mississippi*:

> In the space of one hundred and seventy-six years the Lower Mississippi has shortened itself two hundred and forty-two miles. That is an average of a trifle over one mile and a third per year. Therefore, any calm person, who is not blind or idiotic, can see that ... just a million years ago next November, the Lower Mississippi River was upwards of one million three hundred thousand miles long.

C.R.A.P. DETECTOR III–8

Do not extrapolate regression equations beyond the range of the original data points.

III–5. The following is a true story. Only the names are forgotten to protect the guilty. Several years ago, we came across an article in a reputable, widely read British medical journal. It might have been *Lancelot*, or perhaps it was the *British Magical Journal*. In this article, the authors were examining how physicians performed on a multiple choice test in relation to their year of graduation. They had scores from several hundred physicians, which they grouped by decade of graduation. They calculated the mean score in each decade, then correlated this mean score with the midpoint of the decade of graduation. The correlation was about 0.96. They concluded a nearly perfect relationship existed between performance and year of graduation. Do you agree?

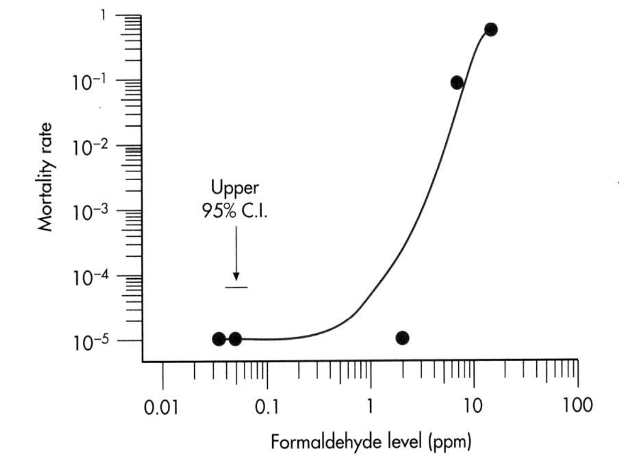

FIGURE III–1 Mortality rate as a function of formaldehyde level in the Cohn (1982) study.

> Heck, no! First of all, a correlation that high should tip you off to something rotten. Very few things in life are that good. But the question is why is it that high? The answer is that they correlated the means in each category, not the original data. As a result, most of the variation of individuals was conveniently lost because the "data" for their correlation had an error equal to the SE of the mean, not the SD (see Chapter 6 if you need reminding of the difference). Goodness knows what the true correlation was, but it was certainly a lot lower.

C.R.A.P. DETECTOR III–9

Beware the SE of the mean. Folks often display data using the SE of the mean because it looks so much better. This is perhaps useful when you want to compare means, but if you want to indicate what the actual data look like, it is deceptive. Some are even dumb enough to analyze their data this way.

III–6. Meedok and Hipokrit attempted to develop an instrument called the TMIADS (Trust Me, I'm a Doctor Scale) to measure patients' feelings about their doc's interpersonal skills.

[1] *At which point the birth rate would rapidly drop.*

TABLE III–1

Rotated factor loading matrix

Item	Factor 1	Factor 2	Factor 3	Factor 4
1	.54	−.02	.18	.21
2	.48	.18	.29	.15
3	.39	.24	.19	.16
4	.25	.35	.12	.08
5	.20	.37	.27	.26
6	.52	.33	.14	.17
7	.26	.28	.36	.22
8	.33	.27	.31	.28
9	.27	.30	.39	.18
10	.21	.15	.38	.31
11	.29	.41	−.02	.49
12	.11	.36	.24	.38
13	.19	.24	.21	.27
14	.47	.17	.19	.31

After weeding out unusable questions, they ended up with 14 True-False items, which they then administered to 50 patients. They said that the rotated factor loading matrix, which is reproduced in Table III–1, shows that the TMIADS is tapping four different areas—Openness, Trust, Empathy, and Looking Like Dr. Kildare. Can you spot any problems with what they did?

Actually, there are more problems than we can mention. Here are some of them:

1. **The Subject-to-Variable ratio.** With 14 items, there should have been an absolute minimum of 70 subjects (5 subjects per variable), and 140 would be preferable (a 10:1 ratio). Only 50 patients for 14 items just doesn't cut it.

C.R.A.P. DETECTOR III–10

The same as III–5 and many others: the subject-to-variable ratio should be 5:1 *at a minimum*, and it should be closer to 10:1.

2. **Eigenvalues.** It's usual to report the eigenvalue for each factor at the bottom of the column. The authors thought they could pull a fast one on us by not giving them. However, you now know that you can figure them out yourself by simply squaring each loading in the column and adding them up. What we get is that the four eigenvalues are 1.5737, 1.1067, 1.0675, and 1.0039. Sure enough, they're all above 1, but we wouldn't get too excited by them.
3. **Percent of variance explained.** Meedok and Hipokrit also didn't report how much variance each factor explained. Again drawing on our vast knowledge of arcane lore, we know that the total variance is 14 because we have that many items. So, the first factor accounted for $1.5737 \div 14 = 11.24\%$, and the four factors together expressed a total of $(1.5737 + 1.1067 + 1.0675 + 1.0039) \div 14 = 33.94\%$ of the variance. If our results were this bad, we'd also be too embarrassed to make them public. Especially with so few items, we'd hope that the first four factors would explain at least 60% or 70% of the variance.
4. **Factorial complexity.** Even after rotation, many of the items load about equally on two or more factors (e.g., items 8, 12, and 13). This makes it hard to argue that these are independent factors.
5. **Number of factors.** Factor 4 has only two items (11 and 12) that load higher on it than on the other factors. We would say that two items don't constitute a factor, and we really have a three-factor solution (accounting for 26.8% of the variance).

C.R.A.P. DETECTOR III–11

The retained factors should (a) comprise at least three items; (b) show minimal factorial complexity; (c) have their eigenvalues reported; (d) have eigenvalues considerably greater than 1.0; and (e) account cumulatively for at least 60% of the variance.

6. **Analyzing binary data.** Don't!

C.R.A.P. DETECTOR III–12

Binary data should not be factor analyzed.

SECTION

THE FOURTH

NON-PARAMETRIC STATISTICS

CHAPTER THE TWENTIETH

Tests of Significance for Categorical Frequency Data

Here we introduce statistical methods used to deal with categorical frequency data of the form, "The number of individuals who..." We begin with the simplest case, the chi-squared test, and then consider some special cases: small numbers (the Fisher Exact Test), paired data (McNemar's chi-squared), two factors (Mantel-Haenszel chi-squared), and finally the general case involving many factors (log-linear analysis).

SETTING THE SCENE

A few years ago, a report (Eidson et al., 1990) indicated that several people in New Mexico had succumbed to a rare but particularly nasty disease, eosinophilia-myalgia syndrome (EMS).[1] The only circumstance they appeared to have in common was that they were health food freaks and had all been imbibing large quantities of an amino acid health food called ***tryptophan****, which is supposed to be good for everything from insomnia to impotence. You, Hercules Parrot, have been assigned to the case by your masters at CDC Atlantis. You scour the countryside far and wide and locate 17 other poor souls who have succumbed under mysterious circumstances. Did tryptophan do it, and how will you prove it? In particular, how do you perform statistics on counts of bodies?*

We confess to a deviation from our tradition. In this case the story, however unlikely, happens to be true (at least true enough to end up in a law court). It is now fairly well accepted by everyone except the manufacturers and distributors of tryptophan that this innocent-appearing stuff actually bumped off about 200 unfortunate folks in the U.S.[2] It did start with a few suspicious cases in New Mexico and grew rapidly from there.

This is the stuff of *real* epidemiology. None of this touchy-feely research based on "How do you feel on a 7-point scale?" questions. Here it is a matter of life and death, and our data are body counts.[3] The question is, of course, how do you analyze bodies, because they don't usually follow a normal distribution unless you pile them that way. But first a small diversion into research design.

You may have heard that the best of all research designs is a **randomized controlled trial**, whereby subjects are assigned at random to a treatment or control group and no one knows until it's over who was in what group. What you heard is true, but it's also impossible to apply in this situation. If we really thought people might die from tryptophan exposure, it's unlikely (we hope) that any ethics committee would let us expose folks to the stuff just for the sake of science. The next best design is a **cohort** study. Here, you assemble cohorts of folks who, of their own volition (smoking), or from an accident of nature (radon) or their jobs (Agent Orange), have been exposed to a substance, match them up as best you can to another group of folks who are similar in every way you can think of but exposure, and then check the frequency of disease occurrence in both. That might work here, except that probably hundreds of thousands of health food freaks are gobbling up megavitamins and all sorts of other stuff, and (1) very few of them actually appear to have come down with EMS, and (2) it would be hard to trace all of them. So you end up at a third design, a **case-control** study, in which you take a bunch of folks with the disease (the cases) and without the disease (controls), and see how much of the exposure of interest each group has had. Although this approach has its problems, it is about the only practical approach to looking at risk when the prevalence is very low.

Off you go, Mr. Parrot, to find cases and controls. You scour hospital records and death certificates around the country, and you eventually locate 80 people with EMS. You also locate some controls, who were hospitalized for something else or died of

[1] *Eosinophilia-myalgia syndrome (EMS) is a very nasty multisystem disease. As well as causing crippling muscle pain and high eosinophil counts, it has many other manifestations (e.g., fever, weakness, nausea, dyspnea, tachycardia), and it occasionally kills.*

[2] *Actually it was a contaminant, but we don't want to get ahead of ourselves.*

[3] *Just like Vietnam, only we'll tell you when we're lying.*

TABLE 20–1

Association between tryptophan exposure and eosinophilia-myalgia syndrome (EMS)

		EMS	Normal	TOTAL
Tryptophan	Yes	42	34	76
	No	38	166	204
	TOTAL	80	200	280

TABLE 20–2

Expected values of the table assuming no association

		EMS	Normal	TOTAL
Tryptophan	Yes	21.7	54.3	76
	No	58.3	145.7	204
	TOTAL	80	200	280

something else. Because there are lots of the latter, you stop at 200. You then administer a detailed questionnaire to their next of kin or by way of seance, ascertaining exposure to all sorts of noxious substances—vitamins, honey, ginseng root, lecithin, and (of course) tryptophan. After the dust settles, 42 of the EMS group and 34 of the control group took tryptophan regularly.[4] Is this a statistically significant difference? That, of course, is what this chapter is all about.

The dilemma is that, like our dummy variables in Chapter 14, this variable has only two values—0 or 1, dead or alive—so it is not normally distributed. (If it were, we would just do a *t*-test.) We might bring logistic regression to the rescue, but that would be overkill (no pun intended) and would ignore the large body of research called **nonparametric statistics**, which antedated logistic regression and big computers by many decades. To explain why this is nonparametric statistics, we have to explain why the other type isn't. ANOVA, regression, and all those other techniques are based on calculated means and SDs, the *parameters* of the normal distribution. By contrast, nonparametric statistics make no assumptions about the nature of the distribution, so it is free of assumed parameters.

Actually, in keeping with note 3, we just lied to you. As we'll see in the next few paragraphs, some of the "nonparametric" tests actually do make some assumptions about the underlying distribution. The chi-squared test is based on a distribution but one that isn't normal. For some reason, it's lumped together with other tests that actually don't make any assumptions. This is probably due to the fact that many of the tests used with categorical data are truly nonparametric and, through laziness, or sloppiness (or more likely ignorance), all statistics designed for categorical data were called nonparametric. So, the next test we'll talk about is a parametric, "nonparametric" statistic.

THE CHI-SQUARED TEST

To begin to tease out a strategy for approaching the data, we'll put the data into a form dearly beloved by clinicians and statisticians alike: a **2 × 2 contingency table**. It's called "2 × 2" because it has two rows and two columns and "contingency" because the values in the cells are contingent on what is happening at the marginals (be patient; we'll get to that in a minute) (Table 20–1).

Now, what we are trying to get at is whether any association exists between tryptophan use and EMS. As usual, the starting point is to assume the null hypothesis (no association) and then try to reject it. The question is, "What would the 2 × 2 table look like if there were no association?" One quick, and wrong, response is that the 280 people are equally divided among the four cells; that is, there would be 280 ÷ 4 = 70 people in each one. Not at all. We began with 80 patients and 200 controls. Were there no association, we would expect that exactly the same *proportion* of patients as controls would have gobbled tryptophan. Our best guess at the proportion of tryptophan users is based on the *marginal* totals, and it equals 76 ÷ 280 = .271. So, the number of EMS patients who ate tryptophan, under the null hypothesis of no association, is 80 × (76 ÷ 280) = 21.7; and the number of controls is 200 × (76 ÷ 280) = 54.3. In a similar manner, the number of EMS folks who abstained is 80 × (204 ÷ 280) = 58.3, and the number of control abstainers is 145.7. If there were no association, then, Table 20–2 would result.

The extent to which the observed values differ from the expected values is a measure of the association, our *signal* again. But if you work it out, it equals zero, just as it did when we determined differences from the mean in the ANOVA case. So we do the standard statistical game and square everything. The signal now looks like:

$$\text{Signal} = (42 - 21.7)^2 + (34 - 54.3)^2 + (38 - 58.3)^2 + (166 - 145.7)^2$$

If we were to follow the now familiar routine, the next step would be to use the individual values within each cell to estimate the noise. Unfortunately, we have only one value per cell. Fortunately, Mother Nature comes to the rescue. It turns out that frequencies follow a particular distribution, called a **Poisson distribution**,[5] which has a very unusual property: the variance is exactly equal to the mean.[6] Thus, for each one of the squared differences in the equation above, we can guess that it would be expected to have a variance equal to the expected mean value. So, the ratio of the squared difference

[4] *These data aren't real. Later on, we'll show you some data that are.*

[5] *To which any regular stats book would devote about 20 pages, just so we could get to the next equation.*

[6] *If you need some convincing, imagine some frequency generator, such as a radioactive source. We know that in the long term it has an average of, say, 100 counts per minute. But any individual count for a minute will differ from this by some amount. It turns out the distribution is about bell-shaped (skewed if the mean is very low), and has an SD of $\sqrt{100}$, = 10, or a variance of 100.*

between the observed and expected frequency to the expected mean is a signal-to-noise ratio. It's called **chi-squared**, for reasons now lost in antiquity. Formally, then:

$$\chi^2 = \sum \frac{[O_i - E_i]^2}{E_i} \tag{20–1}$$

where O_i is the observed frequency and E_i is the expected frequency. And in this case, it equals:

$$\chi^2 = \frac{(42 - 21.7)^2}{21.7} + \frac{(34 - 54.3)^2}{54.3} + \frac{(38 - 58.3)^2}{58.3} + \frac{(166 - 145.7)^2}{145.7} = 36.4 \tag{20–2}$$

That looks like a big enough number, but it's not clear where we should go looking to see whether it's big enough to be statistically significant. As it turns out, chi-squared has a table all to itself (Table F in the appendix). Once again, it's complicated a bit by the *df*. For this table, the *df* is 1. To demonstrate this, keep the marginal frequencies fixed and put the number in one cell. Now, for the cells to add up to the correct marginal totals, all other cells are predetermined: the marginal total minus the filled-in value. So you have only one cell free to vary; hence, one *df*. In the general case of an $(r \times c)$ contingency table (r rows and c columns), there are $(r - 1) \times (c - 1)$ *df*. This particular value is highly significant (the value of chi-squared needed for significance at $p = .05$ for one *df* is 3.84), proving conclusively that health food is bad for your health.[7]

A nice rule of thumb is that the value of χ^2 needed for significance at the .05 level is equal to the number of cells. This approximation becomes more accurate as the number of cells increases. In this case, we would have said 4, which differs a bit from the true value of 3.84. For a 5 × 2 table, our guess would be 10; the actual value is 9.49.

But that's not the end of the story. Careful tracking of EMS cases showed that many were turning up all over the U.S. but virtually none in Canada or Europe. Although Americans were more likely to junk out on health foods and other "alternative" therapies than were staid Brits, it was well known that Canadians were also popping the stuff with gay abandon. So perhaps the illness was caused by a contaminant that snuck into one batch from one manufacturer, not by the stuff itself. This cause was pinpointed in a study by Slutsker et al. (1990). They located 46 cases with EMS who also took tryptophan; 45 of the 46 ate stuff from one manufacturer in Japan (sold through 12 wholesalers and rebottled under 12 different brand names). There were 41 controls who took tryptophan but didn't get EMS; 12 of them ate tryptophan from the Japanese manufacturer, the other controls from other manufacturers. This difference (45/46 to 12/41) is so significant that only a sadist or a software manufacturer would demand a statistical test.

TABLE 20–3 Source of tryptophan by group

Group	Manufacturer: Showa Denko K.K.	Other	Unknown	TOTALS
Cases	29	1	22	52
Random controls	5	4	15	24
Volunteers	16	10	7	33
TOTALS	50	15	44	109

TABLE 20–4 Expected values for Table 20–3

Group	Manufacturer: Showa Denko K.K.	Other	Unknown	TOTALS
Cases	23.8	7.2	21.0	52
Random controls	11.0	3.3	9.7	24
Volunteers	15.1	4.5	13.3	33
TOTALS	50	15	44	109

Another study, this time in Minnesota (Belongia et al., 1990), managed to track the nasties down to a single contaminant. To do this, they first located 52 cases with a high eosinophil count and myalgia. They then formed two control groups: (1) a volunteer group of folks who had been taking tryptophan but weren't sick, located by public announcements ($n = 33$), and (2) a control group of people who had also taken tryptophan and were located by a random telephone survey ($n = 24$). They then interviewed everybody to see what brand of tryptophan they were using. Only 30 cases, 26 volunteers, and 9 random controls could locate the bottle. They rapidly focused the problem down to a single manufacturer. The data are presented in Table 20–3.

And once again, Mr. Parrot, the time has come to crunch numbers. The approach is just the same as with the 2 × 2 table. This time the analysis is analogous to a one-way ANOVA for parametric statistics. First you estimate the expected value in each cell by multiplying the row and column marginal totals and dividing by the grand total. So for row 1 and column 1, this equals (50 × 52) ÷ 109 = 23.8. Working through the expected values results in Table 20–4.

From this we can calculate a chi-squared as we did before, simply by taking the difference between observed and expected values, squaring it, dividing by the expected value, and adding up all 9 terms. The answer is 22.40, and the *df* is (3 − 1) × (3 − 1) = 4; moreover, the result is highly significant, at $p < .0001$. To close the corporate noose around Showa Denko, the investigators then showed that (1) the manufacturer cut back on the amount of activated

[7] *As a brief aside, the astute reader might have noted that the value of chi-squared, 3.84, is just the square of the corresponding z-test (1.96). A general observation, offered without proof, is that chi-squared on 1 degree of freedom, is just z^2. Pay attention, it might be on the exam.*

charcoal at one filtration stage, (2) bypassed some other filter, and (3) 17 of the 29 cases had consumed tryptophan out of one particular batch. The contaminant also showed up on liquid chromatography. In short, the goose was neatly fried.

Another way of looking at the chi-squared test of association is that it is a test of the null hypothesis that the *proportion* of EMS cases among tryptophan users (usually abbreviated as π_t) was the same as the proportion among nonusers (π_n); that is, it is a test of two or more proportions. There is, in fact, a *z*-test of the significance of two independent proportions. We haven't bothered to include it for the simple reason that z^2 is exactly the same as chi-squared. However, it's easier to figure out sample size requirements based on proportions, so we'll come back to this concept when we tell you how to figure them out.

That's the story for chi-squared—almost. Things work out well as long as (1) you only have four cells, and (2) the frequencies are reasonably good. When you have more cells, as in the example of myalgia, or when the numbers are small, then some fancier stuff is required.

DECONSTRUCTING CHI-SQUARED

In the previous example, we showed that the overall chi-squared was significant, but we skipped over a problem[8]—what was significantly different from what? This is similar to the situation after finding a significant *F*-ratio in a one-way ANOVA: we still don't know which groups differ from the others. In the case of ANOVA, we would use one of the post-hoc tests, such as Tukey's or Scheffé's, but we don't have that option with χ^2 . What we have to do is **decompose** the χ^2 table into a number of smaller subtables, and see which ones are significant.

At first glance, it would seem as if we could have nine 2 × 2 subtables: Showa Denko versus Other for Cases versus Controls, Cases versus Volunteers, and Controls versus Volunteers; Showa Denko versus Unknown for the same three group comparisons; and the three group comparisons for Other versus Unknown. But the rules of the game say otherwise; the number of subtables we are allowed and how they are constructed must follow certain conventions. These were summarized by Iversen (1979)[9]—kind of like the Hoyle of χ^2 decomposition. The rules are as follows:

1. The degrees of freedom, summed over all of the subtables, must equal the degrees of freedom of the original table.
2. Each frequency (i.e., cell count) in the original table must be a frequency in one and only one subtable.
3. Each marginal total, including the grand total, must be a marginal total in one and only one subtable.

Are you sufficiently confused yet? Let's work this out for the data in Table 20–3 and see what it looks like. This should either dispel the confusion or make you turn to hard liquor.[10] According to the first rule, we can have subtables with a total of four degrees of freedom, since the original table had *df* = 4. This means we can have four 2 × 2 tables, or one 3 × 2 and two 2 × 2 tables. The easiest breakdown to interpret (although not necessarily the most informative) is a series of 2 × 2 tables, so that's what we'll do; it's easiest to start with the upper left corner. For reasons that will become obvious as we go along, we'll arrange the table so that the row and column we're most interested in (Showa Denko and Cases) are the last ones, rather than the first. The first subtable, then, is shown in Table 20–5A. Now, according to the second rule, none of those four cell counts can appear in any other table, and the marginal totals must appear in some other table. We can satisfy these rules (in part, so far), by combining columns 1 (Other) and 2 (Unknown) and comparing this to column 3 (Showa Denko) for the Controls and Volunteers; this is shown in Table 20–5B. Notice also that some of the row marginals from the original table are now frequencies in 20–5B, in accordance with rule 3. In Table 20–5C, we do the same thing, only this time we're combining the first two rows (Controls and Volunteers) and comparing them with the Cases; in Table 20–5D, we combine both rows and columns. If you want to check, you will see that each of the 16 numbers in Table 20–3 (nine cell frequencies, three row totals, three column

[8]Our next book will consist simply of problems we have skipped over in this one.

[9]And translated into English by Agresti (1990), from the original Statisticalese.

[10]Or both, if we're lucky.

TABLE 20–5

Decomposing the data from Table 20–3 into a series of 2 × 2 subtables

A

Group	Other	Unknown	Total
Controls	4	15	19
Volunteers	10	7	17
TOTALS	14	22	36

B

Group	Other + unknown	Showa Denko	Total
Controls	19	5	24
Volunteers	17	16	33
TOTALS	36	21	57

C

Group	Other	Unknown	Total
Controls + volunteers	14	22	36
Cases	1	22	23
TOTALS	15	44	59

D

Group	Other + unknown	Showa Denko	Total
Controls + volunteers	36	21	57
Cases	23	29	52
TOTALS	59	50	109

totals, and the grand total) appears once and only once in one of the subtables. Now, that wasn't hard, was it?[11]

This is one of a number of possible sets of 2 × 2 tables. It may make more sense, for example, to combine Unknown with Showa Denko than with Other, or to combine Controls with Cases rather than Volunteers. Because you're limited in the number of tables you can make, it's important to think about which comparisons will provide you with the most useful information.

Strictly speaking, the subtables shouldn't be analyzed with the usual method for χ^2, because it does not account for the frequencies in the cells that aren't included. But, people do it anyway, and the results are relatively accurate, if the total *N* in the subtable is close to the *N* of the original table. If you want to be extremely precise, the equation you need is in Agresti (1990). But be forewarned—it's formidable! We'll take the easy route and just do χ^2 on the four of them. It turns out that all four subtables are significant. This tells us that the Other and Unknown preparations produce different results for the Cases and Volunteers. The most important analysis, from our perspective, however, is of Table 20–5D—Showa Denko versus everything else and Cases versus everyone else. The numbers in this table really clinch the case.

SMALL NUMBERS, YATES' CORRECTION, AND FISHER'S EXACT TEST

Yates' Correction for Continuity

When the *expected* value for any particular cell is less than 5, then the usual chi-squared statistic runs into trouble. Part of this is simply instability. Because the denominator is the expected frequency, addition or subtraction of one body can make a big difference when the expected values are small. But the chi-squared tends towards liberalism because it approximates categories with a continuous distribution. However popular this is politically, it is anathema to statisticians. One quick and dirty solution is called **Yates' Correction**. All you do is add or subtract .5 to each difference in the numerator to make it smaller before squaring and dividing by expected values. So the **Yates' corrected chi-squared** is:

$$\chi^2 = \sum \frac{[|O_i - E_i| - .5]^2}{E_i} \tag{20-3}$$

The vertical lines around the *O* and *E* are "absolute value" signs, so you make the quantity positive, then take away half and proceed as before.

Having said all this, it turns out that Yates' correction is a bit too conservative. So half the world's statisticians recommend using it all the time, and half recommend never using it. In any case, the impact is small unless frequencies are very low, in which case an exact alternative is available.

Fisher's Exact Test

Imagine that we have proceeded with the original investigation of whether tryptophan causes the disease and we're using a stronger design—a *cohort* study.[12] You put more signs up in the health food stores, this time asking for people who are taking tryptophan, not people who are sick. You then locate a second group of folks who weren't exposed to the noxious agent tryptophan, perhaps by hitting up the local greasy spoon.

Now fortunately for the populace, but unfortunately for you, tryptophan isn't all that nasty, so very few people actually come down with EMS. If we had 100 of each group, the data might look like Table 20–6. The expected value for both cells in the first column is 5, so we can't use chi-squared on this. The alternative is called **Fisher's exact test**, which is as follows. Instead of calculating a signal-to-noise ratio and then looking it up in the back of the book, we dredge up some of the basic laws of probability to calculate the exact[13] probability of the data under the hypothesis of no association.

To understand how this one works, we'll really stretch the analogy. Cast your mind back to the Civil War, when families were torn asunder, etc. You remember from your history books the famous Battle of Bull Roar, don't you? Let's just briefly remind you.

Bull Roar was a small town in West Virginia. One hot summer night, recruiters from both the Union and the Confederacy descended on the town, hit all the local pubs, stuffed the boys into uniforms, and handed them all muskets. The next day, they assembled in a field on the edge of town. Thirteen wore the blue of the North, and 11 wore the gray of the South. They opened fire, and when the smoke blew away, four Union men and three Confederates lay dead on the ground. At this point, the survivors all took off their uniforms, went into the pubs in their underwear, and got thoroughly sozzled.

The statistical question is, "Given only the information in the marginals—that is, there were 24 able-bodied males, of whom 13 were in blue uniforms and 11 in gray; and 7 ended up dead and 17

[11] *The only acceptable answer is, "No sirs, that wasn't hard at all. May I have more, please?"*

[12] *If you expect us to define this further, forget it; this is a* statistics *book. Read* PDQ Epidemiology.

[13] *That's why it's called the* exact *test.*

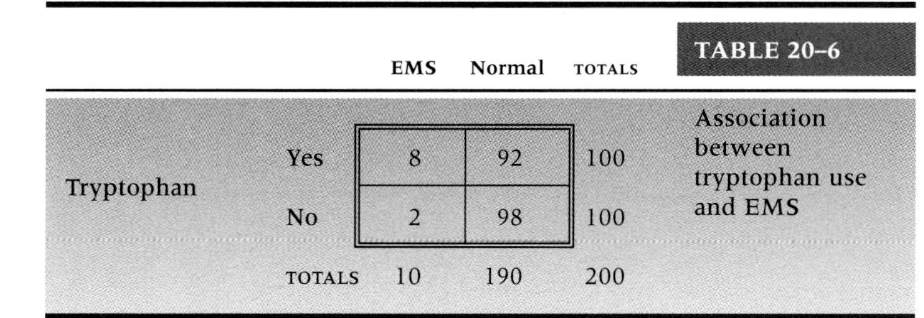

TABLE 20–6 Association between tryptophan use and EMS

		EMS	Normal	TOTALS
Tryptophan	Yes	8	92	100
	No	2	98	100
	TOTALS	10	190	200

Data from Slutsker L et al (1990). Eosinophilia-myalgia syndrome associated with exposure to tryptophan from a single manufacturer. *Journal of the American Medical Association,* **264**:213–217.

TABLE 20–7

Statistics from the Battle of Bull Roar

		Alive	**Dead**	TOTALS
Army	Union	9 (a)	4 (b)	13 (a + b)
	Confed	8 (c)	3 (d)	11 (c + d)
	TOTALS	17 (a + c)	7 (b + d)	24

alive—what is the chance that things could have turned out the way they did?" We might, as we are wont to do in this chapter, put it all into a 2 × 2 table (Table 20–7).

To make things easier, we'll begin by illustrating the field of battle graphically (Figure 20–1). Now let's look at the Union men first. How many ways can 4 of the 13 die? Think of it one shot at a time. The first fatal bullet might have taken out any 1 of 13 men, so there are 13 ways that the first bullet could have done its dirty work. Now one man is dead—what about the second bullet? There are 12 men to choose from, so 12 possibilities. Similarly, there are 11 possibilities for the third bullet, and 10 for the fourth. So in the end, there are 13 × 12 × 11 × 10 possible ways the bullets could have found their mark. However, once the lads are dead on the field, we no longer care in which order they were killed. Again, by the same logic, any one of the four could have been taken out by the first bullet, then three possibilities for the second bullet, and so on. So the overall number of ways that 4 of the 13 Union men could have been killed is (13 × 12 × 11 × 10) ÷ (4 × 3 × 2 × 1). A convenient way of writing this algebraically is through the use of factorials.[14] So, the number of ways to bump off 4 men out of 13 is:

[14]*Remember that $n! = n \times (n-1) \times (n-2) \times (n-3) \times \ldots 3 \times 2 \times 1$.*

$$nC_k = \frac{n!}{(n-k)!k!} = \frac{13!}{9!4!} = \frac{13 \times 12 \times 11 \times 10}{4 \times 3 \times 2 \times 1} = 715 \qquad \textbf{(20–4)}$$

where k is the number of events (deaths) and n is the total number of individuals. Similarly, the number of ways losses on the Confederate side could have occurred are equal to 11! ÷ (8! 3!) = 165.

However, we are ultimately interested in the association between Union/Confederate and Alive/Dead. To get at this, we have to begin with the nonassociation and figure out how many ways a total of 7 men could have been shot out of the 24 who started. We put them all in one long row, regardless of the color of their uniform, and do the same exercise. The answer, using the same logic as before, is 24! ÷ (7! 17!) = 346,104.

That means that there were a total of 346,104 ways of ending up with 7 dead souls out of the 24 with which we began. That is, if we were to line up all the soldiers in a row with the 13 Union guys on the left and the 11 Rebels on the right and fire 7 rounds at them, there are a total of 346,104 ways to take out 7 soldiers. Some of these possibilities are 6 dead Union soldiers and 1 dead Confederate soldier, 5 dead Confederate soldiers and 2 dead Union soldiers, and on and on.

Now, how many possibilities have the right combination of 4 and 3? Well, imagine the Confederates shoot first, so there are just 4 Unionists on the ground. We have already worked out that the number of ways you can kill 4 Union people of 13 is 715. Now, for each one of these possibilities, we can now let the Union guys open fire, and kill 3 Confederates—as before, 165 possible configurations. So the total number of patterns which correspond to 4 and 3 dead, after we have got everyone back in a single line, is just 715 × 165 = 117,975. This is out of the total number of possibilities of 346,104. So, the overall probability of getting the distribution of deaths which occurred at Bull Roar is 117,975/346,104 = .34. Now, if we put all the factorials together, we can see that the formula for the probability that things came out as they did is:

$$p = \frac{\frac{13!}{9!4!} \times \frac{11!}{8!3!}}{\frac{24!}{17!7!}} \qquad \textbf{(20–5)}$$

FIGURE 20–1
Aftermath of the Battle of Bull Roar.

TABLE 20–8

More extreme associations between tryptophan use and EMS

P(1)		EMS	Normal	TOTALS
Tryptophan	Yes	9	91	100
	No	1	99	100
	TOTALS	10	190	200

P(2)		EMS	Normal	TOTALS
Tryptophan	Yes	10	90	100
	No	0	100	100
	TOTALS	10	190	200

More generally, this can be expressed in terms of *a*s and *b*s as:

$$p = \frac{\dfrac{(a+b)!}{a!b!} \times \dfrac{(c+d)!}{c!d!}}{\dfrac{N!}{(a+c)!\,(b+d)!}} \tag{20–6}$$

This simplifies to:

$$p = \frac{(a+b)!\,(c+d)!\,(a+c)!\,(b+d)!}{N!\,a!\,b!\,c!\,d!} \tag{20–7}$$

This then is the probability of a particular configuration in a 2 × 2 table. So going back to our original EMS example, the probability of occurrence of the events in Table 20–6 is:

$$\text{Prob }(2) = \frac{10! \times 190! \times 100! \times 100!}{200! \times 8! \times 2! \times 92! \times 98!} = .0410 \tag{20–8}$$

where the '(2)' means that the count in the cell with the fewest number of subjects is two. We'll see why that's important in a moment.

We're not quite done. The probability used in the statistical test is the entire probability in the tail (i.e., the likelihood of observing a value *as* extreme or even *more* extreme than the one observed). In the discrete case we are considering, this corresponds to tables with stronger associations, which means more extreme values in the cells. There are only two possibilities with more extreme values: 1 case in the control group and 0 cases in the control group.[15] The corresponding 2 × 2 tables are shown in Table 20–8. For one occurrence the formula is:

$$\text{Prob }(1) = \frac{10! \times 190! \times 100! \times 100!}{200! \times 9! \times 1! \times 91! \times 99!} = .0085 \tag{20–9}$$

And for no occurrences this probability equals:

$$\text{Prob }(2) = \frac{10! \times 190! \times 100! \times 100!}{200! \times 10! \times 0! \times 90! \times 100!} = .0008 \tag{20–10}$$

Putting it all together, the overall probability of observing this strong an association is .041 + .0085 + .0008 = .0503. If we find that the first term exceeded .05 (instead of .041), and we didn't want to figure out the exact probability, we could stop right there. This follows since if the first probability is greater than .05, and all subsequent steps can only increase the *p* level, the latter two calculations were unneccessary. As a general rule, we can stop calculating when the probability reaches .05. So, this particular investigation doesn't make it to the *New England Journal of Medicine*.

To summarize, Fisher's exact test is used when the expected frequency of any cell in a 2 × 2 table is less than 5. You construct the 2 × 2 tables for the actual data and all more extreme cases, then work out the probability for each contingency table using the binomial theorem, shown above. The exact probabilities are then added together to give the probability of the observed association or any more extreme.

PAIRED AND MATCHED OBSERVATIONS: McNEMAR CHI-SQUARED

Perhaps you noticed that we began this chapter by telling you that we were going to use a real example, and we then went back to some imaginary data. There was a good reason for this peculiar action.[16] The original study that implicated tryptophan (Eidson et al., 1990) used a slightly more complicated design—complicated in the sense of analysis at any rate. They located 11 individuals who had EMS based on objective criteria and then matched them with 22 controls on the basis of age and sex. The magical word "match" means that we have to try another approach to analysis, equivalent to a paired *t*-test. The approach is called the **McNemar chi-squared**; it is logically complex but computationally trivial. Because the logic is tough enough with simple designs, we will pretend that the investigators just did a one-on-one matching and actually located 22 cases. If we ignored the matching, the data could be displayed as usual (Table 20–9).

Matched or not, clearly this is one case where the *p*-value is simply icing on the cake; however, we will proceed. The logic of the matching is that we frankly don't care about those instances where both case and control took tryptophan, or about those instances where neither took it. All that interests us is the circumstances where either the case took it

[15] *What happened to the top row with 8 cases? It turns out that the binomial distribution, as shown in the formula, is symmetrical. So we could have worked out the probability of observing 8 and 9 and 10 and 11 … and 99 and 100 cases. But it would have taken a bit more time and resulted in the same answer anyway. The two probabilities are not added together because that would amount to counting everything twice.*

[16] *In contrast to many of our peculiar actions.*

TABLE 20–9

Study 3—matched design association between tryptophan use and EMS (shown unmatched)

		Tryptophan use		
		Yes	No	TOTALS
EMS	Yes	22	0	22
	No	2	20	22
	TOTALS	24	20	44

Data from Eidson M et al (1990). L-tryptophan and eosinophilia-myalgia syndrome in New Mexico. *Lancet*, **335**:645–648.

TABLE 20–10

Study 3—matched design association between tryptophan use and EMS (shown matched)

		Control (without EMS) used tryptophan		
		Yes	No	TOTALS
Case (with EMS) used tryptophan	Yes	2	20	22
	No	0	0	
	TOTALS	2	20	22

Data from Eidson M et al (1990). L-tryptophan and eosinophilia-myalgia syndrome in New Mexico. *Lancet*, **335**:645–648.

and the control didn't, or vice versa. So we must construct a different 2 × 2 table reflecting this logic (Table 20–10). The first thing to note is that the total at the bottom right is only 22; the analysis is based on 22 *pairs*, not 44 people. Second, note that the four cells display the four possibilities of the pairs—both used it, both didn't use it, cases did but controls did not, and controls did but cases did not.

Finally, as we said, we're interested in only the two off-diagonal cells because those are where the action will be. The reason is that if *no* association existed between being a case or a control and tryptophan exposure, we would expect that there would be just as many instances where cases used tryptophan and controls didn't as the opposite. We have 20 instances altogether, so we would expect 10 to go one way and 10 the other. In short, for the McNemar chi-squared, the expected value is obtained by totaling the off-diagonal pairs and dividing by two. It is now computationally straightforward to crank out a chi-squared based on these observed and expected values. There is one wrinkle—McNemar recognized that he would likely be dealing with small numbers most of the time, so he built a Yates'-type correction into the formula:

$$\chi^2_M = \frac{(|20 - 10| - 0.5)^2}{10} + \frac{(|0 - 10| - 0.5)^2}{10} = 18.05 \tag{20–11}$$

with one *df*. To no one's surprise, this is significant at the .0001 level. Because of the particular form of the expected values, the McNemar chi-squared takes a simpler form for computation. If we label the top left cell *a*, the top right *b*, the bottom left *c*, and the bottom right *d*, as we did in Table 20–6, the McNemar chi-squared is just:

$$\chi^2_M = \frac{(|b - c| - 1)^2}{(b + c)} \tag{20–12}$$

To summarize, then, the McNemar chi-squared is the approach when dealing with paired, matched, or pre-post designs. Unfortunately, despite its computational simplicity, it is limited to situations with only two response categories and simple one-on-one matching. That's why we modified the example a bit. To consider the instance of two controls to each case, we must look at more possibilities (e.g., one exposed case and one control, case and both controls). It's possible, but a bit hairier. You don't get something for nothing.

TWO FACTORS: MANTEL-HAENSZEL CHI-SQUARED

Well, we're making some progress. We have dealt with all the possibilities where we have one independent categorical variable. Chi-squared, with a 2 × 2 table, is equivalent to a *t*-test, and with more than two categories is like one-way ANOVA. The McNemar chi-squared is the analogue of the paired *t*-test. The next extension is to consider the case of two independent variables, the parallel of two-way ANOVA. The strategy is called a **Mantel-Haenszel chi-squared** (hereafter referred to as M-H chi-squared. Guess why?)

Unfortunately, none of the real data from the EMS studies are up to it, so we'll have to fabricate some. Stretch your biochemical imagination a bit and examine the possibility (admittedly remote) that some other factors interact with tryptophan exposure from the bad batch to result in illness. For example, suppose EMS is actually caused by a massive allergic response to mosquito bites that occurs only when excess serum levels of tryptophan are present. Well now, gin and tonic was originally developed by the British Raj to protect the imperialist swine from another mosquito-borne contagion (malaria) while concurrently providing emotional support (in the form of inebriation). Maybe it would work here as well.

To test the theory (and to deal with the possible response bias resulting from folks in the G and T group saying they are feeling great when they are past feeling anything), we create six groups by combining the two independent variables: Gin and Tonic,

Tonic Only, or No Drinks, with half of each group having taken tryptophan and the other half a placebo (in ANOVA terms, a 3 × 2 factorial design). We can't afford any lab work so we use symptoms as dependent variables: insomnia, fatigue, and sexual dysfunction. When we announced the study in the graduate student lounges, we had no trouble recruiting subjects and got up to 500 per groups, despite the possible risk. However, the dropout rates were ferocious. The No Drink group subjects were mad that they didn't get to drink; the G and T group got so blotto they forgot to show up; and the Tonic Only group presumed they were supposed to be blotto and didn't come either. In the end, the data resulted in Table 20–11.

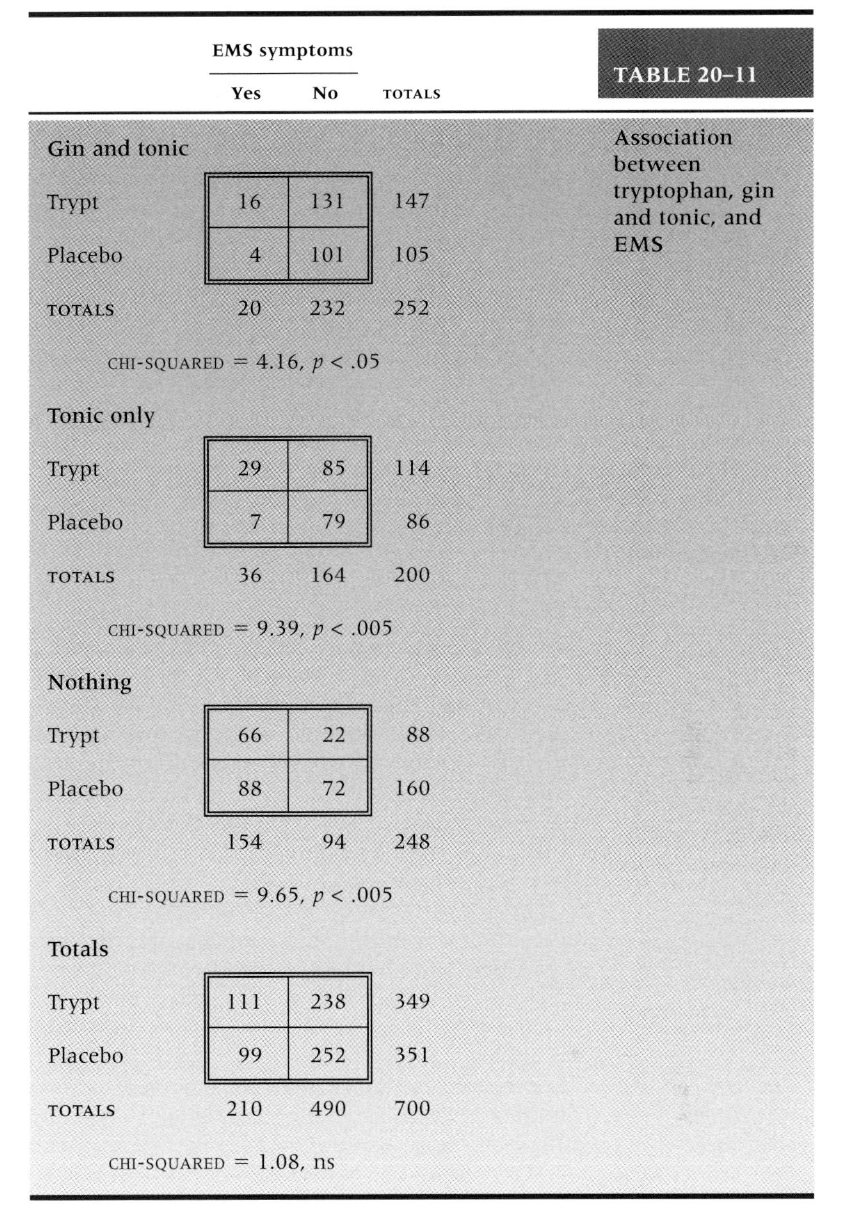

TABLE 20–11

Association between tryptophan, gin and tonic, and EMS

	EMS symptoms: Yes	EMS symptoms: No	TOTALS
Gin and tonic			
Trypt	16	131	147
Placebo	4	101	105
TOTALS	20	232	252
CHI-SQUARED = 4.16, $p < .05$			
Tonic only			
Trypt	29	85	114
Placebo	7	79	86
TOTALS	36	164	200
CHI-SQUARED = 9.39, $p < .005$			
Nothing			
Trypt	66	22	88
Placebo	88	72	160
TOTALS	154	94	248
CHI-SQUARED = 9.65, $p < .005$			
Totals			
Trypt	111	238	349
Placebo	99	252	351
TOTALS	210	490	700
CHI-SQUARED = 1.08, ns			

Before we plunge into the statistical esoterica, take a really close look at the table. Within each 2 × 2 subtable there is a strong association between tryptophan use and symptoms, with about three times as many people with symptoms per 100 in each stratum. The risk of symptoms among tryptophan-exposed individuals in the Tonic Only group is 29 ÷ 114 = 25.4/100; in the Placebo and Tonic Only group it's 8.14/100. So the relative risk is 25.4 ÷ 8.14 = 3.12. But because of the peculiarities of the data—mainly the excess of symptoms in the "Nothing" group, and the factor of two between Tryptophan and Placebo in those who stayed in the trial (160 vs. 88)—when they are combined (shown at the bottom of Table 20–11), the association disappears.

Clearly, one way *not* to examine the association between tryptophan and symptoms, when there are strata with unequal sample sizes, is to add it all together, which makes the effect completely disappear.[17] Instead, we must use some strategy that will recognize the interaction between the two factors, and so must stay at the level of the individual 2 × 2 tables.

We can start as we have before, by considering the expected value of an individual frequency, contrasting this with the observed value, and squaring the lot up. For example, the expected frequency in the G and T, Tryptophan, YES cell is:

$$\text{Exp} = \frac{(a + b)\ (a + c)}{N} = \frac{20 \times 147}{252} = 11.67 \tag{20–13}$$

You probably thought that a reasonable way to proceed now is simply to calculate a chi-squared by doing as we have already done—summing up all the $(O - E)^2 \div E$ for all 12 cells. We thought so too, but Mantel and Haenszel didn't.[18] First, the variance in this situation is *not* just the expected value, as it was when we did the original chi-squared. Here, the variance of the expected value of each frequency is:[19]

$$\text{Var (Exp)} = \frac{(a + b)(a + c)\ (c + d)(b + d)}{N^2(N - 1)} \tag{20–14}$$

The next step is to add up the $(O - E)$ differences for all the individual frequencies in the a (top left) cells across all subtables[20] and square the resulting total. We then throw in a Yates' correction, just for the heck of it. This is the numerator for the M-H chi-squared and is an overall measure of the *signal*, the difference between observed and expected values, analogous to the mean square (between) in ANOVA.

$$\begin{aligned}\text{Numerator} &= [\textstyle\sum(O_i - E_i) - 0.5]^2 \\ &= [(16 - 11.67) + (29 - 20.52) \\ &\quad + (66 - 54.64) - 0.5]^2 \\ &= 23.67^2 = 560.26\end{aligned}$$

Similarly, the variances of the values in each subtable are added together to give the *noise* term, analogous to the mean square (within) in ANOVA:

[17] *The situation is completely analogous to the problems of estimating main effects for ANOVA when there are unequal samples and interactions.*

[18] *Another one of those "It is just so" situations. We're honestly not quite sure why you do all the steps that follow, but that's the way it is.*

[19] *Although this seems strange, it actually is related to more general equations. The general formula for the variance of a proportion is $\pi(1 - \pi) \div n$, where n is the total number of objects and π is the proportion. After a great deal of algebra, this is equal to the formula shown, except for an $n \div (n - 1)$ "fiddle factor" favored by statisticians. See also the section on the phi coefficient in Chapter 21.*

[20] *We just use the "a" cells because all the (O – E) differences in each subtable are the same, and all the variances in each subtable are the same, so this would amount to multiplying both numerator and denominator by four, which changes nothing.*

[21] *Presumably there must be some positive interaction—that's why bartenders put them together.*

$$\sum \frac{(a + b)(a + c)(c + d)(b + d)}{N^2(N - 1)} = 4.49 + 7.27 + 13.41 = 25.17 \qquad (20\text{–}15)$$

where N is the total sample size for each subtable (252, 200, and 248). Finally, the ratio of the two sums is the M-H chi-squared, with $(k - 1)$ *df*, where k is the number of subtables in the analysis—in this case, three. This M-H chi-squared equals 560.26 ÷ 25.17 = 22.25, and it is significant at the .001 level.

Although useful for analyzing stratified data, the M-H chi-squared also appears in the analysis of life-tables because at one level, a life-table is nothing more than a series of 2 × 2 tables (e.g., treatment/control by alive/dead) on successive years over the course of the study. This is described in more detail in Chapter 24.

MANY FACTORS: LOG-LINEAR ANALYSIS

We must still deal with the equivalent of factorial ANOVA—the situation where imaginations and budgets run rampant, and we end up swimming in variables. This frequently occurs on "fishing expeditions" but can also arise when folks do randomized trials, insist on gathering demographic data by the pile, and then make the fatal mistake of analyzing them to show the groups are equivalent. Occasionally, it even happens by design.

In particular, in the last example we examined the combined effects of tryptophan and gin and tonic on EMS symptoms. But the astute ANOVA'er might have noticed that we could have, but didn't, look at the effects of gin and tonic separately. As a result, we cannot separate out the main effect of gin from the interaction between gin and tonic.[21] A better design would be to have four groups—Gin and Tonic, Gin only, Tonic only, and Nothing, with half of each group exposed to tryptophan and half to placebo.

TABLE 20–12 Association between tryptophan, gin, tonic, and EMS

Tonic	Gin		EMS symptoms Yes	No	TOTALS
Yes	Yes	Trypt	16	131	147
		Placebo	4	101	105
		TOTALS	20	232	252
Yes	No	Trypt	29	85	114
		Placebo	7	79	86
		TOTALS	36	164	200
No	Yes	Trypt	32	33	65
		Placebo	23	64	87
		TOTALS	55	97	152
No	No	Trypt	66	22	88
		Placebo	88	72	160
		TOTALS	154	94	248

We had a good reason for not doing it this way. This would have introduced three factors in the design (Tryptophan, Gin, Tonic) and the M-H chi-squared, like the parametric two-way ANOVA, is capable of dealing only with two independent variables. To deal with multiple factors, we must move up yet again in the analytical strategy. The approach to analysis is called **log-linear analysis**. We work out a way to predict the expected frequency in each cell by a product of "effects"—main effects and interactions—and then take the logarithm of the effects to create a linear equation (hence, **log-linear**). It ends up, yet again, as a regression problem using estimates of the regression parameters. Everything seems to be a linear model, or, if it's not, we poke it around until it becomes one!

For relative simplicity, we'll add an extra group to Table 20–11 to separate out the two drinking factors (Table 20–12). In log-linear analysis, we first collapse the distinction between independent and dependent variables. You and I know that Symptoms of EMS is the dependent variable, but from the vantage point of the computer, it's just one more factor leading to vertical or horizontal lines in the contingency table. Table 20–12 could be displayed with any combinations of factors on the vertical and horizontal axis, and it is only logic, not statistics, that distinguishes between independent and dependent variables. Ultimately we care about the association between EMS and Tryptophan, Gin, and Tonic, but this, like a correlation, has no statistical directionality.

We begin by determining what an *effect* is. Let's start by assuming there was *no* effect of any of the variables at all. In this case, the expected value of each cell is just the total divided by the number of cells, 852 ÷ 16 = 53.25.

The next level of analysis presumes a main effect of each factor; this explains the different marginals. This is introduced by multiplying the expected value by a factor reflecting the difference in marginal totals. We would begin by determining the marginal proportion with Gin present, (252 + 152) ÷ 852 = 0.47, and the proportion with Gin absent (0.53). If Gin had no marginal effect, these proportions would be .50 and .50, so we multiply the Gin present cells by .47 ÷ .50 = .94, and the Gin absent cells by .53 ÷ .50 = 1.06.

Working this through for the top left cell, where all effects are present, to account for all the marginal totals, the initial estimate must also be multiplied by the overall probability of Tonic (252 + 200) ÷ 852 = 0.53 ÷ .50 = 1.06; the overall probability of Tryptophan (147 + 114 + 65 + 88) ÷ 852 = 0.48 ÷ .50 = .96; and the overall probability of Symptoms (20 + 36 + 55 + 154) ÷ 852 = .31 ÷ .50 = .62. So, the expected value in this cell is 53.25 × .94 × 1.06 × .96 × .62 = 31.58. If we call β_{G1} (there is no logical reason to call these things βs—that's just what everybody calls them), the main effect of the Gin factor, where the subscript (1) indicates the first level;

β_{P1} the effect of the Pop (Tonic) factor; β_{S1} the main effect of EMS at the first level; and β_{T1} the main effect of tryptophan, then algebraically the expected value of cell (1,1,1,1) with no association is:

$$f_{1111} = N \times \beta_{G1} \times \beta_{P1} \times \beta_{T1} \times \beta_{S1} = 53.25 \times .94 \times 1.06 \times .96 \times .62 = 31.58 \qquad (20\text{–}16)$$

where N is the expected frequency in each cell assuming no main effects, just the total count divided by the number of cells ($852 \div 16 = 53.25$). Going the next step, if we assume that there *is* an association between Gin and Symptoms, but there *is not* an association between Pop and Tryptophan and Symptoms, then this would amount to introducing another multiplicative factor to reflect this interaction, a factor that we might call β_{GS11}. We won't try to estimate this value because there is a limit to our multiplication skills, but algebraically the expected value in the top left cell of such a model would look like:

$$f_{1111} = N \times \beta_{G1} \times \beta_{P1} \times \beta_{T1} \times \beta_{S1} \times \beta_{GS11} \qquad (20\text{–}17)$$

There is no reason to stop here. Several models could be tested, including No Effects (the expected value in each cell is 53.25), then one or more main effects only, then one or more two-way interactions, then the three-way interactions, and finally the four-way interaction.

However, as yet, we have not indicated how we test the models. Here is the chicanery. Recall once again your high school algebra, where you were told (and then forgot) that the logarithm of a *product* of terms is the *sum* of the logarithms of the terms. So if we take the log of the above equation, it becomes:

$$\begin{aligned}\log[f_{1111}] &= \log N + \log\beta_{G1} + \log\beta_{P1} + \log\beta_{T1} + \log\beta_{S1} + \log\beta_{GS11} \\ &= \theta + \lambda_{G1} + \lambda_{P1} + \lambda_{T1} + \lambda_{S1} + \lambda_{GS11}\end{aligned} \qquad (20\text{–}18)$$

Again, unfortunately, there isn't much rationale for the Greek symbols. The first thing looking like an "O" with a bird dropping in the center is called *theta*. The others are called *lambda* and are the Greek "L"—for **l**og-**l**inear, we suppose.

We have now reduced the beast to a regression problem. The usual analytical approach is to fit the models in hierarchical fashion, so that first the main effects model is fitted, then the two-way interactions model, then the three-way interactions model, and on to the full model. Of course, just as in regression, when new terms are introduced into the model, the magnitudes of all the estimated parameters change. One additional constraint is imposed on the analysis: all the λs for a particular effect must add to zero. Thus, when an effect has two levels, as is the case in our example, the λs will be something like +0.602 and −0.602. In turn, because each of the estimated parameters is the logarithm of a factor that multiplies the initial expected cell frequency, it is also possible to determine the expected cell frequencies at any stage by listing the parameter estimates, taking antilogs, and then multiplying the whole lot together. Computer packages that run log-linear analysis will do this for you, of course.

TABLE 20–13

Test of individual interactions in log-linear analysis

Effect	df	Partial association Chi-squared	Partial association p	Marginal association Chi-squared	Marginal association p
S	1	124.77	<.0001		
T	1	0.68	.41		
G	1	2.27	.13		
P	1	3.18	.07		
ST	1	31.27	<.0001	4.44	<.05
SG	1	39.85	<.0001	57.94	<.0001
SP	1	165.54	<.0001	163.99	<.0001
TG	1	5.39	<.05	4.64	<.05
TP	1	52.88	<.05	32.51	<.0001
GP	1	2.45	0.11	32.51	<.0001
STG	1	0.00	0.98	26.99	<.0001
STP	1	0.63	0.43	0.53	.46
SGP	1	0.25	0.61	.23	.27
TGP	1	1.34	0.24	0.12	.73
STGP	1	0.13	0.71	0.78	.38

At each stage of the analysis, a chi-squared is calculated, based on the differences between the observed frequencies and the frequencies estimated from the model. If the model fits the data adequately, we get a nonsignificant chi-squared, indicating no significant differences between the predicted and the observed data. Where do the *df* come from? Two effects. First, note that in this case all variables are at two levels, so each effect is a 2×2 or a $2 \times 2 \times 2$ table, and any combination of 2×2 tables has one *df*. Second, there are 4 main effects, 6 two-way interactions, 4 three-way interactions, and 1 four-way interaction (see Table 20–13), so these are the total *df*, 15.

For the present data, the analysis of zero-, first-, and higher-order interactions results in Table 20–14. It is clear that the test of first-order interactions (i.e., main effects) is significant (chi-squared = 102.15), simply implying that the marginals are not equal; the two-way interactions are also highly significant (chi-squared = 291.44). However, fortunately for us, no evidence of a significant three-way or four-way interaction is found (fortunate because we wouldn't know how to interpret it if it was there). So we conclude that the model with two-way interactions fits the data (i.e., it is the model with the lowest order significant interactions, and no significant chi-squareds exist beyond it), so we stop.

TABLE 20–14 Test of interactions in log-linear analysis

Level	*df*	Chi-squared	*p*
1	4	102.15	<.0001
2	6	291.44	<.0001
3	4	3.03	.55
4	1	0.13	.71

The next step is to examine the individual terms to determine which of the main effects and interactions are significant. For the present data, these are shown in Table 20–13. Looking at the main effects only, we see that all are significant, but this simply says that the frequencies in the Gin and No Gin cells, for example, are not equal. Who cares? More interesting is that all the two-way interactions with Symptoms are significant, so an association does exist between symptoms and tryptophan, gin, and tonic. Tryptophan makes you sicker, tonic makes you better, and gin makes you better. The remaining two-way interactions are not of any particular interest, indicating only that there happen to be interactions among the independent variables. Finally, none of the three-way or four-way interactions are significant.

Note that, in Table 20–13, we show both a *marginal* and a *partial* association. The marginal association is based on frequencies at the marginals and is analogous to a test of a simple correlation. Conversely, the partial association takes into account the effect of the other variables at this level, so it is analogous to the test of the partial correlation.

Not surprisingly, at a conceptual level, the analysis resembles multiple regression, in that it reduces to an estimation of a number of fit parameters based on an assumed linear model, with the exception that in log-linear analysis, you generally proceed in hierarchical fashion, fitting all effects at a given level. For those with an epidemiological bent, there is one final wrinkle. The estimated effect is exactly equal to the log of the *odds ratio*. Thus an effect of −1.5 for G and T implies that the odds ratio (the odds of disease with G and T present to the odds of disease with G and T absent) is equal to exp(−1.5) = 0.22. Similarity to factorial ANOVA also exists in the unique ability of the log-linear analysis to handle multiple categorical variables.

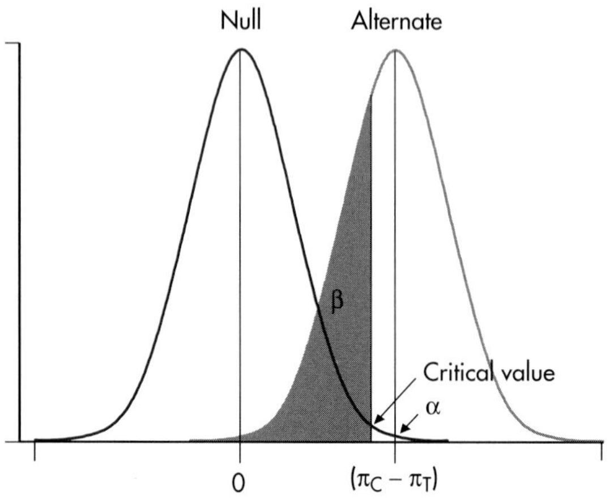

FIGURE 20–2 Visualizing the sample size calculation for two independent proportions.

SAMPLE SIZE ESTIMATION

As we found in earlier situations, sample size procedures are worked out for the simpler cases such as those with two proportions, but not for any of the more advanced situations. The method for two proportions is a direct extension of the basic strategy introduced in Chapter 6. Imagine a standard RCT where the proportion of deaths in the treatment group is π_T and in the control group is π_C. (With a few sad exceptions, π_T is less than π_C.)

We consider two normal curves, one corresponding to the null hypothesis that the two proportions are the same ($\pi_T - \pi_C = 0$), and the second corresponding to the alternative hypothesis that the proportions are different ($\pi_C - \pi_T = \delta$). We're almost set. However, we first have to figure out the SD of the two normal curves. You may recall that the SD of a proportion is related to the proportion itself. In this case, the SD of the proportion π is equal to:

$$\text{SD}(\pi) = \sqrt{\frac{\pi(1-\pi)}{n}} \tag{20–19}$$

and the variance is just the square of this quantity. Now the two bell curves are actually derived from a *difference* between two proportions, so the variances of the two proportions are added. For the H_1 curve on the right, then, the SD is:

$$\text{SD}(\delta) = \sqrt{\frac{\pi_T(1-\pi_T) + \pi_C(1-\pi_C)}{n}} \tag{20–20}$$

Finally, the H_0 curve is a little simpler because the two proportions are the same, just equal to the average of π_T and π_C.

$$\text{SD}(O) = \sqrt{2\frac{\pi(1-\pi)}{n}} \tag{20–21}$$

The whole lot looks like Figure 20–2, which of course bears an uncanny resemblance to the equivalent figure in Chapter 6 (Figure 6–7). We can then do as we did in Chapter 6, and solve for the critical value. The resulting sample size equation looks a little horrible:

$$n = \left[\frac{Z_\alpha \sqrt{2\,\pi\,(1-\pi)} + Z_\beta \sqrt{\pi_T(1-\pi_T) + \pi_C\,(1-\pi_C)}}{(\pi_T - \pi_C)}\right]^2 \tag{20–22}$$

What a miserable mess this is! Now the good news. If you would like to forget the whole thing, that's fine with us because we have furnished tables (Table K) that have performed all this awful calculation for you. These tables are based on a slightly different, and even more complicated, formula, so they will not yield exactly the same result.

For the situation where you wish to test the significance of a single proportion, the formula is a bit simpler. One good example of this is the paired design of the McNemar chi-squared, where the null hypothesis is that the proportion of pairs in each off-diagonal cell is .5. In this case, the SDs are a bit simpler, and the formula looks like:

$$n = \left[\frac{Z_\alpha \sqrt{\pi_1(1 - \pi_1)} + Z_\beta \sqrt{\pi_O(1 - \pi_O)}}{(\pi_1 - \pi_O)} \right]^2 \qquad \textbf{(20–23)}$$

where π_1 is the proportion under the alternative hypothesis, and π_0 is the proportion under the null hypothesis (in this case, .5). Unfortunately, there is no table for this, so get out the old calculator.

To show you how it's done, we refer to an ad we recently saw on TV where it was loudly proclaimed that, "In a recent survey, 57% of consumers preferred Brand *X* to the leading competitor." Pause a moment, and warm up the old C.R.A.P. Detectors. This means that 43% preferred the competitor, and the split is not far from 50–50. They also don't say how many times they did the "study." More particularly, we might ask the essential statistical question, "How large a sample would they need to ensure that the 57–43 split did not arise by chance alone?"

Looking at the formula above, π_1 is .57, and π_0 is .50, so the equation looks like (assuming $\alpha = .05$ and $\beta = .10$):

$$n = \left[\frac{1.96 \sqrt{.57(1 - .57)} + 1.28 \sqrt{.50(1 - .50)}}{(.57 - .50)} \right]^2$$

$$= 410 \qquad \textbf{(20–24)}$$

Any bets on how many consumers they really used?

SUMMARY

We have considered several statistical tests to be used on frequencies in categories. The ubiquitous chi-squared deals with the case of two factors (one independent, one dependent) only, as long as no frequencies are too small. In the case of low frequencies, you use the Fisher exact test. For 2 × 2 tables with paired or matched designs, the McNemar chi-squared is appropriate. Finally, we considered the M-H chi-squared for three factor designs, and log-linear analysis for still more complex designs.

EXERCISES

1. In a small randomized double-blind trial of attar of eggplant for acne, the ZR (medical talk for "zit rate") in the treated group was half that of the control group. However, a chi-squared test of independent proportions showed that the difference was not significant. We can conclude that:
   a. The treatment is useless
   b. The reduction in ZR is so large that we should start using the treatment immediately
   c. We should keep adding cases to the trial until the test becomes significant
   d. We should do a new trial with more subjects
   e. We should use a *t*-test instead of the chi-squared
2. The data below are from a study of previous failure in school, academic or behavioral problems, and dropout.

		Dropout	
Previous failure	**Problems**	**Yes**	**No**
Yes	Yes	32	45
	No	18	99
No	Yes	75	181
	No	84	832

   How would you analyze it?
3. A case-control study was performed to examine the potential effect of marijuana as a risk factor for brain cancer. A total of 75 patients with brain cancer were matched to 75 controls. All subjects were questioned about previous marijuana use.

   Of the cases, 50 said they had used marijuana, and 35 of their matched controls reported marijuana use. No use of marijuana was reported by 25 cases and 20 controls.

   If the data were analyzed with a McNemar chi-squared, what would be the observed frequency in the upper right corner (cell B) in the table below?

	Control	
Case	**Yes**	**No**
Yes	(A)	(B)
No	(C)	(D)

4. Is it really true that "If you don't wear your long underwear, you'll catch your death of cold, dearie!"? We know colds are caused by viruses, but surely all those grannies all those years couldn't have all been wrong.

Let's put it to the test. One cold, wintry week in February, half the kids in the student residence have their longjohns confiscated for science. After a week, the number of colds looks like this:

	Colds		
Longjohns	**Yes**	**No**	
Yes	1	19	20
No	5	15	20
	6	34	40

Analyse the data with:

a. Chi-squared
b. Yates' corrected chi-squared
c. Fisher exact test

How to Get the Computer to Do the Work for You

Chi-squared and Fisher's Exact Test

It may look as if the appropriate place to find these tests would be in **Analyze**, **Nonparametric Tests → Chi-Square**. Well, resist the temptation to do what appears logical. The best way to get chi-squared and Fisher's exact test is:

- From **Analyze**, choose **Summarize → Cross-Tabs**
- Click on the variable you want for the rows and click the arrow to move it into the box labeled **Row(s)**
- Do the same for the second variable, moving it into the **Column(s)** box
- Click the **Statistics** button and choose **Chi-square**, then **Continue**
- Click the **Cells** button and click on **Row**, **Column**, and **Total** in the **Percentages** box, then **Continue**
- Click **OK**

McNemar's Chi-squared

- Simply select it after you click the **Statistics** button

CHAPTER THE TWENTY-FIRST

Measures of Association for Categorical Data

This chapter reviews several measures of association used for contingency tables. The phi coefficient and Cramer's V are directly related to the chi-squared test of significance. Kappa and weighted kappa are popular measures of "agreement beyond chance," the former for nominal scales where no gradation of disagreement is found, and the latter for when degrees of disagreement do exist. The equivalence between kappa and the intraclass correlation (ICC) coefficient is discussed.

SETTING THE SCENE

In an effort to reform the public schools and catch up with education in the rest of the world, a study is initiated to see if school psychologists can detect potential criminals so that taxpayers' dollars won't be wasted in the schools and can be diverted directly to the prisons.

Having succeeded in deriving several approaches to do significance testing for categorical data, the next step is to work out some measures of association. In parametric statistics, once statistical significance was established, we examined nondimensional measures indicating how much association was present. Pearson's correlation did nicely for two variables and simple regression, the multiple R handled multiple independent variables, and the eta-squared did the same for ANOVA situations. All were based on the underlying concept of proportion of variance in Y accounted for by the independent variables.

All is not so straightforward in nonparametric statistics. Just as with the tests of significance, which were an inventors' paradise with two-man teams all over the countryside striving for immortality, nonparametric measures of association are similarly littered with surnames, although these tend to be of solo practitioners. We will mention only a few of the more common ones, attributable to Cohen, Yule, and Cramer. Left for obscurity are the dozens of more esoteric tests.[1]

MEASURES OF ASSOCIATION FOR 2 × 2 AND HIGHER TABLES

Returning to our opening scenario, we must apologize for such a pessimistic attitude. In fact, terms such as "criminal" or "juvenile delinquent" have acquired a pejorative meaning, as if the bearer had actually done something wrong rather than just finding himself or herself in unfortunate circumstances. This labeling gets in the way of rehabilitation. Clearly, in these politically correct times, it's an occasion for a new, neutral label for such unlucky folks. How about "legally challenged?" And for the kids, "youthful legally challenged," or YLC for short.

Along these lines, if we could only identify these kids early, perhaps they might never stray at all. School psychologists should be in an ideal position to do this (and we haven't picked on them yet). Let's do a study to see if they are good at primary prevention.

The design is straightforward. We locate a sample of a couple of hundred kids, both YLCs from the local reformatory and normals (oops, there we go again. Calling them "normals" implies that the YLCs aren't normal. Let's call them "others"). We ask the psychologists at their schools (former or current, depending on the kid) to review the files and predict whether they were likely to end up on the other side of the bars. The data can be arranged in a 2 × 2 table (Table 21–1).

Now, it would be easy enough to apply a statistical test to determine if the relationship is significant. The appropriate test is the chi-squared, which equals 17.61, significant at the .001 level. But a larger question is involved: namely, is it worth putting a lot of effort into attempting to catch these kids early and do counseling, handholding, or whatever is necessary to keep them off the streets if the association is not all that strong? In short, we would

[1] *Who can forget Goodman's Gamma and Lambda or Somer's d? We can, and so can you.*

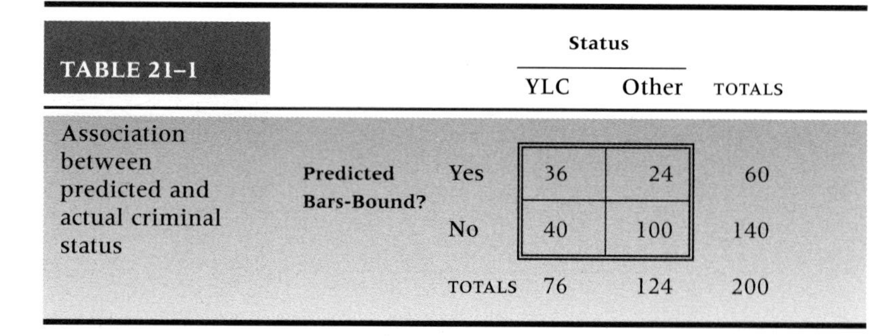

TABLE 21–1 Association between predicted and actual criminal status

		Status: YLC	Status: Other	TOTALS
Predicted Bars-Bound?	Yes	36	24	60
	No	40	100	140
	TOTALS	76	124	200

like a measure of the *strength* of the association, equivalent to a correlation coefficient, before we decide to throw taxpayers' dollars at this social problem.[2]

[2] *We could also treat this as a screening test for YLCs ("If the test is positive, it's highly likely that the kid will end up in the can") and determine the sensitivity (36 ÷ 76 = .47), specificity (100 ÷ 124 = .81), positive predictive value (36 ÷ 60 = .60), and negative predictive value (100 ÷ 140 = .71). We could but we won't—this is a statistics book. For more details see* PDQ Epidemiology.

The Phi Coefficient, Contingency Coefficient, Yule's Q, and Cramer's V

The most obvious approach is to pretend that the data are actually interval and go ahead and calculate a Pearson correlation. If the kid is identified by the psychologist as a troublemaker, he gets a "1," if not, a "0"; if he ends up a YLC, he gets a "1," if not, a "0." And so we stuff 200 (x,y) pairs into the old computer, where each pair looks like (1,1), (1,0), (0,1), or (0,0), and see what emerges. As it turns out, this results in some simplifications to the formula. We won't go through all the dreary details, but will just give you a glimpse. Remember that the numerator of the Pearson correlation was:

$$\text{Numerator} = N\sum XY - \sum X \sum Y$$

Now for the shenanigans. If we call the top left cell a, the top right one b, the bottom left c, and the bottom right cell d, then the first observation is that the sum of XY is equal only to a because this is the only cell where both X and Y are equal to 1. Second, the sum of X (the rows) is just $(a + b)$, and the sum of Y is $(a + c)$, again because this is where the 1s are located. Finally, N is equal to $(a + b + c + d)$, the total sample size. The equation now becomes:

$$\begin{aligned}\text{Numerator} &= (a + b + c + d)(a) - (a + b)(a + c) \\ &= (a^2 + ab + ac + ad) - (a^2 + ab + ac + bc) \\ &= ad - bc\end{aligned}$$

Similar messing around results in a simplication of the denominator, so that the final formula is equal to:

$$\phi = \frac{ad - bc}{\sqrt{(a + b)(a + c)(c + d)(b + d)}} \quad \textbf{(21–1)}$$

We have taken the liberty of introducing yet another weird little Greek symbol, which is called *phi*. The coefficient is, as you may have noticed from the title, the **phi coefficient**. For completeness, we'll put the numbers in:

$$\phi = \frac{36 \times 100 - 24 \times 40}{\sqrt{(60)(76)(140)(124)}} = .297 \quad \textbf{(21–2)}$$

We'll let you be the judge whether this correlation is high enough (or low enough) to merit trying to inspire a change in behavior.

Because the phi coefficient falls directly out of the 2 × 2 table, if the associated chi-squared is significant, so is the phi coefficient (and vice versa). In fact, there is an exact relationship between phi and chi-squared:

$$\phi = \sqrt{\frac{\chi^2}{N}} \quad \textbf{(21–3)}$$

This relationship, and some variations, is the basis of several other coefficients. Pearson's **contingency coefficient**, not to be confused with the product-moment correlation, looks like:

$$\text{Contingency Coefficient} = \sqrt{\frac{\chi^2}{(N + \chi^2)}} \quad \textbf{(21–4)}$$

Cramer's V is based on the chi-squared as well, but it is a more general form for use with I × J contingency tables. It is written as:

$$\text{Cramer's V} = \sqrt{\frac{\chi^2}{[N \times \min(I - 1, J - 1)]}} \quad \textbf{(21–5)}$$

where the denominator means "N times the minimum of (I − 1) or (J − 1). For a 2 × 2 table, this is the same as phi.

Yule's Q is another measure based on the cross-product of the marginals, and it has a particularly simple form:

$$Q = \frac{ad - bc}{ad + bc} \quad \textbf{(21–6)}$$

Choice among these alternatives can be made on cultural or esthetic grounds as well as any other because they are all variations on a theme that give different answers, with differences ranging from none, through slight, to major.

Cohen's Kappa

A second popular measure of association in the biomedical literature is **Cohen's kappa** (Cohen, 1968). Kappa is usually used to examine inter-observer agreement on diagnostic tests (e.g., physical signs, radiographs) but need not be restricted to such purposes. However, to show how it goes, we'll create a new example.

One clear problem with our above study is that we were left with a number of prediction errors, which may be due either to individual psychologists' inability to agree on their predictions (an issue of reliability); or because they may agree based on the evidence available at the school, but this evidence is simply not predictive of future behavior (an issue of validity). Disagreement among observers will reduce the association, so it might be useful to examine the extent of agreement on this classification. This is straightforward. We assemble the files for a bunch of kids (e.g., $n = 300$), and get two psychologists to independently classify each kid as rotten or not. We then examine the association between the two categorizations, which are displayed in a 2 × 2 table, as in Table 21–2.

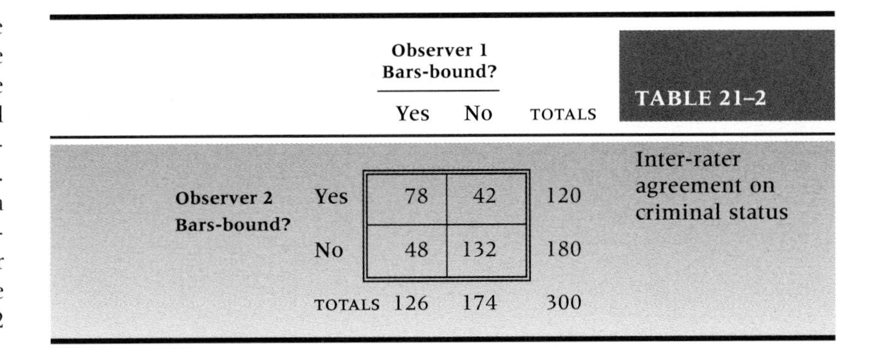

TABLE 21–2

Inter-rater agreement on criminal status

		Observer 1 Bars-bound? Yes	No	TOTALS
Observer 2 Bars-bound?	Yes	78	42	120
	No	48	132	180
	TOTALS	126	174	300

We could use phi here; however, kappa is a more popular choice as a measure of agreement in clinical circles for reasons which are likely more stylistic than substantive. To begin, let's take a closer look at the data. How much agreement would we expect between the two shrinks just by chance? That is, if the psychologists don't know beans about the students' behaviors, they would still agree with each other a number of times, just by chance. Thus, the proportion of agreements would be non-zero.

The chance agreement is calculated by working out the expected proportions for the *a* and *d* cells, using the product of the marginals as we did with the chi-squared test. This equals ($120 \times 126/300^2$) = .168 for the *a* cell and ($174 \times 180/300^2$) = .348 for the *d* cell, a total chance agreement of 0.516.[3]

Next, we must determine the *observed* proportion of times when there is agreement (p_0)—simply the frequency in the *a* and *d* cells, divided by the total frequency. In this case, it equals $(78 + 132)/300 = .70$. The final step is to express the agreement beyond chance as a ratio to the maximum possible agreement beyond chance:

$$\kappa = \frac{(p_o - p_e)}{(1.0 - p_e)} = \frac{(.70 - .516)}{(1.0 - .516)} = .380 \tag{21-7}$$

So, even though the observed agreement was a fairly impressive 70%, much of this was due to chance agreement and the kappa is a less impressive .380.

Although kappa appears to start from a different premise than the phi coefficient, there are more similarities than differences after the dust settles. The numerator of kappa turns out also to equal ($ad - bc$), the same as phi. The denominators are different, but this amounts to a scaling factor. In fact, in this situation, phi is also equal to .380.

Standard error of kappa and significance test. To test the significance of kappa, it is first necessary to derive the standard error (or the variance) of kappa, assuming that it is equal to zero. In its most general form, including multiple categories and multiple raters, this turns out to be a fairly horrendous equation. However, for a 2 × 2 table, it is a lot easier:

$$var(\kappa) = \frac{p_o - p_o^2}{N[1 - p_e]^2} \tag{21-8}$$

In the present case, p_o is .70, p_e is .516, and N is 300, so the variance equals:

$$var(\kappa) = \frac{.70 - .70^2}{300(1 - .516)^2} = .003 \tag{21-9}$$

Once the variance has been determined, the significance of kappa can be determined through a *z*-statistic:

$$z = \frac{\kappa}{\sqrt{var(\kappa)}} \tag{21-10}$$

which in this case equals $.380/\sqrt{.003} = 6.952$—which is significant. In turn, the confidence interval about kappa is just 1.96 times the square root of the variance, or ±.107.

Generalization to multiple levels and dimensions. Kappa, unlike phi, can be generalized to more complex situations. The first is multiple levels; for example, we might have decided to get the counsellors to identify what kind of difficulty the kids would end up in—violent crimes, "white collar" crimes, drugs, and so on. Kappa can be used; it is just a matter of working out the expected agreement by totaling all the cells on the diagonal, then the expected agreement by totaling all the expected values, obtained by multiplying out the marginals. The ratio is then calculated according to the above formula.

Kappa can also be used for multiple observers, which amounts to building a 2 × 2 × 2 table for three observers, a 2 × 2 × 2 × 2 table for four observers, and so on. You can still work out the observed and expected frequencies on the diagonal (only this time in 3-dimensional or 4-dimensional space) and calculate the coefficient. Beware, though, that this is now a measure of *complete* agreement among three, four, or more observers and ignores agreement among a majority or minority of observers.

Finally, kappa can be used for ordinal data, without resorting to ranking. For this we go to the next section.

[3] *Notice that we're dividing by* N^2 *rather than by* N, *as we did previously. The reason is that with chi-squared, we were estimating the number of counts in a cell; whereas, here, we're looking at a proportion.*

TABLE 21–3

Inter-rater agreement on criminal status

		Observer 2				
		Saintly	Slightly crooked	Street thug	Serial killer	TOTALS
Observer 1	Saintly	1	5	8	1	15
	Slightly crooked	8	10	29	5	52
	Street thug	4	14	21	20	59
	Serial killer	1	4	27	12	44
	TOTALS	14	33	85	38	170

PARTIAL AGREEMENT AND WEIGHTED KAPPA

Let's continue to unfold the original question. In the first analysis, we found a relatively low but significant relationship between the prediction and the eventual status. In the next analysis, we explored the agreement on observer rating of criminal tendencies, which was only moderate. One way we might improve agreement is by expanding the categories to account for the degree of criminal tendency. Criminality, like most biomedical variables (blood pressure, height, obesity, rheumatoid joint count, serum creatinine, extent of cancer),[4] is really on an underlying continuum. Shoving it all into two categories throws away information.[5] We should contemplate at least four categories of prediction, for example "Saintly," "Slightly Crooked," "Street Thug," and "Serial Killer." If we again employ two observers, using the same design, the data would take the form of Table 21–3.

The first thing to note is that the overall agreement, on the diagonal, is now 44 ÷ 170, or 26%, which is pretty awful. If we went ahead and calculated a kappa on these data, using the previous formula, it would be less than zero. But there is actually a lot of "near agreement" in the table; 103 additional observations (8 + 14 + 27 + 5 + 29 + 20) agree within one category; combining these would yield an agreement of (44 + 103) ÷ 170 = .865, which is much better. The challenge is to figure out some way to put all these instances of *partial* agreement together into some overall measure of agreement.

Cohen (1968) dealt this problem a body blow with the idea of a **weighted kappa**, whereby all the cells are assigned a weight related to the degree of disagreement. Full agreement, the cells on the diagonal, are weighted zero. (This does not mean that these very important cells are ignored. Stay tuned). The weights on the off-diagonal cells are then varied according to the degree of disagreement. The weights can be arbitrary and assigned by the user. For example, we might decide that a disagreement between Slightly Crooked and Street Thug is of little consequence, so this disagreement gets weighted 1; a difference between Saintly and Slightly Crooked gets a weight of 2; and a difference between Serial Killer and Street Thug is as severe as any of the greater disagreements (e.g., Serial Killer and Saintly) and all get weighted 4. We might do that—but we had better marshal up some pretty compelling reason why we chose these particular weights because the resulting kappa coefficient will not be comparable with any other coefficients generated by a different set of weights. (There is one exception. If the sole reason is to do comparisons *within* a study—for example, to show the effects of training on agreement—this is acceptable.)

The alternative is to use a standard weighting scheme, of which there are two: Cicchetti weights, which apparently are used only by Cicchetti (1972); and **quadratic weights**, which are used by everybody.[6] For obvious reasons,we focus our attention on the latter. Actually the scheme is easy—the weight is simply equal to the square of the amount of disagreement. So, cells on the diagonal are weighted 0; one level of disagreement (e.g., Serial Killer vs. Street Thug) gets a weight of $1^2 = 1$; two levels of disagreement (e.g., Serial Killer vs. Somewhat Crooked) gets weighted $2^2 = 4$, and so on up.

To see how this all works, we begin with the formula for kappa, in Equation 21–7, and then substitute $q = (1 - p)$ for everything. In other words, the formula is rewritten in terms of *disagreement* instead of agreement. The revised formula is now:

$$\kappa = \frac{q_e - q_o}{q_e} = 1 - \frac{q_o}{q_e}$$

(21–11)

It is now a matter of incorporating the various weighting schemes into the qs. No problem—just sum up the weighted disagreements, both observed and expected (by taking the product of the related marginals divided by the total), over all the cells (i,j), which are off the diagonals: where w_{ij} are the weights for the cells. These are then popped back into the original equation, and that gives us weighted kappa.

$$q_o = \sum w_{ij} \times p_{o_{ij}}$$

(21–12)

[4] *We used to think that the only real dichotomous variables were pregnancy and death. However, with life-support technology, death is now up for grabs.*

[5] *For an elaboration, see* Health Measurement Scales: A Practical Guide to Their Development and Use *(Streiner and Norman, 1995).*

[6] *Strictly speaking, the formula is (Everybody – 1); Cicchetti doesn't.*

TABLE 21–4

Expected frequencies for rater agreement

		Observer 2			
		Saintly	Slightly crooked	Street thug	Serial killer
Observer 1	Saintly	—	2.91	6.61	4.23
	Slightly crooked	4.28	—	26.00	11.62
	Street thug	4.85	11.45	—	16.65
	Serial killer	3.62	8.54	19.41	—

TABLE 21–5

Quadratic weights for rater agreement

		Observer 2			
		Saintly	Slightly crooked	Street thug	Serial killer
Observer 1	Saintly	—	1	4	9
	Slightly crooked	1	—	1	4
	Street thug	4	1	—	1
	Serial killer	9	4	1	—

$$q_e = \sum w_{ij} \times p_{e_{ij}} \quad (21\text{–}13)$$

To demonstrate how this all works, let's calculate the example in Table 21–3. In Table 21–4, we have worked out the expected frequencies by taking the product of the marginals and dividing by the total. (Note that we did this calculation only for the off-diagonal cells. Why make work for ourselves when we don't use the data in the diagonal cells?) In Table 21–5 we have shown what the quadratic weights for each cell look like.

Now we can put it all together. Keep in mind that the tables show the *frequencies* and we need the *proportions*, so we will have an extra "170" kicking around in the summations. Now the **observed weighted disagreement**, going across the rows and then down the columns, is:

$$\begin{aligned} q_o &= 1 \times \frac{5}{170} + 4 \times \frac{8}{170} + 9 \times \frac{1}{170} + 1 \times \frac{8}{170} \\ &+ 1 \times \frac{29}{170} + \ldots + 1 \times \frac{27}{170} \\ &= 1.260 \end{aligned} \quad (21\text{–}14)$$

and the **expected weighted disagreement** is:

$$\begin{aligned} q_e &= 1 \times \frac{2.91}{170} + 4 \times \frac{6.61}{170} + 9 \times \frac{4.23}{170} \\ &+ 1 \times \frac{4.28}{170} \ldots + 1 \times \frac{19.41}{170} \\ &= 1.634 \end{aligned} \quad (21\text{–}15)$$

So the weighted kappa in this case is:

$$\kappa = 1 - \frac{q_o}{q_e} = 1 - \frac{1.260}{1.634} = .254 \quad (21\text{–}16)$$

Although this is not terribly impressive, it is an improvement over the unweighted kappa for these data, which would equal −.018. The general conclusion is that the weighted kappa, which takes partial agreement into account, is usually larger than the unweighted kappa.[7]

[7] *It is also a general, although counter-intuitive, finding that increasing the number of boxes on the scale will improve reliability, as assessed by weighted kappa or an ICC, even though the raw agreement is reduced. For an elaboration, see Streiner and Norman (1995).*

RELATION BETWEEN KAPPA AND THE INTRACLASS CORRELATION

One reason why Cicchetti was fighting a losing battle is that the weighted kappa using quadratic weights has a very general property—it is mathematically (i.e., exactly) equal to the ICC. We must be pulling your leg, right? Nope. We know that the ICC comes out of repeated-measures ANOVA (see Chapter 11) and is useful only for interval-level data, and kappa is based on frequencies and nominal or ordinal data.

But just suppose we didn't tell the computer that. We call *Saintly* a 4, *Slightly* a 3, *Street* a 2, and *Serial* a 1. We then have a whole bunch of pairs of data, so the top left cell gives us one (4,4) and the bottom right cell gives us a total of 12 (1,1)s. There are, of course, 170 points in all. We then do a repeated-measures ANOVA where Observer is the within-subject factor with two levels. We calculate an ICC, just as we did in Chapter 11. The result is *identical* to weighted kappa. It also follows that if we were to analyze a 2 × 2 table with ANOVA, using numbers equal to 0 and 1, unweighted kappa would equal this ICC when calculated like we did above (Cohen, 1968).

Who cares? Well, this eases interpretation. Kappa can be looked on as just another correlation, explaining some percent of the variance. And there is another real advantage. If we have multiple observers, we can do an intraclass correlation and report it as an average kappa[8] instead of doing a bunch of kappas for Observer 1 vs. Observer 2, Observer 1 vs. 3, etc.

[8] *This also accommodates apparently religious differences among journals. Some journals like ICCs, some like kappas. We have on occasion calculated an ICC and called it a kappa, and vice versa, just to keep the editor happy.*

SAMPLE SIZE CALCULATIONS

Sample size calculations for phi, kappa, and weighted kappa are surprisingly straightforward. To test the significance of a phi coefficient (i.e., to determine whether phi is different from zero), we simply use the sample size formula for the equivalent chi-squared because both are based on the same 2 × 2 table. This was outlined in Chapter 20, so we won't repeat it.

For kappa, you must first consider a bit of philosophical decision making. If the point of the study is to determine whether kappa is significantly different from zero, we can use the formula in Equation 21–10 to derive an SE for kappa and then insert this into the usual formula for sample size:

$$N = \left[\frac{(z_\alpha + z_\beta)\text{SE}(\kappa)}{\kappa_{est}} \right]^2 \tag{21–17}$$

where κ_{est} is the estimated value of kappa. However, this philosophical stance presumes that a kappa of zero is a plausible outcome. If you are looking at observer agreement, an agreement of zero is hopefully highly implausible (although it happens only too often). In this case, you are really hoping that your estimate of the agreement is somewhere near the true agreement. In short, you want to establish a *confidence interval* around your estimated kappa. The formula for the SE of kappa (Equation 21–10 again) is a likely starting point, and it is necessary only to decide what is a reasonable confidence interval, δ (say .1 on either side of the estimate or .2), then solve for N. Of course, the fact that you have to guess at the likely value for both p_o and p_e in these equations gives you lots of freedom to come up with just about any sample size you want. The formula is now:

$$N = \frac{z_\alpha^2}{\delta^2} \frac{p_o(1 - p_o)}{(1 - p_e)^2} \tag{21–18}$$

The sample size for weighted kappa would require too many guesses, so a rule of thumb is invoked: the minimum number of objects being rated should be more than $2c^2$, where c is the number of categories (Soeken and Prescott, 1986; Cicchetti, 1981). So in our example, with four categories, we should have $2 \times 4^2 = 32$ objects.

SUMMARY

This chapter has reviewed four popular coefficients to express agreement among categorical variables. The phi coefficient and Cramer's V are measures of association directly related to the chi-squared significance test. Kappa is a measure of agreement particularly suited to 2 × 2 tables; it measures agreement beyond chance. Weighted kappa is a generalization of kappa for multiple categories, used in situations where partial agreement can be considered. Unless there are compelling reasons, weighted kappa should use a standard weighting scheme. When quadratic weights are used, weighted kappa is identical to the intraclass correlation, which was discussed in Chapter 11.

EXERCISES

1. Consider a study of inter-rater agreement on the likelihood that psychiatric patients have von Richthofen's disease (characterized by the propensity to take off one's shirt in the bright sunlight—the "Red Barin' Sign"). Two psychiatrists indicate whether or not patients have VRD, rated as Present or Absent.
   Suppose we now did a second study where they did the same rating, only this time on a four-point scale, from "Definitely Present" to "Definitely Absent." What would happen to the following quantities?

	SMALLER	SAME	LARGER	UNDEFINED
a. Raw agreement	_____	_____	_____	_____
b. Unweighted kappa	_____	_____	_____	_____
c. Weighted kappa	_____	_____	_____	_____
d. Phi coefficient	_____	_____	_____	_____

2. The following 2 × 2 table displays agreement between two observers on the presence or absence of the dreaded "Red Barin' Sign" (see above for explanation):

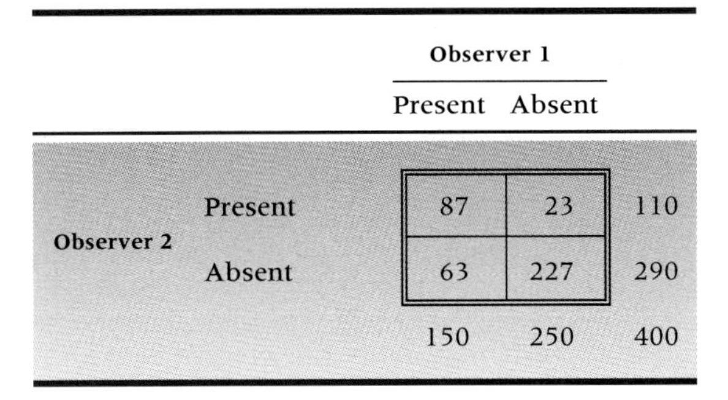

		Observer 1		
		Present	Absent	
Observer 2	Present	87	23	110
	Absent	63	227	290
		150	250	400

Calculate
a. Phi
b. Contingency coefficient
c. Cramer's V
d. Cohen's kappa

How to Get the Computer to Do the Work for You

Phi and Kappa

- From **Analyze**, choose **Summarize** → **Cross-Tabs**
- Click on the variable you want for the rows and click the arrow to move it into the box labeled **Row(s)**
- Do the same for the second variable, moving it into the **Column(s)** box
- Click the **Statistics** button and choose **Phi and Cramer's V** and/or **Kappa**, then **Continue**
- Click **OK**

Data that can be ranked (ordinal data) should be treated differently from categorical data. This chapter reviews several ranking tests—the Mann-Whitney test (or Wilcoxon Rank Sum test) for two independent groups, the Kruskal-Wallis one-way ANOVA for more than two groups, and the Wilcoxon Signed Rank test and Friedman two-way ANOVA for paired data.

CHAPTER THE TWENTY-SECOND

Tests of Significance for Ranked Data

SETTING THE SCENE

You heard that clam juice works wonders for psoriasis. Going one better, you arrange a randomized trial of Bloody Caesars (reasoning that the booze will ease the physical and psychic pain while the clam juice works its miracles). At the end, a bunch of dermatologists examine photographs of the patients and put them in rank order from best to worst. How, Dr. Skinflint, will you analyze this lot?

All academics are slaves of the publish or perish syndrome,[1] and in some the illness is more acute than in others. One easy way to get big grant money (thereby ingratiating yourself to the administration) as well as publication, is to do trials of look-alike drugs or combination drugs for companies. In the present chapter, we discuss one such trial.

Dr. Skinflint, a locally renowned dermatologist, recalls reading somewhere that clam juice works wonders for the misery of psoriasis.[2] He speculates that a combination of clam juice and ethyl alcohol might ease the symptoms while reducing the lesions. So he arranges a randomized trial of Bloody Caesars against Virgin Marys.[3] At the conclusion of the trial, he photographs all the patients and places the pictures together in random order, then he distributes the set of photos to a group of dermatologists, who are asked to simply rank order the pictures from best to worst. The idea then is to examine the ranks of the patients in the BC group against the patients in the VM group.

A comment on the rank ordering. We know a couple of possible alternative approaches to measurement. The photographs could be placed on an interval scale by, for example, measuring the extent of body surface involvement. However, this might not adequately capture other aspects, such as the severity of involvement. Moreover, this could lead to a badly skewed distribution: many patients with only a few percent of body surface area involved, and a few patients in which nearly all the skin is involved can bias parametric tests. Alternatively, some objective staging criteria, such as is used for cancer, might be devised, but this would simply lead to another ordinal scale, which must be analyzed by ranking individual subjects. Similarly, rating individuals on 7-point scales would be regarded by some (but not us) as ordinal level measurement, thus requiring nonparametric statistics. For all these reasons, proceeding directly to a subjective ranking may well represent a viable approach to measurement.

The question is now how to analyze the ranks, which are clearly ordinal-level measurements. We cannot use statistics that employ means, SDs, and the like. To clarify the situation, if 20 patients were in the trial, the ranks would extend from 1 (the best outcome) to 20 (the worst). If the treatment were successful, we would expect that, on the average, patients in the BC group would have higher ranks (lower numbers) than those in the VM group. Suppose Table 22–1 shows the data. It is evident that treatment has some effect. If no effect occurred, the BCs and the VMs would be interspersed, and the average ranks of the BCs and the VMs would be the same. This does not seem to be happening; the BCs appear to be systematically higher in the table than do the VMs. The question is how one puts a p-value on all this.

[1] *One disciple to another while taking Christ from the cross: "He was a great teacher, but he didn't publish."*

[2] *"Somewhere" was in the preeminent international oeuvre,* PDQ Statistics.

[3] *For the teetotalers in our midst (both of them!), a Bloody Caesar contains tomato juice, tabasco, clam juice, and vodka. A Virgin Mary is missing the clam juice and alcohol.*

TABLE 22–1

Ranks of 20 psoriasis patients treated with BC and VM

Rank	Treatment
1	BC
2	BC
3	BC
4	BC
5	VM
6	BC
7	VM
8	BC
9	VM
10	VM
11	BC
12	VM
13	VM
14	BC
15	BC
16	VM
17	BC
18	VM
19	VM
20	VM

TABLE 22–2

Ranks of 20 psoriasis patients treated with BC and VM

Rank	Treatment	Rank of BC group	Rank of VM group
1	BC	1	
2	BC	2	
3	BC	3	
4	BC	4	
5	VM		5
6	BC	6	
7	VM		7
8	BC	8	
9	VM		9
10	VM		10
11	BC	11	
12	VM		12
13	VM		13
14	BC	14	
15	BC	15	
16	VM		16
17	BC	17	
18	VM		18
19	VM		19
20	VM		20
	SUMMED RANK	81	129

TWO INDEPENDENT GROUPS

For this simple case with two groups, several approaches are possible. Characteristically, as with many tests based on ranks, they are "cookbook-ish," and it is nearly impossible to synthesize individual tests into a coherent conceptual whole, as we did (or think we did) with parametric tests. Fortunately, many, such as the Median test and the Kolmogorov-Smirnov test, have now faded into well-deserved obscurity. We will examine only one test, the **Mann-Whitney U** test. However, to make things more confusing in this particular situation, the Mann-Whitney test is also called the **Wilcoxon Rank Sum** test, abbreviated *W*.

The test focuses on the sum of the ranks for the two groups separately, which explains *Rank Sum*, but not *U*. Anyway, as you see in Table 22–2, the calculation is the essence of simplicity. The summed rank for the BC group is 81 and for the VM group is 129. (Actually we didn't have to calculate the second one for two reasons: (1) the sum of all ranks is $N(N + 1) \div 2 = 210$, so we can get it by subtraction, and (2) we don't need it anyway.) The larger the difference between the two sums the more likely the difference is real. The next step is easier still—you turn to the back of the book[4] (as long as the sample size per group is less than 10) and look up 81. You find that the probability of a rank sum (*W* or *U*, depending on where your allegiance lies) as *low* as 81 is .0376 (using a one-tailed test), so Dr. Skinflint can get his publication after all.

What do you do if there are more than 10 per group? Believe it or not, parametric statistics rear their heads yet again. It turns out that the normal distribution and *z*-test can be used as an approximation. If the total sample size is N and m are in the higher ranked group, then the expected value of the rank sum is $m(N + 1) \div 2$, or 105. So we can construct a *z*-test with a numerator of the observed rank sum minus the expected rank sum. The question is the form of the denominator, the SE of the difference. This turns out, after much boring algebra, to equal $\sqrt{mn(N + 1)/12}$, where m and n are the two group sizes, with $m + n = N$. The *z*-test then equals:

$$z = \frac{U + 0.5 - m(N + 1)/2}{\sqrt{mn(N + 1)/12}} = 1.776 \qquad \textbf{(22–1)}$$

Looking this value up in Table A (in the appendix) of the normal distribution, we find that a z of 1.776 results in a one-tailed probability of .0379, very close to the tabulated value up above. As is frequently the case, nonparametric tests are devised because of a concern for bias in the parametric tests. However, except for some limiting cases, the parametric test turns out to be quite a precise approximation.

MORE THAN TWO GROUPS

Following our usual progression, we can next consider the extension to three groups—the equivalent of one-way ANOVA. The strategy is a lot like the Wilcoxon (Mann-Whitney) test. However, instead of examining the total rank in each group, we calculate the average rank. And instead of doing a *t*-test on the ranks, we do a one-way ANOVA on the ranks. Once again, terms such as $N\,(N + 1)$ and fac-

[4] *Some books, not this one. We recommend Siegel and Castellan (1988) for all nonparametric tests. As we'll see in a minute, though, we really don't need the tables.*

TABLE 22–3 Ranks of patients in the BC, VC, and VM trial

	Bloody Caesar	Virgin Caesar	Virgin Mary
	1	4	8
	2	7	12
	3	9	15
	5	11	16
	6	14	21
	10	17	22
	13	20	27
	18	23	28
	19	25	29
	24	26	30
SUM	101	156	208
MEAN	10.1	15.6	20.8

tors of 12 start kicking around, simply because of the use of ranks.

The test is another two-man team like Mann and Whitney, Kolmogorov and Smirnov, or Rimsky and Korsakoff.[5] This time it's the **Kruskal-Wallis one-way ANOVA**. To illustrate, suppose we throw an intermediate group into our original design. They are fed just clamato juice, what might be called a Virgin Caesar (VC), to see whether the vodka is having any real beneficial effect on the skin of the BC group as opposed to that group's souls. We now have 30 patients, and in Table 22–3 we have shown the ranks of the patients in each group. It is clear that these contrived data are working according to plan: the BC group has a mean rank of 10.1; the VC group, a mean rank of 15.6; and the VM group a rank of 20.8. If no difference existed, we would have expected that the average rank of each group would be about halfway, or 15. (Actually, it's $(N + 1) \div 2 = 15.5$, because the ranks start counting at 1.) But is it significant?

To address the question, cued by the title of the test, obviously the first thing to do is to calculate a Mean Square (between groups), exactly as we have been doing since Chapter 8. This looks like:

$$MS_{\text{bet}} = \sum n_j(\bar{R}_j - \bar{R}_{..})^2 = \sum n_j\left(\bar{R}_j - \frac{N+1}{2}\right)^2 \quad \textbf{(22–2)}$$

where n_j is the sample size in each group; $\bar{R}_j$ the group mean; and $\bar{R}_{..}$ the overall mean. So in this case it equals:

$$10(10.1 - 15.5)^2 + 10(15.6 - 15.5)^2 + 10(20.8 - 15.5)^2 = 572.6$$

In the normal course of events, we would now have to work out a Mean Square (within), but because of the use of ranks again, this takes a particularly simple form: $N(N + 1)/12$. So, the final ratio (the Kruskal-Wallis test, or K-W) is equal to:

$$\text{K-W} = \frac{12}{N(N+1)}\sum n_j(\bar{R}_j - \bar{R}_{..})^2 \quad \textbf{(22–3)}$$

For small samples, the K-W test statistic has to be retrieved from the back of someone else's book. However, if more than five subjects are in each group, then it looks like a chi-squared distribution with $(k - 1)$ *df*, where k is the number of groups. Because you shouldn't have fewer than five subjects per group anyway, you don't need the special table.

[5] *Of course, these days a hyphenated name stands for one married woman, not two men. Probably reflects the observation that one woman can do about as much as two men anyway. Certainly, because most of these seem to be a simple adaptation of a parametric test developed by one man, one wonders why it took two to do it.*

[6] *Lest you are offended by the labels, these are all legitimate drinks, and can be purchased in any reputable bar (and many disreputable ones).*

TWO GROUPING FACTORS

We're getting there. Not only have we extended the rank sum tests to include multiple groups, we have also proven, more or less, that booze is good for psoriasis (that's the beauty of fictitious data!). However, we are still limited to one factor only. In any case, if we were setting out to examine the independent and possible interactive effects of vodka and clam juice, a much better design from the outset would be a two-factor one. To be precise, we would have four groups—clam juice and vodka (Bloody Caesar), clam juice only (Virgin Caesar), vodka only (Bloody Mary), and neither (Virgin Mary).[6] We lay on the pepper and tabasco so no one can tell which is which anyway, and we again have dermatologists rank the outcomes, this time for 40 patients.

Unfortunately, it is at this point that tests on ordinal data grind to a screeching halt. Given the simple strategy used by all tests in this chapter, it would seem such a simple trick to take the equations of Chapter 8 and diddle them a bit for rankings. We can see it all now ... "The Streiner-Norman two-way ANOVA for Independent Samples." Better still, why don't you do it, and we'll get back to writing books?

REPEATED MEASURES: WILCOXON SIGNED RANK TEST AND FRIEDMAN TWO-WAY ANOVA

The final step in this walk through the ranked clones of the parametric tests is to consider the issue of matched or paired data—the equivalent of the paired *t*-test and repeated-measures ANOVA. For this excursion, let's take the issue of clones to heart.

Suppose some cowboy scientist, Gene Auful, was let loose in the university molecular biology lab and managed to create some clones of graduate students from samples of blood they unwittingly donated to the Red Cross. The little darlings were raised by foster parents, and in due course, 20 years later, the clones end up as graduate students in the same labs (the experiment is working!). Recognizing that here are the makings of the ultimate nature-nurture experiment, one of the measures we put in place is a measure of achievement and likelihood of success, arrived at by getting the graduate faculty to sit around a table with all the files of both original and clone students and rank order them.[7] One measure of successful clones would be that they, on the average, are ranked just as highly on ability to succeed. The data are in Table 22–4.

There are 15 pairs, and we have listed the rank order of the original and the clone, ranging from 1 to 30, where this time 1 is best and 30 is worst. It seems that the clones are actually a bit inferior because their average rank appears higher than that of the originals. This is confirmed in the fourth column, which is the first step to the **Wilcoxon Signed Ranks test** (wasn't he a busy little lad!), where we have calculated the difference of ranks for each pair. Next, we rank the rank differences in column five, so that the smallest differences have the first rank (ignoring the sign, but carrying it through). You will notice some funny-looking numbers in the right column. We have three 6s, two 2.5s, and two 1.5s, but no 5, 7, or 1 scores. The problem is caused by having three differences of 3, which should take up the ranks of fifth, sixth, and seventh; two differences of 2, which should be third and fourth; and two differences of 1, which should be first and second. Because we don't know which is which, to avoid any infighting, we give them all (or both) the average rank: 6, 3.5, or 1.5.

TABLE 22–4

Ranking of graduate students and their clones on success

Pair	Rank of original	Rank of clone	Signed difference	Signed ranked difference
A	1	3	+2	+3.5
B	8	2	−6	−10.0
C	6	14	+8	+12.0
D	9	10	+1	+1.5
E	5	12	+7	+11.0
F	20	17	−3	−6.0
G	7	26	+19	+15.0
H	4	13	+9	+13.0
I	11	15	+4	+8.0
J	22	27	+5	−9.0
K	25	23	−2	−3.5
L	24	21	−3	−6.0
M	16	19	+3	+6.0
N	18	30	+12	+14.0
O	29	28	−1	−1.5

Finally we sum the ranks of the positive and negative differences. The positive sum is (3.5 + 12 + 1.5 + 11 + 15 + ... + 14) = 84, and the negative sum turns out to be 36; as before, they sum to $N\ (N + 1) \div 2 = 120$.

Now, under the null hypothesis that no difference exists between original and clone, we would anticipate that the average rank of the originals and the clones would be about the same. If so, then the differences between rankings would all be small, and the sum of the rankings for both the positive and the negative differences would be small. If either of the summed differences is large, this indicates a substantial difference between the average original rank of the individuals in the matched pairs and so would lead to rejection of the null hypothesis.

Once again we rush expectantly to the back of the book, only to be disappointed. However, in Siegel and Castellan (1988), a T of 84 (that's what the sum refers to) is not quite significant ($p = .094$). And once again, the table stops at a sample size of 15 matched pairs. For larger samples there is (you guessed it) an approximate z-test, based on the fact that, once again, this statistic is approximately normally distributed, with a mean and SD based on the number of pairs, N. The formula is:

$$z = \frac{T - N(N + 1)/4}{\sqrt{N(N + 1)(2N + 1)/24}} \quad \textbf{(22–4)}$$

In the present case, z equals 1.842, and the associated p-value is .066, not quite the same as the exact value calculated from the table, but close enough.

The extension of this test to three or more groups, the equivalent of repeated-measures ANOVA, is the **Friedman test**. We won't spell it out in detail because (1) it follows along a familiar path of summed ranks, and (2) the applications are rare. Briefly, it considers matched groups of three, four, or however many, each of which is assigned to a different treatment. It calculates the rank of each member of the trio or quartet. If one treatment is clearly superior, then that member of the group would be ranked first every time. If another treatment is awful, the member receiving that treatment would always come in last. And if the null hypothesis were true, all the ranks would be scrambled up. You then calculate the total of the ranks under each condition and plug the average ranks into a formula, again involving sum of squares and Ns. For a small sample, you look it up in a table; for a large sample, you approximate it with an F distribution. For furthermore information, see Siegel and Castellan, yet again.

SAMPLE SIZE AND POWER

We did Medline, Statline, PsycINFO, and Edline searches, and we even called a few 1-900 numbers, but we were unable to come up with any formulas for sample size calculations on rank tests.[8] What should we do if the granting agency demands them? Determine the sample size from the equivalent parametric test (e.g., t-test, one-way ANOVA, paired t-test), and leave it at that. For a long time, it was assumed that nonparametric tests were less powerful than parametric ones, so some people (us, for example, in the first edition) suggested adding a "fudge factor." However, more recent work has shown that you don't lose any power with these tests and, when the data aren't normal, they may even be more powerful than their parametric equivalents.[9]

[7]You know now this is a fictitious example. Professors never agree on anything. Clark Kerr (UCSF) once said that, "A university is a collection of scholars joined by a common heating system."

[8]However, a few of the 1-900 numbers gave us suggestions for other things, which can't be mentioned in this family-oriented book.

[9]If you need some references for the grant and are loath to cite us, use Blair and Higgins (1985) or Hunter and May (1993).

SUMMARY

In this chapter, we dealt with several ways to do statistical inference on ranks. We should remind you that, although the examples used rankings as a primary variable, in circumstances where the distributions are very skewed or the data are suspiciously noninterval, such as staging in cancer, the data can often be converted to ranks and analyzed with one of these nonparametric tests.

Why not use ranking tests all the time and avoid all the assumptions of parametric statistics? The main reason is simply that the technology of rank tests is not as advanced (there really is no Streiner-Norman two-way ANOVA by ranks), so the rank tests are more limited in potential application. A second reason is that they tend to be a *little* bit conservative (i.e., when the equivalent parametric test says $p = .05$, the rank test says $p = .08$); however, they are not nearly as conservative as are tests for categories such as chi-squared when applied to interval-level data.

For two groups, we used the Wilcoxon rank sum test, also called the Mann-Whitney U test. For more than two groups, we used the Kruskal-Wallis one-way ANOVA by ranks. For matched or paired data, Wilcoxon arrived on the scene once more, with the Wilcoxon matched pair signed rank test,[10] and the Friedman test was briefly described as an extension to more than two groups.

[10] *By the time you finish the title, you know what the test does.*

EXERCISES

1. Is it really true that "Gentlemen Prefer Blondes?" To test this hypothesis, we assemble 24 Playboy playmate centerfolds from back issues—8 blondes, 8 brunettes, and 8 redheads. To avoid bias from extraneous variables, we use only the top third of each picture. We locate some gentlemen (with great difficulty) and get them to rank order the ladies from highest to lowest preference.

   The data look like this:

Rank		
Blondes	**Brunettes**	**Redheads**
1	4	5
2	7	6
3	9	13
8	10	16
11	15	20
12	18	21
14	19	23
17	22	24

   Proceed to analyze it with the appropriate test.

2. In retaliation, the ladies decide to do their own pin-up analysis to address another age-old question related to the encounters between the sexes (oops—genders). Is it true that bald men are more sexy? However, to improve experimental control over the sloppy study done by the gents, they work out a way to control for all extraneous variables. They go to one of those clinics that claim to make chromedomes into full heads of hair and get a bunch of before-after pictures. They get some ladies to rank the snapshots from most to least sexy and then analyze the ranks of the boys with and without rugs. It looks like this:

	Ranking	
Subject	**Bald**	**Rug**
A	3	1
B	12	15
C	11	6
D	8	4
E	19	13
F	5	2
G	20	7
H	10	9
I	17	14
J	16	18

   Go ahead and analyze this one too.

3. One last kick at the cat. The gentlemen, most of whom are predictably thinning, express displeasure at the results of the ladies' study, and assault it on methodological grounds (naturally). They claim that men who would go and buy rugs are not representative of all bald men. So the ladies proceed to repeat the study, only this time ripping out Playgirl centerfolds (top third again), and getting ranks. Now the data look like:

Bald	Hairy
5	1
6	2
8	3
9	4
12	7
14	10
15	11
18	13
19	16
20	17

   Analyze appropriately.

How to Get the Computer to Do the Work for You

All of the procedures follow the same format, so we won't bother to go into detail for each of them. The statistic you want, in every case, is the default option. They all start off the same way:

- From **Analyze**, choose **Nonparametric Tests**
- Then select the appropriate option for the desired test

Mann-Whitney U (Wilcoxon Rank Sum Test)

- **2 Independent Samples**

Kruskal-Wallis one-way ANOVA

- **K Independent Samples**

Wilcoxon Signed Rank Test

- **2 Related Samples**

Friedman two-way ANOVA

- **K Related Samples**

This chapter reviews several measures of association for use with ranked data. Spearman's rho is perhaps the most frequently used and is derived from the Pearson correlation coefficient; Kendall's tau is another approach. Kendall's W can be used for multiple observations.

CHAPTER THE TWENTY-THIRD

Measures of Association for Ranked Data

SETTING THE SCENE

The midwives in your community are actively encouraging prenatal classes, of which a major component is the Amaze breathing exercise. They are frustrated by the observation that, when push comes to shove (so to speak), all the mothers from the class appear to abandon their lessons and scream "Epi-epi-epidural!" Thus the midwives set out to find real scientific data to show that the Amaze method really does lead to shorter and easier labor. They rank moms in the class on their mastery of Amaze and then measure the duration of labor. Now they dump it all on your desk and ask for an analysis. How do you proceed?

The problem we now face is hopefully a familiar one: establishing a correlation between two sets of data. If we didn't know any differently, we might proceed with a Pearson correlation. However, on reflection, the data obviously are not at all normal, on two counts.

First, the rankings of Amaze proficiency. Only one person is ranked 1, only one ranked 2, and so on, so that data have a *rectangular* distribution. Duration of labor might be better because it is measured in hours and minutes, except for one tiny detail. We have all heard tales of women who delivered in the taxi[1] 47 seconds after they went into labor. Conversely, middle-aged women seem to take perverse delight in regaling expectant first-time mothers with stories of Aunt Maude, who was in labor for 17 days and nights and then delivered triplets, unbeknownst to all, including the doctor. If there ever was a skewed distribution, duration of labor is likely it, and this is exactly the situation for which nonparametric statistical methods were invented. So it makes sense to begin by converting the raw data on duration to a ranking as well. An example of how the data might look after the exercise of ranking for 15 happy (hah!) moms is shown in Table 23–1.

[1] One might speculate on the association between duration of labor and the driving status of husbands, as women never seem to deliver in their family cars.

[2] If you want to verify our calculations (always a good idea), Chapter 13 lists several forms of the Pearson correlation.

We will now spend the next few pages delighting you with a few ways to approach the business of generating measures of association with these ranked data. But before we do, just to keep a perspective on the whole thing, we have proceeded to calculate a Pearson's correlation on these data as an anchor point for what follows. It equals .89; keep that in mind.[2]

SPEARMAN'S RHO

The most common, most ancient, and most straightforward approach to measuring association was developed by Spearman, a contemporary of Pearson and Fisher, many moons ago. Like much in this game, the process really wasn't very profound. He simply took Pearson's formula for the correlation and figured out what would happen if you used ranks instead of raw data. As with many statistical techniques that predated the birth of computers, the major impetus was to simplify the calculations. This value is called *rho*, which is the Greek letter for *r* but looks like a *p*. However, it's often written as r_s (the correlation due to Spearman), which is unaccountably straightforward.

We won't inflict the derivation on you, but we will give you some sense of what is likely to happen. For example, the *total* of all ranks in a set of data must be related only to the number of data points and unrelated to the actual values. It turns out that the total is always $N(N + 1) \div 2$, where N is the number of data points (and therefore the highest rank). So it then follows that the mean rank must be this quantity divided by N, or $(N + 1) \div 2$. Similar simplications emerge by diddling around with the formulae for SDs.

The long and short of it is that Spearman's formula for the rank correlation, based on simply substituting ranks for raw data in Pearson's formula, is:

$$r_s = 1 - \frac{6\sum d_i^2}{N^3 - N} \qquad (23\text{–}1)$$

where d_i is the difference in ranks of a particular subject on the two measures. In the fourth column on Table 23–1 we have taken the liberty of doing this complex calculation for you. The sum of the squared differences looks like:

$$(+1)^2 + (-1)^2 + (0)^2 + (+2)^2 + (0)^2 + (-2)^2 + \ldots + (-3)^2 = 62$$

So the **Spearman rank correlation** is:

$$r_s = 1 - \frac{6 \times 62}{15^3 - 15} = .89 \qquad (23\text{–}2)$$

Because the formula came directly from the one for the Pearson r, it stands to reason that it equals .89, which is what we found earlier. Note that this does not mean that the two correlations always yield identical results. When both are calculated on ranked data (as in this case), they give the same answer. However, if the original data can be analyzed with a Pearson correlation (normality and all that other stuff), then converting these interval-property data to ranks results in the loss of information, and rho is lower than, rather than equaling, r.

We haven't dealt with the issue of tied ranks, which is an inevitable consequence of using real data, something that has not yet constrained us. Our resource books tell us that if the number of ties is small, we can ignore them. If it is large, we must correct for them.[3] The formula involves messy corrections to both numerator and denominator of the Spearman formula using the number of ties in X and Y. We'll let this one pass us by and let the computer worry about it.

Significance of the Spearman Rho

We have already approached the issue of significance testing of a product-moment correlation. Because rho is so intimately related to the Pearson correlation, so is its significance test. It's a t-test, yet again, where the numerator is the value of rho and the denominator is, as in Chapter 13, simply related to the value of rho and the sample size. So:

TABLE 23–1

Ranking of moms on Amaze prowess and labor duration

Mum	Rank of prowess	Rank of duration	Signed difference
A	1	2	+1
B	2	1	−1
C	3	3	0
D	4	6	+2
E	5	5	0
F	6	4	−2
G	7	11	+4
H	8	10	+2
I	9	7	−2
J	10	9	−1
K	11	8	−3
L	12	15	+3
M	13	13	0
N	14	14	0
O	15	12	−3

1 is highest proficiency; and 1 = briefest labor.

$$t = \frac{r_s}{\sqrt{\dfrac{1.0 - r_s^2}{N - 2}}} \qquad (23\text{–}3)$$

In this case, it equals:

$$t = \frac{.89}{\sqrt{\dfrac{1.0 - .89^2}{15 - 2}}} = \frac{.89}{.1265} = 7.04 \qquad (23\text{–}4)$$

with $df = N - 2$, or $(15 - 2) = 13$. This value is, of course, wildly significant, so we can close up shop for the day.

We could stop there, but then Kendall (keep reading) would be left high and dry and would have to make his fame and fortune in motor oil. So, we'll carry on a bit further.

THE POINT-BISERIAL CORRELATION

Just to ensure that you have a well-rounded statistical education, we will briefly mention another quaint historical piece, derived simply for simplicity in calculation. The **point-biserial correlation** was used in the situation where one variable was continuous and the other was dichotomous.[4] For example, does any association exist between gender (two categories) and height (continuous)? It was calculated by starting with the Pearson formula, inserting a *1* and *0* for one variable, and then simplifying the equation. The resulting form is:

$$r_{pbi} = \frac{\bar{X}_p - \bar{X}_q}{s_x}\sqrt{pq} \qquad (23\text{–}5)$$

[3] *What they don't say, of course, is how small is "small," and how large is "large."*

[4] *Why did this coefficient end up in a chapter on ranked data? Because one variable is continuous, we couldn't put it in Chapter 21. Because the other is categorical, we couldn't put it in Chapter 13. So we averaged continuous and categorical and ended up here.*

TABLE 23–2

Ranking of moms on Amaze prowess and labor duration

Mom	Rank on Amaze	Rank on duration	Order							
A	1	2	−1							
B	2	1	+1	+1						
C	3	3	+1	+1	+1					
D	4	6	+1	+1	+1	−1				
E	5	5	+1	+1	+1	−1	−1			
F	6	4	+1	+1	+1	+1	+1	+1		
G	7	8	+1	+1	+1	+1	+1	+1	−1	
H	8	7	—	—	—	—	—	—	—	
		TOTALS	+5	+6	+5	0	+1	+2	−1	**18**

where p is the proportion of individuals with a 1, q is $(1 - p)$, and s_x is the SD of the scores. Because the same result can be obtained simply by stuffing the whole lot—ones, zeros, and all—into the computer and calculating the usual Pearson correlation, there is little cause for elevating this formula to special status.

However, this formula is still applied with great regularity in one situation. In calculating test statistics for multiple choice tests, one measure of the performance of an individual item is the **discrimination**—the extent to which persons who perform well on the rest of the test and get a high score (the continuous variable) pass the item (the dichotomous score), and vice versa. This index is regularly calculated with (or at least expressed as) the point-biserial coefficient.

KENDALL'S TAU

Kendall created two approaches to measuring correlation among ranked data. The first, called **tau** (yet another Greek letter with no meaningful interpretation), generally underestimates the correlation when compared with other measures such as rho.

Calculation of Tau

The calculation involves a bit of bizarre counting, but no fancy stuff. To get the ball rolling, we have copied the data of Table 23–1 into a new table, Table 23–2. However, the new table has two changes. First, all the ranks of prowess are in "natural" order (i.e., from lowest to highest). They started out that way in Table 23–1, so we didn't have to do anything. But if they hadn't, this would be the first step. The second part is that we have copied only the first eight moms; this is for both our sanity and yours, as you shall soon see.

Once we have ordered one of the variables in ascending ranks and placed the ranks of the second variable alongside, the game now shifts entirely to the second variable. We start at each rank of this variable and count the number of occasions in which subsequent ranks occur in natural (i.e., ascending) order (add 1) or reversed (subtract 1) order. So looking at the first rank, 2, it is followed by a 1, which is in the wrong order, and therefore contributes a −1 to the running total. It is then followed by 3, 6, 5, 4, 8, and 7, all of which are greater than 2, so all contribute +1s to the total. We then go to the next rank in the durations, 1, and find (naturally) that all subsequent ranks—3, 6, 5, 4, 8, and 7—are in the right order, so this column contributes six +1s to the total. And so it goes, and eventually we end up with a total of +18. Clearly this method could drive you bananas if you had more than about 10 cases, but then no one ever said that this stuff was easy.

Now then, what's going on? If no association existed between the two ranks, then (1) the second row of numbers would be distributed at random with respect to the first, (2) there would be as many −1s and +1s, and (3) the total would come out to zero. Conversely, if a perfect relationship existed, all the ranks on the second variable would be in ascending order, and because there are $N(N - 1) \div 2$ comparisons, the running total would be $N(N - 1) \div 2$ + 1s; in this case, 28.

This then leads to the final step in the calculation. You take the ratio of the total to $N(N - 1) \div 2$ and call it "tau."

$$\tau = \frac{S}{\frac{N(N-1)}{2}} = \frac{2S}{N(N-1)} \tag{23–6}$$

where S is the sum of the +1s and −1s. In our example, tau is equal to $18 \div 28 = .64$. This is a whopping lot less than the Spearman correlation, but of course we deleted the last seven cases. Never fear, your intrepid authors took a night off to calculate the ruddy thing. The total S is 71, and the maximum is $15(14) \div 2$, so tau is equal to $71 \div [15(14) \div 2] = .68$.[5] This is still substantially lower than the Spearman rho of .89, which is a general problem with tau. However, we will go the last step and see if it is significant anyway.[6]

[5] *We invite you to check our calculations!*

[6] *We will not bother with the corrections for tied ranks, as we see little point in using the ruddy things.*

TABLE 23–3 Ranking of orthopedic surgeons by three judges

Surgeon	Judge: Clairtete	Judge: Klerkopf	Judge: Kromdome	Mean rank	Squared sum
A	1	4	2	2.33	49
B	2	3	3	2.67	64
C	3	1	6	3.33	100
D	4	8	1	4.33	169
E	5	2	4	3.67	121
F	6	9	5	6.57	400
G	7	6	11	8.00	576
H	8	12	9	9.67	841
I	9	5	7	7.00	411
J	10	7	12	9.67	841
K	11	11	8	10.00	900
L	12	10	10	10.67	1,024
SUM					5,526

1 = highest rank on proficiency.

Significance Testing for Tau

As usual, significance testing involves constructing a *t*-test whose numerator is the coefficient itself and whose denominator is the SD of the coefficient. (Actually the numerator is [tau − 0] because we are trying to see if differs from 0.) These SDs are really messy to derive and are the one place where real statisticians (unlike ourselves) earn their bread, so we'll just take the answer on faith:

$$z_\tau = \frac{\tau}{\frac{\sqrt{2(2N+5)}}{3\sqrt{N(N-1)}}} \quad (23\text{–}7)$$

We have one small caveat; it doesn't work for samples of less than 10, and, of course, only a fool would try to calculate τ for samples greater than 10. However, tables are around for looking up these SDs directly, including, of course, the proverbial bible of nonparametric statistics, Siegel and Castellan (1988). Tau has one redeeming quality; you can use it to calculate a partial correlation coefficient (the correlation between X and Y, controlling for the effect of Z). It is calculated in exactly the same manner as any other partial correlation (see Chapter 14). However, as we can never recall seeing it done, this is probably of marginal benefit.

KENDALL'S *W*

Not surprisingly, even Kendall had the good sense to figure that this one would be unlikely to put him in the history books. On the other hand, all the measures to date for ordinal data are like the Pearson correlation, in that they are limited to considering two variables at a time. It would be nice, particularly since having learned the elegance of the intraclass correlation coefficient, if you could use an equivalent statistic for ordinal data.

Calculation of *W*

Remember the Olympic gymnastic championships, where emaciated little nymphettes paraded their incredible prowess in front of the judges while their doting mothers watched entranced from the sidelines? Remember the finale, where a bunch of 9.4 − 9.5 − 9.4s appeared magically across the TV screen? One score, from the home country, was always a bit higher, and one (from the communist or capitalist country, whichever was oppositely inclined politically), was lower.

Wouldn't it be neat if we could do the same thing for surgery? Well, now we can! Welcome to the first annual Orthopedic Olympics. The surgeons are in the basement doing warm-up exercises, and the judges—Dr. Clairtete from McGill University in the country of Quebec, Herr Dr. Prof. Klerkopf from Heidelberg, and Sam Kromdome from Hawvawd University—are in their booths. The first candidate presents herself and, in the flash of an eye, bashes off a double hip replacement, to the wild applause of all.[7]

After all 12 surgeons display their wares, the data look like those in Table 23–3. The fourth column shows the calculated average rank of each surgeon. So, the first sawbone rates (1 + 4 + 2) ÷ 3 = 2.33. Now, what would happen if the judges were in perfect agreement?[8] The first would get a mean rank of (1 + 1 + 1) ÷ 3 = 1, and the last would get a rank of (12 + 12 + 12) ÷ 3 = 12. Conversely, if there were no agreement—a far more likely proposition—the rank of everybody would be about (6.5 + 6.5 + 6.5) ÷ 3 = 6.5. So the extent of agreement is related to the dispersion of individual mean ranks from the average mean rank. This is analogous to the intraclass correlation coefficient, where agreement was

[7] *All except the son and heir, who hoped the old lady would croak.*

[8] *We would all be absolutely incredulous, that's what.*

captured in the variance between subjects. It just remains to express this dispersion as a sum of squares, as usual, and then divide this by some expression of the maximum possible sum of squares. The latter turns out to equal $N(N^2 - 1) \div 12$ (where N is the number of subjects), so Kendall's W (the name of the new coefficient, for no reason we can figure[9]) equals:

$$W = \frac{\sum (\overline{R}_j - \overline{R})^2}{N(N^2 - N)/12} \quad \textbf{(23–8)}$$

where $\overline{R}_j$ and $\overline{R}$ are mean ranks.

As with the formulae for the SD and other tests, this one is easy to understand conceptually but difficult computationally. We have to figure out the mean rank for each person, the overall mean rank, subtract one from the other, and so on, with rounding error introduced at each stage. An easier formula to use is:

$$W = \frac{12\Sigma R_j^2 - 3k^2N\,(N + 1)^2}{k^2N\,(N^2 - 1)} \quad \textbf{(23–9)}$$

where R_j is the summed rank and k is the number of judges. It looks more formidable, but it's actually quite a bit simpler to use.

Backing up, we have calculated all the squared sums of the ranks in the far right column, which total up to 5,526. So W now equals:

$$W = \frac{12\,(5526) - 3\,(3^2)\,12\,(13)^2}{3^2\,(12)\,(12^2 - 1)} = \frac{11{,}566}{15{,}444}$$

$$= 0.748 \quad \textbf{(23–10)}$$

[9] We do have a theory, however. Perhaps Kendall had (a) a lisp or (b) an Oriental mother, so that when he tried to say "R," it came out "AWA," and he just retained the W. Yes, we know it's farfetched.

Significance Testing for *W*

A simple approach for significance testing for W is available. If N is small (less than seven), you look it up in yet another ruddy table. For larger Ns, as it turns out (again for obscure reasons known to only real statisticians), a little jimcrackery on W gives you a chi-squared with $(N - 1)\ df$:

$$\chi^2 = k(N - 1)\ W = 3 \times 11 \times .749 = 24.71$$

So in this case, the chi-squared is ridiculously significant.

SUMMARY

That completes our little tour of agreement measures for ranked data. The most common by far is the Spearman correlation, rho, which is a reasonable alternative to the Pearson correlation. The advantage of Kendall's W is that it can be used for multiple observers and thus is analogous to the intraclass correlation and is useful for agreement studies. As far as tau is concerned, the less said the better.

EXERCISES

1. For the following designs, indicate the appropriate measure of association:
   a. Agreement between two observers on presence/absence of a Babinski sign.
   b. Agreement between two observers on knee reflex, rated as 0, +, ++, +++, ++++.
   c. Association between income of podiatrists and patient satisfaction (measured on a 7-point scale).
   d. Agreement between two observers on religion of patients (Protestant, Catholic, Jewish, Muslim, Other).
   e. Agreement among four observers on rating of medical-student histories and physical exams, using a 25-item checklist (both individual items and overall percent score).
   f. Association between height and blood pressure.
   g. Association between presence/absence of an elevated jugular venous pressure and cardiomegaly (present/absent on X-ray film).
   h. Association between number of siblings and graduation honors.
   i. Association between gender and graduation honors.
2. When any of us seek professional advice, whether from a plumber, mechanic, or statistician, our satisfaction is usually guided by the "three C's"—correct, cheap, and cwick (sorry!).[10]

   Imagine a descriptive study of the association between time from initial contact with the statistician to the delivery of the analysis. To assess stability, each statistician is consulted twice, once with a simple problem and once with a hard one. Time from contact to delivery is measured in minutes, hours, days, or weeks, as appropriate. The data look like this:

Statistician	Short problem	Long problem
A	32 min	4 days
B	3.7 days	6 days
C	14 min	8.6 hr
D	4.2 days	3.7 months
E	18 min	7.5 days
F	38 sec	2.2 days
G	8.2 hr	1.7 wk
H	3.3 hr	3.9 days

   Analyze the data with the appropriate measure of association.
3. Evaluation of medical residents is notoriously unreliable. One way out of the swamp might be to get evaluators to rank individual residents, rather than rate them.

[10] Except for docs, because (a) you find out if they were wrong only after you die, (b) no doc is cheap, but then no one (hardly) pays for a doc out of his or her pocket anyway, and (c) you want the doc to be quick with everyone ahead of you and really slow with you.

Here are the results from a study involving ranking of 10 residents by (a) peers, (b) nurses, and (c) staff.

Resident	Peers	Nurses	Staff
A	1	4	3
B	2	3	5
C	3	1	6
D	4	8	2
E	5	2	1
F	6	5	4
G	7	7	10
H	8	6	7
I	9	10	9
J	10	9	8

Analyze with the appropriate measure of association.

How to Get the Computer to Do the Work for You

Spearman's rho

- From **Analyze**, choose **Correlate → Bivariate**
- Click the variables you want from the list on the left, and click the arrow to move them into the box labeled **Variables**
- In the section named **Correlation Coefficients**, check the box to the left of **Spearman**
- Click **OK**

Kendall's tau

- From **Analyze**, choose **Correlate → Bivariate**
- Click the variables you want from the list on the left, and click the arrow to move them into the box labeled **Variables**
- In the section named **Correlation Coefficients**, check the box to the left of **Kendall's tau-b**
- Click **OK**

Point-Biserial Correlation

- Use Pearson's r

Kendall's *W*

- From **Analyze**, choose **Nonparametric Tests → K Related Samples**
- Click on the box for **Kendall's W**

Survival, or life-table, analysis allows us to look at how long people are in one state (e.g., alive), followed by a discrete outcome (e.g., death). It can handle situations in which the people enter the trial at different times and are followed for varying periods; it also allows us to compare two or more groups.

CHAPTER THE TWENTY-FOURTH

Life-Table (Survival) Analysis

SETTING THE SCENE

An upstart pharmaceutical company has come up with a new wonder drug, called Hairgro. *Just one injection will give even a bald politician that blow-dried, Kennedy-look hair, good for at least 10 percentage points in the next election, irrespective of political affiliation or strongly held beliefs (if any). Unfortunately, it has one serious side effect: it also causes the politicians to tell the truth, thus shortening their political lives. The company must find out how severe this effect is and wants to do a study. Needless to say, the number of willing participants is severely limited, so the company has to enroll these willing candidates over the course of time. Some of the candidates retire while in office, and the company, for financial reasons, has to stop collecting data after 10 years so they can begin marketing the drug. How can they maximize the use of the data they have collected?*

WHEN WE USE SURVIVAL ANALYSIS

Under ideal circumstances, a study would enroll all of the subjects simultaneously and follow them for either a fixed period or until they all reach some end point, such as recovery or death. However, the situation we just described is not unusual. Studies that require a large number of subjects or that investigate relatively rare conditions must enter subjects over a period of several months or even years. The Multiple Risk Factor Intervention Trial (MRFIT), for instance, involved nearly 13,000 men recruited over a 27-month period (MRFIT, 1977). When our study finally ends (as all trials must, at some time), the subjects will have been followed for varying lengths of time, during which several outcomes could have occurred:

1. Some subjects reach the designated end point. In this example, this means that the politician is defeated (i.e., dies, politically) and is forced to take a cushy job chairing a commission that oversees the saltwater ports in Oklahoma or Alberta. In other types of trials, such as chemotherapy for cancer, the end point may be truly tragic—death or the reappearance of a malignancy.
2. Some subjects drop out of sight: they move without leaving a forwarding address; refuse to participate in any more follow-up visits; or, in our case, retire before the next election.
3. The study ends before all of the subjects reach the end point. When the company shuts down the trial after 10 years, some politicians who started the trial may still be in office. They may be defeated the next day or last for another 20 years, but we won't know because data collection has ended.

Figure 24–1 shows how we can illustrate these different outcomes, indicating what happened to the first 10 politicians in the study. Subjects A, C, D, and F were defeated during the course of the trial; they're labeled *D* for Dead.[1] Subjects B, G, and I retired while undefeated and so were lost to follow-up study (hence the label *L*) at various times after they started the drug. The other subjects, E, H, and J, (labeled *C*) were still in office at the time the trial ended.[2] These last three data points are called "right censored."[3] To be more quantitative about the data, Table 24–1 shows how long each person was in the study and what the outcome was for each pol.

[1]Because defeat is tantamount to death for a politician, we'll call this outcome Death. This will also simplify the discussion because death is the outcome of interest in most survival analyses.

[2]It's called "in office," but they were probably out of their offices and on a junket to examine the garbage disposal facilities in Bali or Paris (coincidentally, in the middle of winter).

[3]This means the data toward the right of the graph were cut off because the study ended before the subjects reached the designated end point; it doesn't mean being silenced by conservative moralists.

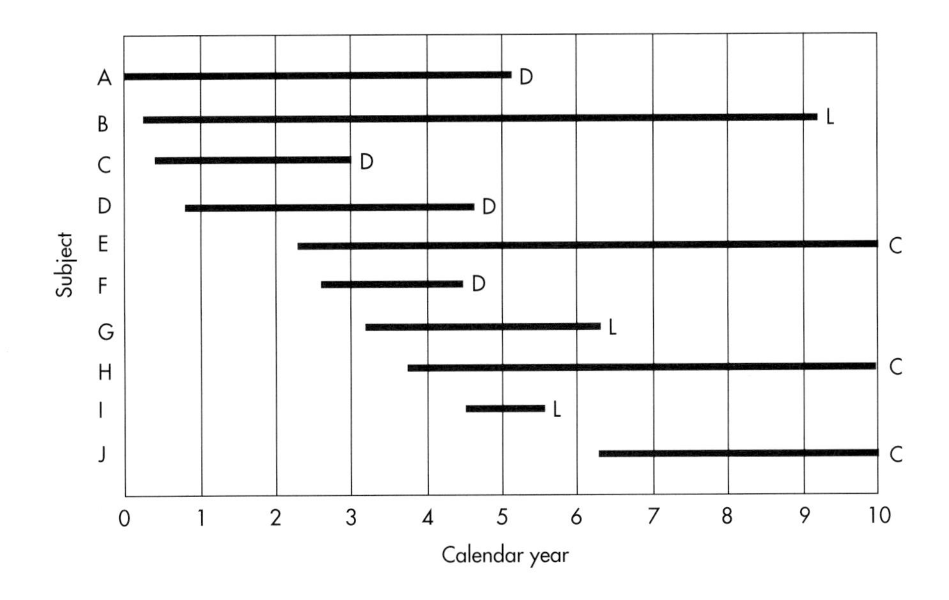

FIGURE 24–1
Entry and withdrawal of subjects in a 10-year study.

SUMMARIZING THE DATA

So, what conclusions can we draw from these data regarding the survival time of pols following a shot of *Hairgro* (and being forced to tell the truth)? What we need is a method of summarizing the results that uses most, if not all, of the data and isn't overly biased by the fact that some of the data are censored. What we'll do is approach the "right" answer in stages. The first two ways of summarizing the data (Mean Survival and Survival Rate) are intuitively appealing but have some problems associated with them, which the third method (using Person-Years) neatly sidesteps.

TABLE 24–1
Outcomes of the first 10 subjects

Subject	Length of time in trial (months)	Outcome*
A	61	Died
B	111	Lost
C	29	Died
D	46	Died
E	92	Censored
F	22	Died
G	37	Lost
H	76	Censored
I	14	Lost
J	45	Censored

*In this and subsequent tables in this chapter, political defeat is referred to as "death."

Mean Survival

One tactic would be to look at only those subjects for whom we have complete data, in that we know their outcome exactly. These subjects would be only those who died—subjects A, C, D, and F.

$$\text{Mean Survival} = \frac{\text{Time to Outcome}}{\text{Number of Subjects Who Reached the Outcome}}$$

Their **mean survival** in office was 39.5 months after taking the drug. The major problem with this approach is that we've thrown out 60% of the subjects. Even more seriously, we have no guarantee that the six people we eliminated are similar to the four whose data we analyzed; indeed, it is most likely that they are *not* the same. Those who dropped out may have been the ones who would have been defeated in the next election in any case. Those who were censored were, by definition, still undefeated and in office. Similarly, those who retired had not been defeated, although some of them might have been, had they decided to run again. Ignoring the data from these subjects would be akin to studying survival rates following radiation therapy but not including those who were still alive when the study ended; any conclusion we drew would be biased by not including these subjects. The extent of the bias is unknown, but it would most likely operate in the direction of underestimating the effect of radiation therapy on the survival rate.

We could include the censored subjects by using the length of time they were in the study. The effect of this, though, would again be to underestimate the survival rate because these people are likely to continue in office for varying lengths of time beyond the study period.

Survival Rate

Another way to summarize the data is to see what proportion of politicians continued in office (i.e., "survived," in the terminology of survival analysis). The major problem is "survived" *as of when*? The survival rate after 1 month would be pretty close to 100%; after 50 years, it would probably be 0%.[4] One way around this is to use a commonly designated follow-up time. Many cancer trials, for instance, look at survival after 5 years. Any subjects who were still around at 5 years would be called "survivors" for the 5-year **survival rate**, no matter what subsequently happened to them.

[4] *Actually, given how long a lot of these codgers hang around, we should probably have used 75 years.*

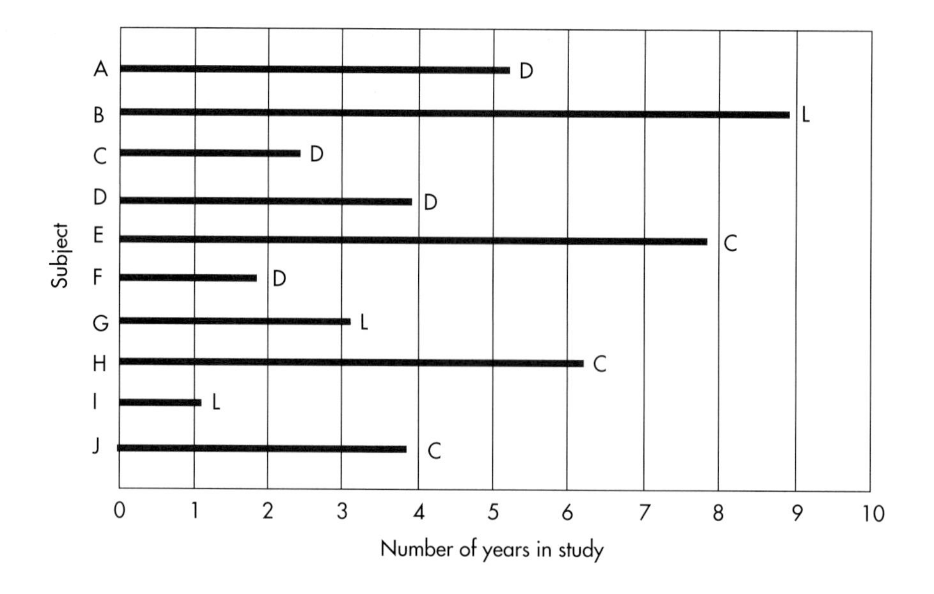

FIGURE 24–2
Figure 24–1 redrawn so all subjects have a common starting date.

$$\text{Survival Rate} = \frac{\text{Number of Subjects Surviving at Time } (t)}{\text{Total Number of Subjects}}$$

This strategy reduces the impact of "the censored ones," although it doesn't eliminate it. Those subjects who were censored *after* 5 years don't bother us any more because their data have already been used to figure out the 5-year survival rate. Now the only people who give us any trouble are those who had been followed for less than 5 years when the study ended. However, the major disadvantage is still that a lot of data aren't being used.

Using Person-Years

In (unsuccessfully) trying to use the mean duration of survival or the survival rate to summarize the data, it was necessary to count *people*. This led us to the problem of choosing which people to count or not count when the data are censored. Because we divide the length of survival or the number of survivors by whichever number we finally decide to include, this has been referred to as the "denominator problem."

A different factor to use for the denominator would be (rather than individual people) the *length of time*, in years, each person was in the study; that is, the total number of **person-years** of follow-up study. But it doesn't have to be measured in years; we can use any time interval that best fits our data. In the *Hairgro* example, we'd use person-*months*. If we were looking at how quickly a new scrotonin reuptake inhibitor drug reduced depressive symptomatology, we could even talk in terms of person-*days*. The major advantage of this approach is that it uses the data even from people who are lost for one reason or another. If we added up all the numbers in the middle column of Table 24–1, we would find a total of 503 person months, during which time 4 pols died, politically speaking. This means that the political mortality rate is (4 ÷ 503) = 0.0080 deaths per month. The major problem with this approach is its assumption that the risk of death is constant from one year to the next. We know that, in this case at least, it isn't. The longer these rascals have been in office, the harder it is to throw them out.

THE LIFE-TABLE (SURVIVAL) TECHNIQUE

What we *can* do is figure out how many people survive for at least 1 year, for at least 2 years, and so on. We're not limited to having equal intervals; they could be days for the first week, then weeks for the next month, and then months thereafter. This approach, called either the **survival table** or, more commonly, the **life-table** technique, has all the advantages of the person-years method (i.e., making maximum use of the data from all subjects), without its disadvantage of assuming a constant risk over time. Two ways to go about calculating a life-table are the **actuarial approach** and the **Kaplan-Meier approach**. They're fairly similar in most details, so we'll begin with the more traditional, actuarial way.

The Actuarial Approach

The first step for both approaches involves redrawing the graph so that all of the people appear to start at the same time. Figure 24–2 shows the same data as Figure 24–1; however, instead of the *X*-axis being *Calendar Year*, it is now *Number of Years in the Study*. The lines are all the same length as in Figure 24–1; they've just been shifted to the left so that they all being at Time 0.

From this figure, we can start working out a table showing the number of people **at risk** of death each year, and the probability of their still being around (surviving) at the end of each year. To begin with, let's summarize the data in Figure 24–2, listing for each year of the study (1) the number of subjects still up and kicking (those at risk), (2) the number who "died," and (3) the number lost to follow-up. We've done this in Table 24–2.

Getting from the graph to the table is quite simple. No lines terminate during the interval 0 to 1 year, so we know no one died and no one was lost. Between years 1 and 2, one line ends with a *D* and one with an *L*, so we enter one Death and one Loss in the table. This means that two fewer subjects began the next time interval, so we subtract 2 from 10, leaving 8 at risk. We continue this until either the study ends or we run out of subjects. Note that we treat people who dropped out of the study and people who were "censored" in the same way; we call them both "lost." The reason is that, from the viewpoint of the researcher, they *are* similar—both groups were still alive (or undefeated) at the time we stopped gathering any more information about them.

The next step is to figure out the probability of dying each year. This would be relatively simple to do if all we had to deal with were subjects who were still in office (alive) at the start of each study year and the number who died. In that case, the probability of death each year is simply:

$$\text{Pr(Death)} = \frac{\text{Number Who Died}}{\text{Number at Risk of Death}} \quad \textbf{(24–1)}$$

To simplify writing our equation, let's use the symbols:

q_i = Probability of death in Year i
$p_i = 1 - q_i$ (i.e., the probability of survival in Year i)
D_i = Number of persons who died in Year i
R_i = Number of subjects at risk starting Year i

So, we can rewrite Equation 24–1 as:

$$q_i = \frac{D_i}{R_i} \quad \textbf{(24–2)}$$

The technical term for this expression is the **hazard.**

The *hazard* is the probability of occurrence of the outcome for people who began in that interval.

Some texts differentiate between the *hazard* and the *hazard function*, which is formally defined as:

$$\lambda(t_{mi}) = \frac{2\, q_i}{h_i(1 + p_i)} \quad \textbf{(24–3)}$$

where h_i is the width of the interval (in this case, 1). The fatality rate is at the middle of the interval (hence the subscript, t_{mi}); q_i is the risk at the *end* of the interval.

TABLE 24–2

Number of subjects at risk, who died, and were lost each year

Number of years in study	Number of subjects at risk	Number of subjects who died	Number of subjects lost
0–1	10	0	0
1–2	10	1	1
2–3	8	1	0
3–4	7	1	2
4–5	4	0	0
5–6	4	1	0
6–7	3	0	1
7–8	2	0	1
8–9	1	0	1
9–10	0	0	0

But, back to our machinations. What do we do with the people who were lost *during* the year? If we look at the data only at discrete intervals, we don't know exactly when they were lost and thus aren't sure for what length of time they were at risk. Do we say that they were at risk for the whole year, or should we drop them entirely at the beginning of the year? For example, Subject G in Figure 24–2 dropped out of sight some time between the start of Year 3 and the start of Year 4. We can either attribute a full year of risk to this subject (Year 3 to Year 4), or limit his time at risk to the end of Year 3. To say he was at risk the whole year assumes that he was in office and undefeated for all 12 months. In reality, he may have stood for election in March and have been defeated. In this case, we would have "credited him" with nine extra months of political life. This therefore underestimates the death rate. On the other hand, to drop him entirely from that year throws away valid data; we know that he at least made it to the beginning of Year 3, if not to the end.

What we do is make a compromise. If we don't know exactly when a subject dropped out, but we know it was sometime within the interval, we count him as half a person-year (or whatever interval we're using). That is, we say that half a person got through the interval, or, putting it slightly differently, the person got through half the interval. In a large study, this compromise is based on a fairly safe assumption. If deaths occur randomly throughout the year, then about half the people will die during the first 6 months, and the other half during the second part of the year. On average, then, giving each person credit for half the year balances this out. So, if we abbreviate the number of people lost each year (i.e., truly lost plus censored) as L_i, we can rewrite Equation 24–1 as:

$$\text{Pr(death)} = \frac{\text{Number Who Died in Year } i}{\text{Number at Risk at Start of Year } i - \dfrac{\text{Lost or Censored}}{2}} \quad \textbf{(24–4)}$$

TABLE 24–3

Data for all subjects in the *Hairgro* trial

Number of years in study	Number of subjects at risk	Number of subjects who died	Number of subjects lost
0–1	100	5	2
1–2	93	4	4
2–3	85	5	1
3–4	79	3	3
4–5	73	7	2
5–6	64	5	4
6–7	55	7	3
7–8	45	10	5
8–9	30	8	2
9–10	20	7	5

TABLE 24–4

Life-table based on data in Table 24–3

Number of years in study	Probability of death (q_i)	Probability of surviving (p_i)	Cumulative probability of surviving (P_i)
0–1	.0505	.9495	.9495
1–2	.0440	.9560	.9078
2–3	.0592	.9408	.8541
3–4	.0387	.9613	.8210
4–5	.0972	.9028	.7412
5–6	.0806	.9194	.6814
6–7	.1308	.8692	.5922
7–8	.2353	.7647	.4529
8–9	.2759	.7241	.3280
9–10	.4000	.6000	.1968

or, in statistical shorthand:

The Hazard

$$q_i = \frac{D_i}{R_i - \frac{L_i}{2}} \tag{24–5}$$

To give ourselves more data to play with, let's assume that these 10 subjects were taken from a larger study involving 100 politicians. The (fictitious) data for the full study are in Table 24–3. Now, using Equation 24–5 with the data in Table 24–3, we can make a new table (Table 24–4), giving (1) the probability of death occurring during each interval, (2) the converse of this, the probability of *surviving* the interval, and (3) the *cumulative probability* of survival (also referred to as the **survival function**). Let's walk through a few lines of Table 24–4 and see how it's done.

The first line of Table 24–3 (0–1 years in study) tells us there were five deaths and two losses. Using Equation 24–5, then, we have:

$$q_{(1)} = \frac{5}{100 - \frac{2}{2}} = \frac{5}{99} = .0505 \tag{24–6}$$

and the probability of death during Year 1 is 5%. Therefore the probability of surviving that year, p_i, which is $(1 - q_1)$, is 0.9495. The second year began with 93 subjects at risk; 4 died during that year, and 4 either retired or were censored. Again we use Equation 24–5:

$$q_{(2)} = \frac{4}{93 - \frac{4}{2}} = \frac{4}{91} = .0440 \tag{24–7}$$

The probability of surviving Year 2 is (1 − .0440) = .9560. The **cumulative probability** of surviving Year 2 (P_2) is the probability of surviving Year 1 (P_1, which in this case is .9495) times the probability of surviving Year 2 (p_2, or .9560), or .9078. If you remember the discussion on probability (Chapter 5), you'll recognize this as an example of a **conditional probability**. The probability of surviving from the beginning of the study until the end of Year 2 is the probability of surviving Year 2, *conditional* on having survived Year 1. The cumulative probability at the end of Year 3 is the probability of surviving Year 3 times the cumulative probability of surviving Year 2, and so on.

What is the difference between p_2 and P_2? The first term tells us that the probability of making it through Year 2 is 95.60% *for those subjects who were around at the beginning of the year*. However, not all people made it to the start of the year; five were defeated during the previous interval and two were lost. Hence the *cumulative* probability, P_2, gives us the probability of surviving the second year *for all subjects who started the study*, whereas p_2 is the probability of surviving the second year *only for those subjects who started Year 2*.

Now we continue to fill in the table for the rest of the intervals. If some of the intervals have no deaths, it's not necessary to calculate q_i, p_i, or P_i. By definition, q_i will be 0.0, p_i will be 1.0, and P_i will be unchanged from the previous interval. Once we've completed the table, we can plot the data in the P_i column (the *survival function*), which we have done in Figure 24–3. This is called, for obvious reasons, a **survival curve**.

The Kaplan-Meier Approach to Survival Analysis

The Kaplan-Meier approach (Kaplan and Meier, 1958) is similar to the actuarial one, with four exceptions. First, rather than placing death within some arbitrary interval, the *exact* time of death is used in the calculation. Needless to say, this presupposes that we know the exact time. If all we have is the fact that the patient died after the 2-year follow-up visit but before the 3-year visit, we're limited to using the actuarial approach.[5] Second, instead of calculating the survival function at fixed times (i.e., every month or year of the study), it's done only when an outcome occurs. This means that some of the data points may be close together in time,

[5] *Patients are notorious for not letting investigators know when they die. This is one of the hazards of clinical research and explains why investigators such as B.F. Skinner preferred using pigeons.*

whereas others can be spread far apart. This also leads to the third difference; the survival curve derived by the actuarial method changes only at the end of an interval, whereas that derived from the Kaplan-Meier method changes whenever an outcome has occurred. What this means is that, with the actuarial approach, equal steps occur along the *time* axis (X). But with the Kaplan-Meier technique, the steps are equal along the *probability* (Y) axis. You can always tell what type of graph you're looking at—if the steps along the X-axis are equal, it's an actuarial graph; if the steps aren't of equal length, then it's a Kaplan-Meier type.

Last, subjects who are lost to follow-up because of retirement or censoring are considered to be at risk up to the time they drop out. This means that if they withdraw at a time between two events (i.e., deaths of others), their data are used in the calculation of the survival rate for the first event but not for the second. If we go back to Figure 24–2, one event occurred when Subject C was defeated; the next when Subject D went down in flames. Between these two times, Subject G dropped out of sight. So, Subject G's data will be used when we figure out the survival rate at the time of C's death, but not D's.

To show how this is done, let's go back and use the data for the 10 subjects in Table 24–1. The first step is to rank order the length of time in the trial, flag which ones reflect the outcome of interest (defeat, in this case), and mark those caused by withdrawal or censoring. We've done this by putting an asterisk after the datum for subjects who were lost to follow-up study because they retired or were censored by the termination of the study.

14* 22 29 37* 45* 46 61 76* 92* 111*

Our life table (Table 24–5) would thus have only four rows; one for each of the four politicians who bit the dust. As a small point, notice that we used the subscript i in each column of Table 24–4. In Table 24–5 we've used t to indicate that we're measuring an exact *time*, rather than an *interval*.

One person was lost before the first person died, so the number at risk at 22 months is only 9. At 46 months, 2 people had died and 3 were lost, so the number at risk is 5, and so on. Because we know the exact time when people were lost to the trial, we don't have to use the fancy correction in Equation 24–5 to approximate when they dropped out of sight. We can use Equation 24–2 to figure out the Death Rate, q_i, as we did in Table 24–5.

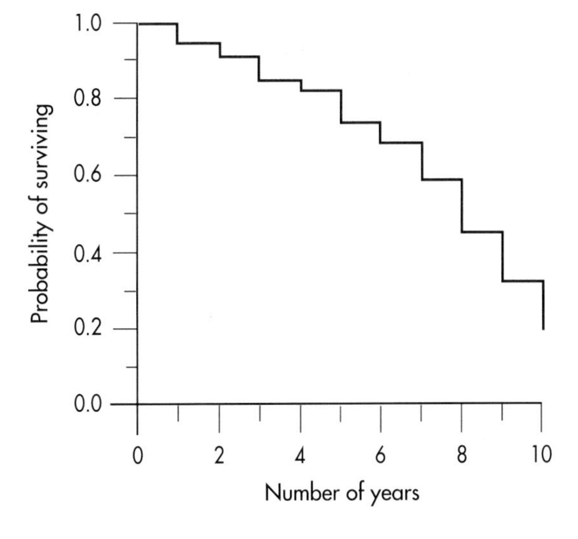

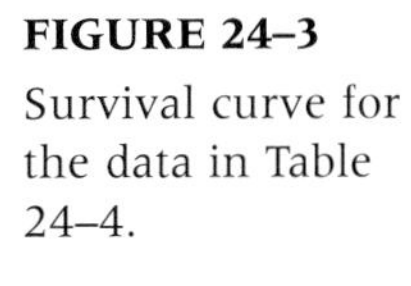

FIGURE 24–3
Survival curve for the data in Table 24–4.

So, which technique do we use, the actuarial or the Kaplan-Meier? When you have fewer than about 50 subjects in the group, the Kaplan-Meier approach is likely more efficient, from a statistical point of view, because you're using exact times rather than approximations for the outcomes. The downside of Kaplan-Meier is that withdrawals occurring between outcomes are ignored; this is more of a problem when $N > 50$. However, in most cases the two approaches lead to fairly similar results, so go with whichever one is on your computer.[6]

[6]*How's that for pragmatism?*

The Standard Error

It's possible to calculate an SE for the survival function, just as we can for any other parameter. (Just to remind you, the survival function consists of the data in the P_i column of Table 24–4, which are plotted as a curve in Figure 24–3.) However, we're limited to estimating it at a specific time, rather than for the function as a whole; that is, there are as many SEs as there are intervals (with the actuarial method) or times (with the Kaplan-Meier approach). There are also several formulae, all of which are approximations of the SE and some of which are quite complicated. A simple approximation, which gives comparable results to the more complex ones, was proposed by Peto et al (1977).

TABLE 24–5
Kaplan-Meier life-table analysis of the data in Table 24–1

Time (months) (t)	Number at risk (R_t)	Number of deaths (D_t)	Death rate (q_t)	Survival rate (p_t)	Cumulative survival rate (P_t)
22	9	1	.1111	.8889	.8889
29	8	1	.1250	.8750	.7778
46	5	1	.2000	.8000	.6222
61	4	1	.2500	.7500	.4667

[7]*Most physicians spend up to 8 years in medical school and residency to be able to recognize this state—it's marked by the patient not paying his or her bill.*

The SE of the Survival Function

$$SE(P_i) = P_i \sqrt{\frac{1 - P_i}{R_i}} \quad \textbf{(24–8)}$$

The equation is exactly the same for the Kaplan-Meier approach; just use terms with the subscript t rather than i. Let's go back to Tables 24–3 and 24–4 to figure out the numbers. For Year 0 − 1, $P_i = .9495$ and $R_1 = 100$, so:

$$SE(P_1) = 0.9495 \sqrt{\frac{1 - 0.9495}{100}} = 0.0213 \quad \textbf{(24–9)}$$

For Year 9 − 10, $P_{10} = 0.1968$ and $R_{10} = 20$:

$$SE(P_{10}) = 0.1968 \sqrt{\frac{1 - 0.1968}{20}} = 0.0394 \quad \textbf{(24–10)}$$

$SE(P_{10})$ is larger than $SE(P_1)$ because the sample size is smaller. In general, then, as the intervals or times increase, so do the SEs, because the estimates of the survival function are based on fewer and fewer subjects.

Assumptions of Survival Analysis

During our discussion so far, we've made several assumptions; now let's make them explicit.

1. *An identifiable starting point.* In this example, the starting point was easily identifiable: when the politicians got their injection of *Hairgro*. If the study looks at survival following some intervention under the experimenter's control, there's usually no problem in identifying the start for each subject. However, if we want to use this technique to look at the natural history of some disorder, such as how long a person is laid up with low-back pain, we may have a problem of specifying when the problem actually began. Is baseline when the person first came to the attention of the physician; when he or she first felt any pain; or when he or she did something presumably injurious to the back? There are difficulties with each of these. For instance, some people run to their family docs at the first twinge of a gluteus maximus, whereas others avoid them at all costs. The other proposed starting points rely on the patients' recall of events, which we know is notoriously inaccurate. The important point is that, whatever starting point is chosen, it must be applied uniformly and reproducibly for all patients.
2. *The end point.* Survival analysis requires a dichotomous and well-defined outcome. Again, this usually isn't a problem if the end point is death.[7] However, we have problems similar to those in identifying a starting point if the outcome isn't as "hard" as death (e.g., the reemergence of symptoms or the reappearance of a cancerous growth). If we rely on a physician's report or the patient's recall, we face the prospect that a multitude of other factors affect these, many of which have nothing to do with the disorder. The more we have to rely on recall or reporting, the more error we introduce into our identification of the end point and hence into our measurement of survival time.
   Another problem occurs if the end point can occur numerous times for the same subject. Hospitalization is one example. In this case, the usual rule is to take the *first* occurrence of the outcome. Finally, deciding which events to count as reflecting the outcome is yet one more problem. This is a thorny issue that isn't always as easy to resolve as it first appears. If we're studying the effectiveness of a combination of chemotherapy and radiation therapy for cancer, with the end point being a reappearance of a tumor, what do we do with patients who commit suicide? They could be counted simply as withdrawals because their deaths were not caused by the cancer; or were they? If the patients believed that they were again becoming symptomatic and took their own lives rather than face the prospect of a lingering death, then they should actually be considered treatment failures and included in the numerator. This issue is dealt with in more depth by Sackett and Gent (1979).
3. *Loss to follow-up study should not be related to the outcome.* We've been assuming that the reason people were lost to follow-up study is because they dropped out of the study and that this had nothing to do with the outcome. If the reasons *are* related, then our estimation of the survival function will be seriously biased, in that we'd underestimate the death rate and overestimate the survival rate. In our example, if politicians retire because they fear that their uncontrollable urge to tell the truth jeopardizes their chances for reelection, they likely would have had a greater probability of defeat than did those remaining in the study. However, because they dropped out before we could determine what actually happened to them, they never appear in the numerator of the equations; they're only subtracted from the denominator.[8] In medical or surgical trials, if the patient is "lost" because he or she dies of the disease, unknown to the investigators, the effect will be the same. This isn't a problem with people whose data are censored —only with those who withdraw or drop out of sight.

[8]*How would you like this as your epitaph: "He was only deducted from the denominator." On second thought, compared with what we could say about politicians, this may be a blessing.*

TABLE 24–6 Data for both groups in the *Hairgro* trial

Number of years in study	Experimental group			Control group		
	At risk (R_1)	Died (D_1)	Lost (L_1)	At risk (R_2)	Died (D_2)	Lost (L_2)
0–1	100	5	2	250	0	2
1–2	93	4	4	248	3	5
2–3	85	5	1	240	5	4
3–4	79	3	3	231	1	5
4–5	73	7	2	225	10	4
5–6	64	5	4	211	8	6
6–7	55	7	3	197	9	5
7–8	45	10	5	183	12	9
8–9	30	8	2	162	15	5
9–10	20	7	5	142	13	8

4. *There is no secular trend.* When we construct the life table we start everyone at a common time, t_0. In studies that recruit and follow patients for extended periods, there could be up to a 5-year span between the time the first subject actually enters and leaves the trial and when the last one does. We assume that nothing has happened over this interval that would affect who gets into the trial, what is done to them, and what factors influence the outcome. If changes have occurred over this time (referred to as **secular changes**[9] or **trends**), then the subjects recruited at the end may differ systematically, as may their outcomes, from those who got in early. This may result from changes in diagnostic practices (e.g., the introduction of a more sensitive test), different treatment regimens, or even a new research assistant who codes things differently. Therefore we wouldn't be able to assume that the group was homogeneous and thus could be combined in the manner in which we consolidated them.

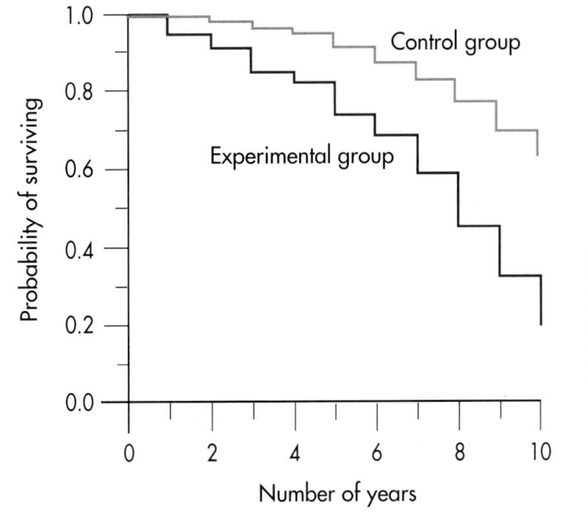

FIGURE 24–4 Survival curves for both groups in the *Hairgro* study, from Table 24–6.

COMPARING TWO (OR MORE) GROUPS

Although the survival curve shown in Figure 24–3 tells us what happened to politicians who were unfortunate enough to have to tell the truth, there's still an important, unanswered question: How do they compare with pols who weren't so burdened? Were they in office for a shorter period, or are voters so cynical that they don't listen anyway, and the pols' terms in office were totally unaffected? To answer questions such as these, we naturally need at least two groups, so we'll compare these "experimental" subjects to 250 "control" politicians who didn't use *Hairgro*. The data are presented in Table 24–6. The first four columns are the same as in Table 24–3, and the last three columns give the data for the new subjects.

The first thing we should do is draw the survival curves for the two groups on the same graph so we can get a picture of what (if anything) is going on (Figure 24–4). This shows us that truth-telling (or Kennedy-like hair) is fatal for politicians. For those who received a shot of *Hairgro*, the survival curve dropped at a faster rate than did that of the control group. But is the difference statistically significant?

The z-Test

One approach to answering this question would be to compare the two curves at a specific point. To do this, using our old standby (the z-test), we have to assume that the cumulative survival rates of the two curves are normally distributed. So:

The z-test

$$z = \frac{P_{i_1} - P_{i_2}}{\sqrt{[SE(P_{i_1})]^2 + [SE(P_{i_2})]^2}} \qquad \textbf{(24–11)}$$

where P_{i1} and P_{i2} are the values of P (the cumulative probability of surviving) for groups 1 and 2 at some arbitrarily chosen interval i (or time t, if we used the Kaplan-Meier approach), and SE are the standard errors at those times, calculated using Equation 24–8.

[9] *We presume as opposed to "ecclesiastical changes," which affect only members of the clergy.*

TABLE 24–7

Calculating a log-rank test on the data in Table 24–6

Number of years in study	At risk Exper (R_{i_1})	At risk Cont (R_{i_2})	At risk Total (R_i)	Dead Exper (D_{i_1})	Dead Cont (D_{i_2})	Dead Total (D_i)	Expected Exper (E_{i_1})	Expected Cont (E_{i_2})
0–1	100	250	350	5	0	5	1.43	3.57
1–2	93	248	341	4	3	7	1.91	5.09
2–3	85	240	325	5	5	10	2.62	7.38
3–4	79	231	310	3	1	4	1.02	2.97
4–5	73	225	298	7	10	17	4.16	12.84
5–6	64	211	275	5	8	13	3.03	9.97
6–7	55	197	252	7	9	16	3.49	12.51
7–8	45	183	228	10	12	22	4.34	17.66
8–9	30	162	192	8	15	23	3.59	19.41
9–10	20	142	162	7	13	20	2.47	17.53
TOTALS				$O_1 = 61$	$O_2 = 76$		$E_1 = 28.06$	$E_2 = 108.93$

This method is quite easy to calculate and is very useful if we are interested in differences in survival rates at one specific time, such as 5-year survival in cancer. An added advantage is that it is simple to determine the **Relative Risk (RR)** at this point. The RR is the ratio of the probability of having some outcome occur among subjects in Group 1 as opposed to it occurring among those in Group 2. In this example, it would be the risk of defeat for people in the *Hairgro* group, relative to the controls. The formula for determining the RR is:

The Relative Risk at Interval i

$$\mathrm{RR}_i = \frac{1 - P_{i_1}}{1 - P_{i_2}} \tag{24–12}$$

The z-test can also be applied to test the significance of the RR.

The Mantel-Cox Log-Rank Test

However,[10] this approach has two problems. The first involves intellectual honesty: you should pick your comparison time *before* you look at the data, ideally before you even start the trial. Otherwise, there is a great temptation to choose the time that maximizes the differences between the groups. The second problem is more substantive; we've ignored most of the data and focused on only one point. A better approach would be to use all of the data. This is done by using the **Mantel-Cox log-rank** (or logrank) **test**, which is a modification of the Mantel-Haenszel chi-squared test we ran into earlier. Although it is a nonparametric test, it is more powerful than the parametric z-test because it makes use of more of the data.

As with most chi-squared tests, the log-rank test compares the *observed* number of events with the number *expected*, under the assumption that the null hypothesis of no group differences is true. That is, if no differences existed between the groups, then at any interval (or time), the total number of events should be divided between the groups roughly in proportion to the number of subjects at risk. For example, if Group A and Group B have the same number of subjects, then each group should have about the same number of events. On the other hand, if Group A is twice as large as B, then it should experience two times the number of outcomes.

If we go back to Table 24–6, we see that there were 350 people at risk during the first interval; 100 in the experimental group and 250 in the control group. Because 28.6% of the subjects were in the *Hairgro* condition, and there were a total of 5 deaths during this interval, we would expect that 5 × .286 = 1.43 deaths would have occurred in this group, and 3.57 among the controls.[11] The shortcut formula for calculating the expected frequency for Group k(k = 1 or 2) at interval i is:

$$\mathrm{E}_{i_k} = D_i \times \frac{R_{i_k}}{R_{i_1} + R_{i_2}} \tag{24–13}$$

where D_i is the total number of deaths.

Using this in the example we just worked out:

$$E_{i_1} = 5 \times \frac{100}{100 + 250} = 1.43$$

$$E_{i_2} = 5 \times \frac{250}{100 + 250} = 3.57 \tag{24–14}$$

Doing this for each interval, we get a new table listing the observed and expected frequencies at each interval, as in Table 24–7.[12] As a check on our (or the computer's) math, the total of the observed deaths (61 + 76) should equal the sum of the expected ones (28.06 + 108.93), within rounding error. The last step, then, is to figure out how much our *observed* event rate differs from the *expected* rate. To do this, we use (finally) the **Mantel-Cox chi-squared**:

The Mantel-Cox Chi-Squared

$$\chi^2 = \frac{(O_1 - E_1)^2}{E_1} + \frac{(O_2 - E_2)^2}{E_2} \tag{24–15}$$

[10] *There's always a "however" when anything is simple.*

[11] *We'll not deal with the existential question of how there can be a fraction of a death.*

[12] *In reality, we let the computer do this for us. After all, that's why they were placed on this earth.*

TABLE 24–8

Political survival time of the two groups stratified by political party

Party	Experimental: Observed deaths	Experimental: Expected deaths	Control: Observed deaths	Control: Expected deaths
Old Deadbeats	28	12.01	34	49.99
New Do-Nothings	33	13.91	42	61.09
TOTALS	61	25.92	76	111.08

with 1 *df*. (Some texts subtract ½ from the value |(O − E)| before squaring. As we discussed earlier, though, we doubt the usefulness of this correction for continuity.) If we had more than two groups, we would simply extend Equation 24–15 by tacking more terms on the end and using $k - 1$ *df* (where k is the number of groups). For some strange reason, this is called the log-rank test, although nowhere did we use ranks or take the logarithm of anything. Let's apply it to our data in Table 24–7:

$$\chi^2(1) = \frac{(61 - 28.06)^2}{28.06} + \frac{(76 - 108.93)^2}{108.93} = 38.67 + 9.95 = 48.62$$

(24–16)

which is highly significant.

The RR of using *Hairgro* can be figured out by using the formula:

The Overall Relative Risk

$$RR = \frac{O_1 / E_1}{O_2 / E_2}$$

(24–17)

For our data, this works out to be:

$$RR = \frac{61 / 28.06}{76/108.93} = 3.12$$

(24–18)

Because the chi-squared value was significant, we can go ahead and look at the RR. By convention, we disregard any RR under 2 as not being anything to write home about. This RR of 3.12 tells us that politicians who tell the truth (or have blow-dried hair) are three times as likely to be voted out of office as are controls—let's all go out and buy some *Hairgro* for our favorite pols!

ADJUSTING FOR COVARIATES

Having gone to all this trouble to demonstrate the log-rank test, it would be a pity if we could use it only to compare two (or more) treatment groups. In fact, it does have more uses, mainly in testing for the possible effects of covariates. If we thought, for example, that telling the truth was more of a liability for politicians in one party than those in another,[13] then we could divide the *Hairgro* group by political affiliation and do a survival analysis (either actuarial or Kaplan-Meier) on these two (or more) strata. (If the covariate were continuous, such as age or length of time in office, we could dichotomize the covariate by splitting it at the median or some other logical place.) *Taking the covariate into consideration involves an "adjustment" which takes place at the level of the final chi-squared, where we use the strata-adjusted expected frequencies.* Let's assume that we divided each group by political party and had the computer redo the calculations for Table 24–7 two times: once for the first party (the Old Deadbeats) divided by experimental condition (*Hairgro* versus control), and again for the second party (the New Do-Nothings) split the same way. Table 24–8 shows what we found. Using these new figures in Equation 24–15 gives us:

$$\chi^2 = \frac{(61 - 25.92)^2}{25.92} + \frac{(76 - 111.08)^2}{111.08} = 58.59$$

(24–19)

which is larger than the unadjusted log-rank test, telling us that *Hairgro* did indeed affect politicians from the two parties differentially.

There are a few problems with this way of going about things, though. First, each time we split the group into two or more strata, our sample size in each subgroup drops. Unless we have an extremely large study, then, we're limited as to the number of covariates we can examine at any one time. The second problem is that we may be taking perfectly good interval or ratio data and turning them into nominal categories (e.g., converting length of time in office into <10 years and ≥10 years). This is a good way to lose power and sensitivity. Last, although we can calculate the statistical significance of adjusting for the prognostic factor(s) (i.e., the covariates), we don't get an estimate of the *magnitude* of the effect.

What we need, then, is a technique that can (1) handle any number of covariates, (2) treat continuous data as continuous, and (3) give us an estimate of magnitude of the difference; in other words, an equivalent of an analysis of covariance for survival data. With this build-up, it's obvious that such a statistic is around and is the next topic we tackle. This technique is called the **Cox proportional hazards model** (Cox, 1972).

[13] *We won't say for which party it's more of a handicap for fear that we won't insult half the readers.*

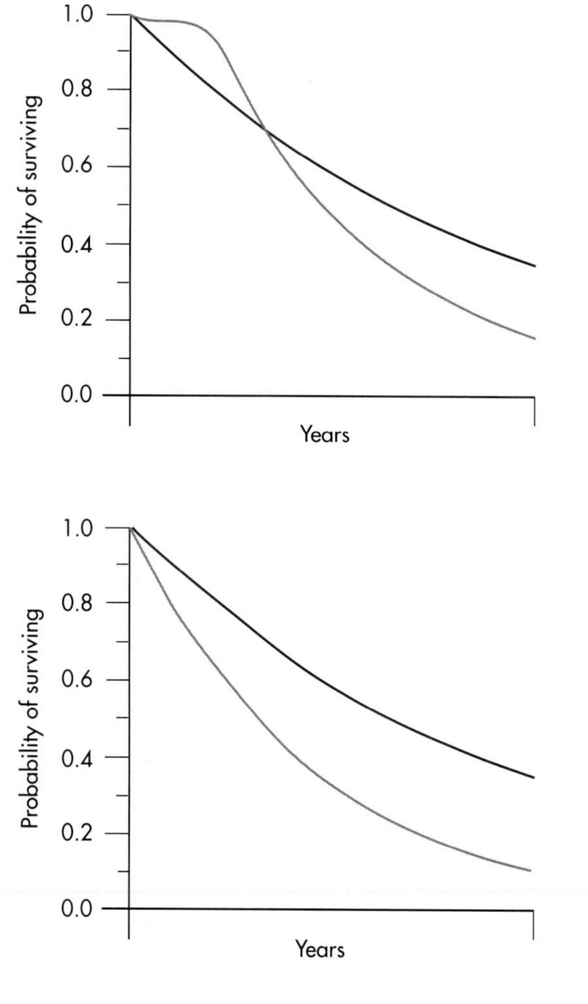

FIGURE 24–5

Two survival curves that do not meet the assumption of no change in the effect of the prognostic variable over time.

FIGURE 24–6

These two curves meet the assumption of no change in the effect of the prognostic variable over time.

[14]*Now there's a euphemism, if we ever heard one.*

Let's go back to the definition of the hazard, which we defined in Equation 24–2 as:

$$q_t = \frac{D_t}{R_t} \tag{24–20}$$

Putting this into English:

> **The hazard at time t, q_t, is the probability of an event at time t, given survival (no event) up to time t.**

The proportional hazards model extends this to read:

> **The proportional hazard at time t is the probability of an event at time t, given survival up to time t, *and for a specific value of a prognostic variable, x.***

In the example we did illustrating the Mantel-Cox chi-squared, it would be the probability of defeat at time t (or interval i), given that the person belonged to one political party or another. With our new, enhanced technique, the prognostic variable can be either discrete (e.g., political party) or continuous (such as age or number of years in office), and we can have several of them (age *and* years in office *and* political party). So, to be more precise, instead of having just one x, we can have several, x_1, x_2, and so on, with each x representing a different prognostic variable. However, let's stick to just one variable for now to simplify our discussion.

The major assumption we make is that the effect of the prognostic variable depends on the *value* of that variable *and does not depend on time*. That is, we assume that if political party plays a role in determining survival or defeat, then that effect is constant and doesn't change over time. If the popularity of the different parties changes over time, and if this affects how a specific politician will do in an election, then we can't use this model. We can use the graph of the survival curves to see if our data meet the assumption. First, if the two survival curves cross at anytime, as in Figure 24–5, then this immediately tells us that our data do *not* meet this criterion. Second, not only must the curves not cross, they should get further and further apart over time, as in Figure 24–6.

Another way of stating this assumption is that the RR at a specific value of x doesn't change over time. If we're looking at tenure in office as the covariate, and if the RR resulting from *Hairgrō* is 1.5 for people who have served[14] for 10 years, then we assume that it is 1.5 whether the politicians were enrolled during the first year of the study or 5 years later. Putting this into the form of an equation, we can say:

> **The proportional hazard at time t for some specific value of x = (some constant that depends on t) times (some function dependent on x)**

Writing this in mathematical shorthand, we get:

The Proportional Hazard

$$h(t \mid x) = c(t) \times f(x) \tag{24–21}$$

where c is the constant dependent on t, and f is the function dependent on x. What the term c tells us is how fast the curve drops; are most of the pols booted out of office within the first few years, or do they keep hanging around, year after year?

Now, let's start using it with some data. To keep the number of subjects manageable, we'll use the 10 pols we first met in Table 24–1. In Table 24–9, we've added one covariate, Duration in Office, and rank ordered the subjects by their time in the study because we'll use a Kaplan-Meier approach.

The first death occurred at 22 months and was Subject F (let's call him the **index case** for this calculation). All of the other politicians were in the study at least 22 months, with the exception of Subject I, who was lost to follow-up study after 14 months. The next step is to figure out the probability of Subject F being defeated at 22 months, versus the probabilities for the other people at risk. He had been in office for 36 months, so his probability of death at 22 months after starting *Hairgro* is $c(22) \times f(36)$.

We[15] now repeat this procedure in turn for all of the other people who died, each of them in turn becoming the index case. In each calculation, we include in the denominator only those people who were still in the study at the time the index case was killed off by the voters; that is, those still at risk. We don't include those who were already dead or those whose data were censored before the time the index case died.

When we're finally done with these mind-numbing calculations, what we've got[16] for each person is some expression involving the term *f*. Multiplying all of the expressions together gives us the overall probability of the observed defeats. Now the fun begins.

Those of you who are still awake may have noticed that we've been talking about the term *f* without ever really defining it. Based on both fairly arcane statistical theory as well as real data,[17] the distribution of deaths over time can best be described by a type of curve called **exponential**. The two curves in Figure 24–6 are of this type; more events occur early on when there are more people at risk, and then the number tapers off as time goes on. In mathematical shorthand, we write the equation for an exponential curve as:

$$y = e^{-kt}$$

where e is the base of the natural logarithm and is roughly equal to the value 2.71828. Another way of writing this to avoid superscripts is: $y = \exp(-kt)$. What this equation means is that some variable y (in this case, the number of deaths), gets smaller over time (that's why the minus sign is there). The k is a constant; it's what makes the two curves in Figure 24–6 differ from one another. All of this is an introduction for saying that the f in our equation is really $\exp(-kx)$; x is the specific value of the covariate we're interested in (time in office, in this case), and we've gone through all these calculations simply to determine the value of k.

The computer now goes through its gyrations and comes up with an answer. Let's say it tells us that k is .02, with an associated p level of .03. First, the p tells us that the effect of time in office is significant; politicians with longer tenures have a different rate of dying that have those who haven't been fooling the public as long. Knowing the exact value of k, we can compare the RRs at any two times. For example, to compare those in office for 40 months with those in for 20 months, we simply calculate:

$$RR = \frac{\exp(-0.02 \times 40)}{\exp(-0.2 \times 20)} = \frac{e^{-0.8}}{e^{-0.4}} = 0.67 \quad (24\text{–}22)$$

TABLE 24–9

Outcomes of the first 10 subjects

Subject	Length of time in trial (months)	Duration in office (months)	Outcome
I	14	24	Lost
F	22	36	Died
C	29	58	Died
G	37	60	Lost
J	45	48	Censored
D	46	111	Died
A	61	105	Died
H	76	134	Censored
E	92	228	Censored
B	111	120	Lost

This would indicate that tenure confers protection against death; those in office for 40 months had only two-thirds the risk of being defeated as had those in office for only half as long.

So, to reiterate, the proportional hazards model allows us to adjust for any number of covariates, whether they are discrete (e.g., gender or political party) or continuous (e.g., age or tenure in office).

SAMPLE SIZE AND POWER

As is usual in determining the required sample size for a study, we have to make some estimate of the magnitude of the *effect size* that we wish to detect. For the t-test, the effect size is the ratio of the mean difference between the groups divided by the SD. In survival analysis, the effect size is the ratio of the hazards, q, at a given time, such as 5 or 10 years. If we call this ratio δ (delta), then the number of events (deaths, defeats, and so on) we need in each group can be figured out with an equation proposed by George and Desu (1974):

$$d = \frac{2(Z_\alpha + Z_\beta)^2}{(\ln \delta)^2} \quad (24\text{–}23)$$

where the term "ln δ" means the natural logarithm of δ. To save you the hassle of having to work through the formula, we've provided sample sizes for various values of δ in Table L in the book's appendix.

Remember that these aren't the sample sizes at the *start* of the study; they're how many people have to have *outcomes*. To figure out how many people have to enter the trial, you'll have to divide these numbers by the proportion in each group you expect will have the outcome. So, if you're planning on a two-tailed α of .05, a β of .20, and δ of 2, Table L says you'll need 33 events per group. If you expect that 25% of the subjects in the control group will experience the outcome by the time the study ends, then you have to start with $(33 \div .25) = 132$ subjects. A different approach to calculating sample

[15] *Or, more accurately, the computer, as no rational being would ever want to do this by hand, except to atone for some otherwise unpardonable sin, such as reading another stats book.*

[16] *Apart from a headache.*

[17] *For a change, both theory and facts give the same results.*

sizes, based on the *difference* in survival rates, is given by Freedman (1982), who also provides tables.

To determine the power of a trial after the fact, we take Equation 24–23 and solve for z_β. For those who care, this gives us:

$$z_\beta = \frac{(\ln \delta)\sqrt{d} - z_\alpha\sqrt{2}}{\sqrt{2}} \qquad \text{(24–24)}$$

A minor problem arises if the number of outcomes (d) is different in the two groups. If this happens, the best estimate of the average number of events in the groups can be derived using the formula for the harmonic mean:

$$d = \frac{2}{\frac{1}{d_1} + \frac{1}{d_2}} \qquad \text{(24–25)}$$

For example, if group 1 had 13 events at the end, and group 2 had 20, we would have:

$$d = \frac{2}{\frac{1}{13} + \frac{1}{20}} = 15.76 \qquad \text{(24–26)}$$

so we would use 15.76 for d.

EXERCISES

Television executives are becoming worried that, at one point or another, all TV talk-show hosts become afflicted with a case of terminal megalomania and think they are as powerful as the Assistant Junior Vice President in Charge of Washroom Keys. To slow the onset of this insidious condition, the executives try an experiment. They hire a group of television psychologists[18] to give half of the 20 hosts a course in Humbleness 101[19] and have the other half serve as the controls. The outcome is any 5-minute interval where the host says "I," "me," or "my" more than 15 times; a sure sign that the course did not work or that its effects are wearing off.

Because the course is a grueling one, the company can take in only one or two people each month over the 2 years of the study. Also, some of the hosts are killed off by irate viewers, crashes in their Lamborghinis, or enraged Assistant Junior Vice Presidents in Charge of Washroom Keys. The data for the experiment are shown in the accompanying table.

1. Draw an actuarial curve for these data.
2. What test would you use to determine whether your treatment works?
3. What would your data look like if you simply approached it with a contingency table chi-squared?
4. What is the SE at 6 to 7 months for the control group?
5. What is the relative risk at 18 months? (If you cheated on your homework and didn't work out the table for the experimental group, p_E = .567.)

[18] *They used to be pet psychologists until the market went to the dogs.*

[19] *The motto of this company is, "The most important thing is sincerity. Once you learn to fake that, the rest is easy."*

TV host	Treatment/control	Month of entry	Month of conclusion	Outcome	
1	T	2	24	Still humble	Outcomes for subjects in both groups
2	C	3	18	Megalomaniacal	
3	T	4	24	Still humble	
4	T	4	20	Megalomaniacal	
5	C	5	8	Megalomaniacal	
6	T	5	23	Dead	
7	C	6	22	Still humble	
8	C	6	9	Megalomaniacal	
9	T	8	23	Megalomaniacal	
10	T	8	24	Still humble	
11	C	9	22	Still humble	
12	C	11	24	Still humble	
13	T	11	20	Megalomaniacal	
14	C	13	15	Dead	
15	T	14	24	Still humble	
16	C	16	24	Still humble	
17	T	19	23	Still humble	
18	T	21	24	Still humble	
19	C	20	21	Megalomaniacal	
20	C	22	24	Still humble	

How to Get the Computer to Do the Work for You

Actuarial Method

- From **Analyze**, choose **Survival** → **Life Tables**
- Click on the variable in the left column that indicates the time since entry into the study, and click the arrow to move it into the **Time** box
- Fill in the boxes in the **Display Time Intervals** area. In the first box (**0 through** ❑), fill in the last interval you want analyzed. In the second box (**by** ❑), enter 1 if you want to analyze every interval, 2 for every other interval, and so on
- Click on the variable in the left column that indicates the outcome for that case, and move it into the **Status** area with the arrow button
- Click the **Define Event** button
- The default is **Single Value.** Enter the number that indicates that an outcome has occurred (all other values will be treated as censored data)
- If you have two or more groups, you can analyze each group separately by moving the variable that defines group membership into the **Factor** box
- By clicking the **Options** button, you can elect to display the survival function or the hazard function, or both.
- Click **Continue** and then **OK**

Kaplan-Meier Method

- From **Analyze**, choose **Survival** → **Kaplan-Meier**
- The rest is identical to the Actuarial Method

Cox Proportional Hazards Method

- From **Analyze**, choose **Survival** → **Cox Regression**
- Then follow the instructions for the Actuarial Method, except now you can enter the name of the covariate(s) into the **Covariates** box or analyze categorical data by strata by entering the name of the variable into the **Strata** box

C.R.A.P. DETECTORS

IV–1. As well as banner headlines proclaiming the phenomenon of the "carcinogen of the week," North American media in the 1980s have mounted a continual barrage of study results with exhortations to eat less of this, do more of that, raise our serum rhubarb level, lower our urine asparagus level, and so on, ad nauseam. One such study, reported in the health affairs section of the *National Prevaricator*, attempted to lower serum cholesterol levels through ingestion of daily doses of pine needles, the reasoning being that the natural solvents (turpentine comes from pine) would dissolve the clots away. The investigators randomized 3,000 men with screamingly high cholesterols to sprinkle either pine needles or a plastic imitation on their dinners every night. Six months later, 1,000 men had died of perforations of the GI tract, but 1,000 per group remained to have their cholesterols measured. After the dust settled, 22% of the men in the pine group had cholesterols in the normal range, versus 20% in the plastic group. That was good enough for the *Nat Prev*, and the headline screamed *"Don't Pine Your Life Away, Don't Stand for any More Needling from your Spouse—Pine Needles Make You Live Forever."* However, the chi-squared was not significant ($\chi^2(1) = 1.20$, $p = .27$). Might you have done it differently?

Any sane researcher wouldn't do it at all. But if you had, among the many sins committed by this study was the cardinal one of taking a perfectly good ratio variable such as cholesterol (or blood pressure, or even depression) and collapsing it into a 2 × 2 table. The excuse offered usually goes something like, "Well, yah, I know it's a ratio variable but clinicians must, after all is said and done, make binary decisions about whether to treat or not." True, but this confounds statistics with decision making. The table can be constructed after the event, but the statistics should be done on the original data. Lest you think this happens only in the popular press, read on.

C.R.A.P. DETECTOR IV–1

Taking a ratio variable such as serum cholesterol and putting it in a 2 × 2 table is an incredible waste. Typically, this may require an increase of sample size by a factor of 5 to 10 to maintain the same power. If you are concerned about skewed distributions, as might be the case in the present example, then use a rank test such as the Mann-Whitney U, which usually results in little or no loss of power.

IV–2. The study we are about to describe is true; however, the names and some of the numbers are changed to protect the guilty. These folks reported a trial of "Critical Appraisal Skills." A group of residents had one session a week for a number of weeks where they did critical appraisal of journal articles. A control group had some other unrelated "placebo" treatment. Both groups then took a multiple choice post-test, which showed a tiny difference between the two groups. For obvious reasons, they never quite got around to analyzing these results, but we did ($t = 1.20$, ns).

Now things got really interesting. The subjects then crossed over, so the original control group now got the treatment. After it was over, they had a second knowledge test, and analysis was on the change scores from the first administration. To everyone's relief, the new treatment group had a gain of a few percent and the new control (old treatment) group had a loss of a few percent, ($p < .05$). Finally, they reported that "21% of the group B residents versus 5% of the group A residents showed an improvement of 18% or more. We think this is a significant improvement for this subset of residents."

On the basis of this, would you now introduce a critical appraisal course where you live?

A finer demonstration of "Do as I say, not as I do" would be hard to find. This is an exercise in design detecting as much as anything. The only clean comparison in the study is the post-test scores after the first trial period, which only we analyzed. The comparison they did, which involved change scores, actually compared one group right after the treatment with another group some time after treatment. Education is not like a vaccine that confers permanent immunity, nor is it like a drug with a 24-hour washout. On reflection, over 8 weeks it is likely that some of the knowledge of the first group (and some of their motivation) will be lost, so the comparison confounds the loss of knowledge of one group with the gain of the other and means nothing.

But that has nothing to do with what we've discussed in Section 4. The last sentence of the description sure does. Recognize that there was no accepted "clinically important change" on their little post-test. Also keep in mind that, in *every* group, a few subjects will increase a lot, a few will decrease a lot, and many will be in the middle. Consequently, it is as easy as pie to look at the data after the fact and establish a cut-point between "clinically important" and "not clinically important": a placed somewhere out on the end of the distribution, where a few extra souls in one group (21% of 29, or 6 residents) or the other (5% of 41, or 2 residents) sneak over the cut-point and give you statistical significance. (In fact, this difference is just significant by the Fisher Exact Test.) We are not saying the authors did this; we *are* saying that they did not provide any evidence in the paper that they didn't.

C.R.A.P. DETECTOR IV–2

As a corollary to IV–1, any collapsing into 2 × 2 tables using arbitrary cut-points (in particular, arbitrary and after the fact) is a license to steal and is absolutely contraindicated.

IV–3. The incidence of spouse beating is growing, which the perpetrators are explaining with statements such as, "I just couldn't stand his nagging any longer." Concerned, the National Irritating Larynx Syndrome (ILS) Society has mounted a television campaign featuring the co-chairs, Wendell Winer and Sally Druthers, making an appeal for tolerance and dollars. They mounted a study where they survey people with the question, "Would you stay married to a nagging, whining spouse?" They analyzed responses by gender and exposure to the appeal and found that (a) fewer men would stay married than women; (b) among men, exposure to the commercial resulted in more negative responses to the question ($\chi^2 = 3.98$, $p < .05$); and (c) similarly among women, exposure to the commercial resulted in more negative responses to the question ($\chi^2 = 4.20$, $p < .05$). However, collapsing the data across gender revealed no significant effect of exposure to the commercial. Confused, they dumped the whole lot on your desk and whined off into the sunset. What would you do now?

A Mantel-Haenszel chi-squared, of course. When multiple sub-tables are collapsed, strange things can result. In any case, having gone to the effort of gathering the data by both gender and exposure, they should analyze the data appropriately.

C.R.A.P. DETECTOR IV–3

When dealing with two independent categorical variables, the appropriate analysis is the Mantel-Haenszel chi-squared (or log-linear analysis).

IV–4. A recent review article in a major psychological journal addressed the issue, "Are adolescents slaves to their hormones?" The answer is a resounding "Yes!" But this is of little consolation to parents, who need an objective test of hormone bondage. One such test is based on reaction time: a series of statements, such as, "Son, would you. . . .", or "Jane, did you. . . ." are interspersed with adolescent-neutral phrases and displayed on the screen while sensors monitor the teen's time to become apoplectic. One study did this for a sample of 14-year-olds and a control group of 12-year-olds. Because reaction time is generally horribly skewed, with a few really long stretches where the subject apparently fell asleep at the switch, a nonparametric test is appropriate. The investigators opened the stats book at the median test, wherein all the reaction times for the 12- and 14-year-olds are ranked in one long line, and the median was established. A 2 × 2 table was then constructed based on the number of 12-year-olds and 14-year-olds above and below the median, and a chi-squared test was performed. Why did you find the median test here, instead of in the middle of Chapter 22?

For a now familiar reason. Once again, we are throwing away information by reducing all the ranked data to two categories.

C.R.A.P. DETECTOR IV–4

The median test is a very conservative test of ranked data, and other methods should be used instead.

IV-5. Dr. Dreikopf, a shrink at the Mesmer School of Health Care and Tonsorial Trades, has taken seriously the dictum, "Behind every twisted mind is a twisted molecule." He believes that severe depression is genetically based, and because the DNA strands coil counterclockwise, the best cure would be to spin patients around in a clockwise direction at 45 r.p.m.[1] so the strands could realign themselves. He decides to do a study by comparing time to relapse in twirled patients to similar patients on another ward who were not iatrogenically twisted. He sends a questionnaire to 50 patients he has treated in the past decade and to 50 control patients, asking them how long after discharge they experienced depression (if ever). Of the spin group, 28 patients replied, and 42 control patients replied. Eight of the 28 (25%) had experienced a relapse, compared with 28 of the 42 (67%) controls. The test is significant, and he begins the headlong rush to publication. How do you stop him, and should you?

[1] *Someone finally found a use for all those old turntables gathering dust since the CD revolution.*

All sorts of things are wrong with this one. A short list:

a. *No well-defined dichotomous end point.* Forgive us for sounding repetitive, but depression is more of a continuous variable. In any case, getting patients to say when it began guarantees that everybody has a different criterion. One way to avoid this is to use a depression scale, which creates a continuous score and which usually has well-defined thresholds.
b. *Loss to follow-up study is related to outcome.* One risk of depression is suicide; thus, some patients may be lost to follow-up study because they have departed this vale of tears. Making matters worse, a much higher attrition is evident in the treatment group, so they are still dizzy or they may just be dead.

C.R.A.P. DETECTOR IV–5

If you are going to analyze survival data, either you must account for all patients in the follow-up period or be absolutely certain that no relationship exists between the outcome and attrition.

c. *A substitution game.* He began looking at time to relapse but then substituted a simple measure of prevalence of relapse, which again has less information. Far better would be to follow the patients' time to relapse using a life-table and then do a Mantel-Haenszel or Cox Model analysis. Admittedly, the numbers in this study may be too small to do any meaningful analysis.

C.R.A.P. DETECTOR IV–6

Survival data, particularly in situations where there is variable length of follow-up study, are better analyzed with life-table analysis than with simple contingency tables.

SECTION

THE FIFTH

REPRISE

In this chapter, we discuss ways of locating anomalous data values, how to handle missing data, and what to do if the data don't follow a normal distribution.

CHAPTER THE TWENTY-FIFTH

Screwups, Oddballs, and Other Vagaries of Science

Locating Outliers, Handling Missing Data, and Transformations

SETTING THE SCENE

You've carefully planned your study and have estimated that you need 100 subjects in each of the two groups, with each subject tested before and after the intervention. With much effort, you're able to locate these 200 patients. But, at the end of the trial, you find that 8 subjects didn't show up for the second assessment; 2 subjects forgot to bring in urine samples; and you lost the sheet with all the demographic data on 1 subject. Your printout also tells you that your sample includes 2 pregnant men, a mother of 23 kids, and a 187-year-old woman. To add insult to injury, some of the data distributions look about as normal as the Three Stooges.

The situation we just described is, sad to say, all too common in research. Despite our best efforts, some data always end up missing, entered into the computer erroneously, or are accurate but reflect someone who is completely different from the madding crowd. Sometimes the fault is ours; we lose data sheets, punch the wrong numbers into the computer, or just plain screw up in some other way. Other times, the fault lies with the subjects;[1] they "forget" to show up for retesting, put down today's date instead of their year of birth, omit items on questionnaires, or are so inconsiderate that they up and die on us before filling out all the necessary paperwork. Last, what we've learned about the normal curve tells us that, although most of the people will cluster near the mean on most variables, we're bound to find someone whose score places him or her somewhere out in left field.

Irrespective of the cause, though, the results are the same. We may have a few anomalous data points that can screw up our analyses, we have fewer valid numbers and less power for our statistical tests than we had initially planned on, and some continuous variables look like they cannot be analyzed with parametric tests. Is there something we can do with sets of data that contain missing, extreme, and obviously wrong values?

Of course there is, otherwise we wouldn't have a chapter devoted to the issue. We have two broad options: grit our teeth, stiffen our upper lip, gird our loins, take a deep breath, and simply accept the fact that some of the data are fairly anomalous, wrong, or missing, and throw them out (and likely all of the other data from that case). Or, we can grab the bull by the horns and "fake it"—that is, try to come up with some reasonable estimates for the missing values. Let's start off by trying to locate extreme data points and obviously (and sometimes not so obviously) wrong data. This is the logical first step because we would usually want to throw out these data, and we then end up treating them as if they were missing.

FINDING ANOMALOUS VALUES

Ideally, this section would be labeled "Finding Wrong Values," because this is what we really want to do—find the data that eluded our best efforts to detect errors before they became part of the permanent record.[2] For instance, if you washed your fingers this morning and can't do a thing with them,

[1] At least that's what we tell the granting agency.

[2] We know of several ways to do this, such as entering the data twice and looking for discrepancies. But if you're reading this book to find out other ways, you've picked up the wrong volume, so go look somewhere else.

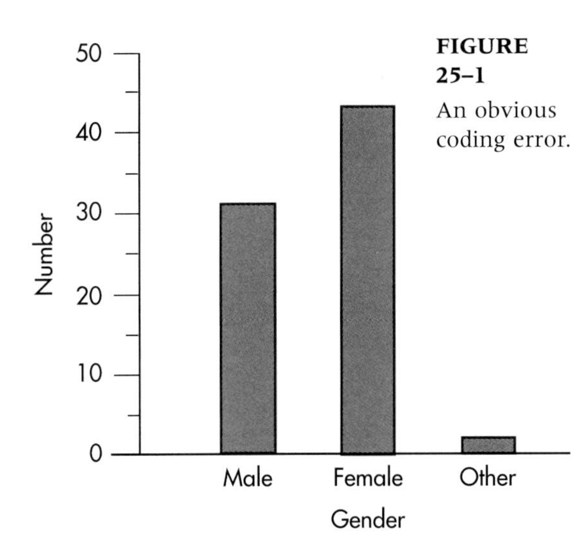

FIGURE 25–1 An obvious coding error.

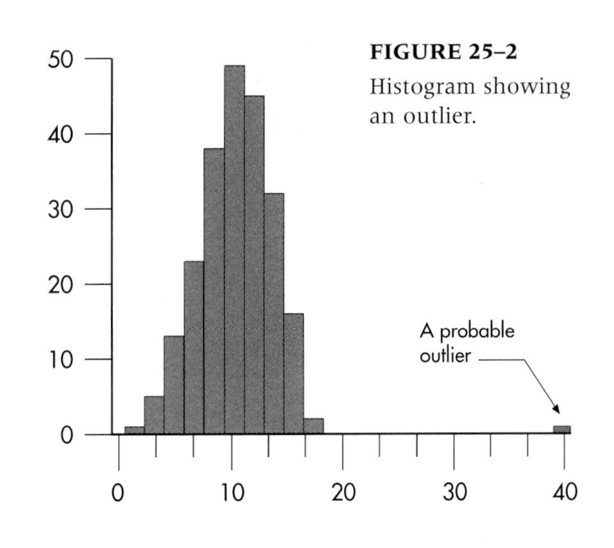

FIGURE 25–2 Histogram showing an outlier.

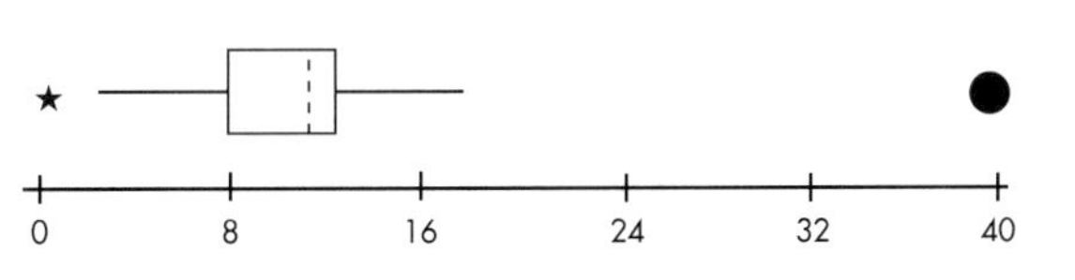

FIGURE 25–3 Box plot with one, possibly two, outliers.

and entered a person's age as 42 rather than 24, you may never find this error. Both numbers are probably within the range of legitimate values for your study, and there would be nothing to tell us that you (or your research assistant) goofed. The best we can do is to look for data that are outside the range of expected values or for where there are inconsistencies within a given case.

The easiest type of anomaly to spot is where a number falls either into a category that shouldn't exist or above or below an expected range. For example, we can make a histogram of the subjects' gender, using one of the computer packages we mentioned in Section 1. If we got the result shown in Figure 25–1, we'd know we've got problems.[3]

With continuous data, the two primary ways of spotting whether any data points are out of line are (1) visually and (2) statistically. The visual way involves plotting each variable and seeing if any oddballs are way out on one of the tails of the distribution. You can use a histogram, a frequency polygon, or a box plot; with each of them, the eyeball is a good measurement tool. Figure 25–2 shows what an outlier looks like on a histogram, and Figure 25–3 shows the same data displayed in a box plot. The solid circle on the right of Figure 25–3 is a far outlier, corresponding to the blip on the right of the histogram in Figure 25–2; and the asterisk is a run-of-the-mill outlier.

Notice that the histogram did not identify this low value as an outlier. The difference is that, with a histogram, we rely on our eyeballs only to detect outliers. With box plots, outliers are defined statistically, and this may pick up some of the buggers we would otherwise have overlooked. So, box plots combine visual detection of outliers with a bit of statistics.

You get a purely "statistical" look when you ask most computer packages to summarize a variable (show the mean, SD, and the like); they will also give the smallest and largest value for each variable. So if you're studying the fertility patterns of business women, a minimum value of 2 or a maximum of 99 for age should alert us to the fact that something is awry, and you should check your data for outliers. Quite often, values such as 9 or 99 are used to indicate missing data. Again, check to see if this is the case.

A more sophisticated approach looks at how much each score deviates from the mean. You no doubt remember that the easiest way of doing this is to transform the values into z-scores. Each number now represents how far it is from the mean, in SD units. The cut-off point between what's expected and what's an outlier is somewhat arbitrary, but usually anything over $+3.00$ or under -3.00 is viewed with suspicion. Doing this, we find that the highest value is 7.33—definitely an outlier that should be eliminated from further analysis. The lowest value has a z-score of -2.54, so even though one program[4] flagged it as suspicious, we'd probably keep it.

By eliminating the outlier(s), we've changed the distribution a bit. In this case, the mean dropped from 10.28 to 10.15, and the SD naturally got smaller (going from 4.06 to 3.54). Consequently, values that weren't extreme previously may now have z-scores beyond ± 3.00.[5] So it makes sense to go through the data a few times, eliminating the outliers on each pass, until no more come to light.

More difficult to spot are "multivariate" errors. These occur when you've got two or more variables, each of which looks fine by itself, but some combinations are a bit bizarre. Imagine that we surveyed the incoming class of the Mesmer School of Health Care and Tonsorial Trades and got some basic demo-

[3] *Actually, our problems are not as serious as those of the two people labeled "Other," if the data* aren't *wrong.*

[4] *We used Minitab in this case.*

[5] *We can actually figure out which, if any, values would be "revealed." Using all the original data, a z-score of 3 corresponds to a raw score of 22.46 (i.e., 10.28 + 3 × 4.06); after we eliminated the outlier, a z of 3 corresponds to a score of 20.77. So any score between these two values would not be detected the first time but would be extreme on the second pass through the data.*

TABLE 25–1

Data with some problems

Subject	Age	Gender	Number of pregnancies
1	27	M	99
2	18	F	5
3	19	F	0
4	32	M	99
5	21	F	3
6	23	M	2

graphic information. Take a look at the data in Table 25–1, which summarizes what we found for the first six students. If we used all of the tricks we just outlined, none of the variables would look too much out of line: the ages go from 18 to 32, which is reasonable; the only genders listed are male (M) and female (F); and once we realize that 99 means "Not Applicable," the number of pregnancies looks okay. But wait a minute—we've got an 18-year-old female who had 5 pregnancies, and a 23-year-old man who had 2! Even in these days of more liberal attitudes toward sex, and a blurring of the distinctions between the genders, we would hazard a guess that these are, to use the statistical jargon, boo-boos.

The important point is that neither of these errors would have been detected if we restricted our attention to looking at the variables one at a time; they were spotted only because we took two into consideration at the same time—age and number of pregnancies, and gender and number of pregnancies. One problem, though, is that if we have N variables, we have $N \times (N - 1) \div 2$ ways of looking at them two at a time. For these 3 variables, there are 3 combinations; 10 variables would have 45, and so on. Although it may not make sense to look at all of these pairs, you should still examine those where being in a certain category on one of the variables limits the range of possible categories on the other. For example, age imposes limits on marital status (few people under the age of 17 have entered into the state of matrimonial bliss), number of children, income (not too many teenagers gross over $1,000,000 a year, although they all spend money as if their parents do), and a host of other factors.

Checking the data for integrity[6] is a boring job that can best be compared to being forced to listen to politicians. But it has to be done. The only saving grace is that we can hire research assistants to do the work for us; you can't find anybody who'll listen to politicians, for love or money.

[6] *Its, not yours.*

[7] *That's one of the saving graces about being short.*

FILLING IN THE BLANKS

Just Forget about It

Once data are missing or have been eliminated as wrong or too anomalous, they are gone for good. Some statistical purists may say that any attempt to estimate the missing values either introduces a new source of error or results in biased estimators. Their solution would be to acknowledge the fact that some data are missing and then do the best with what is at hand. In fact, this is likely the most prudent path to take, especially when only a small amount of the data are missing. As in other areas of statistics, the definition of "small" is subjective and arbitrary, but it probably hovers around 5% of the values for any one variable.

Even so, we still have a choice to make: to use all the available data that are left, or to eliminate all of the data associated with a subject who is missing at least one data point. To illustrate the difference, let's do a study testing a hypothesis based on our years of clinical observation working in a faculty of health sciences: the major criterion used to select deans (at least for males) is height. You can be the head of the largest clinical department, pull in the most grant money, and be responsible for a scientific advance that reduced suffering among thousands of patients, but if you ain't over 6′ tall, you won't become a dean.[7] To test this hypothesis, we'll collect five pieces of datum on former chairmen from several schools: whether or not they became a dean (coded 0 = No, 1 = Yes); the number of people in their department; the number of grants received during the last 5 years of their chairmanships; a peer rating of their clinical competence, on a 7-point scale (1 = Responsible for more Deaths than Attila the Hun, 7 = Almost as Good as I Am); and, of course, their height. The data for the first 10 people are shown in Table 25–2.

Each of these 10 people was supposed to have 5 scores. As you can see, though, 5 subjects have some missing data: variable X_1 (whether or not the person became a dean) for Subject 7; variable X_2 for Subject 3; variable X_3 for Subject 5, variable X_4 for Subject 4, and Subject 8 has variable X_5 missing. Assuming we want to correlate each variable with the others, how much data do we have to work with?

If we use as much data as possible, then the correlation between variables X_2 and X_3 is based on 8 subjects who have complete data for both variables (Subjects 1, 2, 4, 6, 7, 8, 9, and 10), as is the correlation between variables X_2 and X_4 (Subjects 1, 2, and 5 through 10) and similarly for all other pairs of variables. Intuitively, this approach is the ideal one to take because it makes maximum use of the existing data and makes no assumptions regarding what is missing. This way of analyzing missing data is sometimes referred to as **pairwise** deletion of data.

In *pairwise* deletion of data, a subject is eliminated from the analysis only for those variables where no data are available.

Needless to say, if anything seems logical, easy, and sensible in statistics, there must be something dreadfully wrong, and there is. Note that each of the 10 possible correlations is based on a different sub-

TABLE 25–2

Data set with missing values

	Variable				
Subject	Dean X_1	Department size X_2	Number of grants X_3	Competence X_4	Height (in.) X_5
1	0	22	12	2	70
2	0	49	5	6	69
3	1	—	8	5	76
4	0	44	10	—	68
5	1	47	—	1	72
6	0	45	15	4	66
7	—	27	9	3	67
8	0	13	1	4	—
9	1	42	12	5	71
10	0	32	7	6	68

set of subjects. This makes it difficult to compare the correlations, especially when a larger proportion of cases have missing data. Moreover, techniques that begin with correlation matrices (and this would include all the multivariate procedures, along with ordinary and logistic regression) may occasionally yield extremely bizarre results, such as *F*-ratios of less than 0 or correlations greater than 1.0.

The other way of forgetting about missing data is to eliminate any case that has any data missing; this is referred to as **casewise** or **listwise** deletion of data. All of the statistics are then based on the same set of subjects.

In *casewise* data deletion, cases are eliminated if they are missing data on any of the variables.

The trade-off is the potential loss of a large number of subjects. In our example, fully 50% of the subjects have some missing data and so would be dropped from all analyses. Although admittedly a bit extreme, the example does serve as a warning: if values are missing throughout the data set, casewise deletion can result in the elimination of a large number of subjects.

When in Doubt, Guess

The second way of handling missing data is by **imputing** what they should be. This is simply a fancy way of saying "taking an educated guess." Several techniques have been developed over the years, which in itself is an indication of the ubiquity of the problem and the lack of a totally satisfactory solution.[8]

1. *Deduction* (the Sherlock Holmes technique). Sometimes it is possible to deduce a logical value for a missing data point. For instance, if a person's race was missing, but we had data on the person's parents, it's a safe bet that the data would be the same. This approach is not always possible but is actually quite useful in the cases where it can be used. It does work well in one common situation, where one too many (or too few) spaces were added during data entry. If an adult has an age of 5.2 years or 520 years, it's pretty safe to assume that the correct age is 52, but the number got moved in one direction or the other. An "age" of 502 is a bit more tricky; should it have been 50 or 52?
2. *Replace with the mean.* The most straightforward method is to replace the missing data point with the mean of the known values for that particular variable. For example, the mean of the nine known values for variable X_2 is 35.7, so we could assume that the value for Subject 3 is 36. Note that this hasn't changed the value of the mean at all; it still remains 35.7 (plus or minus a tiny bit of error introduced by rounding). However, we reduced the variance somewhat; in this case, from 12.71 for the 9 values to 11.98 when we impute a value of 36. The reason is that it would be highly unusual for the missing value to have actually been the same as the mean value, so we've replaced the "real" (but lost) value with one which is closer to the mean—in fact, it is the mean. If only a small number of items are missing, the effect is negligible; once we get past 5% to 10%, however, we start to dramatically underestimate the actual variance. A good approximation of how much the variance will be reduced is n/N, where n is the number of nonmissing values, and N is the total number of subjects.

Replacing the missing value with the mean would still result in an unbiased estimate of the numerator in statistical methods such as the *t*-test. However, the denominator may be a bit smaller, leading to a slightly optimistic test. On the other hand, correlations tend to be more conservative, by the amount:

$$\frac{\sqrt{n_x n_y}}{N} \qquad (25\text{–}1)$$

where n_x is the number of nonmissing values for variable X, and n_y is the number of non-

[8] *Where do all the data go when they go missing? Is there some place, equivalent to the elephants' burial ground, filled with misplaced 1s, 32s, and 999s?*

missing values for variable Y. The distribution of scores is also affected by mean replacement, tending to become more leptokurtic.

Sometimes we can be even more precise. For example, departments of medicine are usually much larger than departments of radiology. So if we were missing the number of faculty members for a chairman of medicine, we'd get a better estimate by using the mean of only departments of medicine, rather than a mean based on all departments. However, because replacing the missing values with the mean lowers the SD, changes the distribution, and distorts correlations with other variables, this type of imputation should be avoided.

3. *Use multiple regression*. The next step up the ladder of sophistication is to estimate the missing value using the other variables as predictors. For example, if we were trying to estimate the missing value for the number of grants, we would run a multiple regression, with X_3 as the dependent variable and variables X_1, X_2, X_4, and X_5 as the predictors. Once we've derived the equation, we can plug in the values for subjects for whom we don't know X_3 and get a good approximation (we hope).

A few problems are associated with this technique. First, it depends on the assumption that we can predict the variable we're interested in from the others. If there isn't much predictive ability from the equation (i.e., if the R^2 is low), then our estimate could be way off, and we'd do better to simply use the mean value. The opposite side of the coin is that we may predict *too* well; that is, the predicted value will tend to increase the correlation between that variable and all the others, for the same reason that substituting the mean decreases the variance of the variable. The usual effect is to bias correlations toward +1 or −1. A better tactic is to add some random error to the imputed value. That is, instead of replacing the missing value with $\hat{y}$, it's replaced by $\hat{y} + N(0, s^2)$,[9] where s^2 is the Mean Square (Residual) of the multiple regression. This substitution preserves the mean, the variable's distribution, and its correlations with other variables. Last, multiple regressions are usually calculated using casewise deletion. Because several variables may be used in the regression equation, we may end up throwing out a lot of data and basing the regression on a small number of cases (i.e., we're shafted by the very problem we're trying to fix)!

Assumptions

All of these imputation techniques make certain assumptions about the missing data. Replacement with the mean and with an estimate based on regression (with or without some random error built in) assume that the data are Missing at Random (MAR).[10] Missing at Random means that the probability that the value of a given variable is missing does not depend on that variable, although it may be related to some other variable. In human language, the assumption is that missingness may not be completely at random but it is equivalent across groups. Casewise deletion of data assumes that the data are Missing Completely at Random (MCAR; that is, missingness is unrelated to any other variable), which is almost never the case, and is yet another reason that it should be avoided whenever possible. The last thing to note is that it is impossible to determine from the data whether MAR or MCAR is true; we just assume it's true and go on from there.[11]

TRANSFORMING DATA

To Transform or not to Transform

In previous chapters, we learned that parametric tests are based on the assumption that the data are normally distributed.[12] Some tests make other demands on the data; those based on multiple linear regression (e.g., MLR itself, ANOVA, and ANCOVA), as the name implies, assume a straight-line relationship between the dependent and independent variables. However, if we actually plot the data from a study, we rarely see perfectly normal distributions or straight lines. Most often, the data will be skewed to some degree or show some deviation from mesokurtosis, or the "straight line" will more closely resemble a snake with scoliosis. Two questions immediately arise: (1) can we analyze these data with parametric tests? and, if not, (2) is there something we can do to the data to make them more normal? The answers are: (1) it all depends, and (2) it all depends.[13]

Let's first clarify what effect (if any) non-normality has on parametric tests. The concern is not so much that deviations from normality will affect the final value of t, F, or any other parameter testing the difference between means (except to the degree that extreme outliers affect the mean or SD); it is that they *may* influence the p-value associated with that parameter. For example, if we take two sets of 100 numbers at random from normal distributions and run a t-test on them using an α level of .05, we should find statistical significance about 5% of the time. The concern is that if the numbers came from a distribution that wasn't normal, we'd find significance by chance more often than 1 time in 20. However, several studies have simulated non-normal distributions on a computer, sampled from these distributions, and tested to see how often the tests were significant. With a few exceptions that we discuss below, the tests yielded significance by chance about 5% of the time (i.e., just what they should have done). In statistical parlance, most parametric tests (at least the univariate ones) are fairly *robust* to even fairly extreme deviations from normality. This would indicate that, in most situations, it's not

[9] *The term* N *(0,* s^2*) is statistical shorthand for a normally distributed variable with a mean of 0 and a standard deviation of* s^2*. Most statistical packages allow you to do this quite easily.*

[10] *No, this doesn't mean that the data are missing completely at random; for that, stay tuned. Don't you just love statistical jargon?*

[11] *Yet another nail in the coffin of the "objectivity" of statistics.*

[12] *Some tests assume other distributions, such as the Poisson or exponential. However, because we've been successful so far in ignoring them, we'll continue to pay them short shrift.*

[13] *It's been rumored that graduate students receive their Ph.D.s in statistics when they reflexively answer, "It all depends" to any and all questions.*

necessary to transform data to make them more normal.

There's a second argument against transforming data, and that has to do with the interpretability of the results. For example, one transformation, called the "arc sine" and sometimes used with binomial data, is:

$$X' = 2\ sin^{-1}\sqrt{X + .5} \tag{25–2}$$

A colleague of ours once told us that his master's thesis involved looking at the constipative effects of medications used by geriatric patients. He reasoned (quite correctly) that because his dependent variable—whether or not the patient had a bowel movement on a given day—was binomially distributed, he should use this transformation. Proud of his deduction and statistical skills, he brought his transformed data to his supervisor, who said, "If a clinician were to ask you what the number means, are you going to tell him, 'It is two times the angle whose sine is square root of the number of patients (plus ½) who shat that day'?" Needless to say, our friend used the untransformed data.[14] The moral of the story is that, even when it is statistically correct to transform the data, we pay a price in that lay people (and we!) have a harder time making sense of the results.

Having said that, there are still some instances when transforming the data makes sense. Four examples we discuss are when (1) the data are J-shaped, (2) we're calculating correlations, (3) transforming the data makes them easier to understand, and (4) the SD is related to the mean.

As the name implies, J-shaped data are highly skewed, either to the right or to the left, as in Figure 25–4. Data such as this occur when there's a limit at one end of the values that can be obtained, but not at the other end. For example, several studies have tried to puzzle out what is disturbed in the thought processes of people with schizophrenia by seeing how quickly they react to stimuli under various conditions. The lower limit of reaction time is about 200 ms, reflecting the time it takes for the brain to register that a stimulus has occurred, decide whether or not it is appropriate to respond, and for the action potential to travel down the nerves to the finger. However, no upper limit exists; the person could be having a schizophrenic episode or be sound asleep at the key when he or she should be responding. When data like this are analyzed with parametric tests, the p-values could be way off, so it makes sense to transform them.

A second situation in which transforming data is helpful is when we're calculating Pearson correlations or linear regressions. Recall that these tests tell us the degree of *linear* relationship between two or more variables. It's quite possible that two variables are strongly associated with one another, but the shape of the relationship is not linear. Around the turn of the century, Yerkes and Dodson postulated that anxiety and performance are related to each other in an ∩-shaped (called an inverted U) fashion: not enough anxiety, and there is no motivation to do well; too much, and it interferes with the ability to perform. Who studies 10 weeks before a big exam, and who *can* study the night before? As Figure 25–5 shows, a linear regression attempts to do just what its name implies: draw a straight line through the points.

As you can see, the attempt fails miserably. The resulting correlation is 0. Although this is an extreme example, it illustrates the fact that doing correlations where the relationship is nonlinear underestimates the degree of association; in this case, fairly severely. It would definitely help in this situation to transform one or both of the variables so that a straight line runs through most of the data points.

The third situation where transformations help is similar to the previous one, when, because of the nature of the data, they are *expected* to follow a nonlinear pattern, such as logarithmic or exponential. This assumption can be tested by doing the correct transformation and seeing if the result is a straight line. For example, if the relationship between the variables is exponential, a logarithmic transformation should make the line appear straight, and vice versa.[15] Even if it isn't necessary to transform the variable for statistical reasons, simply seeing that the line is straight confirms the nature of the underlying relationship (Figure 25–6).

[14] *And later went on to become the head of Statistics Canada.*

[15] *AHA! Finally, an explanation of the phrase, log linear. It appears linear when we take the log of one variable.*

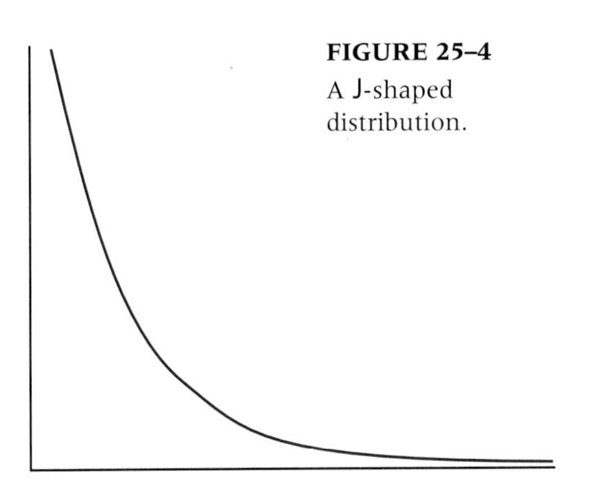

FIGURE 25–4
A J-shaped distribution.

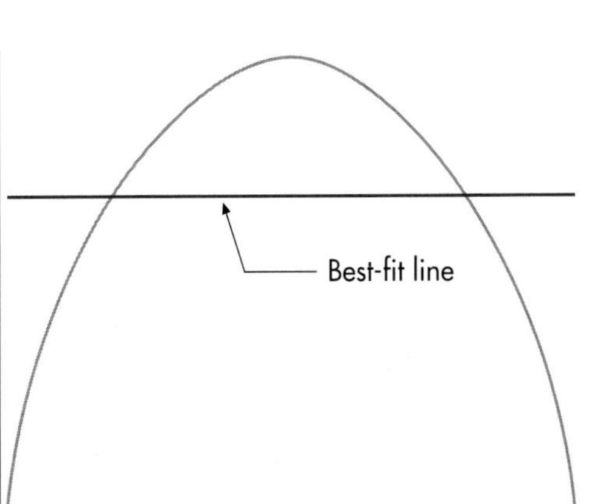

FIGURE 25–5
A straight-line fit through an ∩-shaped distribution.

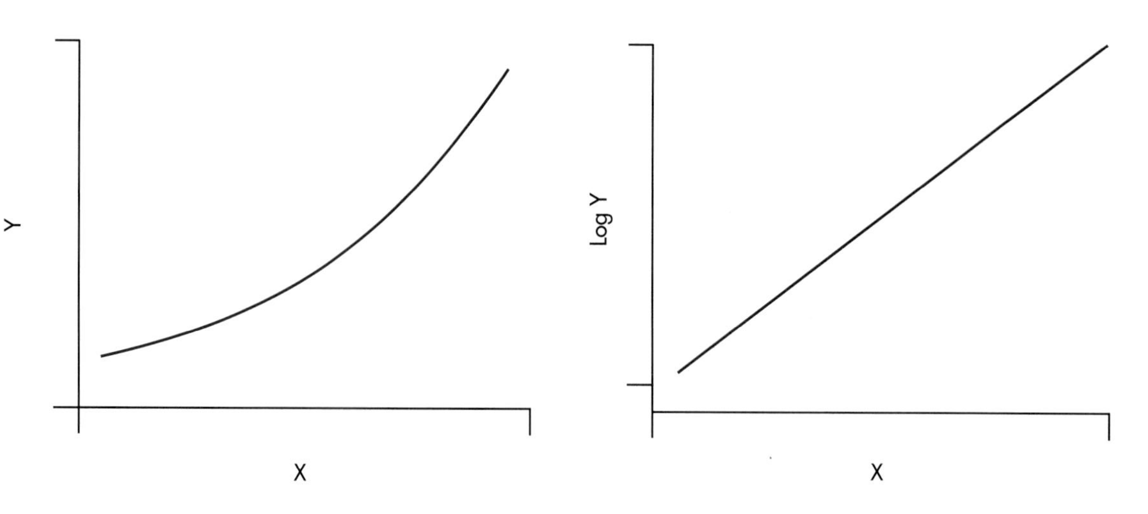

FIGURE 25–6
An exponential curve straightens out with a logarithmic transformation.

The last situation where transformations may be warranted is when the SD is correlated with the mean across groups. Way back in Chapter 4, we mentioned that one of the desirable properties of the normal distribution is that the variance stays the same when the mean is increased. In fact, that's one of the underlying assumptions of the ANOVA; we change the means of some groups with our interventions, but homogeneity of variance is (in theory, at least) maintained. This independence of the SD from the mean sometimes breaks down when we're looking at frequency data: counts of blood cells, positive responses, and the like. If the correlation between the mean and the variance is pronounced,[16] a transformation is the way to go. A good way to check this out is visually: plot the mean along the X-axis and the variance along the Y-axis; if the line of dots is heading toward the upper right corner, as in Figure 25–7A, you've got heteroscedasticity. If the line is relatively flat, as in Figure 25–7B, there's no relationship between the two parameters.

So let's get down to the bottom line: should we transform data or shouldn't we? We would propose the following guidelines:[17]

Don't transform the data if:

1. The deviation from normality or linearity is not too extreme.
2. The data are in meaningful units (e.g., kilos, mm of mercury, or widely known scales, such as IQ points).
3. The sample size is over 30.
4. You're using univariate statistics, especially ones whose robustness is known.
5. The groups are similar to each other in terms of sample size and distribution.

Transform the data if:

1. The data are highly skewed.
2. The measurements are in arbitrary units (e.g., a scale developed for the specific study or one that isn't widely known).
3. The sample size is small (usually under 30).
4. You'll be using multivariate procedures because we don't really know how they do when the assumptions of normality and linearity are violated.
5. A large difference exists between the groups in terms of sample size or the distribution of the scores.

[16] Yet another precise term to which we can't assign a number.

[17] Bear in mind that these are just guidelines. Any statistician worth his or her salt can think of a dozen exceptions, even before the first cup of coffee.

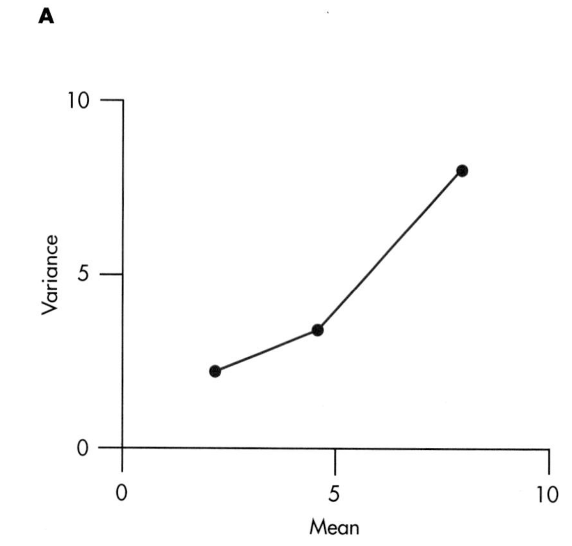

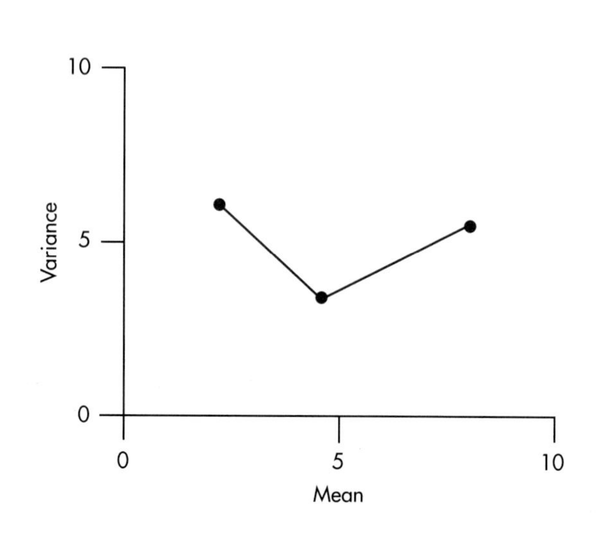

FIGURE 25–7

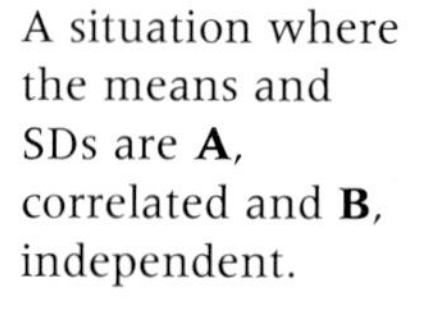
A situation where the means and SDs are **A**, correlated and **B**, independent.

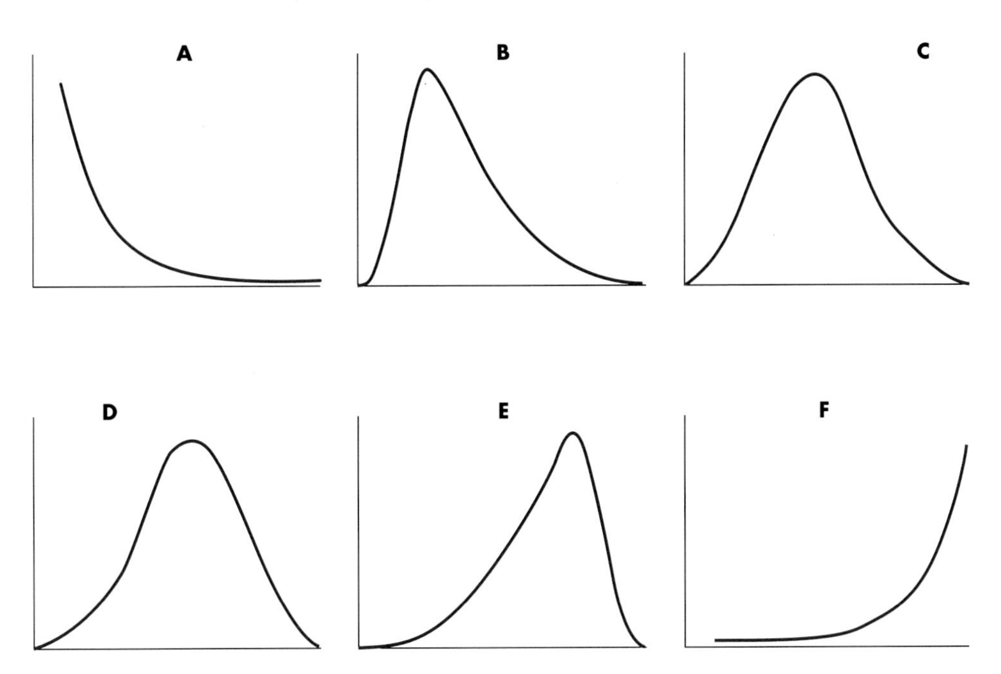

FIGURE 25–8
The "family" of distributions.

6. A moderate-to-strong correlation exists between the means and the SDs across groups or conditions.

So You Want to Transform

You've made the momentous decision that you want to transform some variables. Now for the hard question: which transformation to use? We can think of distributions as ranging from extremely skewed to the right (sort of a backward J), through normal, to extremely leftward skewed, as in Figure 25–8. In the same way, a range of transformations can be matched to the shapes almost one-to-one:

Shape	Figure	Transformation
Reverse J	25–8 *A*	$1 \div X$
Severe skew right	25–8 *B*	$\text{Log}(X)$
Moderate skew right	25–8 *C*	$\sqrt{X}$
Moderate skew left	25–8 *D*	$-1 \div \sqrt{X}$
Severe skew left	25–8 *E*	$-1 \div \text{Log}(X)$
J-shaped	25–8 *F*	$-1 \div X$

The first transformation we'll do is on these terms, by turning them into English. In fact, we can make this task even easier for ourselves; although it looks like we have six transformations here, we really have only three.[18] The -1 term in the last three rows serves to "flip" the curve over, so the skew left curves become skewed right, allowing us to use the top three transformations. Let's finish talking about this flip[19] before explaining the transformations themselves. It's obvious that if we started with all positive numbers (such as scores on some test), we'll end up with all negative ones.

Although the statistical tests don't really mind, some people have trouble coping with this. We can get around the "problem" in a couple of ways. First, before the data are transformed, we can find the maximum value, add 1 to it (to avoid too many zeros when we're done), and subtract each raw value from this number. For example, if we started out with the numbers:

1 1 2 4 8 9

then we would subtract each number from 10 (the maximum, 9, plus 1), yielding:

9 9 8 6 2 1

We would then use the transformations for right-skewed data, rather than left-skewed; that is, this reflection takes the place of dividing into -1.

The other method of eliminating the negative numbers is fairly similar[20] but takes place after we divide the appropriate denominator into the -1 term. First, find the smallest number (i.e., the biggest number if we ignore the sign); subtract 1 from it (again to avoid too many zeros); and then add the absolute value of the number to all the data points. So, if our transformed data were:

-.1 −2 −3.7 −5 −6 −11

we would subtract 1 from -11, giving us -12; the absolute value is $+12$, and the result of the additions would be:

11.9 10 8.3 7 6 1

Now to explain the transformations. The first one, $1 \div X$, is simply the **reciprocal** of *X*; if *X* is 10,

[18] *There are actually many more possible transformations, including the arc sine one, but they're rarely used, so we'll ignore them.*

[19] *The correct word would be "reflect," as in a mirror—not meaning to ponder (we never do that in statistics).*

[20] *For once, we don't get rid of the minus sign by squaring!*

the transformed value is ($1 \div 10 = 0.1$).[21] The last transformation, $-1 \div X$, is exactly the same, except that 10 now becomes -0.1 (i.e., $-1 \div 10$).

The second (and fifth) transformation involves taking the **logarithm** of the raw data. It really does not matter if you use logs to the base 10 or to the base *e*;[22] in fact, Cleveland (1984) often uses base 2 because the resulting numbers are in more easily understood units. When you use a log transformation, *be careful*; don't have any zeros or negative numbers among your raw data, or the computer will have a major infarct. If you have zeros or negative values for some variable, add a constant to each number so the smallest one is now over zero.

The **square root** transformation is similar to the log transformation in that zeros and negative numbers are taboo. Use the same technique to eliminate them.[23]

These rules may seem to imply that you look at your data, pick the right transformation, and you're off and running. Unfortunately, reality isn't quite like that. The curves we get in real life don't look like these idealized shapes;[24] they always fall somewhere in between two of the models. What you have to do is try out a transformation and actually see what it does to the data (perhaps by looking at the figures for skewness and kurtoses, or at a box plot). It's possible that you chose a transformation that overcorrected and turned a moderate left skew into a moderate right one. This gains you nothing except heartache. So, if this has happened, go back and try a less "powerful" transformation—perhaps square root rather than log, or log rather than reciprocal.

[21] *Don't get too worried that you'll have to do all these transformations by hand; at the end of the chapter, we'll show you how to get the computer to do the work for you.*

[22] *If you don't know the difference, it matters even less.*

[23] *Needless to say, this presupposes that you've looked at your data beforehand and know if you have any zeros or minus signs. If you haven't looked, go back to Chapter 2 and start reading all over again (and miss your dessert, too, as extra punishment).*

[24] *This is beginning to sound like a commercial for unmentionable undergarments.*

How to Get the Computer to Do the Work for You

Finding Cases that are Outliers

- From **Analyze,** choose **Descriptive Statistics → Explore**
- Select the variables in the left column, and move them into the **Dependent List**
- Click **OK**

Imputing Missing Values

- From **Transform,** choose **Replace Missing Values**
- Select the **Method** [try Linear Trend at Point or Series Mean]
- Select the variable from the left column and click the arrow key
- Repeat this for all the variables [you can use different transformations for each variable]
- Click **OK**

  [New variables are created with the imputed values replacing the missing data.]

Transforming Data

- From **Transform,** choose **Compute**
- Type in a new variable name in the **Target Variable** box
- Select or type in what you want to do in the **Numeric Expression** box; for example, to take the natural logarithm of variable **VAR1**:
  - Type **VAR2** in **Target Variable**
  - Choose **LN(numexpr)** from the **Functions:** box and click the up arrow
  - Select the variable to be transformed from the list, and click the arrow

CHAPTER THE TWENTY-SIXTH

Putting It All Together

In this chapter, we provide some final signposts: (1) flow charts to help you select the right test, (2) simplified sample size calculations, and (3) names of some software available for doing sample size calculations.

SETTING THE SCENE

As a result of reading this book to the end, you are fired up with enthusiasm for the arcane delights of doing statistics. You rush out to the local software house, drop piles of your hard-earned shekels on the table, and buy the latest version of SPSS. You cram it into your PC, sacrificing some neat computer games along the way. And there you sit, like the highwayman of yore, ready to pounce on the next unsuspecting data set that passes your way. In due course it arrives, and suddenly you are faced with the toughest decision of your brief career as a statistician, "What test do I use???"

Every professional has his or her top problems on the hit parade. For family docs, it's snotty-nosed kids and high blood pressure; for neurologists, it's migraines and seizure disorders; for respirologists, it's asthma and COPD; and for psychiatrists, it's depression and schizophrenia. Routine is a depressing fact of the human condition. As one psychologist put it, "An expert doesn't have to solve problems any more."

And so it is for statisticians. Ninety percent of the lost souls who enter our offices come with one of two questions. If they have bits of ragged paper covered in little numbers, it's, "What test do I use?" And if they come with a wheelbarrow full of grant proposals,[1] it's "How big a sample do I need?"

In this last chapter, we hope to help you answer these questions all by yourself. The chapter is admittedly self-serving, because unlike some health professionals, we rarely charge for our advice. If we do this right, some of you may learn enough that you need not bother us or other members of our clan with one of these questions, so we can stay home, write books, and make royalties.

DESCRIPTIVE STATISTICS

The flow chart for descriptive stats is shown in Table 26–1. The first decision point is between one variable and two—whether you are looking at distributions or associations. The next step, in either case, is to dredge out some definitions. Decide if the variable is **nominal**—a frequency count in one of several named categories, **ordinal**—ranked categories or actual rankings, or **interval or ratio** (the distinction is unimportant)—a measured quantity on each subject. Some judgement calls must be made along the way, of course. Will you treat the responses on the 7-point scales as ordinal or interval data? The answer depends, at least in part, on the journal you are sending your results to. Of course, as you move to extremes, it becomes clearer. A 2- or 3-point scale really should be treated as frequencies in categories; conversely, a sum of 10 or 20 ratings, regardless of whether they comprise 2-point scales (e.g., "Can you climb the stairs?") or 7-point Likert scales, can justifiably be treated as interval data.

From here on in, it's easy. Let's deal with the description of single variables first. If the variable is interval or ratio, the appropriate statistics are the mean and SD (and additional measures of skewness and kurtosis, if it suits your fancy). Several graphing methods are suitable—stem-leaf plots for information from the raw data, histograms or frequency polygons to show the data graphically, and box plots to summarize the various statistics.

[1] *Why is it that it takes more text to describe the study you're going to do than to describe the one you did? Typically, granting agencies allow 20 pages, or 5000 words, for the proposal, but journals allow only 2500 to 3500 words for the finished product.*

TABLE 26–1 Descriptive statistics and graphs

Level of measurement	What are you looking for?	Statistic	Graph
Number of variables—1			
Ratio or interval	Central tendency	Mean	Histogram Stem-leaf plot Frequency polygon Box plot
	Dispersion	Standard deviation	
	Deviation from normality	Skew, kurtosis	
Ranks (ordinals)	Central tendency	Median, mode	Bar chart Box plot
	Dispersion	Range, IQR, Index of Dispersion	
Frequencies categories (nominal)	Central tendency	Mode	Bar chart Dot plot
	Dispersion	Number of categories	
Number of variables—2			
Interval/Ratio	Association	Pearson's r	Scatter plot
Nominal/Ordinal	Comparison	Kappa, phi, rho, weighted kappa	Paired bar chart Box plot

For ordinal data, means and SDs are replaced with medians, modes, and ranges or interquartile ranges. In displaying the data, connecting the dots is out of order, so we use bar charts and box plots only.

Finally, for nominal data, about all we can use to summarize the data is the mode (most commonly occurring category) to indicate central tendency and the number of filled categories to show dispersion. The data are displayed as a bar chart or dot plot (point graph).

What about showing the association between variables? For interval and ratio variables, the Pearson correlation is the only accepted measure. For categorical nominal variables, there are several contenders, but leading the pack are phi and Cohen's kappa. For ordinal data in categories, weighted kappa would be used; if the data are ranked, then Spearman's rho is the most useful measure.

The association between interval/ratio variables is illustrated with a scatter plot. With nominal variables, we can use a paired bar chart to display frequencies or a box plot when one variable is nominal and the other is interval/ratio (i.e., two groups).

UNIVARIATE STATISTICS

Now we get on to the bread and butter of stats—inferential statistics. The tables are organized more or less as was Table 26–1 on descriptive stats.

Once again, we begin by deciding whether the dependent variable is a measured quantity—an **interval** or **ratio** variable, a rank or **ordinal** variable, or a frequency or **nominal** variable. Interval and ratio variables are analyzed with parametric statistics, as in Chapters 7 through 19, and illustrated in Tables 26–2 and 26–3. Ranks and frequencies are analyzed with nonparametric statistics in Chapters 20 to 24 and covered in Table 26–4. Once this separation is made, we spell out the specifics for the two forms of statistics.

Parametric Statistics

The next major concern is with the independent variable(s), as shown in Table 26–2. If it is (or they are) also measured (interval or ratio) variables, then you are getting into examining the association among the variables with some form of regression analysis. If you have one variable, it's simple regression, and the measure of association is the Pearson correlation, r. If you have more than one independent variable, then the game is multiple correlation, with all its complexities, and the overall measure of association is the multiple correlation, R.

By contrast, independent variables that are categorical lead to tests of differences among means, t-tests, and ANOVA methods. To sort out all these complexities, look at Table 26–3.

The first issue in arriving at the right test of differences among means is to examine the design. The two classes of simple designs are (1) those which involve **independent** samples, where subjects are randomly assigned to groups, and (2) those which involve **related** samples, where one measure is dependent on another. That is, studies that involve matched controls, pretest and post-test measure-

TABLE 26–2 Parametric statistics (ratio or interval data)

Independent variable	Number of independent variables	Method	Measure of association	Chapter
Ratio or interval	1	Simple regression	*r*	13
	>1	Multiple regression	*R*	14
		Logistic regression		
Categorical	1	*t*-test and ANOVA (see expansion)	eta^2	8
	>1	ANOVA (see expansion)	eta^2	8
Interval, ratio, and categorical	Any	ANCOVA		16

TABLE 26–3 Analysis of variance

Independent/ related samples	Number of variables	Number of levels	Method	Chapter
Independent	1	2	*t*-test	7
		>2	One-way ANOVA	8
	≥2	≥2	Two-way ANOVA	9
		≥2	Factorial ANOVA	
Related	1	2	Paired *t*-test	10
		>2	Repeated-measures ANOVA	11
	>1		Repeated-measures ANOVA	11

ments, or other situations with more than one measurement on each case, are called related samples.

The simplest of the independent sample tests is the *t*-test, which involves only one grouping variable with only two levels (in simple language, two groups). If you have one independent variable but more than two groups, then you use one-way ANOVA.

Next in complexity is the consideration of more than one independent grouping factor. If you have just two factors (regardless of how many levels of each), it's two-way factorial ANOVA. If you have more than two factors, then you are doing (generic) factorial ANOVA, which is a label attached to any number of wild and woolly[2] designs (and also includes the two-way case).

Finally, the most general methods, which apply to all mixtures of interval/ratio and categorical independent variables, use analysis of covariance (ANCOVA) (back to Table 26–2).

Having spelled out all these intricacies, keep in mind that most of these methods are rapidly becoming historical oddities. Any of the simpler tests are also special cases of the more complicated ones. Factorial ANOVA programs can do two-way ANOVA, one-way ANOVA, and *t*-tests. So bigger fishes continue to eat littler fishes all the way down the line.

Why bother with all this, "What test do I use" nonsense when actually one test will do? Several reasons are listed below:

1. It shows that you are well grounded in the folklore of statistics.
2. It helps you understand what less erudite people did when they analyzed their data.
3. The general programs, because they are more general, often require many more set-up specifications.
4. Last, you may just find yourself somewhere in the middle of a campground with no electric-

[2] *Time for one last joke before we leave you. A train was crossing the Scottish border on its way from London to Edinburgh. In the compartment were three professors—a physician, a statistician, and a philosopher. They spied a herd of black sheep, whereupon the conversation went as follows:*

TABLE 26–4	Independent variable	Number of levels	Method	Measure of association	Chapter
Nonparametric statistics	**Ranks (ordinal)**				
		2	Wilcoxon rank sum		22
			Mann-Whitney U		22
	Categorical independent				
		>2	Kruskal-Wallis		22
		2	Wilcoxon signed rank		22
	Categorical related				
		>2	Friedman test		22
		2		Spearman's rho	23
				Kendall's tau	23
	Ranks				
		>2		Kendall's W	23
	Categories (nominal)				
		1	Chi-squared		20
		(independent)	Fisher's exact		20
	Categorical				
		1	McNemar	Phi	20/21
		(related)	chi-squared	Cramer's V	21
				Kappa	21
		1	Life-table		24
		2	Mantel-Haenszel chi-squared		20
		2	Life-table		24
		>2	Log-linear		20
	Any	>2	Logistic regression		15
		≥2	Cox model (life-table)		24

Physician: "In my experience, all sheep in Scotland are black."
Statistician: "You know, on statistical grounds, you can't really conclude that. All you can really say is that some *sheep in Scotland are black."*
Philosopher: "No, dear boy, that is incorrect. Logically all you can conclude is that one side *of some sheep in Scotland is black."*

ity and only a solar calculator. It's really handy to remember some of the simple strategies in this situation.

Nonparametric Statistics

On to nonparametric statistics. The major division facing you is between ranked data, which are ordinal and categorical data, which are usually nominal (but may be ordinal, such as Stages I, II, and III) (Table 26–4). If they're ranked data, then the next distinction is between independent and related samples, as in the parametric tests. For independent samples and ranked data, we are looking at the ordinal equivalent of the *t*-test (two groups) and one-way ANOVA (more than two groups, which are the Wilcoxon rank sum test [Mann-Whitney U] for two samples and the Kruskal-Wallis one-way ANOVA by ranks for more than two groups. If the samples are related (matched pairs or repeated observations), then the equivalent of the paired *t*-test is the Wilcoxon signed rank test. And for more than two groups or observations, we use the Friedman test for significance testing. Finally, if we are examining the relationship between two ranked variables, we can use Spearman's rho or Kendall's *W* (the former is preferred). If we want an overall measure of association among three or more rankings, analogous to an intraclass correlation coefficient, Kendall's *W* does the trick.

For categorical variables (see Table 26–4), the nonparametric tests concentrate on cross-classification and contingency tables (i.e., both independent and dependent variables are categorical). Note that on several occasions such as the discussion of log-linear models, we have collapsed this distinction. In fact, the distinction between independent and dependent variables is more of a **design** consideration than an analysis decision. For example, what nonparametric test do we use when we are exploring the relationship between height of professors and tenure status (knowing the bias of at least one of the authors)? The dependent variable is tenure status (yes/no), and the independent variable is height. The appropriate test is a *t*-test, contrasting the mean heights in the tenured and untenured groups. In short (for a change, no pun intended), statistics is indifferent to the causal direction of the variables; only interpretation cares.

Now let's run through the cookbook. If you have two categorical variables, the standard line of defense is the chi-squared test. If the **expected** frequency in any cell of the continency table is less than five, then the Fisher exact test should be used, but unfortunately this works only for 2 × 2 tables. If you have a larger table and a low expected frequency, it may be possible to collapse some cells to get the counts up without losing the meaning of the analysis. If you have three categorical variables, use the Mantel-Haenszel test. And if you have more than three, then the log-linear heavy artillery emerges. Finally, logistic regression, treated in Chapter 15, is a general strategy when the dependent variable is dichotomous and the independent variables are mixed.

Two quick detours: (1) If the samples are related or matched, with two variables, then a McNemar chi-squared is used. With more than two variables, no approach is available. (2) With survival data, you first construct a life table, then do a Mantel-Haenszel chi-squared. Then, to examine predictors of survival, the Cox proportional hazards model is appropriate.

MULTIVARIATE STATISTICS

Insofar as multivariate ANOVAs are concerned, we'd prefer not to repeat ourselves. So, just re-read the section on ANOVAs, and stick an M in front of every mention of the term ANOVA; it's that simple. For every ANOVA design (one-way, factorial, repeated-measures, factorial with repeated measures, fixed factors, random factors, ad infinitum), there's an equivalent MANOVA for those situations when you have more than one dependent variable. And now, the fish gets even bigger—a two-way MANOVA can do a one-way MANOVA, which can do a Hotelling's T^2, which can do a t-test.

If you just have independent variables with no dependent variable and want to see what's going on, use exploratory factor analysis. If you already have some idea of what's happening and want to check out your hypotheses, a more powerful test (in terms of the theory, not necessarily in terms of statistical power) would be confirmatory factor analysis. Structural equation modeling allows you to simultaneously do a number of CFAs to define latent variables and then look at the partial regressions among (or amongst, for our British readers—both of them) the latent variables.

QUICK AND DIRTY SAMPLE SIZES

We have spent an inordinate amount of time describing one approach after another to get a sample size, all the while emphasizing that nearly all the time the calculated sample size, despite its aura of mathematical precision, was a rough and ready approximation—nothing more. Well, someone has called our bluff and, along the way, made the whole game a lot easier. Lehr (1992) invented the "Sixteen s-squared over d-squared" rule, which should never be forgotten.

It goes like this. Recall that the sample size for a t-test is as follows:[3]

[3] *Actually don't bother recalling; we're giving it to you anyway.*

$$n = 2\frac{[(Z_\alpha + Z_\beta)\sigma]^2}{\delta^2} \quad \textbf{(26–1)}$$

where σ is the joint SD and δ is the difference between the two means. Now if we select $\alpha = .05$ (as usual), $Z_\alpha = 1.96$. If we pair this up with $\beta = .20$ (a power of .80), then $Z_\beta = .84$. And $2\,(1.96 + .84)^2 = 15.68$, which is near enough to 16. So the whole messy equation reduces to something awfully close to:

$$n = \frac{16\,s^2}{d^2} \quad \textbf{(26–2)}$$

if we just abandon the Greek script and call δ "d" and σ "s." Say it together now, class: "Sample size equals 16 ess squared over dee squared."

"Ah, yes," sez you, "But what about all the other esoteric stuff in the other chapters?" Well, continuing in the rough and ready (R and R) spirit, let's deal with them in turn. Here we go.

Difference between Proportions

The SD of a proportion is related to the formula $\sqrt{p(1-p)}$, where p is the proportion. If you want to be sticky about it, there are different SDs for the two groups, but they usually come out very close. So the R and R formula is:

$$n = 16\frac{p(1 - p)}{(p_1 - p_2)^2} \quad \textbf{(26–3)}$$

where p_1 and p_2 are the two proportions and p is the average of the two.

Difference among Many Means

One-way ANOVA. Pick the two means you really care about then apply the formula, and this tells you how many you need for each group. If you have a previous estimate of the Mean Square (Residual), use this for s^2.

Factorial ANOVA. Same strategy. Pick the difference that matters the most, and work it out accordingly. If you are nit-picky, add "1" per group for each other factor in the design, but this is not in the spirit of R and R calculations.

Correlations. We told you the fancy formula already, in Chapter 13. But to test whether a correlation is significantly different from zero, you can use this formula with the knowledge that the SE of the correlation is about equal to $1 \div \sqrt{(n-2)}$. The formula then becomes:

$$n = 4 + \frac{8}{r}$$

(26–4)

So perhaps you (and we) can relegate the high-powered formulae to the back burner. Certainly one of the beauties of the rule of 16 is that it brings into sharp focus some of the properties of the relation between sample size and differences. Everything is squared, so if you double the difference you want to detect, you cut the sample size by a factor of four. If you double the SD, the sample size goes up by a factor of four. Incidentally, that also explains how we statisticians are so successful at making the calculated sample size exactly equal the number of available patients. All you need do is make plausibly small adjustments in the initial estimates, and they can have big effects on the calculation.

How to Get the Computer to Do the Work for You

This time, you can forget about SPSS. We have come across two different sample size programs for the PC. They are written as "shareware," so you are at liberty to copy the program. If you like it, you donate a fixed sum ($15) to the author. The one we know well (and goodness knows, there may be many more) comes in two versions. Both are called PC-SIZE, and the author is Gerald E. Dallal, 53 Baltran Street, Malden, MA, 02148. It is described in the *American Statistician* (Dallal, 1986, 1990).

The earlier version does sample size calculations for complex designs: one-factor, two-factor, and randomized-blocks ANOVAs, paired *t*-tests, correlation coefficients, and proportions. The 1990 version does sample size and power calculations for paired and unpaired *t*-tests and chi-squared on two independent proportions. For once, we can honestly say that no instructions are needed; the programs really are user-friendly and self-explanatory. Just type in "SIZE" or "PC-SIZE" and follow orders. There are also a number of more sophisticated (read "more expensive") commercial programs. One is called PASS, which is distributed by NCSS, 329 North 1000 East, Kaysville, UT, 84037. Another is SOLO, which is now distributed by SPSS.

Test Yourself

Being a Compendium of Questions and Answers

The purpose of this section is for you to see how well you've picked up the material so far. It consists of two types of problems: abstracts of real articles, and studies we have made up for the occasion.[1] *With the abstracts, we've deleted all the irrelevant stuff and any mention of the statistics they used. Your job is to figure out the correct statistical test to use to analyze the data. The answer section gives what* we *think should have been done, which in some cases is not what actually was done.*[2] *Of course, this is only our opinion, and you are at liberty to disagree.*[3]

If you pass this test, it's obviously a testament to our superb skills as educators. If you fail, though, it's just as obviously your fault for not paying close enough attention, so go out and buy another copy of this book right now!

[1] *You can tell which are real articles—a reference is given, and the clinical questions are far more mundane than the ones we made up.*

[2] *If no comments are made, we go along with how the authors handled the data.*

[3] *You are also at liberty to write your own statistics book.*

QUESTIONS

Problem 1. Andersen et al (1990) compared the excess mortality rate following transurethral resection of the prostate versus the more traditional, open resection in men with benign hypertrophy. They used hospital data, following 38,067 cases for up to 10.5 years.[4] However, this was not a randomized trial, and the two groups differed in terms of age and previous health status. How did they do it?

[4] *"Following" is an epidemiological term, meaning that they kept track of the cases through medical records; the authors did not hire Bulldog Drummond or Sam Spade to shadow the patients.*

Problem 2. It was found that more male patients with ocular rectitis (OR) had a family history of hemorrhoids than did the healthy controls. In addition, it was noted that such individuals tended to wear tight underwear more than did the controls. How would you analyze these data?

Problem 3. A retrospective study looked at risk factors for Chronic Fatigue Syndrome. Fifteen CFS sufferers were matched by age and sex to 15 controls in the same company. They examined three predictor variables—Life stress score (0–64), Locus of control (Internal [0] or External [1]), and White collar (1) or Blue collar (2)—to see what best predicted CFS. What analysis would you do?

Problem 4. Patients with chronic obstructive pulmonary disease were randomly assigned to either a comprehensive rehabilitation program or an educational control program. The primary outcome variable was exercise endurance, as determined by treadmill time, measured monthly over a 6-month follow-up period (Toshima, Kaplan, and Ries, 1990).

Problem 5. A scotchophile wanted to see whether other connoisseurs could really discriminate single malt from blended scotches, or well-aged from relatively young scotches. He assembled 4 scotches of each type (8 years old/12 years old, and single malt/blended) and had them rated for quality by a panel of 5 judges, each judge rating all 16 samples. Compliance exceeded 100%, although some of the later ratings were nearly indecipherable.

Problem 6. To judge the effect of this book, your intrepid authors gave away free copies to a bunch (n = 34) of undergraduates on the condition that they take the test you are now taking (a) before they left the bookstore, and (b) after they read the book. Mean percent score was 23% (SD 14%) in the pretest and 45% (SD 12%) at the post-test.

Problem 7. A palm reader (of the hands, not the dates) hears of the success of ear creases in predicting coronary artery disease and wonders if it generalizes to other body parts, specifically those she can exploit. She assembles a bunch of heart attack victims (n = 12) from an old folks' home and also a control sample. She counts the number of wrinkles on the middle knuckle of each finger and each toe (excluding thumb and big toe) and determines whether any of these can differentiate between heart disease patients and healthy people.

Problem 8. To determine if the prevalence of phobias is different among older men and women, and if the prevalence changes with age, 512 people between the ages of 50 and 89 were given a telephone-administered questionnaire (Liddell, Locker, and Burman, 1991). Then what did the researchers do?

Problem 9. Marshall et al (1991) compared recidivism rates among male exhibitionists who received or did not receive therapy.[5] Recidivism was a binary variable—occurring or not occurring within the follow-up period.

[5]Fortunately for the patients, treatment was not vivisection therapy, which is based on the biblical injunction, "If thine eye offend thee, pluck it out."

Problem 10. In a study of 50 mammograms, two observers classified each film as "Normal," "Suspicious—Repeat Test," or "Likely Malignant." How well do they agree with each other?

Problem 11. To test the hypothesis that sinistrality[6] is associated with decreased "survival fitness,"[7] Coren and Halpern (1981) compared the longevities of right-handed versus left-handed baseball players; how did they?

[6]That's a fancy term for left-handedness.

[7]Another fancy euphemism meaning they die at an earlier age.

Problem 12. To see if attractiveness has any bearing on performance on oral examinations, class photographs of 50 final-year dental students were ranked by 2 patients from most to least attractive. These rankings were pooled and then compared with their class standing on the final oral examination.

Problem 13. Because "Hispanics, being a Mediterranean people, tend to put statements in relatively strong terms," Hui and Triandis (1989) hypothesized that they would be more likely to use the extreme ends of 5-point rating scales than would non-Hispanics. They also hypothesized that this difference would disappear when the scale had 10 points, rather than 5. Each subject completed 165 items, using either a 5- or 10-point rating scale.

Problem 14. In another trial of the wonder drug Clamazine, patients evaluated their itchiness on an 11-point scale before and after using the medication. For the 208 patients, their itchiness before Clamazine was 9.5 (SD = 4.2) and 5.7 (SD = 2.8) after the drug.

Problem 15. In the previous problem (number 14, if you've lost track), is there sufficient information to proceed to calculate the test statistic?

a. Yes, we know the means, SDs, and sample size
b. No, we don't know the comparable data for the control group
c. No, we don't know the SD of the differences
d. No, we don't know the *df*

Problem 16. The local dermatologist, building on the growing interest in clam juice for psoriasis, did a study of varying dosage regimens. He looked at 30 ml b.i.d. (twice a day), 60 ml daily, and 20 ml t.i.d. (three times a day). Twelve patients were assigned to each cohort for a 2-month period, and extent of lesions measured at the end of the trial.

Problem 17. Weiss and Larsen (1990) hypothesized that scores on the four subscales of the multidimensional Health Locus of Control (HLC) scale and a Health Value Index would individually and together predict participation in "health-protective behaviors" (HPB), such as using seat belts and undertaking vigorous exercise. HPB was measured on a 10-point scale.

Problem 18. Leary and McLuhan investigated whether any association existed between pot smoking in the 1960s and cocaine addiction in the 1990s. They did a case control study involving 50 coke addicts from the Bay area and 50 normal controls, and they inquired whether these folks were potheads in the 1960s (Never, Occasional, Continuous). The association was significant (chi-squared = 4.56). How could they measure the strength of association?

Problem 19. Minsel, Becker, and Korchin (1991) looked at whether "mental health" was seen the same way in four different cultures (United States, France, Germany, and Greece). The questionnaire consisted of 186 items answered on a 5-point scale. The primary aim was to look at the *relative* importance of each item, rather than the absolute value. Unfortunately, there were cultural differences, in that Greeks used the higher end of the scale more than did the other groups, and the Germans and Americans used the lower end more often.

a. How did they eliminate this "cultural bias?"
b. How did they see if the four groups had similar concepts of "mental health?"

Problem 20. A local clinician became convinced that, among other evils, smoking causes cirrhosis of the liver, because "if you go into any bar, they are all smoking." Smoking causes drinking, which causes liver damage. On reflection, it might be desirable to look at the effects of both smoking and drinking (categorized as smoker/nonsmoker, drinker/nondrinker), on a sample of cirrhosis patients and controls.

Problem 21. To see if proxy assessments by relatives could substitute for patient assessments of physical and psychosocial health status, Rothman et al (1991) had 275 patient-proxy pairs complete the Sickness Impact Profile. How did they evaluate the similarity?

Problem 22. College fraternities and sororities traditionally run "Dog Pools" on prom nights. All contribute and the one who ends up with the most unattractive mate wins the pool. More often than not, it seems that the winner of the fraternity dog pool is paired with the winner of the sorority dog pool. To test this scientifically, two frat rats used the graduation pictures of all concerned and rank ordered them by male or female pulchritude, as the case may be. They then determined who was paired with whom on prom night and looked for a measure of association.

Problem 23. Bennett et al (1987) used a randomized trial to improve students' knowledge of critical appraisal in two areas: diagnosis and therapeutics. They administered a pretest and a post-test in each area to the treatment group (which received training in critical appraisal) and the control group (which did not). For the treatment group, the paired t-test was highly significant ($p < .001$) for diagnosis, and it was significant for therapy ($p < .01$). For the control group, neither t-test was significant. They concluded that critical appraisal works. Would you have analyzed the study this way?

Problem 24. Feighner (1985) ran an RCT of patients on the new wonder drug Prozac against the old stand-by, amitriptyline. Each group had 22 patients, who were assessed at baseline on three measures—the HAM-D, the Raskin depression scale, and the Covi anxiety scale—and then weekly thereafter for 5 weeks. A one-tailed Wilcoxon signed rank test was used to compare improvement from week 0 to week 5.

Problem 25. Physicians in Ontario are all subjected to regular peer review of records, and those who have problems identified are sent for a further 2-day assessment that includes various measures—simulated patients, oral examinations, chart review, O.S.C.E. (Objective Structured Clinical Examination), and written tests. The question posed was whether some identifiable underlying components are assessed by all these measures. Can statistics help?

Problem 26. A concern was that Medicare beneficiaries who join health maintenance organizations (HMOs) were sicker than people not on Medicare. To check this out, Lichtenstein et al (1991) looked at a 9-level functional health status measure in patients in 23 HMOs. How did they determine if health status differed between recipients and nonrecipients of Medicare?

Problem 27. To see whether reaction times in traffic situations deteriorate in low-light conditions, a simulator was set up in which the same traffic situations could be displayed under high- and low-light conditions. Subjects were tested with a series of 20 videotaped traffic situations, where 10 were in daylight and 10 were at night, and response times were measured. Because RT has a severely skewed distribution, it was analyzed with a nonparametric test.

Problem 28. Thomas and Holloway (1991) investigated whether unplanned hospital readmission was related to hospital size, length of stay, discharge to home versus an organized care facility, teaching status of the hospital, and so on.

Problem 29. Sorenson et al (1991) wanted to compare the age of onset of any depressive disorder in the population for (a) men versus women and for (b) non-Hispanic whites versus Mexican-Americans born in the United States versus Mexican-Americans born in Mexico.

Problem 30. Does the high school yearbook have any predictive validity? To investigate this, students in the graduating class of one midwestern school were rank ordered by teachers on their likelihood of success. Ten years later, their success was assessed by educational attainment—no completed postsecondary education, bachelor's degree, or graduate degrees (this is a measure of success?).

ANSWERS

Problem 1. They used a survival analysis, with the Cox proportional hazards approach.

Problem 2. There are three variables—one dependent (OR yes/no) and two independent variables (Jockey/boxer shorts; family history yes/no). Mantel-Haenszel chi-squared or log-linear analysis; the choice is yours.

Problem 3. To examine variables individually, you could do a paired t-test on the life stress score and the McNemar chi-squared on locus of control and white/blue (remember that it is a matched design). To see what combination best predicts CFS, use logistic regression. But the matching creates real problems, as these procedures are really for independent samples. Also note that, in doing this, we have interchanged the independent and dependent variables for the purpose of analysis. Now CFS/Normal is acting like a dependent variable. This happens often in statistics and is of no consequence, as the computer doesn't know which is which.

Problem 4. The data were analyzed using a repeated-measures ANOVA, with Treatment (rehab versus education) as a between-subjects factor and Time as a within-subjects factor.

Problem 5. The analysis is a factorial ANOVA. There are two grouping (between subject) factors: (1) young/old, and (2) single malt/blended; and one within subject factor (rater). The real trick is that "subject," in this case, is the scotch, which is the object of measurement. If you like, the design looks like this:

			Rater				
		Scotch	**A**	**B**	**C**	**D**	**E**
		1					
		2					
	Young	3					
Single		4					
malt		5					
	Old	·					
		8					
		9					
	Young	·					
Blended		12					
		13					
	Old	·					
		16					

Problem 6. Because it's a pretest–post-test design, a paired *t*-test would do. Repeated-measures ANOVA gives the same answer.

Problem 7. Repeated-measures ANOVA. There is one between-subject factor (heart attack/normal) and three repeated-measures (hand/foot, left/right, and first, second, third, little) factors, so there are 2 × 2 × 4 measures on each subject. By the way, there again we have flipped the independent and dependent variables, treating heart attack/normal as an independent variable. No one cares.

Problem 8. They broke the subjects down into three age groups (50 to 64, 65 to 74, and 75+), and then used an Age × Sex ANOVA. Although correct, another approach would be to use a multiple regression, with Age and Sex as the predictors. This would preserve the ratio nature of Age and not force it into arbitrary categories.

Problem 9. The data were analyzed with a 2 × 2 chi-squared.

Problem 10. Because the films are classified in three ordinal categories, a weighted kappa is appropriate.

Problem 11. Coren and Halpern argued that a *t*-test wouldn't be appropriate because the data are highly skewed. They used the Wald-Wolfowitz Runs test, which is a nonparametric test of differences between groups. You could also do a life-table analysis on the data.

Problem 12. Two rankings on each of 50 students were compared. Use a Spearman rank correlation.

Problem 13. They used a 2 (Hispanic versus non-Hispanic) × 2 (5- or 10-point scale) factorial between-subjects ANOVA, with the dependent variable being the number of times an extreme response category was chosen.

Problem 14. Because the data are continuous, forget about χ^2s, of any flavor. Because both values are collected from the same subject, we need a paired, as opposed to an unpaired, *t*-test. (If you said a repeated-measures ANOVA, give yourself ½ a point; a paired *t*-test is a form of ANOVA, but it's much easier to calculate if you've got only two values.)

Problem 15. The correct answer is *c*. The SDs that are given are *between* subjects. The whole point of the paired *t*-test is that it uses the *within*-subject SDs—that is, the SD of the differences, which we don't have.

Problem 16. The quantitative doses suggest a regression problem; however, the total daily dose is the same in all schedules, so the differences are qualitative. A straight one-way ANOVA is appropriate. However, the sensitivity of the experiment would be enhanced by measuring at baseline and doing ANCOVA with baseline measure as covariate.

Problem 17. They used a multiple regression, with the HLC subscales and the health value as predictors and the HPB score as the dependent variable. The correct way to see if any interaction between HCL and health value exists would be to create a new variable that is the product of these two (i.e., an interaction term). The authors state that, "because of the unusually high multicollinearity," this could not be done; they ended up dichotomizing health value and running separate regressions for the two groups. This study was very well analyzed.

Problem 18. Several epidemiologic measures of strength of association exist, such as odds ratio or log likelihood ratio. On the statistical side, we could use one of the measures based on chi-squared (phi, Cramer's *V*, contingency coefficient). Note that we cannot use a kappa or any of the other measures that depend on a 2 × 2 table, as this is a 2 × 3 table.

Problem 19. (*a*) The data for each subject were transformed into standard scores, with a mean of 100 and an SD of 10. (*b*) The data were factor analyzed for each group separately, and the factor matrices were examined for comparability.

Problem 20. Because there are three factors and all are categorical, the choice is between Mantel-Haenszel and log-linear analysis.

Problem 21. They first correlated the two sets of scores to see if they were associated with one another and then did a paired *t*-test to look for any systematic bias. They could not have used an independent (unpaired) *t*-test because, although the two scores came from different people (patient and proxy), they were about the same person (the patient). But if you wanted to look at agreement, it would be better to use an intraclass correlation coefficient.

Problem 22. Use a measure of association for ranked data—Spearman rho or one of the alternatives.

Problem 23. Approaching the analysis this way essentially ignores the control group. They should have done an *unpaired t*-test on the difference scores, contrasting the treatment with the control group. In fact, the investigators reported both analyses.

Problem 24. First, using a nonparametric procedure on these data is unnecessary. There probably isn't much loss of power, but it does limit the analysis. Second, they threw out the data from weeks 1 to 4 to do this test. What they should have done was three ANCOVAs (one for each variable), with the baseline as the covariate and weeks 1 to 4 as repeated measures. And, by the way, the one-tailed test is really hard to justify on this occasion.

Problem 25. Factor analysis would determine whether scores group into homogeneous factors.

Problem 26. They ran *t*-tests between the two groups for each of the 23 HMOs. It would have been much better to do one ANOVA, with two between-subject factors: Medicare status (two levels) by HMO (23 levels). This would have allowed them to see if differences existed among the HMOs, as well as avoid the problem of running so many *t*-tests.

Problem 27. This is a within-subject design, with an average RT for day and night for each subject. Use a Wilcoxon signed rank test. Alternatively, as you actually have 20 RTs per subject, you might want to transform using a log transformation to reduce skewness, then do a repeated-measures ANOVA with two factors—Day/Night, and specific scenario (10 levels).

Problem 28. For each of 22 diagnosis-related groups (DRGs), they ran stepwise logistic regressions. This makes sense because admission is a binary outcome, and there is little assurance of multivariate normality among the predictor variables.

Problem 29. They began by simply looking at the median age of onset. Recognizing that, because the subjects could be any age at the time of the interview, and therefore at risk of becoming depressed for varying lengths of time, they also used a survival analysis.

Problem 30. Three groups, ordinal ranks, Kruskal-Wallis one-way ANOVA by ranks.

Answers to Chapter Exercises

CHAPTER 1

1. a. The IV is *drug* (ASA or placebo). The DV is *the number of coronary events.*
   b. The intention was for *cholesterol level* to be the IV, with *cancer* as the DV (i.e., the probability cancer is dependent on your triglyceride level). As a matter of fact, it is more probable that the relationship goes the other way—CA may reduce cholesterol level. So, any time you simply look at the relationship between two variables, it is difficult to categorically state which is the IV and which the DV.
   c. The IV is *group membership*. The DV is the *quality of life score.*
   d. Again, they probably meant for *occupation* to be the IV and *CHD* the DV. However, because it hasn't been shown that a causal link exists between the two (in fact, *body mass* may explain both coronary morbidity as well as opting for a more sedentary job), we can't really call one the IV and the other the DV.
2. a. A number of sessions is a *discrete* variable.
   b. Time is a *continuous* variable.
   c. Money is *discrete* because it is not divisible beyond cents.
   d. Because your after-taxes income will be $0, it doesn't apply.
   e. Weight is *continuous.*
   f. The number of hairs is *discrete.*
3. a. Income is *ratio.*
   b. A list of specialties is *nominal.*
   c. The ranking of specialties is *ordinal.*
   d. This scale is *ordinal*, bordering on *interval.*
   e. ROM is *ratio.*
   f. Strictly speaking, scores on a questionnaire are most often *ordinal;* although the intervals between successive scores on the test are equal, they likely don't reflect equal increments in anxiety. In actuality, though, we'd likely end up treating them as if they were *interval.*
   g. Staging is *ordinal.*
   h. ST depression is *ratio.*
   i. Grouping the data has changed it into an *ordinal* scale.
   j. A list of diagnoses is *nominal.*
   k. BP (systolic, diastolic, or other) is *ratio.*
   l. As with *f*, an *ordinal* scale that we often treat as *interval.*

CHAPTER 2

1. Histogram.
2. Frequency polygon or histogram.
3. Bar chart. A bar chart is preferable to a histogram in this situation because the category "2+" makes the underlying scale more ordinal than interval. If the data weren't grouped at the end, a histogram would be appropriate.
4. Frequency polygon.
5. Bar chart, with the specialties rank-ordered by income. If there are many specialties, it may be better to use a point graph.
6. Frequency polygon.
7. Histogram or frequency polygon.

CHAPTER 3

1. a. Mean, SD.
   b. Because there was no upper limit, the data are probably skewed to the right, so the median and interquartile range would probably be better than the mean and SD.
   c. Mean, SD.
   d. Mode, none.

2. a. The *mean* is (4 + 8 + 6 + 3 + 4) ÷ 5 = 25 ÷ 5 = 5.
   b. The *median* is 4 (3 and 4 are below it, and 6 and 8 above).
   c. The *mode* is also 4 (there are two of them).
   d. The *range* is 8 − 3 = 5.
   e. The *standard deviation* should be computed using equation (3–6) as:

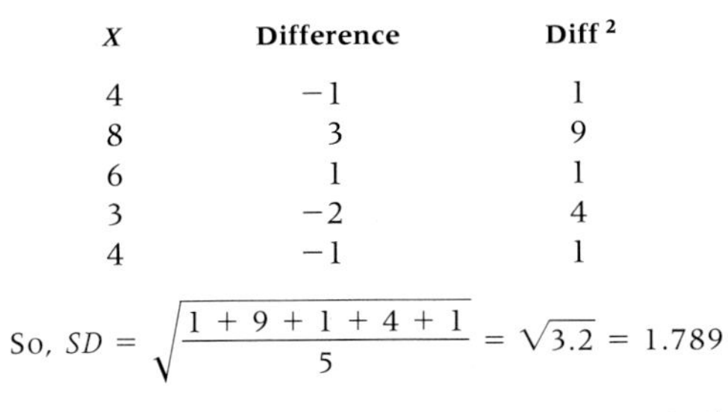

X	Difference	Diff²
4	−1	1
8	3	9
6	1	1
3	−2	4
4	−1	1

$$\text{So, } SD = \sqrt{\frac{1 + 9 + 1 + 4 + 1}{5}} = \sqrt{3.2} = 1.789$$

(Ans 3–1)

3. a. The *mode* should stay the same, unless there are fewer than 5 subjects at that value. In that case, the new mode will be 99.
   b. The *median* will probably increase, except in the unlikely event that all of the missing values were above the median to begin with. If that were so, the median would stay the same.
   c. The *mean* will increase.
   d. and e. The *standard deviation* and the *range* will both increase.

CHAPTER 4

1. $z = \frac{(70 - 60)}{12} = \frac{10}{12} = .833$

(Ans 4–1)

2. $z = \frac{(35 - 40)}{10} = \frac{-5}{10} = -.500$

(Ans 4–2)

3. A score of 78 for males is equivalent to a z of 1.50. What we have to do now is find the raw score for females that yields this value. So,

$$1.50 = \frac{(X - 40)}{10};\ 15 = X - 40;\ X = 55$$

(Ans 4–3)

4. A score of 30 has a z value of −1.00; 45 has a z value of .50. Now, let's look those up in the table of normal curve (Table A in the Appendix.) There's no −1.00; we have to look up +1.00 and then remember that we're talking about the area to the *left* of the mean. It's .3413. The table also shows that 19.14% of the area to the *right* of the mean is between 0 and +.50. Adding these together, we get (.3413 + .1914) = .5327. In other words, 53¼% of women have scores between 30 and 45. We can go even further; because there were 97 women in our sample, 51 or 52 had scores within this range.
5. A score of 68 is equivalent to a z-score of .667. The closest value in the table is .67, which has an "Area Below" value of .2486. Therefore the area above is .5 − .2486 = .2514, or just over 25%.
6. To find the top 10%, we first have to remember that we're looking at the area to the *right* of some z-score. Second, because the table gives the z-scores for only the upper half of the curve, we need to find the value that marks off 40% between it and the mean. Going down the column labeled "Area Below," the closest value to .40 is .3997, equivalent to a z of 1.28. So, 90% of people have scores below this (40% between $z = 0$ and $z = 1.28$, and the remaining 50% between $z = 0$ and $z = -4.0$). Now we have to convert a z of 1.28 back into raw scores, as we did in question 3:

$$1.28 = \frac{(X - 40)}{10};\ 12.8 = X - 40;\ X = 53$$

(Ans 4–4)

CHAPTER 5

1. Actually, it beats us where they got this figure. If they used the formula:

$$\Pr(\text{at least one failure}) = 1 - (1 - \alpha)^n$$

with α of .02 (for a 98% reliability) and n = 30, the probability is actually .4545; half of OTA's estimate. To get a probability of .889 (that's 8 in 9), there would have to be either 109 flights at this level of reliability or 30 flights where the reliability of the shuttle is 93%.
2. The probability is zero—the usual laws of probability don't apply once the stake rises above $200 or so. (The real answer is $\frac{1}{47} + \frac{1}{46} = .043$.)
3. a. Type 1: .40 × .10 = .04 or 4%
   Type 2: .40 × .90 = .36 or 36%
   Type 3: .60 × .10 = .06 or 6%
   Type 4: .60 × .90 = .54 or 54%
   b. $(0.1)^3 = 0.1\%$
   c. $[(1 - .40)(1 - .10)]^3 = 15.75\%$
   d. .36, $.36^2$, $.36^3$, etc.
   e. (.04 + .06); $.10^2$, $.10^3$, etc.
4. The probability is 100%. The laws of probability don't apply on your holidays, except when they can work against you (see question 2).[1] (In fact, the real answer is $0.1^3 = .001$.)

[1] *This is like Damon Runyon's line, "In all human affairs, the odds are always six to five against."*

CHAPTER 6

1. d; the ".05" refers to the null hypothesis.
2. e; because the result was not statistically significant, no substantive conclusion is possible, eliminating a, b, and c. Option d is just bad research design etiquette.
3. Estimates of the parameters don't change systematically with sample size, so the SD will stay the same, more or less, as will the estimates of the means. However, the SE will shrink by the ratio of $\sqrt{100/10}$. In turn, the statistical test will increase by the same amount, and so the associated probability will be reduced.
4. b; because the critical value for $p = .01$ is larger, the β error will increase and the power will decrease. Look at Figure 6–8 and move the "CV" to the right.
5. a. The Type I (α) error rate will decrease from 5% to 1%.
   b. If the Type I rate decreases, then the Type II (β) rate will increase.
   c. By definition, if the Type II error rate increases, then the power must fall, because Power = $1 - \beta$.
   d. There will be no effect on the *df*.
6. a. First, we must calculate the *z*-test. The population mean is 50.0, the SD is 15.0, the sample mean is 56.0, and the sample size is 16. So the statistical test is:

$$z = \frac{56.0 - 50.0}{15/\sqrt{16}} = \frac{6.0}{15/4} = 1.6$$

(Ans 6–1)

The corresponding probability (two-tailed, of course) is .110.

   b. To calculate power, we must first determine the critical value, which is:

$$x = 50 + 1.96 \times \frac{15}{\sqrt{16}} = 57.35$$

Again, look at Figure 6–8. The critical difference is now (57.35−60) = −2.65, so:

$$z = \frac{-2.65}{3.75} = -0.707$$

The corresponding probability, from Table A is .258, and the power is (.258 + .5) = .758.

   c. The sample size calculation is:

$$N = \frac{[(z_\alpha + z_\beta)\ (\mathrm{SD})]^2}{\mathrm{Difference}} = \frac{[(1.96 + 1.28)15]^2}{10^2}$$

$$= 23.62 = 24 \text{ per group}$$

(Ans 6–3)

CHAPTER 7

1. a. True
   b. False
   c. False
   d. False (but if SDs are very different, assumptions are violated)
   e. True (but it is critical only when sample size is small)
2. a. The difference of the means is 15.8 − 9.8 = 6.0
   b. To calculate the SE of the difference, see equation 7–6.

$$\mathrm{SE} = \sqrt{([s_1^2 + s_2^2] \div n)}$$

$$= \sqrt{(8.59^2 + 6.21^2) \div 5} = 4.74$$

   c. Now, $t = 6.0 \div 4.74 = 1.26$
   d. There are (5 + 5 − 2) = 8 *df*, and the critical value of *t* is 2.306, so this is not a significant result.

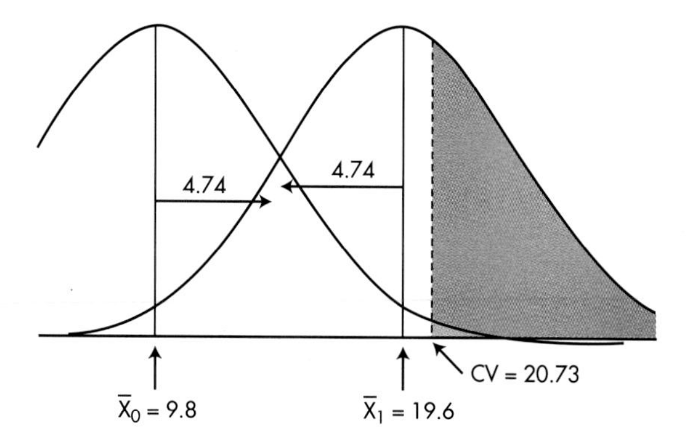

3. a. It helps to reason power calculations out graphically. The data look like:
   So, the critical value falls at t = 2.306, which means that

$$\frac{(CV - 9.8)}{4.74} = 2.306$$

and the critical value is 20.73.

(Ans 7–1)

Then, the distance from the alternative hypothesis mean of 19.6 to the critical value is (20.73 − 19.6) = 1.13. Because the SE is 4.74, then Z_β is 1.13 ÷ 4.74 = 0.238. This is looked up in a table of the normal distribution, which indicates that the area to the right of the mean is .095. So the power is (.500 − .095) = .405.

   b. For $\alpha = .05$, $Z_\alpha = 1.96$, and for $\beta = .10$, $Z_\beta = 1.28$. So, from equation 7–15:

$$n = \frac{2[(1.96 + 1.28) \times 4.74]^2}{4.9^2}$$

$$= 19.6/\text{group} = 20/\text{group}$$

(Ans 7–2)

CHAPTER 8

1. a. a, c, g, h. All are positively related to Sum of Squares (between), which captures the effect.
   b. a, b, c, d, g, h. It is evident that the Sum of Squares and Mean Square (within) are related to the random variation. But the Sum of Squares and Mean Square (between) are also related. See the discussion on expected mean square for clarification. And the *F*-ratio is *inversely* related to within group variation.
   c. a, b, e, f. The Sum of Squares is directly related to the number of terms, which is, in turn, related to the number of groups, assuming the subjects per group remain constant. Degrees of freedom (between) is $(k - 1)$, where k is number of groups, and *df* (within) is $k\,(n - 1)$, where n = subjects per group; thus both are related.
   d. a, b, c, f, g. Sum of Squares related to number of terms. Mean Square (between) contains a factor of "n" multiplying the within variance, but Mean Square (within) does not (see discussion of expected mean squares). This then carries over to the *F*-ratio.
   e. h only. *F*-ratio gets bigger (see d) so probability gets smaller.
   f. h only. See e.
2. The Grand Mean, across all groups, is 7.00. The Sums of Squares can then be calculated. These look like:

$$SS_{bet} = 4[(5.0 - 7.00)^2 + (7.0 - 7.00)^2 + (9.0 - 7.00)^2] = 32.0$$

$$SS_{within} = (4 - 5)^2 + (4 - 5)^2 + (7 - 5)^2 + (5 - 5)^2 + (7 - 7)^2 + (8 - 7)^2 + (6 - 7)^2 + (7 - 7)^2 + (7 - 9)^2 + (9 - 9)^2 + (10 - 9)^2 + (10 - 9)^2 = 14.0$$

$$SS_{total} = 32.0 + 14.0 = 46.0$$

The *df* are:
Between group = (3 − 1) = 2
Within group = 3(4 − 1) = 9
Total = 3 × 4 − 1 = 11

a. In the end, the ANOVA table looks like this:

Source	Sum of Squares	df	Mean square	F	p
Between	32	2	16	10.28	.005
Within	14	9	1.556		
Total	46				

So a significant difference exists in suicide ratings among the roadhouses.

b. The denominator of the Scheffé test equals:

$$1.556 \times [(1 \div 5) + (1 \div 5)] = .6224$$

So the Scheffé test corresponding to each of the differences is:

A − B = 2.0 ÷ .6224 = 3.213
A − C = 4.0 ÷ .6224 = 6.426
B − C = 2.0 ÷ .6224 = 3.213

The critical *F*, on 2 and 9 *df*, is 4.26, which is multiplied by 3 = 12.78.

So according to the Scheffé test, none of the comparisons are significantly different.

Tukey's LSD uses the same denominator, but it takes the square root and multiples by the appropriate *t*-test, in this case on 6 *df*. So LSD equals:

$$\sqrt{.6224} \times 2.45 = 1.235$$

and all the contrasts are significant by the LSD test because all exceed this quantity.

CHAPTER 9

1. a. Independent variables are Maze type (2 levels) and Ulcer Treatment (3 levels). Dependent variable is Lesion size. Maze and Treatment are crossed.
   b. There is now a third IV—Brand—which is crossed with Maze type and nested within Treatment.
   c. There are three independent factors—Beer/Ale (2 levels), Brand (5 levels), and Rater (4 levels)—and one dependent variable, the Rating. Brand is nested within Beer/Ale (Michelob is a beer, Labatt's 50 is an ale), and both are crossed with Rater.
   d. Trick question. Although conceptually, undergraduate grades is the independent variable and success the dependent variable, when it comes to analysis we turn it around. So it's a one-way ANOVA, with Honors/Pass/Fail as independent grouping factor and grades as dependent variable. Crossed versus nested does not apply.
   e. There are three independent factors (Patient, Rater, Bilateral/Lateral); however, Rater is completely nested within Patient (each patient is rated by a different chiropractor), so all you really have is one independent variable (Bilateral/Lateral), and you use a one-way ANOVA or *t*-test.
2. The factors are as follows:
   a. Maze—fixed (likely), Treatment—fixed.
   b. Brand is random.
   c. Beer/Ale is fixed; Brand is random.
   d. Success/Failure is fixed.
   e. Patient/Rater is random, Lateral/Bilateral is fixed.
3. a. The design is a two-way ANOVA and looks like the following table. We have also included the actual cell means in parentheses and the expected cell means in brackets. Factors are Heat Level (Suicide/Mild) and Roadhouse (A/B/C), which are crossed.

	Roadhouse A	B	C
Heat level			
Mild	4, 2, 6, 4	3, 3, 4, 2	1, 3, 2, 2
	(4.0)	(3.0)	(2.0)
	[2.5]	[3.0]	[3.5]
Suicide	4, 4, 7, 5	7, 8, 6, 7	7, 9, 10, 10
	(5.0)	(7.0)	(9.0)
	[6.5]	[7.0]	[7.5]

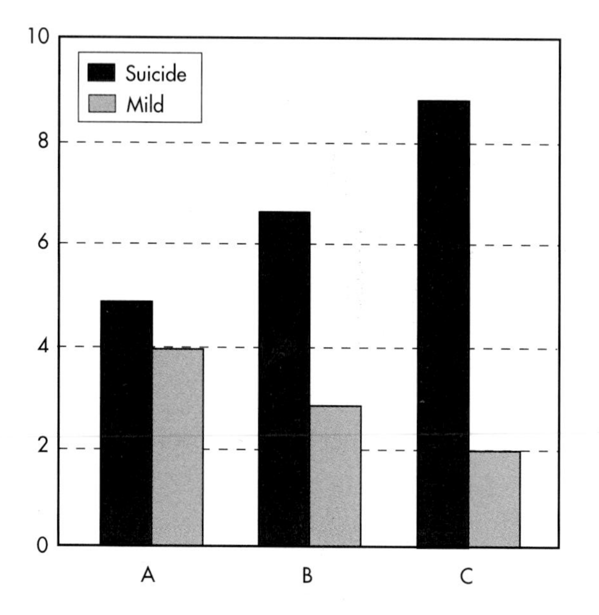

b. The graph of the data is shown in the figure above.

c. The ANOVA table looks like this:

Source	Sum of Squares	*df*	Mean square	*F*	*p*
Heat level	96	1	96	66.46	.0001
Roadhouse	4	2	2	1.38	ns
Level × roadhouse	36	2	18	12.46	.01
Within	26	18	1.444		

As an example, Sum of Squares (level) equals:

$$12[(3.0 - 5.0)^2 + (7.0 - 5.0)^2] = 96.0$$

Sum of Squares (interaction) contains terms such as:

$$4[(4.0 - 2.5)^2 + (3.0 - 3.0)^2 + (2.0 - 3.5)^2 + (9.0 - 7.5)^2] = 36.0$$

CHAPTER 10

1. a. Paired *t*-test. Simple before/after measurement.
   b. Paired *t*-test. Each patient has two measures, Gold and Iron.
   c. Unpaired *t*-test. Each subject is either Only child or With siblings.
   d. Paired *t*-test. Younger child is paired with older child.
   e. Unpaired *t*-test. At one point in time, any child is a member of a one- or two-parent family, not both.
   f. Unpaired *t*-test on difference scores. Take difference between older and younger, then compare for those siblings raised together versus raised apart.
2. The appropriate test is now a paired *t*-test. The differences look like:

Subject	Drug	Placebo	Difference
1	12	5	+7
2	14	10	+4
3	28	20	+8
4	3	2	+1
5	22	12	+10
MEAN	15.8	9.8	6.0
SD	9.60	6.94	3.54

So, the paired $t = \dfrac{6}{3.54 / \sqrt{5}} = 3.79$, $p < .02$.

CHAPTER 11

1. Between-subjects and within-subjects factors

	Subjects	Between subjects (*)	Within subjects
a	Patients	None	Week (12)
b	Patients	None	Pill (3), Headache (6)
c	Patients	Pill (3)	Headache (6)
d	Slide	None	Pathologist (6)
e	Slide	None	Level (3) Pathologist (2)
f	Slide	Cancer/ Normal (2)	Level (3) Pathologist (2)

(*) Number of levels

2. a. This might work. This would improve the estimate of MS (bet) and MS (within), but it won't change them. However, the critical *F* test gets smaller as the number of *df* in the numerator increases.
   b. This might work. This is equivalent to increasing the sample size, and it results in reduction of the critical *F* test.
   c. This might work.
   d. This definitely won't work. Going to categories will result in loss of information and reduced power.

3. a. The Lairds

Source	Sum of squares	*df*	Mean square	*F*
Laird (L)	320	16	20	2.0
Night/Morn (NM)	42	1	42	4.2
NM × L	160	16	10	

b. The Bugs

Source	Sum of squares	*df*	Mean square	*F*
NA/SA (NS)	1300	1	1,300	6.5
Bug (B)	3800	19	200	
Leg (L)	5000	5	1,000	100
L × NS	550	5	110	11.0
L × NS × B	950	95	10	

c. The Clerks

Source	Sum of squares	*df*	Mean square	*F*
Students (S)	950	19	50	
Patient (P)	300	2	150	30
P × S	190	38	5	
Observer (O)	120	1	120	24
O × S	95	19	5	
P × O	34	2	17	17
P × O × S	38	38	1	

CHAPTER 12

1. a. Univariate. There is one independent (or grouping) variable, disease type, and one dependent variable, quality of life.
   b. Univariate. There are now two grouping variables, disease type and gender, but still only one DV.
   c. Univariate, but, it is sometimes better to analyze repeated measures with MANOVA.
   d. Multivariate. There are now eight DVs—the subscales of the quality-of-life scale.
   e. Same as d.
   f. Same as d.
2. The probability is $1 - (1 - .05)^8 = 0.337$.
3. b. *M* is highly significant, and we're not sure from this if the group sample sizes are equal, so proceed with *extreme* caution.
4. a. None of the *F*s are significant, so use the results of the multivariate tests.
5. The intercept is significant, so the variables are doing something, but the effect of setting isn't significant.

CHAPTER 13

1. The net effect of an increase in the sample size is to change all the sums, but there will be no influence on the calculated parameters. The significance of the *p*-value gets smaller because this is related to sample size.

	1 > 2	1 = 2	1 < 2	1 ? 2
Sum of Squares (Regression)			*	
Sum of Squares (Residual)			*	
Coefficient of determination		*		
Correlation		*		
Significance of the correlation	*			
Slope		*		
Intercept		*		

2. Study 3 ends up with a much more homogeneous sample (assuming readers of the Financial Times are likely to have higher incomes than has the general population). This will reduce the correlation and make the line nearer horizontal.

	3 > 2	3 = 2	3 < 2	3 ? 2
Sum of Squares (Regression)			*	
Sum of Squares (Residual)		*		
Coefficient of determination			*	
Correlation			*	
Significance of the correlation			*	
Slope			*	
Intercept	*			

3. a. *No change.* This just improves the precision of the estimate.
   b. *Decrease.* Restricting the range reduces the correlation.
   c. *Increase.* Taking extreme groups inflates the correlation.

CHAPTER 14

1. a. Overall, poor prediction because the variables explain only 2.25% of the variance. Only age is a significant predictor.
   b. Still poor overall prediction. But the sample size is huge, so individual predictors, although accounting for little variance, are all statistically significant.
   c. Very good prediction, because $R^2 = .5625$, but this is based on a sample size of only 5. So none of the individual variables (or for that matter, the overall prediction) is significant.
2. a. No change in *R* or betas, but the significance will go up.
   b. Assuming private school kids have a higher socioeconomic status, at minimum SES will now likely not be significant because range is restricted. If kids are more homogeneous overall, *R* and betas may also drop.
   c. Because income and SES are highly correlated, likely if one goes into the equation; the other won't.

d. Because these kids are still likely more depressed than the average, the range on depression scores will probably decrease. Also, sample size is now smaller. Both changes result in a reduction in *R*, betas, and significance.

CHAPTER 15

1. a. Logistic regression is right, since the dependent variable is categorical—married or not.
   b. Repeated-measures ANOVA or an unpaired *t*-test. Don't let all the talk about laparoscopic versus conventional surgery fool you. It's an independent variable.
   c. Logistic regression is fine here, since the dependent variable is MI or not.
   d. Ordinary *t*-test. Even though conceptually the dependent variable is cancer Yes/No, statistically we turn it around and analyze stick-years as a dependent variable (assuming it's reasonably well behaved distributionally).
   e. Multiple regression. Trick question. Although the dependent variable talks about proportions, for each subject this would be a number like 0, .1, .2, ...9, 1.0. Likely, it will be fairly normally distributed, and it's certainly interval-level. If you're still puzzled, imagine it to be based on 100 written cases, in which the subject would get a percent correct.
2. This involves a bunch of transformations. First, if $\text{Exp}(b_1) = -2.0$, then the odds ratio is 2.0. If the probability of cardiac death in the control group is 10%, this means that we can construct a 2 × 2 table like:

	Dead	Alive
Treat	a	b
Control	.1	.9

Now, the OR is a/b ÷ .1/.9 = 2.0, so a/b = (.1/.9) × 2.0 = 2/9. From this, a = 2/9b, and a/(a + b) = 2/9b ÷ (2/9b + b) = 2/11. So, the risk of cardiac death in the treatment group is 2/11 = .09. The risk in the control group is .20, and the relative risk is .09/.20 = .45.

CHAPTER 16

1. Between-subject and within-subject factors and covariates.

	Between subjects	Within subjects	Covariate
a.	Stats book	None	Undergrad math mark
b.	Calcium/Placebo	None	None

This is a bit of a trick question. The patients made only one assessment of change since beginning of treatment. This is, incidentally, a really dumb idea, and the researchers would do much better to assess present status at the beginning and the end than to do ANCOVA.

c.	None	TENS/none	Power level

Because of the crossover, TENS/Placebo is within subject. Power level, although a treatment effect, would be handled as a covariate because it is a ratio-level variable.

d.	Gender	TENS/none	Power level
e.	Right-/left-handed	None	Reaction time IQ

2. a. "Subjects" in this case is the speaker. We are obtaining rating information on each speaker.
   b. Because six speakers were NN and six were SS, NN/SS is the Between-Subjects factor, with 1 *df*.
   c. Age is the only continuous variable, so it is the covariate. Because only one beta coefficient is associated with the linear relationship to age, there is one *df*.
   d. There are two repeated measures: (1) Gender, with one *df*, and (2) Rater, with 10 levels and 9 *df*.

CHAPTER 17

1. c. In this simple pre-post test design, ANOVA and paired *t*-test are the same test and will give the same *p*-value. In the absence of measurement error, which is pretty well the case for this example, there is no regression to the mean, so ANCOVA will also yield the same result, although you lose a degree of freedom estimating the coefficient.
2. a. The analysis is clearly wrong. Separate analysis of the treatment and control groups amounts to separate tests of the hypothesis that the change is zero. If you're going to do that, you don't need a control group.
   b. i. and ii. are completely equivalent. Since the scores in the control group got worse, either will yield a smaller *p*-value than the initial analysis.
   iii. is preferred for all the reasons outlined in the text. It more appropriately uses the baseline data and all the follow-up data. Depending on the shape of the response over time, it may turn out to be more or less powerful than the test of 12-month values. Also, with orthogonal decomposition, it can reveal something about the shape of the response curve.

CHAPTER 18

1. Answer C would be correct *if* the communalities were high and there were many items for each factor. However, because there are only 12 items, which are divided into (hopefully) 3 scales, then we should be looking at a 10:1 ratio. So, D is the right answer.
2. Only the first three factors have eigenvalues greater than 1. By this criterion, we should drop the fourth factor.
3. A bit of reasoning is required here (and just a soupçon of math). The sum of the eigenvalues of the first 3 factors is 4.412. Because there are 12 items, these factors account for $(4.412 \div 12) = .368$, or 36.8% of the variance. We're not going to become famous at this rate!
4. Item 11 doesn't seem to load on any of the factors and probably should be dropped (or reworded for the next validation study). Item 12 may also have some problems. It is factorially complex, loading on Factors 3 and 4. We should really rerun the analysis, limiting it to 3 factors, and see what happens to this item. If it remains complex by loading on Factor 1 or 2, you may want to again drop it or rewrite it.
5. a. The communality, you'll remember, is the sum of the squared loadings. So, for item 1 (and dropping the fourth factor) it is:

   $$.27^2 + .53^2 + .33^2 = .463$$

   b. The uniqueness is $(1 - \text{communality}) = 1 - .463 = .537$.
   c. This means that 46.3% of the variance of item 1 can be explained by the three factors. Conversely, 53.7% of the variance is *unique* to item 1; that is, not explained by the factors.
   d. No.
   e. Yes. Uniqueness is high when the factor loadings are low. So, if the uniqueness is too high (communality too low), then that variable may be an outlier, not associated with any of the remaining factors.

CHAPTER 19

1. The correlation is $0.20 + (0.50 \times 0.30) = 0.35$.
2. The correlation is $0.30 + (0.50 \times 0.20) = 0.40$.
3. Variables A and B are exogenous; the outcome is endogenous.
4. Only exogenous variables get disturbance terms; hence, A and B.
5. There are three observed variables and, therefore, six observations.
6. The fit is abysmal.

CHAPTER 20

1. d
2. Mantel-Haenszel chi-squared or log-linear analysis.
3. The final table looks like this:

	Control		
Case	**Yes**	**No**	TOTALS
Yes	35 (A)	(B) 15	50
No	5 (C)	(D) 20	25
TOTALS	40	35	75

4. The expected values in cells A and C are 3, and in B and D are 17.
   a. Chi-squared then equals:

$$\frac{(1-3)^2}{3} + \frac{(5-3)^2}{3} + \frac{(19-17)^2}{17} + \frac{(15-17)^2}{17} = 3.12 \quad \textbf{(Ans 20–1)}$$

   $.10 > p > .05$
   b. Yates' corrected chi-squared equals:

$$\frac{[(1-3)-.5]^2}{3} + \frac{[(5-3)-.5]^2}{3} + \frac{[(19-17)-.5]^2}{17} + \frac{[(15-17)-.5]^2}{17} = 1.764 \quad \textbf{(Ans 20–2)}$$

   probability $> .10$
   c. Fisher's exact test is based on:

$$P(1) = (6! \times 34! \times 20! \times 20!) \div (1! \times 19! \times 5! \times 15! \times 40!) = .083$$

$$P(0) = (6! \times 34! \times 20! \times 20!) \div (0! \times 20! \times 6! \times 14! \times 40!) = .010$$

   so the probability is $.083 + .010 = .093$.

(Ans 20–3)

CHAPTER 21

1. a. Smaller. More values would be off the diagonal, hence look like disagreement.
   b. Smaller. See *a*.
   c. Larger. Finer scale divisions result in improved measurement and an increase in weighted kappa.
   d. Undefined. Phi is for 2 × 2 tables.
2. a. Phi = .529.
   b. Contingency coefficient = .468.
   c. Cramer's V = .529.
   d. Kappa = .515.

CHAPTER 22

1. The appropriate analysis is a Kruskal-Wallis one-way ANOVA by ranks because there are three unrelated groups.

   The group means are BLONDES = 8.5, BRUNETTES = 13.0, and REDHEADS = 16; the overall mean rank is just (24 + 1) ÷ 2 = 12.5. We can now proceed with the ANOVA calculation.

$$MS_{bet} = 8[8.5 - 12.5)^2 + (13 - 12.5)^2 + (16 - 12.5)^2] = 228.$$

   and the K=W test equals:

$$\frac{12}{8(8+1)} \times 228. = 38.0 \qquad \textbf{(Ans 22–1)}$$

   which is like a chi-squared with 2 degrees of freedom, and is significant. The few remaining gentlemen on the planet really do prefer blondes.

2. This time, the data are paired ranks, so the appropriate test is the Wilcoxon matched pairs signed rank test. In the third column, we calculated the difference in ranks, then we added up the signed differences.

Subject	Ranking			
	Bald	Rug	Difference	Rank
A	3	1	+2	+2.5
B	12	15	−3	−4
C	11	6	+5	+8
D	8	4	+4	+7
E	19	13	+6	+9
F	5	2	+3	+5.5
G	20	7	+13	+10
H	10	9	+1	+1
I	17	14	+3	+5.5
J	16	18	2	2.5

   The next step is to determine the smallest sum of signed ranks, which is obviously (−2.5 + −4) = −6.5. Because we assume you don't have a copy of Siegal, we will then do the z-test approximation to get the level of significance:

$$z = \frac{6.5 - 10(11)/4}{\sqrt{(10 \times 11 \times 21)/24}} = \frac{(6.5 - 27.5)}{9.81} = 2.14 \qquad \textbf{(Ans 22–2)}$$

   so the test is just significant. And bald men are less sexy. Pity for your authors.

3. The appropriate analysis is the Mann-Whitney U, which uses the differences in summed ranks:

	Bald	Hairy
	5	1
	6	2
	8	3
	9	4
	12	7
	14	10
	15	11
	18	13
	19	16
	20	17
Sum	126	84

   The z test of these ranks equals:

$$z = \frac{126 + 0.5 - (10 \times 21)/2}{\sqrt{10 \times 10 \times 21/12}} = \frac{21.5}{13.22} = 1.626 \qquad \textbf{(22–3)}$$

   so this time the difference is not significant.

CHAPTER 23

1. a. Kappa—two categories, nominal scale.
   b. Weighted kappa. A four-level ordinal scale.
   c. Because income is likely highly skewed, use Spearman rho on the ranked data.
   d. Unweighted kappa because of the nominal categories.
   e. Normally, you would use kappa on individual checklist items and ICC on the total score. However, because there are four observers, do an ICC on individual items as well (then report it as kappa if you want).
   f. Pearson correlation.
   g. You would probably calculate sensitivity and specificity because there is a "gold standard," but phi or kappa are overall measures of agreement. However, JVP is a continuous variable, so it would be better to not categorize it and, instead, use a point-biserial correlation.
   h. Because number of siblings has several categories, use a chi-squared related measure (e.g. Cramer's V).
   i. Phi coefficient.
2. Because the data are horrendously skewed, we must use a measure of association based on ranks. Spearman's rho is appropriate. The following table adds ranks to the data and determines the difference in ranks.

Sub.	Short problem	Rank	Long problem	Rank	*d*	d^2
a	32 min	4	4 days	4	0	0
b	3.7 days	7	6 days	5	2	4
c	14 min	2	8.6 hr	1	1	1
d	4.2 days	8	3.7 months	8	0	0
e	18 min	3	7.5 days	6	−3	9
f	38 sec	1	2.2 days	2	−1	1
g	8.2 hr	6	1.7 wk	7	−1	1
h	3.3 hr	5	3.9 days	3	2	4

So the sum of d^2 is 20, and rho equals:

$$1 - [(6 \times 20) \div (8^3 - 8)] = 1 - [120 \div 504] = .762$$

The test of significance = .786 ÷ $\sqrt{[(1 - .786^2) \div (8 - 2)]}$ = .786 ÷ .253 = 3.11, which is significant at the .05 level.

3. Because there are three rankings, use Kendall's *W*. For this, we need the squared mean rank for each resident, as shown at right:

Resident	Peer	Nurse	Staff	Rank sum	R^2
a	1	4	3	8	64
b	2	3	5	10	100
c	3	1	6	10	100
d	4	8	2	14	196
e	5	2	1	8	64
f	6	5	4	15	225
g	7	7	10	24	576
h	8	6	7	21	441
i	9	10	9	28	784
j	10	9	8	27	729

The sum of the squared mean ranks equals 3279, and *W* equals:

$$W = \frac{12 \times 3279 - 3 \times 3^2 \times 10 \times 11^2}{3^2 \times 10 \times (10^2 - 1)} = \frac{39348 - 32670}{8910} = 0.75 \qquad \text{(Ans 23–1)}$$

and the significance test is:

$$\chi^2 = 3(10 - 1)\ W = 3 \times 9 \times .75, = 20.25$$

on 9 *df*, which is highly significant ($p < .01$).

CHAPTER 24

1. To successfully pull off this analysis, you must first realize that this is a bit of a trick question. Here, *death* amounts to a *loss to follow-up study*. Also, you have to start off by creating a table for the probabilities of "surviving" as humble for each of the two groups. We have worked through the numbers for the control group in the accompanying table; you should do the same for the experimental group. The graph is shown in the accompanying illustration.

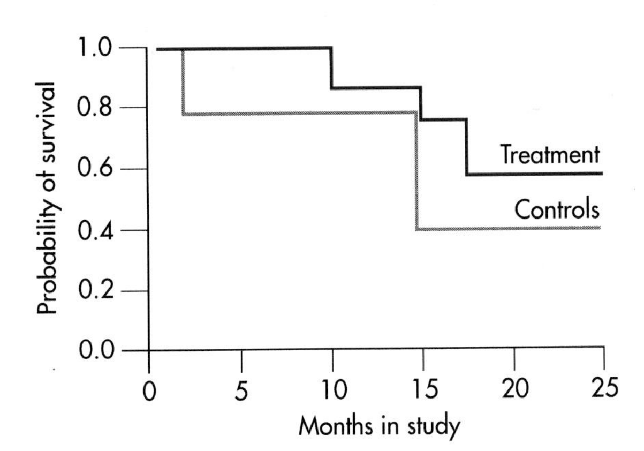

Table for the control group

Number of months in study	Number of TV hosts at risk	Number megalomaniacal	Lost or censored	Probability of surviving	Cumulative probability
0–1	10	0	1	1.000	1.000
2–3	9	2	2	.750	.750
4–5	5	0	0	1.000	.750
6–7	5	0	0	1.000	.750
8–9	5	0	1	1.000	.750
10–11	4	0	0	1.000	.750
12–13	4	0	1	1.000	.750
14–15	2	1	0	.500	.375
16–17	1	0	1	1.000	.375

2. This would be analyzed with a Mantel-Haenszel chi-squared, although the small cell sizes in the demonstration examples are problematic.
3. The contingency table analysis, in its most informative form, would be a 2 × 3 table (Treatment versus Control by Still humble, Megalomaniacal, or Lost to follow-up), which would look like this:

	Still humble	Megalomaniacal	Lost/ censored
Treatment	3	3	4
Control	3	4	3

Note that the differences evident in the curves, amounting to early megalomania in the control group, are virtually obscured in this analysis.

4. The SE is:

$$SE(P_i) = P_i \sqrt{\frac{1 - P_i}{R_i}} = .75 \sqrt{\frac{1 - .75}{5}} = .168$$

(Ans 24–1)

5. The RR is:

$$RR = \frac{1 - P_c}{1 - P_E} = \frac{1 - .375}{1 - .567} = 1.44$$

(Ans 24–2)

References and Further Reading

REFERENCES

Agresti A (1990). *Categorical data analysis.* New York, Wiley & Sons.

Andersen TF, Brønnum-Hansen H, Sejr T, and Roepstorff C (1990). Elevated mortality following transurethral resection of the prostate for benign hypertrophy! But why? *Medical Care,* **28**:870–881.

Andrews DF and Herzberg AM (1985). *Data, a collection of problems from many fields for the student and research worker.* New York, Springer-Verlag.

Arbuckle JL (1997). *AMOS users' guide: version 3.6.* Chicago, SmallWaters Corporation.

Baron RM and Kenny DA (1986). The moderator-mediator variable distinction in social psychological research: conceptual, strategic, and statistical considerations. *Journal of Personality and Social Psychology,* **51**: 1173–1182.

Beck AT, Ward CH, Mendelson M, et al (1961). An inventory for measuring depression. *Archives of General Psychiatry,* **4**:561–571.

Belongia EA, Hedberg CW, Gleich GJ, et al (1990). An investigation of the cause of the eosinophilia-myalgia syndrome associated with tryptophan use. *New England Journal of Medicine,* **323**:357–365.

Bennett KJ, Sackett DL, Haynes RB, et al (1987). A controlled trial of teaching critical appraisal of the clinical literature to medical students. *Journal of the American Medical Association,* **257**:2451–2454.

Bentler PM and Bonett DG (1980). Significance tests and goodness of fit in the analysis of covariance structures. *Psychological Bulletin,* **88**:588–606.

Binzel RP (1990). Pluto. *Scientific American,* **262**(6):50–58.

Blair RC and Higgins JJ (1985). Comparison of the power of the paired samples *t* test to that of Wilcoxon's test under various population shapes. *Psychological Bulletin,* **97**:119–128.

Bloch A (1979). *Murphy's law and other reasons why things go wrong!* Los Angeles, Price/Stern/Sloan.

Borenstein M, Cohen J, Rothstein HR, et al (1990). Statistical power analysis for one-way analysis of variance: a computer program. *Behavior Research, Methods, Instruments, and Computers,* **22**:271–282.

Bryk AS and Raudenbush SW (1987). Application of hierarchical linear models to assessing change. *Psychological Bulletin,* **101**: 147–158.

Cicchetti DV (1972). A new measure of agreement between rank ordered variables. *Proceedings of the American Psychological Association,* **7**:17–18.

Cicchetti DV (1981). Testing the normal approximation and minimal sample size requirements of weighted kappa when the number of categories is large. *Applied Psychological Measurement,* **5**:101–104.

Cleveland WS (1984). Graphical methods for data presentation, full scale breaks, dot charts, and multibased logging. *The American Statistician,* **38**:270–280.

Cohen J (1968). Weighted kappa: nominal scale agreement with provision for scaled disagreement or partial credit. *Psychological Bulletin,* **70**:213–220.

Cohen J (1977). *Statistical power analysis for the social sciences* (2nd ed). New York, Academic Press.

Cohen J (1990). Things I have learned (so far). *American Psychologist,* **45**:1304–1312.

Cohen MS (1982). *Revised carcinogenic risk assessment of urea formaldehyde foam insulation.* Washington, DC, Consumer Product Safety Commission.

Comrey AL (1978). Common methodological problems in factor analytic studies. *Journal of Consulting and Clinical Psychology,* **46**:648–659.

Coren S and Halpern DF (1981). Left-handedness: a marker for decreased survival fitness. *Psychological Bulletin,* **109**:90–106.

Cox DR (1972). Regression models and life tables. *Journal of the Royal Statistical Society,* **34**:187–220.

Croog SH, Levine S, Testa MA, et al (1986). The effects of antihypertensive therapy on the quality of life. *New England Journal of Medicine,* **314**:1657–1664.

Dallal GE (1986). PC-SIZE: a program for sample size determinations. *American Statistician,* **40**:52.

Dallal GE (1990). PC-SIZE Consultant: a program for sample size determinations. *American Statistician,* **44**:243.

de Groot AD (1965). *Thought and choice in chess.* The Hague, Mouton.

Dunlap WP and Cornwell JM (1994). Factor analysis of ipsative measures. *Multivariate Behavioral Research,* **29**:115–126.

Eidson M, Philen RM, Sewell CM, et al (1990). L-tryptophan and eosinophilia-myalgia syndrome in New Mexico. *Lancet,* **335**:645–648.

Feighner JP (1985). A comparative trial of fluoxetine and amitriptyline in patients with major depressive disorder. *Journal of Clinical Psychiatry,* **46**:369–372.

Feinstein AR (1977). *Clinical biostatistics.* St. Louis, Mosby.

Fish LJ (1988). Why multivariate methods are usually vital. *Measurement and Evaluation in Counseling and Development,* **21**:130–137.

Fisher RA (1925). *Statistical methods for research workers.* Edinburgh, Oliver & Boyd.

Fleiss JL (1971). Measuring nominal scale agreement among many raters. *Psychological Bulletin,* **76**:378–382.

Floyd FJ and Widaman KF (1995). Factor analysis in the development and refinement of clinical assessment instruments. *Psychological Assessment,* **7**:286–299.

Freedman LS (1982). Tables of the number of patients required in clinical trials using the log rank test. *Statistics in Medicine,* **1**:121–129.

Friedman LD (1990). World watch. *The Planetary Report,* **10**(5):24–25.

George S and Desu MM (1974). Planning the size and duration of a clinical trial studying the time to some critical event. *Journal of Chronic Disease,* **27**:15.

Glass GV and Stanley JC (1970). *Statistical methods in education and psychology.* Englewood Cliffs, Prentice Hall.

Gorsuch RL (1983). *Factor analysis.* Hillsdale, NJ, Lawrence Erlbaum Associates.

Hardy MA (1993). *Regression with dummy variables.* Newbury Park, CA, Sage.

Hochberg Y (1988). A sharper Bonferroni procedure for multiple tests of significance. *Biometrika,* **75**:800–802.

Holm SA (1979). A simple sequentially rejective multiple test procedure. *Scandinavian Journal of Statistics,* **6**:65–70.

Holmes TH (1978). Life situations, emotions, and disease. *Psychosomatics,* **19**:747–754.

Horn JL and Engstrom R (1979). Cattell's scree test in relation to Bartlett's chi-squared test and other observations on the number of factors problem. *Multivariate Behavioral Research,* **14**:283–300.

Hui CH and Triandis HC (1989). Effects of culture and response format on extreme response style. *Journal of Cross-Cultural Psychology,* **20**:296–309.

Hunter MA and May RB (1993). Some myths concerning parametric and nonparametric tests. *Canadian Psychology,* **34**: 384–389.

Iversen GR (1979). Decomposing chi–squared. *Sociological Methods and Research,* **8**:143–157.

Kaiser HF (1970). A second generation little jiffy. *Psychometrika,* **35**, 401-415.

Kaplan EL and Meier P (1958). Nonparametric estimation from incomplete observations. *Journal of the American Statistical Association,* **53**:457–485.

Kenny DA (1979). *Correlation and causality.* New York, Wiley.

Kleinbaum DG, Kupper LL, and Muller KE (1988). *Applied regression analysis and other multivariable methods* (2nd ed). Boston, PWS-Kent.

Klem L (1995). Path analysis. In: Grimm LG and Yarnold PR (eds). *Reading and understanding multivariate statistics.* Washington, American Psychological Association: 65–97.

LaTour SA and Miniard PW (1983). The misuse of repeated measures analysis of variance in marketing research. *Journal of Marketing Research,* **20**:45–57.

Läuter J (1978). Sample size requirements for the T^2 test of MANOVA (tables for one-way classification). *Biometrical Journal,* **20**: 389–406.

Lehr R (1992). Sixteen S-squared over D-squared: a relation for crude sample size estimates. *Statistics in Medicine*, **11**:1099–1102.

Leigh JP (1988). Assessing the importance of an independent variable in multiple regression. Is stepwise unwise? *Journal of Clinical Epidemiology*, **41**:669–677.

Lichtenstein R, Thomas W, Adams-Watson J, et al (1991). Selection bias in TEFRA at-risk HMOs. *Medical Care*, **29**:318–331.

Liddell A, Locker D, and Burman D (1991). Self-reported fears (FSS-II) of subjects aged 50 years and over. *Behaviour Research and Therapy*, **29**:105–112.

Marshall WL, Eccles A, and Barbaree HE (1991). The treatment of exhibitionists: a focus on sexual deviance versus cognitive and relationship features. *Behaviour Research and Therapy*, **29**:129–135.

McFarlane AH, Norman GR, Streiner DL, and Roy RG (1983). The process of social stress: stable, reciprocal, and mediating relationships. *Journal of Health and Social Behavior*, **24**:160–173.

Meehl P (1990). Why summaries of research on psychological theories are often uninterpretable. *Psychological Reports*, **66**:195–244. (Monograph Supplement 1-V66).

Micceri T (1989). The unicorn, the normal curve, and other improbable creatures. *Psychological Bulletin*, **105**:156–166.

Minsel B, Becker P, and Korchin SJ (1991). Cross-cultural view of positive mental health: two orthogonal main factors replicable in four countries. *Journal of Cross-Cultural Psychology*, **22**:157–181.

Minitab Users' Group (1988). Kaplan-Meier product-limit estimate. *Newsletter*, **9**:2–4.

MRFIT Group (1977). Statistical design considerations in the NHLI Multiple Risk Factor Intervention Trial (MRFIT). *Journal of Chronic Disease*, **30**:261–275.

Norman GR (1989). Issues in the use of change scores in randomized trials. *Journal of Clinical Epidemiology*, **42**:1097–1105.

Norman GR and Streiner DL (1997). *PDQ statistics* (2nd ed). Toronto, B.C. Decker.

Norman GR, Regehr G, and Stratford PW (1997). Problems in the retrospective computation of responsiveness to change: the lesson of Cronbach. *Journal of Clinical Epidemiology*, **50**:869–879.

Olson CL (1976). On choosing a test statistic in multivariate analysis of variance. *Psychological Bulletin*, **83**:579–586.

Peto R, Pike MC, Armitage P, et al (1977). Design and analysis of randomized clinical trials requiring prolonged observation of each patient. II. Analysis and examples. *British Journal of Cancer*, **35**:1–39.

Rosnow RL and Rosenthal R (1991). Statistical procedures and the justification of knowledge in psychological science. *American Psychologist*, **44**:1276–1284.

Rothman ML, Hedrick SC, Bulcroft KA, et al (1991). The validity of proxy-generated scores as measures of patient health status. *Medical Care*, **29**:115–124.

Sackett DL and Gent M (1979). Controversy in counting and attributing events in clinical trials. *New England Journal of Medicine*, **301**:1410–1412.

Sawilowsky SS and Hillman SB (1992). Power of the independent samples *t* test under a prevalent psychometric measure distribution. *Journal of Consulting and Clinical Psychology*, **60**:240–243.

Scailfa CT and Games PA (1987). Problems with step-wise regression in research on aging and recommended alternatives. *Journal of Gerontology*, **42**:579–583.

Schumacker RE and Lomax RG (1996). *A beginner's guide to structural equation modeling*. Mahwah, NJ, Lawrence Erlbaum Associates.

Siegel S and Castellan NJ Jr (1988). *Nonparametric statistics for the behavioral sciences* (2nd ed). New York, McGraw-Hill.

Slutsker L, Hoesly FC, Miller L, et al (1990). Eosinophilia-myalgia syndrome associated with exposure to tryptophan from a single manufacturer. *Journal of the American Medical Association*, **264**:213–217.

Soeken KL and Prescott PA (1986). Issues in the use of kappa to estimate reliability. *Medical Care*, **24**:733–741.

Sorenson SB, Rutter CM, and Aneshensel CS (1991). Depression in the community: an investigation into age of onset. *Journal of Consulting and Clinical Psychology*, **59**:541–546.

Stevens J (1980). Power of the multivariate analysis of variance tests. *Psychological Bulletin*, **88**:728–737.

Stevens J (1986). *Applied multivariate statistics for the social sciences*. Hillsdale, NJ, Lawrence Erlbaum Associates.

Streiner DL (1994). Regression in the service of the superego: the do's and don'ts of stepwise regression. *Canadian Journal of Psychiatry*, **39**:191–196.

Streiner DL (1998). Factors affecting reliability of interpretations of scree plots. *Psychological Reports*, **83**:687–694.

Streiner DL and Norman GR (1995). *Health measurement scales: a practical guide to their development and use* (2nd ed). Oxford, Oxford University Press.

Streiner DL and Norman GR (1996). *PDQ epidemiology* (2nd ed). Toronto, B.C. Decker.

Tabachnick BG and Fidell LS (1989). *Using multivariate statistics* (2nd ed). New York, HarperCollins.

Tabachnick BG and Fidell LS (1996). *Using multivariate statistics* (3rd ed). New York, HarperCollins.

Thomas JW and Holloway JJ (1991). Investigating early readmission as an indicator for quality of care studies. *Medical Care*, **29**:377–394.

Toshima MT, Kaplan RM, and Ries AL (1990). Experimental evaluation of rehabilitation in chronic obstructive pulmonary disease: short-term effects on exercise endurance and health status. *Health Psychology*, **9**:237–252.

Tukey JW (1977). *Exploratory data analysis.* Reading, MA, Addison-Wesley.

Wagner RF Jr, Reinfeld HB, Wagner KD, et al (1984). Ear-canal hair and the ear-lobe crease as predictors for coronary-artery disease. *New England Journal of Medicine*, **311**:1317–1318.

Wainer H (1976). Estimating coefficients in linear models: it don't make no never mind. *Psychological Bulletin*, **83**:213–217.

Wainer H and Thissen D (1976). Three steps towards robust regression. *Psychometrica*, **41**:9–34.

Weiss GL and Larsen DL (1990). Hea alve, health locus of control, and prediction of health protective behaviors. *Social Behavior and Personality*, **18**, 121-135.

Wilkinson L (1979). Tests of significance in stepwise regression. *Psychological Bulletin*, **86**:168–174.

Williams AW, Ware JE, and Donald CA (1981). A model of mental health, life events and social supports applicable to general populations. *Journal of Health and Social Behavior*, **22**:324–336.

Wright S (1934). The method of path coefficients. *Annals of Mathematical Statistics*, **5**: 161–215.

Zung WK (1965). A self-rating depression scale. *Archives of General Psychiatry*, **12**:63–70.

Zwick R (1985). Nonparametric one-way multivariate analysis of variance: a computational approach based on the Pillai-Bartlett trace. *Psychological Bulletin*, **97**:148–152.

TO READ FURTHER

In this section, we've tried to provide you with some texts and articles if you want to delve further into any of these topics. We've omitted ones written in statisticalese and tried to list only those which are comprehensible to normal people. Needless to say, PDQ Statistics *2nd ed. (Norman and Streiner, 1997) and* PDQ Epidemiology, *2nd ed. (Streiner and Norman, 1996) are mandatory readings, so we won't bother to list them under every section.*

Section the First

The Nature of Data and Statistics

Cleveland WS (1985). *The elements of graphing data.* Monterey, CA, Wadsworth.

Tufte ER (1983). *The visual display of quantitative information.* Cheshire, CT, Graphics Press.

Tukey JW (1977). *Exploratory data analysis.* Reading, MA, Addison-Wesley.

Wainer H (1997). Improving tabular displays, with NAEP tables as examples and inspirations. *Journal of Educational and Behavioral Statistics,* **22**:1–30.

Wainer H (1997). *Visual revelations.* New York, Springer-Verlag.

Section the Second

Analysis of Variance

Glass GV and Stanley JC (1970). *Statistical methods in education and psychology.* Englewood Cliffs, Prentice Hall.

Howell DC (1992). *Statistical methods for psychology* (3rd ed). Monterey, CA, Wadsworth.

Kirk RE (1968). *Experimental design: procedures for the behavioral sciences.* Belmont, CA, Wadsworth.

Loftus GR and Loftus EF (1982). *The essence of statistics.* Monterey, Brooks/Cole.

Weinfurt KP (1995). Multivariate analysis of variance. In: Grimm LG and Yarnold PR (eds). *Reading and understanding multivariate statistics.* Washington, American Psychological Association: 245–276.

Winer BJ (1971). *Statistical principles in experimental design* (2nd ed). New York, McGraw-Hill.

Section the Third

Regression and Correlation

Achen CH (1982). *Interpreting and using regression.* Beverly Hills, CA, Sage.

Berry WD and Feldman S (1985). *Multiple regression in practice.* Beverly Hills, CA, Sage.

Bryk AS and Raudenbush SW (1987). Application of hierarchical linear models to assessing change. *Psychological Bulletin,* **101**:147–158.

Francis DJ, Fletcher JM, Stuebing KK, et al (1991). Analysis of change: modeling individual growth. *Journal of Consulting and Clinical Psychology,* **59**:27–37.

Hardy MA (1993). *Regression with dummy variables.* Newbury Park, CA, Sage.

Hosmer DW and Lemeshow S (1989). *Applied logistic regression.* New York, Wiley.

Kleinbaum DG (1994). *Logistic regression: a self-learning text.* New York, Springer-Verlag.

Kline RB (1998). *Structural equation modeling.* New York, Guilford Press.

Kleinbaum DG, Kupper LL, and Muller KE (1988). *Applied regression analysis and other multivariable methods* (2nd ed). Boston, PWS-Kent.

Schroeder LD, Sjoquist DL, and Stephan PE (1986). *Understanding regression analysis, an introductory guide.* Beverly Hills, CA, Sage.

Schumacker RE and Lomax RG (1996). *A beginner's guide to structural equation modeling.* Mahwah, NJ, Lawrence Erlbaum Associates.

Tabachnick BG and Fidell LS (1996). *Using multivariate statistics* (3rd ed). New York, HarperCollins.

Section the Fourth

Nonparametric Statistics

Fienberg SE (1980). *The analysis of cross-classified categorical data* (2nd ed). Cambridge, MIT Press.

Fleiss JL (1981). *Statistical methods for rates and proportions* (2nd ed). New York, Wiley.

Greenhouse JB, Stangl D, and Bromberg J (1989). An introduction to survival analysis, statistical methods for analysis of clinical trial data. *Journal of Consulting and Clinical Psychology,* **57**:536–544.

Peto R, Pike MC, and Armitage P, et al (1976). Design and analysis of randomized clinical trials requiring prolonged observation of each patient, I. Introduction and design. *British Journal of Cancer*, **34**:585–612.

Peto R, Pike MC, and Armitage P, et al (1977). Design and analysis of randomized clinical trials requiring prolonged observation of each patient, II. Analysis and examples. *British Journal of Cancer*, **35**:1–39.

Pierce A (1970). *Fundamentals of nonparametric statistics.* Belmont, CA, Dickenson.

Siegel S and Castellan NJ Jr (1988). *Nonparametric statistics for the behavioral sciences* (2nd ed). New York, McGraw-Hill.

Tibshirani R (1982). A plain man's guide to the proportional hazards model. *Clinical and Investigative Medicine*, **5**:63–68.

Section the Fifth

Reprise

Hartwig F and Dearing BE (1979). *Exploratory data analysis.* Beverly Hills, CA, Sage.

Lepkowski JM, Landis JR, and Stehouwer SA (1987). Strategies for the analysis of imputed data from a sample survey, The National Medical Care Utilization and Expenditure Survey. *Medical Care*, **25**:705–716.

Tabachnick BG and Fidell LS (1996). *Using multivariate statistics* (3rd ed). New York, HarperCollins.

Tukey JW (1977). *Exploratory data analysis.* Reading, MA, Addison-Wesley.

Other Topics

Sample Size and Confidence Intervals

Cohen J (1988). *Statistical power analysis for the behavioral sciences* (2nd ed). Hillsdale, NJ, Lawrence Erlbaum.

Gardner MJ and Altman DG, editors (1989). *Statistics with confidence, confidence intervals and statistical guidelines.* London, British Medical Journal.

Kraemer HC and Thiemann S (1987). *How many subjects?* Beverly Hills, CA, Sage.

An Unabashed Glossary of Statistical Terms

For those who have already experienced the delights of *PDQ Statistics*, closing the whole thing off with an unabashed glossary is nothing new. After all, the whole idea of statistics, and the abuses to which it is put, is a bit ludicrous at the best of times. If it weren't for the fact that journal editors, peer reviewers, and the like take it all so seriously, we might all be able to laugh it off. However, statistics in biomedical sciences is no laughing matter—until now, that is.

What follows is the latest version of the (a trumpet flourish, please) Unabashed Glossary. We apologize to readers of *PDQ*; a few of the entries are repeats but, in our view, are well deserving of repetition. However, most are new, so read on.

If you find some of these are sexist, racist, or otherwise offensive, don't bother to write. Rest assured such crudity was the deliberate intent of the authors and in no way implies that **we** are sexist or racist.[1] And if you are under 18, or have lived a sheltered life, perhaps you should ask your parents' permission before you read on.

Here we go again!

ANCOVA One blanket.

ANOVA (1) Anne Boleyn's favorite position. (2) One egg.

Bar chart A list of local watering holes. In Boston, it's pronounced "Bah chaht," and means "Humbug."

Bartlett's test Used to test the goodness of pears.

Binomial Having two names (e.g., Betty Mae or Jean-Pierre).

Box plot (1) A conspiracy of squares. (2) A cemetery map.

Central limit theorem Nothing gets past Kansas or Manitoba.

Centroids A painful medical condition, relieved by Preparation C.

Communality A living condition adopted by hippies in the 1960s.

Confirmatory factor analysis A test performed on young boys age 12. In Judaism, it's called Bar Mitzvahtory factor analysis.

Correlation A sibling.

Multiple correlation Two or more siblings.

Partial correlation A half-sibling.

Covariance Dressing together in drag.

Cox model Chippendale (male stripper).

Degrees of freedom Stalinism, *glasnost*, democracy, anarchy

Descriptive statistics 36–24–36 (in metric, 90–60–90).

Discrete Mediterranean slang for an island (like "dat Sicily").

Discriminant function Ku Klux Klan ball.

Dot plot Dorothy's final resting place.

Dummy coding Mentally handicapped, Retarded, Challenged, and many other labels through the years.

***F* test** When boys become men.

Factorially complex A psychiatric condition, related to Oedipally complex.

Family-wise A person who has had at least one kid.

Fisher's exact test Under 2 pounds, throw it back in the lake.

Goodness-of-fit test Exercise ECG.

Greenhouse-Geisser Champagne in the Jacuzzi.

Heteroscedasticy A Greek historian (ca. 423 BC to 364 BC).

Histogram A delivered message for historians; for psychologists it is a psychogram, etc.

[1] *As far as our offensiveness, you had better get a second opinion.*

Homogeneous Identical twins.
Inferential statistics Adolescent fantasies (see **Descriptive statistics**).
Interaction The step preceding **Skew** (see below).
Internal consistency The result of eating prunes and bran.
Interrupted time series Cancelled subscription (see **Times Series**).
Kaiser criterion The prerequisite to lead imperial Germany—a pointed head.
Kurtosis Doggie tootsies.
Log line Straight board.
Log rank Mahogony > Cedar > Pine.
MANCOVA (1) A lid for an access hole in a street, now called a "personcova." (2) A blanket for a male.
MANOVA The missionary position.
Matrix Spring Johns.
- **Correlation matrix** Incest for money.
- **Identity matrix** And seeing your name in the local newspaper after.
- **Loading matrix** The Johns are stevedores.
- **Structure matrix** They are engineers.

Mean square Sadistic conformists.
Media Clairvoyant with a cold.
Multiple regression Simultaneous thumb and toe sucking (see **Regression**).
Oblique Part of the French term "noblesse oblique"; in English, "tilted gentry."
Orthogonal Male birth control pill.
Outlier The third person in a two-man tent.
Path analysis Tracking method used by Indian scouts.
Phi Part of what Jack heard on the beanstalk, "Phee, phi, pho, phum."
Platykurtosis Another strange Australian creature.
Polygon Said of an escaped parrot.
Power series Granada, Panama, Libya, and Iraq.
Principal components Job description for a school head. Includes a loud voice and rigidity.
Profile analysis The step before rhinoplasty.
Quartile Two pintiles (in metric, 1.14 litriles).
$p < .05$ End-stage renal disease.
Regression Thumbsucking.
Rho Caviar.
Scedasticity A town in upstate New York.
Secular trend Drifting away from the church.
Simple structure A lean-to.
Skew An impolite term for intercourse.
- **Positive skew** Having done it.
- **Negative skew** "Not tonight dear, I have a headache."

Split-plot Grave site for divorced couples.
Stem-leaf plot Compost heap.
Student's t Pub night at Bethesda Baptist College.
Tau (Greek) Appendage on the phut.
Test of sphericity Use by baseball umps to assess the goodness of the balls.
Time series A magazine subscription.
Tukey's LSD John's acid.
Yule's *Q* Christmas shopping.
X-bar Cocktail lounge for divorcees.

Appendix

Table A Area of the normal curve.

Table B Sample size requirements to show a difference between two means of size σ/δ.

Table C Critical values for the *t*-test.

Table D Sample size requirements for the independent *t*-test.

Table E Power table for the independent *t*-test.

Table F Critical values for the chi-squared test.

Table G Critical values for the Pearson correlation coefficient (r).

Table H Critical values for the F test.

Table I Sample size requirements for the one-way ANOVA.

Table J Power table for the one-way ANOVA.

Table K Sample size requirements for the difference between independent proportions ($\alpha = .05$).

Table L Required number of events per group for survival analysis (two-tailed test).

Table M Critical values of the Studentized Range test.

Table N Sample size requirements for MANOVA.

Table O Power of the two-group MANOVA.

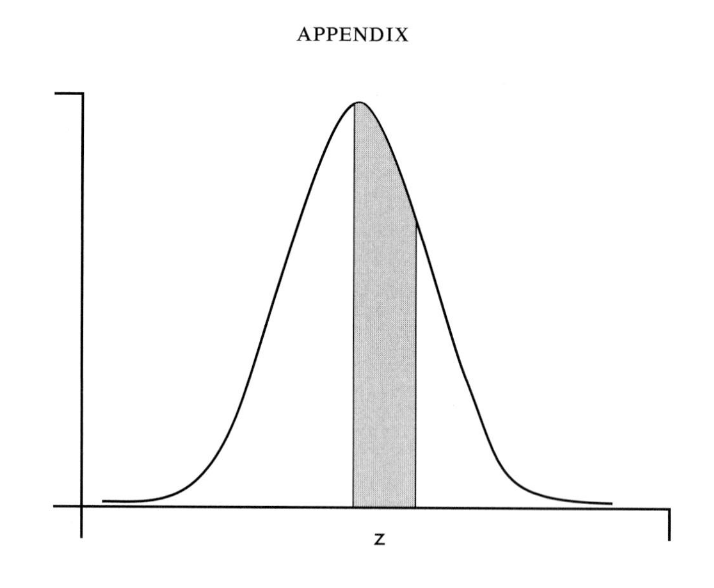

TABLE A

Area of the normal curve

z	Area below	z	Area below	z	Area below	z	Area below
.00	.0000						
.01	.0039	.41	.1591	.81	.2910	1.21	.3869
.02	.0079	.42	.1627	.82	.2932	1.22	.3888
.03	.0119	.43	.1664	.83	.2967	1.23	.3906
.04	.0159	.44	.1700	.84	.2995	1.24	.3925
.05	.0199	.45	.1736	.85	.3023	1.25	.3943
.06	.0239	.46	.1772	.86	.3051	1.26	.3962
.07	.0279	.47	.1808	.87	.3078	1.27	.3980
.08	.0318	.48	.1844	.88	.3106	1.28	.3997
.09	.0358	.49	.1879	.89	.3133	1.29	.4015
.10	.0398	.50	.1914	.90	.3159	1.30	.4032
.11	.0438	.51	.1950	.91	.3186	1.31	.4049
.12	.0477	.52	.1985	.92	.3212	1.32	.4066
.13	.0517	.53	.2019	.93	.3238	1.33	.4082
.14	.0556	.54	.2054	.94	.3264	1.34	.4099
.15	.0596	.55	.2088	.95	.3289	1.35	.4115
.16	.0635	.56	.2122	.96	.3315	1.36	.4131
.17	.0675	.57	.2156	.97	.3340	1.37	.4147
.18	.0714	.58	.2190	.98	.3365	1.38	.4162
.19	.0753	.59	.2224	.99	.3389	1.39	.4177
.20	.0792	.60	.2257	1.00	.3413	1.40	.4192
.21	.0831	.61	.2291	1.01	.3437	1.41	.4207
.22	.0870	.62	.2324	1.02	.3461	1.42	.4222
.23	.0909	.63	.2356	1.03	.3485	1.43	.4236
.24	.0948	.64	.2389	1.04	.3508	1.44	.4251
.25	.0987	.65	.2421	1.05	.3531	1.45	.4265
.26	.1025	.66	.2454	1.06	.3554	1.46	.4279
.27	.1064	.67	.2486	1.07	.3577	1.47	.4292
.28	.1102	.68	.2517	1.08	.3599	1.48	.4306
.29	.1141	.69	.2549	1.09	.3621	1.49	.4319
.30	.1179	.70	.2580	1.10	.3643	1.50	.4332
.31	.1217	.71	.2611	1.11	.3665	1.51	.4345
.32	.1255	.72	.2642	1.12	.3686	1.52	.4357
.33	.1293	.73	.2673	1.13	.3708	1.53	.4370
.34	.1330	.74	.2703	1.14	.3729	1.54	.4382
.35	.1368	.75	.2734	1.15	.3749	1.55	.4394
.36	.1406	.76	.2764	1.16	.3770	1.56	.4406
.37	.1443	.77	.2793	1.17	.3790	1.57	.4418
.38	.1480	.78	.2823	1.18	.3810	1.58	.4429
.39	.1517	.79	.2852	1.19	.3830	1.59	.4441
.40	.1554	.80	.2881	1.20	.3849	1.60	.4452

TABLE A

Area of the normal curve, cont'd

z	Area below	z	Area below	z	Area below	z	Area below
1.61	.4463	2.21	.4864	2.81	.4975	3.41	.4997
1.62	.4474	2.22	.4868	2.82	.4976	3.42	.4997
1.63	.4484	2.23	.4871	2.83	.4977	3.43	.4997
1.64	.4495	2.24	.4875	2.84	.4977	3.44	.4997
1.65	.4505	2.25	.4878	2.85	.4978	3.45	.4997
1.66	.4515	2.26	.4881	2.86	.4979	3.46	.4997
1.67	.4525	2.27	.4884	2.87	.4979	3.47	.4997
1.68	.4535	2.28	.4887	2.88	.4980	3.48	.4997
1.69	.4545	2.29	.4890	2.89	.4981	3.49	.4998
1.70	.4554	2.30	.4893	2.90	.4981	3.50	.4998
1.71	.4564	2.31	.4896	2.91	.4982	3.51	.4998
1.72	.4573	2.32	.4898	2.92	.4982	3.52	.4998
1.73	.4582	2.33	.4901	2.93	.4983	3.53	.4998
1.74	.4591	2.34	.4904	2.94	.4984	3.54	.4998
1.75	.4599	2.35	.4906	2.95	.4984	3.55	.4998
1.76	.4608	2.36	.4909	2.96	.4985	3.56	.4998
1.77	.4616	2.37	.4911	2.97	.4985	3.57	.4998
1.78	.4625	2.38	.4913	2.98	.4985	3.58	.4998
1.79	.4633	2.39	.4916	2.99	.4986	3.59	.4998
1.80	.4641	2.40	.4918	3.00	.4987	3.60	.4998
1.81	.4649	2.41	.4920	3.01	.4987	3.61	.4998
1.82	.4656	2.42	.4922	3.02	.4987	3.62	.4999
1.83	.4664	2.43	.4925	3.03	.4988	3.63	.4999
1.84	.4671	2.44	.4927	3.04	.4988	3.64	.4999
1.85	.4678	2.45	.4929	3.05	.4989	3.65	.4999
1.86	.4686	2.46	.4931	3.06	.4989	3.66	.4999
1.87	.4693	2.47	.4932	3.07	.4989	3.67	.4999
1.88	.4699	2.48	.4934	3.08	.4990	3.68	.4999
1.89	.4706	2.49	.4936	3.09	.4990	3.69	.4999
1.90	.4713	2.50	.4938	3.10	.4990	3.70	.4999
1.91	.4719	2.51	.4940	3.11	.4991	3.71	.4999
1.92	.4726	2.52	.4941	3.12	.4991	3.72	.4999
1.93	.4732	2.53	.4943	3.13	.4991	3.73	.4999
1.94	.4738	2.54	.4945	3.14	.4992	3.74	.4999
1.95	.4744	2.55	.4946	3.15	.4992	3.75	.4999
1.96	.4750	2.56	.4948	3.16	.4992	3.76	.4999
1.97	.4756	2.57	.4949	3.17	.4992	3.77	.4999
1.98	.4761	2.58	.4951	3.18	.4993	3.78	.4999
1.99	.4767	2.59	.4952	3.19	.4993	3.79	.4999
2.00	.4772	2.60	.4953	3.20	.4993	3.80	.4999
2.01	.4778	2.61	.4955	3.21	.4993	3.81	.4999
2.02	.4783	2.62	.4956	3.22	.4994	3.82	.4999
2.03	.4788	2.63	.4957	3.23	.4994	3.83	.4999
2.04	.4793	2.64	.4959	3.24	.4994	3.84	.4999
2.05	.4798	2.65	.4960	3.25	.4994	3.85	.4999
2.06	.4803	2.66	.4961	3.26	.4994	3.86	.4999
2.07	.4808	2.67	.4962	3.27	.4995	3.87	.4999
2.08	.4812	2.68	.4963	3.28	.4995	3.88	.4999
2.09	.4817	2.69	.4964	3.29	.4995	3.89	.4999
2.10	.4821	2.70	.4965	3.30	.4995	3.90	.5000
2.11	.4826	2.71	.4966	3.31	.4995	3.91	.5000
2.12	.4830	2.72	.4967	3.32	.4995	3.92	.5000
2.13	.4834	2.73	.4968	3.33	.4996	3.95	.5000
2.14	.4838	2.74	.4969	3.34	.4996	3.94	.5000
2.15	.4842	2.75	.4970	3.35	.4996	3.95	.5000
2.16	.4846	2.76	.4971	3.36	.4996	3.96	.5000
2.17	.4850	2.77	.4972	3.37	.4996	3.97	.5000
2.18	.4854	2.78	.4973	3.38	.4996	3.98	.5000
2.19	.4857	2.79	.4974	3.39	.4997	3.99	.5000
2.20	.4861	2.80	.4974	3.40	.4997	4.00	.5000

TABLE B

Sample size requirements to show a difference between two means of size σ/δ

σ/δ	$\alpha =$ .10			.05			.01		
	$\beta =$ .20	.15	.10	.20	.15	.10	.20	.15	.10
.5	2	2	2	2	2	3	3	4	4
1.0	6	7	9	8	9	11	12	14	16
1.5	14	17	20	18	21	24	27	30	34
1.6	16	19	22	21	24	28	31	34	39
1.7	18	21	25	23	27	31	35	39	44
1.8	20	24	28	26	30	35	39	43	49
1.9	23	26	31	29	33	39	43	48	55
2.0	25	29	35	32	37	43	48	53	61
2.1	28	32	38	35	40	47	53	59	67
2.2	30	35	38	39	44	52	58	65	73
2.3	33	39	46	42	48	57	63	71	80
2.4	36	42	50	46	53	61	68	77	87
2.5	39	46	54	50	57	67	74	83	95
2.6	42	49	59	54	62	72	80	90	102
2.7	46	53	63	58	67	78	87	97	110
2.8	49	57	68	62	72	83	93	104	118
2.9	52	61	73	67	77	90	100	112	127
3.0	56	66	78	72	82	96	107	120	136
3.1	60	70	83	76	88	102	114	128	145
3.2	64	74	88	81	93	109	121	136	154
3.3	68	79	94	86	99	116	129	144	164
3.4	72	84	100	92	105	123	137	153	174
3.5	76	89	106	97	112	130	145	162	185
3.6	81	94	112	103	118	138	153	172	195
3.7	85	99	118	108	125	145	162	181	206
3.8	90	105	124	114	131	153	171	191	217
3.9	95	110	131	120	138	161	180	201	229
4.0	99	116	138	127	145	170	189	212	241

*NOTE δ = difference between means; σ = standard deviation.

TABLE C

Critical values for the t-test

	α for a 1-tailed test					
	.10	.05	.025	.01	.005	.0005
	α for a 2-tailed test					
df	.20	.10	.05	.02	.01	.001
1	3.078	6.314	12.706	31.820	63.656	636.615
2	1.886	2.920	4.303	6.965	9.925	31.599
3	1.638	2.353	3.182	4.541	5.841	12.924
4	1.533	2.132	2.776	3.747	4.604	8.610
5	1.476	2.015	2.571	3.365	4.032	6.869
6	1.440	1.943	2.447	3.143	3.707	5.959
7	1.415	1.895	2.365	2.998	3.499	5.408
8	1.397	1.860	2.306	2.896	3.355	5.041
9	1.383	1.833	2.262	2.821	3.250	4.781
10	1.372	1.812	2.228	2.764	3.169	4.587
11	1.363	1.796	2.201	2.718	3.106	4.437
12	1.356	1.782	2.179	2.681	3.055	4.318
13	1.350	1.771	2.160	2.650	3.012	4.221
14	1.345	1.761	2.145	2.624	2.977	4.141
15	1.341	1.753	2.131	2.602	2.947	4.073
16	1.337	1.746	2.120	2.583	2.921	4.015
17	1.333	1.740	2.110	2.567	2.898	3.965
18	1.330	1.734	2.101	2.552	2.878	3.922
19	1.328	1.729	2.093	2.539	2.861	3.883
20	1.325	1.725	2.086	2.528	2.845	3.849
21	1.323	1.721	2.080	2.518	2.831	3.819
22	1.321	1.717	2.074	2.508	2.819	3.792
23	1.319	1.714	2.069	2.500	2.807	3.768
24	1.318	1.711	2.064	2.492	2.797	3.745
25	1.316	1.708	2.060	2.485	2.787	3.725
26	1.315	1.706	2.056	2.479	2.779	3.707
27	1.314	1.703	2.052	2.473	2.771	3.690
28	1.313	1.701	2.048	2.467	2.763	3.674
29	1.311	1.699	2.045	2.462	2.756	3.659
30	1.310	1.697	2.042	2.457	2.750	3.646
35	1.306	1.690	2.030	2.438	2.724	3.591
40	1.303	1.684	2.021	2.423	2.704	3.551
45	1.301	1.679	2.014	2.412	2.690	3.520
50	1.299	1.676	2.009	2.403	2.678	3.496
55	1.297	1.673	2.004	2.396	2.668	3.476
60	1.296	1.671	2.000	2.390	2.660	3.460
70	1.294	1.667	1.994	2.381	2.648	3.435
80	1.292	1.664	1.990	2.374	2.639	3.416
90	1.291	1.662	1.987	2.368	2.632	3.402
100	1.290	1.660	1.984	2.364	2.626	3.390

TABLE D

Sample size requirements for the independent *t*-test*

d	α (1-tail) = .05				.025				.005			
	α (2-tail) = .10				.05				.01			
	β = .20	.15	.10	.05	.20	.15	.10	.05	.20	.15	.10	.05
.10	1237	1438	1713	2165	1570	1795	2102	2599	2337	2609	2977	3563
.20	309	359	428	541	393	449	526	650	584	652	744	891
.25	198	230	274	346	251	287	336	416	374	417	476	570
.30	137	160	190	241	174	199	234	289	260	290	331	396
.40	77	90	107	135	98	112	131	162	146	163	186	223
.50	49	58	69	87	63	72	84	104	93	104	119	143
.60	34	40	48	60	44	50	58	72	65	72	83	99
.70	27	31	35	44	32	37	43	53	48	53	61	73
.75	24	28	30	38	30	32	37	46	42	46	53	63
.80	21	24	29	34	27	30	33	41	37	41	47	56
.90	17	20	23	29	21	24	28	32	31	32	37	44
1.0	14	16	19	24	18	20	23	28	25	28	30	36
1.1	12	14	16	20	15	17	19	23	21	24	27	31
1.2	11	12	14	17	13	14	17	20	18	20	23	27
1.3	9	11	12	15	11	13	14	17	16	17	20	23
1.4	8	9	11	13	10	11	13	15	14	15	17	20
1.5	7	8	10	12	9	10	11	14	12	14	15	18
1.6	7	8	9	10	8	9	10	12	11	12	14	16
1.7	6	7	8	9	7	8	9	11	10	11	12	14
1.8	6	6	7	9	7	8	8	10	9	10	11	13
1.9	6	6	7	8	6	7	8	9	8	9	10	12
2.0	6	6	6	7	6	6	7	8	8	9	9	11
2.1	6	6	6	7	6	6	7	8	7	8	9	10
2.2	6	6	6	6	6	6	6	7	7	7	8	9
2.3	6	6	6	6	6	6	6	7	6	7	8	9
2.4	6	6	6	6	6	6	6	7	6	7	7	8
2.5	6	6	6	6	6	6	6	6	6	6	7	8
3.0	2	6	6	6	6	6	6	6	6	6	6	6
3.5	2	2	2	6	2	2	6	6	6	6	6	6
4.0	2	2	2	2	2	2	2	6	2	6	6	6

*NOTE: sample sizes are per group.

TABLE E

Power table for the independent *t*-test

	α (1-tail) = .05				.025				.005			
	α (2-tail) = .10				.05				.01			
N per group *d* =	.20	.40	.60	.80	.20	.40	.60	.80	.20	.40	.60	.80
2	.129	.127	.151	.171	.086	.090	.097	.106	.048	.049	.050	.051
3	.114	.135	.171	.219	.063	.075	.096	.126	.021	.024	.030	.038
4	.114	.147	.202	.275	.058	.079	.114	.164	.015	.020	.030	.046
5	.116	.161	.234	.329	.058	.086	.135	.206	.013	.020	.036	.061
6	.119	.176	.265	.380	.059	.095	.159	.248	.012	.022	.043	.078
7	.122	.190	.296	.428	.060	.105	.182	.290	.011	.024	.051	.099
8	.126	.205	.326	.472	.062	.115	.206	.331	.011	.027	.061	.121
9	.130	.219	.354	.514	.064	.125	.230	.371	.012	.030	.072	.145
10	.133	.233	.382	.553	.066	.135	.254	.410	.012	.033	.083	.171
11	.137	.247	.409	.589	.068	.146	.278	.448	.012	.037	.095	.197
12	.141	.261	.435	.623	.070	.157	.301	.483	.013	.041	.108	.224
13	.144	.275	.460	.654	.073	.167	.324	.518	.013	.045	.121	.251
14	.148	.288	.484	.683	.075	.178	.347	.550	.014	.050	.135	.279
15	.152	.301	.507	.710	.078	.189	.369	.581	.014	.054	.149	.307
16	.156	.314	.529	.735	.080	.199	.391	.611	.015	.059	.163	.336
17	.159	.327	.551	.758	.083	.210	.413	.638	.015	.064	.178	.364
18	.163	.340	.572	.779	.085	.221	.434	.664	.016	.069	.193	.391
19	.167	.353	.591	.798	.088	.232	.455	.689	.017	.074	.208	.419
20	.170	.365	.610	.816	.090	.242	.475	.712	.018	.079	.223	.446
21	.174	.377	.629	.833	.093	.253	.494	.734	.018	.084	.239	.472
22	.178	.389	.646	.848	.095	.263	.513	.754	.019	.090	.255	.498
23	.181	.401	.663	.862	.098	.274	.532	.773	.020	.095	.270	.523
24	.185	.413	.679	.874	.100	.284	.550	.791	.021	.101	.286	.548
25	.189	.424	.695	.886	.103	.295	.568	.807	.021	.107	.302	.571
30	.207	.479	.763	.930	.116	.346	.648	.874	.026	.137	.381	.678
35	.225	.530	.817	.958	.130	.395	.716	.919	.030	.169	.458	.764
40	.242	.576	.860	.975	.143	.442	.773	.949	.035	.203	.530	.831
45	.260	.619	.893	.985	.156	.487	.819	.968	.041	.238	.596	.881
50	.277	.658	.919	.991	.170	.529	.858	.980	.046	.273	.656	.918
55	.293	.693	.939	.995	.183	.569	.888	.988	.052	.308	.710	.944
60	.310	.725	.954	.997	.197	.606	.913	.993	.058	.343	.757	.962
65	.326	.755	.966	.998	.210	.640	.932	.996	.064	.378	.797	.975
70	.342	.781	.974	.999	.224	.673	.948	.997	.071	.413	.832	.984
75	.357	.805	.981	.999	.237	.702	.960	.998	.077	.447	.862	.989
80	.373	.826	.986	1.00	.251	.730	.969	.999	.084	.480	.887	.993
85	.388	.846	.990	1.00	.264	.755	.977	.999	.091	.512	.908	.996
90	.403	.863	.992	1.00	.277	.779	.982	1.00	.098	.543	.926	.997
95	.417	.878	.994	1.00	.290	.800	.987	1.00	.106	.572	.940	.998
100	.431	.892	.996	1.00	.303	.820	.990	1.00	.113	.601	.952	.999

TABLE F

Critical values for the chi-squared test

	α					
df	.10	.05	.025	.01	.005	.001
1	2.706	3.842	5.024	6.635	7.879	10.828
2	4.605	5.992	7.378	9.210	10.597	13.816
3	6.251	7.815	9.348	11.345	12.838	16.266
4	7.779	9.489	11.143	13.277	14.860	18.467
5	9.236	11.071	12.833	15.086	16.750	20.515
6	10.645	12.592	14.449	16.812	18.548	22.457
7	12.017	14.067	16.013	18.475	20.278	24.321
8	13.362	15.507	17.535	20.090	21.955	26.124
9	14.684	16.919	19.023	21.666	23.589	27.877
10	15.987	18.307	20.483	23.209	25.188	29.588
11	17.275	19.675	21.920	24.725	26.757	31.264
12	18.549	21.026	23.336	26.217	28.299	32.909
13	19.812	22.362	24.736	27.688	29.819	34.528
14	21.064	23.685	26.120	29.141	31.319	36.123
15	22.307	24.996	27.488	30.578	32.801	37.697
16	23.542	26.296	28.845	32.000	34.267	39.252
17	24.769	27.587	30.191	33.409	35.718	40.790
18	25.989	28.869	31.526	34.805	37.156	42.312
19	27.204	30.144	32.852	36.191	38.582	43.820
20	28.412	31.410	34.170	37.566	39.997	45.314
21	29.615	32.671	35.479	38.932	41.401	46.797
22	30.813	33.924	36.781	40.289	42.796	48.268
23	32.007	35.172	38.076	41.638	44.181	49.728
24	33.196	36.415	39.365	42.980	45.558	51.178
25	34.382	37.652	40.647	44.314	46.928	52.620
26	35.563	38.885	41.924	45.642	48.290	54.052
27	36.741	40.113	43.195	46.963	49.645	55.476
28	37.916	41.337	44.461	48.278	50.993	56.892
29	39.087	42.557	45.723	49.588	52.336	58.301
30	40.256	43.773	46.980	50.892	53.672	59.703

TABLE G

Critical values for the Pearson correlation coefficient (r)

	α for a 1-tailed test			
	.05	.025	.01	.005
	α for a 2-tailed test			
df	.10	.05	.02	.01
1	.988	.997	.9995	.9999
2	.900	.950	.980	.990
3	.805	.878	.934	.959
4	.729	.811	.882	.917
5	.669	.755	.833	.874
6	.621	.707	.789	.834
7	.582	.666	.750	.798
8	.549	.632	.715	.765
9	.521	.602	.685	.735
10	.497	.576	.658	.708
11	.476	.553	.634	.684
12	.458	.532	.612	.661
13	.441	.514	.592	.641
14	.426	.497	.574	.623
15	.412	.482	.558	.606
16	.400	.468	.543	.590
17	.389	.456	.529	.575
18	.378	.444	.516	.561
19	.369	.433	.503	.549
20	.360	.423	.492	.537
21	.352	.413	.482	.526
22	.344	.404	.472	.515
23	.337	.396	.462	.505
24	.330	.388	.453	.496
25	.323	.381	.445	.487
26	.317	.374	.437	.479
27	.312	.367	.430	.471
28	.306	.361	.423	.463
29	.301	.355	.416	.456
30	.296	.349	.409	.449
35	.275	.325	.381	.418
40	.257	.304	.358	.393
45	.243	.288	.338	.372
50	.231	.273	.322	.354
55	.220	.261	.307	.339
60	.211	.250	.295	.325
70	.195	.232	.274	.302
80	.183	.217	.257	.283
90	.173	.205	.242	.267
100	.164	.195	.230	.254
125	.147	.174	.206	.228
150	.134	.159	.189	.208
175	.124	.147	.175	.193
200	.116	.138	.164	.181
250	.104	.124	.146	.162

TABLE Ha

Critical values for the F test*

df_2	df_1—Numerator degrees of freedom 1	2	3	4	5	6	7	8	9	10
1	161	200	216	225	230	234	237	239	241	242
	4052	**5000**	**5403**	**5625**	**5764**	**5859**	**5928**	**5981**	**6023**	**6056**
2	18.5	19.0	19.2	19.2	19.3	19.3	19.4	19.4	19.4	19.4
	98.5	**99.0**	**99.2**	**99.2**	**99.3**	**99.3**	**99.4**	**99.4**	**99.4**	**99.4**
3	10.1	9.55	9.28	9.12	9.01	8.94	8.89	8.85	8.81	8.79
	34.1	**30.8**	**29.5**	**28.7**	**28.2**	**27.9**	**27.7**	**27.5**	**27.3**	**27.2**
4	7.71	6.94	6.59	6.39	6.26	6.16	6.09	6.04	6.00	5.96
	21.2	**18.0**	**16.7**	**16.0**	**15.5**	**15.2**	**15.0**	**14.8**	**14.7**	**14.5**
5	6.61	5.79	5.41	5.19	5.05	4.95	4.88	4.82	4.77	4.74
	16.3	**13.3**	**12.1**	**11.4**	**11.0**	**10.7**	**10.5**	**10.3**	**10.2**	**10.2**
6	5.99	5.14	4.76	4.63	4.39	4.28	4.21	4.15	4.10	4.06
	13.7	**10.9**	**9.78**	**9.15**	**8.75**	**8.47**	**8.26**	**8.10**	**7.98**	**7.87**
7	5.59	4.74	4.35	4.12	3.97	3.87	3.79	3.73	3.68	3.64
	12.2	**9.54**	**8.45**	**7.85**	**7.46**	**7.19**	**6.99**	**6.84**	**6.72**	**6.62**
8	5.32	4.46	4.07	3.84	3.69	3.58	3.50	3.44	3.39	3.35
	11.3	**8.64**	**7.59**	**7.01**	**6.63**	**6.37**	**6.18**	**6.03**	**5.91**	**5.81**
9	5.12	4.26	3.86	3.63	3.48	3.37	3.29	3.23	3.18	3.14
	10.6	**8.02**	**6.99**	**6.42**	**6.06**	**5.80**	**5.61**	**5.47**	**5.35**	**5.26**
10	4.96	4.10	3.71	3.48	3.33	3.22	3.14	3.07	3.02	2.98
	10.0	**7.56**	**6.55**	**5.99**	**5.64**	**5.39**	**5.20**	**5.06**	**4.94**	**4.85**
11	4.84	3.98	3.59	3.36	3.20	3.09	3.01	2.95	2.90	2.85
	9.65	**7.21**	**6.22**	**5.67**	**5.32**	**5.07**	**4.89**	**4.74**	**4.63**	**4.54**
12	4.75	3.89	3.50	3.26	3.11	3.00	2.91	2.85	2.80	2.75
	9.33	**6.93**	**5.95**	**5.41**	**5.06**	**4.82**	**4.64**	**4.50**	**4.39**	**4.30**
13	4.67	3.81	3.41	3.18	3.03	2.92	2.83	2.77	2.71	2.67
	9.07	**6.70**	**5.74**	**5.21**	**4.86**	**4.62**	**4.44**	**4.30**	**4.19**	**4.10**
14	4.60	3.74	3.34	3.11	2.96	2.85	2.76	2.70	2.65	2.60
	8.86	**6.51**	**5.56**	**5.04**	**4.69**	**4.46**	**4.28**	**4.14**	**4.03**	**3.94**
15	4.54	3.68	3.29	3.06	2.90	2.79	2.71	2.64	2.59	2.54
	8.86	**6.36**	**5.42**	**4.89**	**4.56**	**4.32**	**4.14**	**4.00**	**3.89**	**3.80**
16	4.49	3.63	3.24	3.01	2.85	2.74	2.66	2.59	2.54	2.49
	8.53	**6.23**	**5.29**	**4.77**	**4.44**	**4.20**	**4.03**	**3.89**	**3.78**	**3.69**

*upper number is 5% level, lower (**in bold**) is 1%.

df_1—Numerator degrees of freedom									
11	12	15	20	25	30	40	50	75	100
243	244	246	248	249	250	251	252	253	253
6083	**6106**	**6157**	**6209**	**6239**	**6261**	**6287**	**6302**	**6323**	**6334**
19.4	19.4	19.4	19.4	19.5	19.5	19.5	19.5	19.5	19.5
99.4	**99.4**	**99.4**	**99.4**	**99.5**	**99.5**	**99.5**	**99.5**	**99.5**	**99.5**
8.76	8.74	8.70	8.66	8.63	8.62	8.59	8.58	8.56	8.55
27.1	**27.1**	**26.9**	**26.7**	**26.6**	**26.5**	**26.4**	**26.4**	**26.3**	**26.2**
5.94	5.91	5.86	5.80	5.77	5.75	5.72	5.70	5.68	5.66
14.5	**14.4**	**14.2**	**14.0**	**13.9**	**13.8**	**13.7**	**13.7**	**13.6**	**13.6**
4.70	4.68	4.62	4.56	4.52	4.50	4.46	4.44	4.42	4.41
9.96	**9.89**	**9.72**	**9.55**	**9.45**	**9.38**	**9.29**	**9.24**	**9.17**	**9.13**
4.03	4.00	3.94	3.87	3.83	3.81	3.77	3.75	3.73	3.71
7.79	**7.72**	**7.56**	**7.40**	**7.30**	**7.23**	**7.14**	**7.09**	**7.02**	**6.99**
3.60	3.57	3.51	3.44	3.40	3.38	3.34	3.32	3.29	3.27
6.54	**6.47**	**6.31**	**6.16**	**6.06**	**5.99**	**5.91**	**5.86**	**5.79**	**5.75**
3.31	3.28	3.22	3.15	3.11	3.80	3.04	3.02	2.99	2.97
5.73	**5.67**	**5.52**	**5.36**	**5.26**	**5.20**	**5.12**	**5.07**	**5.00**	**4.96**
3.10	3.07	3.01	2.94	2.89	2.86	2.83	2.80	2.77	2.76
5.18	**5.11**	**4.96**	**4.81**	**4.71**	**4.65**	**4.57**	**4.52**	**4.45**	**4.41**
2.94	2.91	2.84	2.77	2.73	2.70	2.66	2.64	2.60	2.59
4.77	**4.71**	**4.56**	**4.41**	**4.31**	**4.25**	**4.17**	**4.12**	**4.05**	**4.01**
2.82	2.79	2.72	2.65	2.60	2.57	2.53	2.51	2.47	2.46
4.46	**4.40**	**4.25**	**4.10**	**4.01**	**3.94**	**3.86**	**3.81**	**3.74**	**3.71**
2.72	2.69	2.62	2.54	2.50	2.47	2.43	2.40	2.37	2.35
4.22	**4.16**	**4.01**	**3.86**	**3.76**	**3.70**	**3.62**	**3.57**	**3.50**	**3.47**
2.63	2.60	2.53	2.46	2.41	2.38	2.34	2.31	2.28	2.26
4.02	**3.96**	**3.82**	**3.66**	**3.57**	**3.51**	**3.43**	**3.38**	**3.31**	**3.27**
2.57	2.53	2.46	2.39	2.34	2.31	2.27	2.24	2.21	2.19
3.86	**3.80**	**3.66**	**3.51**	**3.41**	**3.35**	**3.27**	**3.22**	**3.15**	**3.11**
2.51	2.48	2.40	2.33	2.28	2.25	2.20	2.18	2.14	2.12
3.73	**3.67**	**3.52**	**3.37**	**3.28**	**3.21**	**3.13**	**3.08**	**3.01**	**2.98**
2.46	2.42	2.35	2.28	2.23	2.19	2.15	2.12	2.09	2.07
3.62	**3.55**	**3.41**	**3.26**	**3.16**	**3.10**	**3.02**	**2.97**	**2.90**	**2.86**

TABLE Hb

Critical values for the *F* test*

df_2	df_1—Numerator degrees of freedom 1	2	3	4	5	6	7	8	9	10
17	4.45	3.59	3.20	2.96	2.81	2.70	2.61	2.55	2.49	2.45
	8.40	**6.11**	**5.18**	**4.67**	**4.34**	**4.10**	**3.93**	**3.79**	**3.68**	**3.59**
18	4.41	3.55	3.16	2.93	2.77	2.66	2.58	2.51	2.46	2.41
	8.29	**6.01**	**5.09**	**4.58**	**4.25**	**4.01**	**3.84**	**3.71**	**3.60**	**3.51**
19	4.38	3.52	3.13	2.90	2.74	2.63	2.54	2.48	2.42	2.38
	8.18	**5.93**	**5.01**	**4.50**	**4.17**	**3.94**	**3.77**	**3.63**	**3.52**	**3.43**
20	4.35	3.49	3.10	2.87	2.71	2.60	2.51	2.45	2.39	2.35
	8.10	**5.85**	**4.94**	**4.43**	**4.10**	**3.87**	**3.70**	**3.56**	**3.46**	**3.37**
21	4.32	3.47	3.07	2.84	2.68	2.57	2.49	2.42	2.37	2.32
	8.02	**5.78**	**4.87**	**4.37**	**4.04**	**3.81**	**3.64**	**3.51**	**3.40**	**3.31**
22	4.30	3.44	3.05	2.87	2.66	2.55	2.46	2.40	2.34	2.30
	7.95	**5.72**	**4.82**	**4.31**	**3.99**	**3.76**	**3.59**	**3.45**	**3.35**	**3.26**
23	4.28	3.42	3.03	2.80	2.64	2.53	2.44	2.37	2.32	2.27
	7.88	**5.66**	**4.76**	**4.26**	**3.94**	**3.71**	**3.54**	**3.41**	**3.30**	**3.21**
24	4.26	3.41	3.01	2.78	2.62	2.51	2.42	2.36	2.30	2.25
	7.82	**5.61**	**4.72**	**4.22**	**3.89**	**3.67**	**3.50**	**3.36**	**3.26**	**3.17**
25	4.24	3.39	2.99	2.76	2.60	2.49	2.40	2.34	2.28	2.24
	7.77	**5.57**	**4.68**	**4.18**	**3.86**	**3.63**	**3.46**	**3.32**	**3.22**	**3.13**
26	4.23	3.37	2.98	2.74	2.59	2.47	2.39	2.32	2.27	2.22
	7.72	**5.53**	**4.64**	**4.14**	**3.82**	**3.59**	**3.42**	**3.29**	**3.18**	**3.09**
27	4.21	3.35	2.96	2.73	2.57	2.46	2.37	2.31	2.25	2.20
	7.68	**5.49**	**4.60**	**4.11**	**3.78**	**3.56**	**3.39**	**3.26**	**3.15**	**3.06**
28	4.20	3.34	2.95	2.71	2.56	2.45	2.36	2.29	2.24	2.19
	7.64	**5.45**	**4.57**	**4.07**	**3.75**	**3.53**	**3.36**	**3.23**	**3.12**	**3.03**
29	4.18	3.33	2.93	2.70	2.55	2.43	2.35	2.28	2.22	2.18
	7.60	**5.42**	**4.54**	**4.04**	**3.73**	**3.50**	**3.33**	**3.20**	**3.09**	**3.00**
30	4.17	3.32	2.92	2.69	2.53	2.42	2.33	2.27	2.21	2.16
	7.56	**5.39**	**4.51**	**4.02**	**3.70**	**3.47**	**3.30**	**3.17**	**3.07**	**2.98**
40	4.08	3.23	2.84	2.61	2.45	2.34	2.25	2.18	2.12	2.08
	7.31	**5.18**	**4.31**	**3.83**	**3.51**	**3.29**	**3.12**	**2.99**	**2.89**	**2.80**
50	4.03	3.18	2.79	2.56	2.40	2.27	2.20	2.13	2.07	2.03
	7.17	**5.06**	**4.20**	**3.72**	**3.41**	**3.19**	**3.02**	**2.89**	**2.78**	**2.70**
75	3.97	3.12	2.73	2.49	2.34	2.22	2.13	2.06	2.01	1.96
	6.99	**4.90**	**4.05**	**3.58**	**3.27**	**3.05**	**2.89**	**2.76**	**2.65**	**2.57**
100	3.94	3.09	2.70	2.46	2.31	2.19	2.10	2.03	1.97	1.93
	6.90	**4.82**	**3.98**	**3.51**	**3.21**	**2.99**	**2.82**	**2.69**	**2.59**	**2.50**

*upper number is 5% level, lower (**in bold**) is 1%.

df_1—Numerator degrees of freedom									
11	12	15	20	25	30	40	50	75	100
2.41	2.38	2.31	2.23	2.18	2.15	2.10	2.08	2.04	2.02
3.52	**3.46**	**3.31**	**3.16**	**3.07**	**3.00**	**2.92**	**2.87**	**2.80**	**2.76**
2.37	2.34	2.27	2.19	2.14	2.11	2.06	2.04	2.00	1.98
3.43	**3.37**	**3.23**	**3.08**	**2.98**	**2.92**	**2.84**	**2.78**	**2.71**	**2.68**
2.34	2.31	2.23	2.16	2.11	2.07	2.03	2.00	1.96	1.94
3.36	**3.30**	**3.15**	**3.00**	**2.91**	**2.84**	**2.76**	**2.71**	**2.64**	**2.60**
2.31	2.28	2.20	2.12	2.07	2.04	1.99	1.97	1.93	1.91
3.29	**3.23**	**3.09**	**2.94**	**2.84**	**2.78**	**2.69**	**2.64**	**2.57**	**2.54**
2.28	2.25	2.18	2.10	2.05	2.01	1.96	1.94	1.90	1.88
3.24	**3.17**	**3.03**	**2.88**	**2.79**	**2.72**	**2.64**	**2.58**	**2.51**	**2.48**
2.26	2.23	2.15	2.07	2.02	1.98	1.94	1.91	1.87	1.85
3.18	**3.12**	**2.98**	**2.83**	**2.73**	**2.67**	**2.58**	**2.53**	**2.46**	**2.42**
2.24	2.20	2.13	2.05	2.00	1.96	1.91	1.88	1.84	1.82
3.14	**3.07**	**2.93**	**2.78**	**2.69**	**2.62**	**2.54**	**2.48**	**2.41**	**2.37**
2.22	2.18	2.11	2.03	1.97	1.94	1.89	1.86	1.82	1.80
3.09	**3.03**	**2.89**	**2.74**	**2.64**	**2.58**	**2.49**	**2.44**	**2.37**	**2.33**
2.20	2.16	2.09	2.01	1.96	1.20	1.87	1.84	1.80	1.78
3.06	**2.99**	**2.85**	**2.70**	**2.60**	**2.54**	**2.45**	**2.40**	**2.33**	**2.29**
2.18	2.15	2.07	1.99	1.94	1.90	1.85	1.82	1.78	1.76
3.02	**2.96**	**2.81**	**2.66**	**2.57**	**2.50**	**2.42**	**2.36**	**2.29**	**2.25**
2.17	2.13	2.06	1.97	1.92	1.88	1.84	1.81	1.76	1.74
2.99	**2.93**	**2.78**	**2.63**	**2.54**	**2.47**	**2.38**	**2.33**	**2.26**	**2.22**
2.15	2.12	2.04	1.96	1.91	1.87	1.82	1.79	1.75	1.73
2.96	**2.90**	**2.75**	**2.60**	**2.51**	**2.44**	**2.35**	**2.30**	**2.23**	**2.19**
2.14	2.10	2.03	1.94	1.89	1.85	1.81	1.77	1.73	1.71
2.93	**2.87**	**2.73**	**2.57**	**2.48**	**2.41**	**2.33**	**2.27**	**2.20**	**2.16**
2.13	2.09	2.01	1.93	1.88	1.84	1.79	1.76	1.72	1.70
2.91	**2.84**	**2.70**	**2.55**	**2.45**	**2.39**	**2.30**	**2.25**	**2.17**	**2.13**
2.04	2.00	1.92	1.84	1.78	1.74	1.69	1.66	1.61	1.59
2.73	**2.66**	**2.52**	**2.37**	**2.27**	**2.20**	**2.11**	**2.06**	**1.98**	**1.94**
1.99	1.95	1.87	1.78	1.73	1.69	1.63	1.60	1.55	1.52
2.63	**2.56**	**2.42**	**2.27**	**2.17**	**2.10**	**2.01**	**1.95**	**1.87**	**1.82**
1.92	1.88	1.80	1.71	1.65	1.61	1.55	1.52	1.47	1.44
2.49	**2.43**	**2.29**	**2.13**	**2.03**	**1.96**	**1.87**	**1.81**	**1.72**	**1.67**
1.89	1.85	1.77	1.68	1.62	1.57	1.52	1.48	1.42	1.39
2.43	**2.37**	**2.22**	**2.07**	**1.97**	**1.89**	**1.80**	**1.74**	**1.65**	**1.60**

TABLE I

Sample size requirements for the one-way ANOVA*

Effect size (f)	Number of groups	$\alpha = .05$, $\beta = .30$	$\alpha = .05$, $\beta = .20$	$\alpha = .05$, $\beta = .10$	$\alpha = .01$, $\beta = .30$	$\alpha = .01$, $\beta = .20$	$\alpha = .01$, $\beta = .10$
.1	3	251	315	415	384	460	578
	4	217	269	351	324	386	481
	5	191	237	307	283	335	415
	6	173	213	274	252	298	368
	7	148	182	235	216	256	315
.2	3	64	80	105	97	116	146
	4	55	68	89	82	99	121
	5	49	60	78	72	85	105
	6	44	54	69	64	76	93
	7	38	46	59	55	65	80
.3	3	29	36	47	44	53	66
	4	25	31	40	37	44	55
	5	22	27	35	33	39	48
	6	20	25	32	29	35	42
	7	17	21	27	25	30	36
.4	3	17	21	27	26	30	38
	4	15	18	23	22	26	32
	5	13	16	20	19	22	27
	6	12	14	18	17	20	24
	7	10	12	16	15	17	21
.5	3	11	14	18	17	20	25
	4	10	12	15	15	17	21
	5	9	11	13	13	15	18
	6	8	10	12	10	13	16
	7	7	8	10	10	12	14
.6	3	8	10	13	13	15	18
	4	7	9	11	11	12	15
	5	7	8	10	9	11	13
	6	6	7	9	9	10	12
	7	5	6	8	8	9	10

*numbers are sample sizes per group.

TABLE J

Power table for the one-way ANOVA

Number of groups	N per group	α = .05, f = .10	.20	.30	.40	α = .01, f = .10	.20	.30	.40
3	5	.059	.088	.140	.218	.012	.202	.037	.067
	10	.068	.140	.272	.453	.013	.036	.095	.206
	15	.080	.199	.405	.647	.015	.060	.177	.380
	20	.093	.259	.527	.785	.019	.090	.272	.548
	25	.107	.320	.632	.875	.023	.124	.372	.689
	30	.121	.378	.719	.930	.028	.162	.469	.795
	35	.135	.435	.789	.962	.033	.203	.560	.871
	40	.149	.490	.844	.980	.039	.246	.642	.921
	45	.164	.541	.886	.990	.045	.289	.713	.954
	50	.179	.588	.918	.995	.051	.334	.773	.973
4	5	.059	.090	.149	.239	.012	.021	.041	.078
	10	.069	.150	.301	.508	.013	.040	.112	.252
	15	.082	.216	.454	.717	.016	.069	.215	.463
	20	.097	.286	.589	.850	.021	.106	.332	.650
	25	.112	.355	.701	.926	.026	.148	.451	.789
	30	.127	.422	.788	.965	.031	.195	.563	.881
	35	.143	.487	.853	.985	.037	.246	.661	.937
	40	.160	.547	.901	.993	.044	.298	.743	.968
	45	.176	.604	.934	.997	.051	.351	.810	.984
	50	.193	.655	.957	.999	.059	.405	.862	.993
5	5	.059	.093	.159	.262	.012	.022	.045	.090
	10	.071	.160	.332	.562	.014	.045	.131	.301
	15	.085	.235	.502	.777	.017	.079	.255	.542
	20	.101	.313	.647	.898	.022	.123	.393	.736
	25	.117	.391	.760	.958	.028	.174	.527	.863
	30	.135	.466	.843	.984	.034	.230	.647	.934
	35	.152	.537	.901	.994	.041	.290	.745	.971
	40	.171	.603	.939	.998	.049	.352	.822	.988
	45	.189	.662	.963	.999	.057	.414	.879	.995
	50	.208	.714	.978	1.00	.066	.474	.920	.998
6	5	.060	.097	.170	.286	.012	.023	.049	.102
	10	.072	.170	.362	.612	.014	.049	.151	.350
	15	.088	.254	.547	.826	.019	.089	.295	.615
	20	.105	.341	.699	.932	.024	.140	.452	.805
	25	.123	.427	.810	.976	.030	.200	.597	.913
	30	.142	.509	.885	.992	.037	.266	.719	.965
	35	.162	.584	.934	.998	.045	.335	.812	.987
	40	.182	.653	.963	.999	.054	.405	.879	.996
	45	.202	.713	.980	1.00	.064	.474	.925	.999
	50	.223	.765	.989	1.00	.074	.540	.966	1.00

TABLE K

Sample size requirements for the difference between independent proportions ($\alpha = .05$)

p_S	Difference between proportions ($p_L - p_S$) .05	.10	.15	.20	.25	.30	.35	.40	.45	.50	.55	.60	.65	.70	.75	.80
.05	424	132	69	44	31	24	19	15	13	11	9	8	7	6	5	4
	485	151	79	51	36	27	21	17	14	12	10	9	8	7	6	5
	567	177	93	59	42	32	25	20	17	14	12	10	9	8	7	6
	702	219	115	73	52	39	31	25	21	17	15	13	11	10	8	7
.10	681	195	96	59	41	30	23	18	15	12	10	9	7	6	5	5
	778	223	110	67	46	34	26	21	17	14	12	10	9	7	6	5
	911	261	129	79	54	40	31	24	20	17	14	12	10	9	7	6
	1127	323	159	98	67	49	38	30	25	20	17	15	12	11	9	8
.15	903	248	119	71	48	34	26	20	16	13	11	9	8	7	5	
	1032	283	136	81	54	39	30	23	19	15	13	11	9	7	6	
	1208	332	159	95	64	46	35	27	22	18	15	12	10	9	7	
	1494	410	196	117	79	57	43	34	27	22	18	15	13	11	9	
.20	1092	292	137	80	53	38	28	22	17	14	12	9	8	6		
	1249	334	156	92	61	43	32	25	20	16	13	11	9	7		
	1462	391	183	108	71	51	38	29	23	19	15	13	10	9		
	1808	483	226	133	88	63	47	36	29	23	19	16	13	11		
.25	1250	328	151	88	57	40	30	23	18	14	12	9	7			
	1429	375	173	100	65	46	34	26	21	16	13	11	9			
	1673	439	202	117	77	54	40	31	24	19	15	12	10			
	2069	542	250	145	95	67	49	38	30	24	19	15	12			
.30	1376	356	162	93	60	42	31	23	18	14	11	9				
	1573	406	185	106	69	48	35	27	21	16	13	10				
	1842	476	217	124	80	56	41	31	24	19	15	12				
	2278	588	268	153	99	69	51	38	30	23	18	15				
.35	1471	375	169	96	61	42	31	23	17	13	10					
	1681	429	193	110	70	48	35	26	20	15	12					
	1969	503	226	128	82	57	41	31	23	18	14					
	2434	621	280	159	101	70	51	38	29	22	17					
.40	1534	387	173	97	61	42	30	22	16	12						
	1753	443	197	111	70	48	34	25	19	14						
	2053	518	231	130	82	56	40	29	22	17						
	2538	641	286	160	101	69	49	36	27	20						
.45	1565	391	173	96	60	40	28	20	15							
	1798	447	197	110	69	46	32	23	17							
	2095	524	231	128	80	54	38	27	20							
	2591	648	286	159	99	67	47	34	25							
.50	1565	387	169	93	57	38	26	18								
	1789	443	193	106	65	43	30	21								
	2095	518	226	124	77	51	35	24								
	2591	641	280	153	95	63	43	30								

NOTE 1: sample sizes calculated using the arcsine formula, with Fleiss' correction for continuity.

NOTE 2: line 1: $\beta = .20$
line 2: $\beta = .15$
line 3: $\beta = .10$
line 4: $\beta = .05$

NOTE 3: p_L = larger probability, p_S = smaller probability.

TABLE L

Required number of events per group for survival analysis (two-tailed test)

	α = .05				α = .01			
	β				β			
δ	.20	.15	.10	.05	.20	.15	.10	.05
1.2	472	542	632	782	704	789	897	1075
1.4	139	159	186	230	207	232	264	316
1.6	71	82	96	118	106	119	135	162
1.8	46	53	61	76	68	76	87	104
2.0	33	38	44	55	49	55	63	75
2.2	26	29	34	42	38	43	48	58
2.4	21	24	28	34	31	35	39	47
2.6	18	20	23	29	26	29	33	40
2.8	15	17	20	25	23	25	29	34
3.0	13	15	18	22	20	22	25	30
3.2	12	14	16	20	18	20	23	27
3.4	11	13	15	18	16	18	20	24
3.6	10	11	13	16	15	16	19	22
3.8	9	11	12	15	14	15	17	21
4.0	9	10	11	14	13	14	16	19
4.2	8	9	11	13	12	13	15	18
4.4	8	9	10	12	11	12	14	17
4.6	7	8	10	12	11	12	13	16
4.8	7	8	9	11	10	11	13	15
5.0	7	7	9	11	10	11	12	14

TABLE M

Critical values of the Studentized Range test

df	Number of steps 2	3	4	5	6	7	8	9	10
1	17.97	26.98	32.82	37.08	40.41	43.12	45.40	47.36	49.07
2	6.08	8.33	9.80	10.88	11.74	12.44	13.03	13.54	13.99
3	4.50	5.91	6.82	7.50	8.04	8.48	8.85	9.18	9.46
4	3.93	5.04	5.76	6.29	6.71	7.05	7.35	7.60	7.83
5	3.63	4.60	5.22	5.67	6.03	6.33	6.58	6.80	6.99
6	3.46	4.34	4.90	5.30	5.63	5.89	6.12	6.32	6.49
7	3.34	4.17	4.68	5.06	5.36	5.61	5.82	6.00	6.16
8	3.26	4.04	4.53	4.89	5.17	5.40	5.60	5.77	5.92
9	3.20	3.95	4.42	4.76	5.02	5.24	5.43	5.60	5.74
10	3.15	3.88	4.33	4.65	4.91	5.12	5.30	5.46	5.59
11	3.11	3.82	4.26	4.57	4.82	5.03	5.20	5.35	5.49
12	3.08	3.77	4.20	4.51	4.75	4.95	5.12	5.26	5.39
13	3.06	3.73	4.15	4.45	4.69	4.88	5.05	5.19	5.32
14	3.03	3.70	4.11	4.41	4.64	4.83	4.99	5.13	5.25
15	3.01	3.67	4.08	4.37	4.59	4.78	4.94	5.08	5.20
16	3.00	3.65	4.05	4.33	4.56	4.74	4.90	5.03	5.15
17	2.98	3.63	4.02	4.30	4.52	4.70	4.86	4.99	5.11
18	2.97	3.61	4.00	4.28	4.49	4.67	4.82	4.96	5.07
19	2.96	3.59	3.98	4.25	4.47	4.65	4.79	4.92	5.04
20	2.95	3.58	3.96	4.23	4.44	4.62	4.77	4.90	5.01
24	2.92	3.53	3.90	4.17	4.37	4.54	4.68	4.81	4.91
30	2.89	3.49	3.84	4.10	4.30	4.46	4.60	4.72	4.82
40	2.86	3.44	3.79	4.04	4.23	4.39	4.52	4.63	4.73
60	2.83	3.40	3.74	3.98	4.16	4.31	4.44	4.55	4.65
120	2.80	3.36	3.68	3.92	4.10	4.24	4.36	4.47	4.56
∞	2.77	3.31	3.63	3.86	4.03	4.17	4.29	4.39	4.47

TABLE Na

Sample size requirements for two-group MANOVA (T^2)

Effect size	Number of variables	$\alpha = .05$			$\alpha = .01$		
		$\beta =$ .30	.20	.10	$\beta =$ .30	.20	.10
Very large (d = 1.5)	2	9	11	13	13	15	18
	3	10	12	15	15	17	21
	4	12	14	17	17	19	22
	5	13	15	18	18	20	24
	6	14	16	19	19	22	26
	8	16	18	22	21	24	28
	10	18	20	24	24	26	31
	15	21	24	28	28	31	36
	20	25	28	32	32	36	42
Large (d = 1)	2	17	21	27	26	31	38
	3	20	24	31	29	34	42
	4	22	27	34	32	37	46
	5	24	29	36	35	40	48
	6	26	31	39	37	44	52
	8	29	35	44	42	48	56
	10	32	38	46	44	52	60
	15	38	46	54	52	60	70
	20	44	52	62	60	68	78
Moderate (d = 0.75)	2	29	36	48	44	52	66
	3	34	42	54	50	58	72
	4	37	46	58	54	64	78
	5	42	50	62	58	68	84
	6	44	52	66	62	72	88
	8	48	58	72	68	80	96
	10	54	64	78	74	86	105
	15	64	74	92	86	100	120
	20	72	84	105	98	110	130
Small (d = 0.5)	2	64	80	105	96	115	145
	3	74	90	120	110	130	160
	4	80	98	130	120	140	170
	5	88	110	135	125	150	180
	6	94	115	145	135	160	190
	8	105	125	160	150	175	210
	10	115	135	170	160	185	230
	15	135	160	195	185	220	260
	20	150	180	220	210	240	290

Abridged with permission from Läuter J (1978). Sample size requirements for the T^2 test of MANOVA (tables for one-way classification). *Biometrical Journal,* **20**:389–406.

TABLE Nb

Sample size requirements for three-group MANOVA

Effect size	Number of variables	α = .05, β = .30	α = .05, β = .20	α = .05, β = .10	α = .01, β = .30	α = .01, β = .20	α = .01, β = .10
Very large ($d = 1.5$)	2	11	13	16	15	17	21
	3	12	14	18	17	20	24
	4	14	16	19	19	22	26
	5	15	17	21	20	23	28
	6	16	18	22	22	25	29
	8	18	21	25	24	28	32
	10	20	23	27	27	30	35
	15	24	27	32	32	35	42
	20	27	31	37	36	40	46
Large ($d = 1$)	2	21	26	33	31	36	44
	3	25	29	37	35	42	50
	4	27	33	42	38	44	54
	5	30	35	44	42	48	58
	6	32	38	48	44	52	62
	8	36	42	52	50	56	68
	10	39	46	56	54	62	74
	15	46	54	66	64	72	84
	20	54	62	74	72	80	94
Moderate ($d = 0.75$)	2	36	44	58	54	62	76
	3	42	52	64	60	70	86
	4	46	56	70	66	78	94
	5	50	60	76	72	82	100
	6	54	66	82	76	88	105
	8	60	72	90	84	98	120
	10	66	78	98	92	105	125
	15	78	92	115	110	125	145
	20	90	105	130	125	140	165
Small ($d = 0.5$)	2	80	98	125	115	140	170
	3	92	115	145	135	155	190
	4	105	125	155	145	170	210
	5	110	135	170	155	185	220
	6	120	145	180	165	195	240
	8	135	160	200	185	220	260
	10	145	175	220	200	230	280
	15	170	210	255	240	270	320
	20	195	230	280	270	300	360

Abridged with permission from Läuter J (1978). Sample size requirements for the T^2 test of MANOVA (tables for one-way classification). *Biometrical Journal,* **20**:389–406.

TABLE Nc

Sample size requirements for four-group MANOVA

Effect size	Number of variables	α = .05 β = .30	α = .05 .20	α = .05 .10	α = .01 β = .30	α = .01 .20	α = .01 .10
Very large (*d* = 1.5)	2	12	14	17	17	19	23
	3	14	16	20	19	22	26
	4	15	18	22	21	24	28
	5	16	19	23	23	26	30
	6	18	21	25	24	27	32
	8	20	23	28	27	30	36
	10	22	25	30	29	33	39
	15	26	30	36	35	39	46
	20	30	34	40	40	44	52
Large (*d* = 1)	2	24	29	37	34	40	50
	3	28	23	42	39	46	56
	4	31	37	46	44	50	60
	5	34	40	50	48	54	64
	6	36	44	54	50	58	70
	8	42	48	60	56	64	76
	10	46	52	64	62	70	82
	15	54	62	76	72	82	96
	20	60	70	86	82	92	110
Moderate (*d* = 0.75)	2	42	50	64	60	70	86
	3	48	58	72	68	80	96
	4	54	64	80	76	88	105
	5	58	70	86	82	94	115
	6	62	74	92	86	100	120
	8	70	84	105	96	115	135
	10	78	92	115	105	120	145
	15	92	110	130	125	145	170
	20	105	125	150	140	160	190
Small (*d* = 0.5)	2	92	115	145	130	155	190
	3	105	130	165	150	175	220
	4	120	145	180	165	195	240
	5	130	155	195	180	210	250
	6	140	165	210	190	220	270
	8	155	185	230	220	250	300
	10	170	200	250	240	270	320
	15	200	240	290	280	320	370
	20	230	270	330	310	350	420

Abridged with permission from Läuter J (1978). Sample size requirements for the T^2 test of MANOVA (tables for one-way classification). *Biometrical Journal,* **20**:389–406.

TABLE O

Power for Hotelling's T^2 at $\alpha = .05$ and $\alpha = .10$*

Number of variables	N in each group	D^2 .25	.64	1.00	2.25
2	15	.26 (.32)	.44 (.60)	.65 (.77)	.95
	25	.33 (.47)	.66 (.80)	.86	.97
	50	.60 (.77)	.95	1.00	1.00
	100	.90	1.00	1.00	1.00
3	15	.23 (.29)	.37 (.55)	.58 (.72)	.91
	25	.28 (.41)	.58 (.74)	.80	.95
	50	.54 (.65)	.93	1.00	1.00
	100	.86	1.00	1.00	1.00
5	15	.21 (.25)	.32 (.47)	.42 (.66)	.83
	25	.26 (.35)	.42 (.68)	.72	.96
	50	.44 (.59)	.88	1.00	1.00
	100	.78	1.00	1.00	1.00
7	15	.18 (.22)	.27 (.42)	.37 (.59)	.77
	25	.22 (.31)	.38 (.62)	.64 (.81)	.94
	50	.40 (.52)	.82	.97	1.00
	100	.72	1.00	1.00	1.00

Modified from Stevens J (1980). Power of the multivariate analysis of variance tests. *Psychological Bulletin*, **88**:728–737, with permission from the author and the American Psychological Association.

*NOTE: power values at $\alpha = .10$ are in parentheses; values of 1.00 are approximately equal to 1.

Index

A

Actuarial approach to survival analysis, 238–240, 249
Additive model, 81
Additive rule, 34–35
Adequacy, Kaiser-Meyer-Olkin Measure of Sampling, 166
Adjusted Goodness-of-Fit Index (AGFI), 191
AGFI. *See* Adjusted Goodness-of-Fit Index
AIC. *See* Akaike's Information Criterion
Akaike's Information Criterion (AIC), 191
Alpha in significance testing, 47–50
Alternative hypothesis, 44
AM. *See* Arithmetic mean
Analysis
  assumptions, 152–153, 156–159
  of covariance,147, 154, 157–159
  factor (*see* Factor analysis)
  log-linear, 212–214
  survival (*see* Survival analysis)
  of variance (*see* ANOVA)
ANCOVA. *See* Analysis of covariance
Anomalous values, 256–258
ANOVA
  advanced topics in, 145–154
  between-subjects factors, 97–99
  crossed factors, 86–87
  factorial (*see* Factorial ANOVA)
  Friedman two-way, 226–227, 229, 268
  Latin square design, 99
  in multiple regression, 128
  nested factors, 86–87
  one-way (*see* One-way ANOVA)
  orthogonal decomposition, 159–160
  repeated-measures, 94–102
  within-subjects factors, 97–99
Antiimage correlation matrix, 166
Area of normal curve, 31
  table for, 296–297
Arithmetic mean, 16–17
  uses of, 24
Association
  for categorical data, 217–223
  for ranked data, 230–235
Assumptions, 260
Average, 17

B

Bar charts, 6–9
Bartlett test of sphericity, 166
BDI. *See* Beck Depression Inventory
Beck Depression Inventory, 29
Bell curve, 22, 27. *See also* Normal distribution
Best fit, 128
Beta
  in multiple regression, 131–132
  in significance testing, 47–50
Between-subjects factor, 97–99
Binomial distribution, 37–40
Binomial expansion, 37
Bonferroni correction in one-way ANOVA, 71–72
Box plots, 20, 22–23
Box's *M*, 106

C

Case-control study, 203
Categorical data
  measures of association for, 217–223
  significance testing for, 203–216 (*see also* Significance testing)
Cattell's scree test, 169
Causality, 181, 196
Central limit theorem, 28
  in ANOVA, 100
Central tendency, measures of, 17, 266
  uses of, 23–24
Centroid, 105
Change, 155, 162
Chi-squared test, 204–215, 216, 268
  critical values for, 302
  deconstructing larger tables, 206–207
  goodness-of-fit, 143
  Mantel-Cox, 244–245
  Mantel-Haenszel, 210–212, 269
  McNemar, 209–210

CI. *See* Confidence intervals
Clinical importance, 52–53
Coefficient
  beta, 131–132
  contingency, 218
  correlation (*see* Correlation coefficient)
  factor score, 175
  phi, 218
  standardized regression, 132
    in factor analysis, 169
Cohen's kappa, 218–219
  standard error, 219
  weighted, 220
Cohort study, 203
Communality, term, 165, 168
Comparative fit indices, 190
Comparing two groups, 62–67
Computer. *See also* SPSS/PC
  in factor analysis, 166–167, 175
  in multiple regression, 133–135
Conditional probability, 35–36
  in survival analysis, 240
Confidence intervals, 51–52, 222
  in correlation, 124
Confirmatory factor analysis, 164
Contingency coefficient, 218
Contingency table, 204
Continuity, Yates' correction for, 207–209
Continuous data, 3
Controlled trial, randomized, 203
Cook's distance, 136
Correction
  Bonferroni, 71–72
  Yates', 207
Correlation
  confidence intervals and significance tests in, 124
  mean differences in, 269–270
  in multiple regression, 130–132
  partial, 166
  point-biserial, 231–232, 235
  Spearman rank, 231
Correlation coefficient, 121–122
  interpretation of, 122–124
  in multiple regression, 128–131
  Pearson's, 122
    critical values for, 303
Correlation matrix in factor analysis, 165–167
Covariance, 122
  analysis of, 147–154
Cox proportional hazards model, 245–247, 249, 268
Cramer's V, 218, 222, 223, 268
Critical ratio, 192
Critical values, 48, 53
  for chi-squared test, 302
  for *F* test, 304–307
  for Pearson's correlation coefficient, 303
  for *t*-test, 299
Crossed factors in ANOVA, 86–87, 98
Cubic, term, 145, 159
Cumulative frequency polygon, 12, 15
Cumulative probability, 240–246
Curve
  bell, 22
  normal, 30–32
    area of, 296–297
  survival, 240

D

Data, 6–14
  assumptions, 260
  categorical, 217–223
  deletion of, 258–259
  in factor analysis, 163–168
  histogram and bar chart, 6–9
  missing, 256–260
  ordinal, 8
  ranked, 230–235
  specific point of, 16
  in survival analysis, 237–238
  transforming, 260–264
  types of, 3–5, 175–176
  usefulness of, 6
Database in multiple regression, 137. *See also* SPSS/PC
Death probability. *See* Survival analysis
Deduction, 259
Degrees of freedom, 69–70
  in multiple regression, 128
Dependent probability, 34
Dependent variable, 3, 105
Descriptive statistics, 2, 265–266
Design, 268
Determination, coefficient of, 121–122
Deviation
  mean, 20
  standard) (*see* Standard deviation
Difference
  between groups, 106–107
  within groups, 106–107
Discrepancy, 135
Discrete data, 3
Discrimination in point-biserial correlation, 232
Dispersion, measures of, 18–21, 266
  uses of, 23–24
Distance, 135–136
Distribution
  binomial, 37–40
  Gaussian, 27
  normal (*see* Normal distribution)
  rectangular, 28
  of variance, 170
Disturbance terms, 183
Dot plots, 7
Dummy coding, 134, 135
Dunnett's test, 71, 73, 74

E

Effect size, 54
  in one-way ANOVA, 76
  in *t*-test, 66
Eigenvalue one test, 167, 169
Empirically derived probability, 33–34

Endogenous variables, 181
End-point, 242
Epsilon (ϵ), 110
Equal sample sizes in *t*-test, 63–64
Error
  random, 42
  in repeated-measures ANOVA, 95, 100
Error variance, 82
ES. *See* Effect size
Estimate, pooled, 65
Eta in one-way ANOVA, 76
Exclusive events, mutually, 34–35
Existential variables, 4
Exogenous variables, 181
Expansion, binomial, 37
Exponential, 247

F

F-ratio
  distribution of, 70
  in multiple regression, 128
F test
  critical values for, 304–307
  partial, 130–132
  in regression analysis, 121
FA. *See* Factor analysis
Factor analysis, 163–177, 187–188
  confirmatory, 164, 191–193
  confirmatory vs. exploratory, 164
  definitions in, 163–165
  dichotomous data,
  factors in, 164
    extracting, 167
    retaining and discarding, 168–169
    rotating, 170–173
  interpretation of, 173–174
  matrix in
    correlation, 165–167
    loading, 169–170
  MSA (Measure of Sampling Adequacy), 166
  PCA versus FA, 163–164, 167–168
  sample size, 176
  types of data to use, 175–176
  use of, 174–175
Factorials, 38
  complexity of, 170
Factorial ANOVA, 79–88
  crossed and nested factors in, 86–87
  graphing data in, 82–85
  mean differences in, 269
  random and fixed factors in, 85–86
  sample size calculations in, 87
  sums of squares and mean squares in, 79, 82
  two-way, 87
Far outliers in box plots, 23
Fences in box plots, 22–23
Fisher's exact test, 207–209, 216, 268–269
Fixed factors in ANOVA, 85–86
Freedom
  degrees of, 69–70
    in multiple regression, 128
Frequency, significance testing for; see Significance testing
Frequency polygon, 10–12, 15
Friedman two-way ANOVA, 226–227, 229, 268

G

Gambler's fallacy, 36
Gaussian distribution, 27. *See also* Normal distribution
General linear model, 151
Geometric mean, 24–25
GLM. *See* General linear model
GM. *See* Geometric mean
Goodness-of-fit
  Adjusted Goodness-of-Fit Index (AGFI), 191
  Akaike's Information Criterion (AIC), 191
  comparative fit indices, 190
  in logistic regression, 142–143
  in path analysis, 185
  in structural equation modeling, 191
  Normed Fit Index (NFI), 190
  parsimony fit indices, 191
Graphing, 6–9, 265–266
  of ANOVA, 82–85
Greenhouse-Geisser adjustment, 110
Group comparison, 62–67
Growth curve analysis, 160–161

H

Harmonic mean, 25, 65
Hazard, term, 239–240, 246
Hazards model, Cox proportional, 245–247
Hierarchical stepwise regression, 132–133
Hierarchical linear modality, 161
Histograms, 6–9, 15
Hochberg method, 71–72
Holm method, 71–72
Homoscedasticity, 100
Honestly significant difference, 71, 73–74
Hotelling's T^2, 107, 316
Hotelling's trace, 107
HSD. *See* Honestly significant difference
Huynh-Feldt adjustment, 110
Hypothesis testing
  constructs in, 164
  in correlation, 124–125
  in inferential statistics, 44

I

ICC. *See* Intraclass correlation
Identity matrix in factor analysis, 166
Imputing data, 259–264
Incremental Fit Index (IFI), 190
Independent samples, 36, 266–267
Independent variable, 3, 105
Index case, 246
Index of dispersion, 20
Individual growth curve analysis, 160–161
Inferential statistics, 42–55
  box plots in, 20, 22–23
  clinical importance in, 52–53
  concepts of, 42–43
  confidence intervals in, 51

elements of, 44
errors in, 47–50
hypothesis testing in, 44
one-and two-tail tests in, 50–51
populations in, 43
samples in, 43
estimations of, 53–55
signal-to-noise ratio in, 46–47
significance testing in, 45
standard deviation and error in, 44–45
statistical significance in, 52–53
z-test in, 44
Influence, 135–136
Inner fence in box plots, 22–23
Interaction, 80–83, 95, 134, 152
Interquartile range, 19–20
in box plots, 22–23
uses of, 23–24
Interval
confidence, 51–52
in correlation, 124
variable of, 4–5, 265–267
graphing of, 8
t-test and, 62
width of, 8–9
Intraclass correlation, 100, 221
Ipsative data,
IQR. *See* Interquartile range

K

Kaiser criterion, 168–169
Kaiser-Meyer-Olkin measure of sampling adequacy, 166
Kaplan-Meier approach to survival analysis, 240–241, 243, 249
Kappa, 223, 268
Cohen's, 218–219
standard error, 219
weighted, 220–222
Kendall's tau, 232–233, 235, 268
Kendall's *W*, 233–234
KMO. *See* Kaiser-Meyer-Olkin measure of sampling adequacy
Kruskal-Wallis one-way ANOVA, 226, 229, 268
Kurtosis, 21–22, 265

L

Lambda. *See* Wilk's lambda
Latent variables, 163
Latin square design, 99
Lawley test, 169
Least significant difference, 71, 72–73
Least-squares analysis, 120, 140–141, 190
Leptokurtic, term, 22
Levene's test, 107
Leverage, 135–136
Life-table analysis. *See* Survival analysis
Likelihood function, 141
unconditional vs. conditional, 141
Line, regression, 120
Log-linear analysis, 212–214, 268
Log-rank test, Mantel-Cox, 244–245
Logarithm in data transformation, 264
Logarithmic transformation, 261–264
Logistic regression, 139–144, 268
unconditional vs. conditional likelihood function, 141
goodness-of-fit, 142–143
likelihood function, 141
odds ratio, 142
stepwise, 143–144
Wald test, 142
Logit function, 140
Lost to follow-up, 239, 242
LSD. *See* Least significant difference

M

Mann-Whitney U test, 225, 229, 268
MANOVA. *See* Multivariate analysis of variance
Mantel-Cox log-rank test, 244–245
Mantel-Haenszel chi-squared test, 210–212, 268
Marginals, 35
Matched observations in significance testing, 209
Matrix
correlation, 165–167
factor, 169–170, 175
Mauchly's *W*, 109
Maximum likelihood, 141
McNemar chi-squared test, 209–210, 215, 216, 268
Mean, 17
arithmetic, 24
in binomial distribution,39–40
confidence interval around, 51
difference among, 269
geometric, 24–25
harmonic, 25, 65
standard error of, 44
uses of, 23–24
Mean deviation, 20
Mean square
in ANCOVA, 150
in ANOVA
factorial, 82
one-way, 70
in multiple regression, 128
Mean survival, 237
Measurement model, 189
Measures of association
for categorical data, 217–223
for ranked data, 230–235
repeated, 90, 95–96
Median, 17–18, 22
in box plots, 22
uses of, 23–24
Mesokurtic, term, 22
Method of least squares, 140–141
Midspread, 19
Missing zero, 57
Mode, 18, 23
Motion, range of, 118–121
MSA (Measure of Sampling Adequacy), 166
Multicollinearity, 135–136, 186
Multiple comparisons in one-way ANOVA, 71

Multiple correlation coefficient, 128–131
Multiple regression, 127–138
beta coefficients in, 131–132
calculations for, 127–129
Cook's distance, 136
diagnostic tests, 136
discrepancy, 135
dummy coding, 135
influence, 135–136
interactions in, 152
leverage, 135–136
missing data and, 260
multicollinearity, 136
partial correlations in, 131
part *r*,
partial *F* tests in, 130–132
partial *r*,
pragmatics of, 137
sample size calculations in, 137
semipartial *r*, 131
standardized regression coefficient in, 132
stepwise, 132–135
tolerance, 136
variables in, 131
Variance Inflation Factor (VIF), 136
Multiplicative law, 35–36
Multivariate analysis rationale,
Multivariate analysis of variance (MANOVA), 107
assumptions, 111
Box's *M*, 106
epsilon (ϵ), 110
errors, 257
Greenhouse-Geisser adjustment, 110
homogeneity of VCV, 106, 111
Hotelling's T^2, 111
Hotelling's trace, 107
Huynh-Feldt adjustment, 110
lambda (*see* Wilk's lambda)
Levene's test, 107
Mauchly's *W*, 109
outliers, 111
Pillai's trace, 107–108, 111
power, 111
repeated-measures, 109
robustness, 111
Roy's largest root, 107
sample size, 111, 313–315
sphericity, 109
sum of squares and cross-products (SSCP), 106–107
Wilk's lambda (λ), 107–108
Mutually exclusive events, 34–35

N

Nested factors in ANOVA, 86–87, 98
Newman-Keuls test, 71, 73, 74
Noise
in statistical inference, 46–47
in *t*-test, 63
Nominal variable, 4–5, 265–266
Nonlinear regression, 150–151
Nonparametric statistics, 201–254, 204, 267–268
Nonrecursive model, 189
Normal curve, 30–32
area of, 296–297
Normal distribution, 27–32
binomial and, 39–40
Normed Fit Index (NFI), 190
Null hypothesis, 44–45, 55, 105

O

Oblique rotation in factor analysis, 172
Odds ratio, 142
log of, 214
One-tailed test, 50–51
One-way ANOVA, 68–78
comparisons in, 68–78
multiple, 68–78
planned orthogonal, 74–75
post-hoc, 71
degrees of freedom in, 69–70
F-ratio distribution in, 70, 75
Kruskal-Wallis, 226
mean differences in, 269
power table for, 309
relationship strength in, 75
sample size for, 76, 77, 308
sum of squares in, 69
Ordinal data, 4–5, 265–268
dispersion of, 18
graphing of, 8
Orthogonal, term, 167
Orthogonal comparisons in one-way ANOVA, 74–75
Orthogonal decomposition, 159–160
Orthogonal rotation in factor analysis, 171–172
Outer fence in box plots, 23
Outliers, 257
in box plots, 23
finding, 264
in MANOVA, 111
in multiple regression, 111, 135

P

p-value in regression analysis, 121
Paired observations in significance testing, 210
Paired *t*-test, 89–93
Parameter
population, 43
in regression analysis, 121
Parametric statistics, 266–267
Parsimony fit indices, 191
Partial *F* tests, 130–132
Path analysis
assumptions, 186
and causality, 181
correlated path model, 182
decomposing correlations, 180
direct effects, 180
direct models, 181
disturbance, 183
endogenous variables, 181
exogenous variables, 181
goodness-of-fit, 185

independent path model, 181
indirect effects, 180
indirect model, 182
mediated model, 182
nonrecursive model, 183
number of observations, 184
number of parameters, 184
recursive model, 183
sample size, 186
PCA. *See* Principle components analysis
Pearson's correlation coefficient, 122, 261
critical values for, 303
Person-years, 238
Phi coefficient, 212, 222, 223, 268
Pillai's trace, 107
Planned comparisons in one-way ANOVA, 71, 74–75
Platykurtic, term, 22
Plots
box, 22–23
stem-leaf, 9–10, 15
Point-biserial correlation, 231–232, 235
Point chart, 7
Poisson distribution, 204
Polygon, frequency, 10–12, 15
Pooled estimate in *t*-test, 65
Populations in inferential statistics, 43
Post-hoc comparisons in one-way ANOVA, 71
Power
in one-way ANOVA, 76–77
in significance testing, 49–50
in survival analysis, 247–248
in *t*-test, 301
Power series regression, 151
Principle components analysis, 163–177. *See also* Factor analysis
Probability, 33–41, 240–241
binomial distribution in, 37–40
conditional, 35–36
in survival analysis, 240
cumulative, 240
death (*see* Survival analysis)
empirical derivation of, 33–34
independent events in, 36
law of "at least one" in, 36–37
mutually exclusive events in, 34–35
theoretical derivation of, 34
Proportional differences, 269, 310
Proportional hazards model, 245–247

Q

Quadratic, term, 145, 159–161
Quadratic weights, 220
Quartic, 145
Quartiles in box plots, 22

R

Random error, 42
Random factors in ANOVA, 85–86
Randomized controlled trial, 203
Range, 8
interquartile, 19–20
in box plots, 22–23
of motion, 118–121
uses of, 23–24
Rank correlation, Spearman, 231
Ranked data
measures of association for, 230–235
significance testing for, 224–229
Ratio, signal-to-noise, 46–47
in *t*-test, 62
Ratio variable, 4–5, 265–267
graphing of, 8
t-test and, 62
Rectangular distribution, 28
Recursive models, 183
Regression, 118–126
advanced topics in, 145–154
in ANCOVA, 145–154
Cook's distance, 136
diagnostic tests, 136
discrepancy, 135
dummy coding, 135
general linear model, 151
influence, 135–136
leverage, 135–136
logistic (*see* Logistic regression)
multicollinearity, 136
multiple, 127–138 (*see also* Multiple regression)
nonlinear, 150–151
power series, 151
statistical significance, 130–131
tolerance, 136
Variance Inflation Factor (VIF), 136
Regression coefficient, standardized
in factor analysis, 169
in multiple regression, 132
Regression toward the mean, 156–157
Reject null hypothesis, 45, 55
Related samples, 266–267
Relative risk, 244
Reliability coefficient, 100
Repeated-measures
ANOVA, 94–102, 95–96
orthogonal decomposition, 159–160
Respecification, 191
Rho, Spearman's, 230–231, 235
Risk in survival analysis, 238–239, 244, 245
ROM. *See* Range, of motion
Roy's largest root, 107
RR. *See* Relative risk

S

Samples,
in ANCOVA, 153
in ANOVA, 87, 101
one-way, 76–77, 308
in correlation, 124
estimation of, 65
in factor analysis, 176
in independent proportion testing, 266–267, 310
in inferential statistics, 43
in logistic regression, 144
in MANOVA, 313–315

- mean differences and, 298
- in measures of association for categorical data, 222
- in multiple regression, 137
- in paired *t*-tests, 92
- related, 266–267
- in significance testing
  - for categorical frequency data, 214–215
  - for ranked data, 227
- size requirements, 298
- in survival analysis, 247–248
- in *t*-test, 300
  - equal, 63–66
  - unequal, 64

Scheffé's method in one-way ANOVA, 71, 74
Scree test, Cattell's, 169
SD. *See* Standard deviation
Secular trend, 243
Self-Rating Depression Scale, 29
Semipartial *r*, 131
Signal
- in statistical inference, 46–47
- in *t*-test, 63

Signal-to-noise ratio, 205
Significance testing, 45–46
- for categorical frequency data, 203–216
  - chi-squared test in, 204–207
  - Fisher's exact test in, 207–209
  - log-linear analysis in, 212–214
  - Mantel-Haenszel chi-squared in, 210–212
  - McNemar chi-squared test in, 209–210, 215
  - sample size estimation in, 214–215
  - Yates' correction for continuity in, 207
- Cohen's kappa in, 218–219
- in correlation, 124
- for ranked data, 224–229
- for tau, 233
- for *W*, 234

Skewness, 21–24, 265
Somaliland Camelbite Fever, 139–144
Spearman's rho, 230–231, 235, 268
Specific data point, 16
Sphericity, Bartlett test of, 166
SPSS/PC, ix–x, 107
- in ANCOVA, 95–96, 100, 154
- in ANOVA
  - factorial, 88
  - one-way, 78
  - repeated-measures, 102
- chi square, 216
- comparing groups, 67
- cumulative, 15
- in data analysis, 15
- in data description, 26
- Fisher's exact, 216
- frequency polygons, 15
- graphing, 26
- histograms, 15
- MANOVA, 113
- in measures of association
  - for categorical data, 223
  - for ranked data, 235
- for missing values, 264
- for outliers, 264
- in paired *t*-test, 93
- in principal components and factor analysis, 177
- in regression analysis, 127–138
  - logistic, 154
  - multiple, 128–131
  - nonlinear, 150–151
  - residual, 120–121, 128
  - simple, 126
  - total, 128
- in significance testing
  - for categorical frequency data, 216
  - for ranked data, 229
- statistical inference, 56
- stem-and-leaf plot, 15
- in survival analysis, 249
- in *t*-test, 67
- for transformations, 264

Square root transformation, 264
Squares
- mean (*see* Mean square)
- sum of (*see* Sum of squares)

SSCP (sum of squares and cross-products)
Standard deviation, 21
- in binomial distribution, 39–40
- in confidence intervals, 51–52
- in inferential statistics, 43
- in normal distribution, 28
- in *t*-test, 63–64
- uses of, 23–24

Standard error
- of mean, 44
- in survival analysis, 241–242
- in *t*-test, 63

Standard scores, 28–29
Standardized regression coefficient
- in factor analysis, 169
- in multiple regression, 132

Starting point, 242
Statistics
- descriptive, 2
- inferential (*see* Inferential statistics)
- multivariate, 269
- needs for, 2
- nonparametric, 201–254, 204, 267–268
- parametric, 266–267
- significance of, 52–53
- univariate, 266–269

Stem-leaf plots, 9–10, 15
Step in box plots, 22
Stepwise regression, 132–135
Structural equation modeling, 186–196
- Adjusted Goodness-of-Fit Index (AGFI), 191
- Akaike's Information Criterion (AIC), 191
- assumptions, 186
- and causality, 181, 196
- confirmatory factor analysis, 191
- constraining parameters, 189
- correlated path model, 182
- decomposing correlations, 180
- direct effects, 180

direct models, 181
disturbance, 183, 189
endogenous variables, 181, 189–191
exogenous variables, 181, 189
and factor analysis, 187–188
fixed parameters, 189
free parameters, 189
goodness-of-fit, 185, 191
independent path model, 181, 193
indirect effects, 180
indirect model, 182
measurement model, 189
mediated model, 182
model specification, 188–189
nonrecursive model, 183, 189
Normed Fit Index (NFI), 190
number of observations, 184–185
number of parameters, 184–185
parsimony fit indices, 191
recursive model, 183
sample size, 186
Studentized range test, 72, 312
Student's *t*-test, 63
Sum of squares
in ANCOVA, 150
in ANOVA
factorial, 82
one-way, 69
repeated-measures, 95–96
in regression analysis, 120–121
multiple, 129–130
Sum of squares and cross-products (SSCP), 106–107
Survival analysis, 236–249
actuarial, 238–240
assumptions, 242
censored, 239
covariate adjustment in, 245–247
data summarizing in, 237–238
group comparisons in, 243–244
life-table, 238
required number of events for, 311
sample size and power in, 247–248
techniques of, 238–243
use of, 236
Survival function, 240
Survival rate, 237–238, 248

T

t-test, 62–67, 105, 131
critical values for, 299
paired, 89–93
power table for, 301
sample size for, 63–66, 300
Tables, 12–14
Tau, Kendall's, 232–233, 235
Theorem, central limit, 28
Theoretically derived probability, 34
Tolerance, 136
Transformations, 260–264
Trial, randomized controlled, 203
Trimodal, term, 18
Tukey's least significant difference, 71–73, 78, 88
2 x 2 contingency table, 204, 206, 213
measures of association for, 217–219
Two repeated observations, 89–93
Two-tailed test, 50–51
Two-way ANOVA, Friedman, 226–227
Type I and II error, 47–50

U

Unequal sample sizes in *t*-test, 64–65
Unipolar factors, 170–171
Uniqueness, 165
Univariate statistics, 266–269
Unweighted least squares, 190

V

Variables, 3, 105
association between, 266
in descriptive statistics, 265–267
endogenous, 181
exogenous, 181
interval, 62, 265–267
latent, 163–164
ratio, 62, 265–267
Variance, 21
analysis of (*see* ANOVA)
in binomial distribution, 39–40
in factor analysis, 170
in *t*-test, 65
Variance-covariance matrix (VCV), 186
Variance Inflation Factor (VIF), 136
Varimax in factor analysis, 171
VCV. *See* Variance-covariance matrix
VIF. *See* Variance Inflation Factor

W

W. *See* Mauchly's *W*
W, Kendall's, 233–235
Wald test, 142
Weight
measurement of, 220–221
sample size, 65
Weighted kappa, 220–222
Weighted least squares, 190
Whiskers in box plots, 22
Wilcoxon rank sum test, 225, 229, 268
signed, 226–227, 229, 260
Wilk's lambda (λ), 107
Within-subjects factor, 97–99

X

X bar, 16. *See also* Mean

Y

Yates' correction for continuity, 207–209
Yule's Q, 218

Z

z-test
calculating, 44
in survival analysis, 243–244
z-score, 29
Zero, missing, 57